Essays on Agricultural Economy

Nonexperimental Writings on Agricultural Policy and Development Administration in Nigeria

Essays on Agricultural Economy

Nonexperimental Writings on Agricultural Policy and Development Administration in Nigeria

G. B. Ayoola

Copyright © 2018 by G. B. Ayoola.

Library of Congress Control Number: 2018907885
ISBN: Hardcover   978-1-5434-0181-3
            Softcover   978-1-5434-0180-6
            eBook   978-1-5434-0179-0

All rights reserved. No part of this book may be reproduced or transmitted in any form or by any means, electronic or mechanical, including photocopying, recording, or by any information storage and retrieval system, without permission in writing from the copyright owner.

Any people depicted in stock imagery provided by Getty Images are models, and such images are being used for illustrative purposes only.
Certain stock imagery © Getty Images.

Print information available on the last page.

Rev. date: 10/30/2018
Reprint date: 01/03/2020

To order additional copies of this book
Contact: Topdave Multimedia Ltd
Tel: +2348132187221 / 07019281916
topdavemultimedia@gmail.com
topdavemultimedia.com

CONTENTS

Foreword to Part 1 ix
Preface xiii
Acknoweldegement: xv

PART 1

FIRST PHILOSOPHICAL DECADE (1988 – 1998)

1. The Policy Template For Food And Agricultural Strategies In Nigeria 3
2. Self-Sufficiency in Food Production: An Essential Foundation for Economic and Industrial Development in Nigeria* 16
3. The Impact of Government Policies On Rural Life in Nigeria 41
4. Sectoral Planning of the National Economy in Perspective: The Case of Agriculture 59
5. From Food Deficiency to Food Sufficiency: Strategies to Overcome The Food Crisis in Nigeria 82
6. The National Question in Nigerian Agriculture: A Proximate Analysis 95
7. The Unknown Variables in Nigeria's Food Self-Sufficiency Equation 107
8. Food Self-Sufficiency and Mass Production: What Approach? 117
9. The Institutional Dimensions of Unifying Agricultural Extension System in Nigeria..133
10. Effective Communication with Women for Agricultural Development in Nigeria* 144

11. Agricultural Education In Nigeria: Problem And Prospects* 157
12. Technology and Nigerian Agricultural Development 170
13. Socio-Economic Problems in Commercializing Agricultural Research Findings and Technological Breakthrough in Nigeria 196
14. The Policy and Rural Infrastructure Factors in Stimulating Maize and Sorghum Production 219
15. The Nigerian Grain Market and Industrial Utilization of Sorghum: A Perspective 238
16. Accelerated Soyabean Production in Nigeria: Strategies and State of the Art* 247
17. Nigeria's Rural Infrastructure: Past and Present* 257
18. The Institutions and Processes of Rural Infrastructural Development in Nigeria 277
19. Proposals for Alternative Mode of Market Intervention* 297
20. Management Issues in Financing the Graduate Farmers Scheme in Nigeria* 312
21. The Marketing of Agricultural Pesticides in Nigeria: Organization and Efficiency* 323
22. The World Bank-Assisted Agricultural Development Project as a Model for Public Sector Financing in Nigeria 338
23. Deregulation and Decontrol of Nigeria's Fertilizer in the Context of Structural Adjustment Programme 349
24. A Synthesis of the Nigerian Livestock Policy 360
25. The New Land Development Strategy for Accelerating Food Production in Nigeria* 370
26. The Structural Adjustment Programme (SAP) and the Nigeria Food Crop Sub-Sector* 380
27. The Food Sub-Sector and Structural Adjust Programme in Nigeria: Some Matters Arising 396
28. Of Privatization and Public Agricultural Enterprises 408
29. Dilemmas of Sustaining Increased Crop Production in a Deregulated Economy 420

PART 2

SECOND PHILOSOPHICAL DECADE
(1998 – 2008)

30. Micro And Macro Farm Business Data Requirements For Agricultural Planning And Policymaking In Nigeria 431
31. Emerging Issues For The Formulation Of Policy For Agri-Input Delivery System In Nigeria 442
32. Analysis Of Budget Policy On Agriculture Under Different Governance Regimes 454
33. Agriculture And National Development 465
34. Policy Advocacy: The Missing Link In Nigeria's Quest For Agricultural Transformation 481
35. Trends And Observations In The Funding Of Agricultural Research In The Naris And Fcas 490
36. The Role Of Agricultural Commodities In Nigeria's Economic Development 505
37. The Nexus Of Food, Drug, Water, And Agricultural Development In Nigeria: The Role Of An Agricultural University 510
38. Relevance Of Horticulture For Economic Empowerment 518
39. Social Safety Nets As A Means Of Delivering The Right To Food In Nigeria 534

PART 3

THIRD PHILOSOPHICAL DECADE
(2008 - 2018)

40. Transparency, Corruption, and Sharp Practices: The Policy Analyst's Viewpoint 547
41. Agricultural Rebirth in Nigeria: Issues of Policy and Institutional Realignment 555
42. Rethinking Africa's Food Policy in Terms of Rights 573
43. Using Smart Subsidies to Support Small Scale Farmers in Africa—Lessons from Nigeria's Growth Enhancement Scheme 586

44. Academic Corruption and the Role of NNMA in Curbing It........ 606
45. A Multisectoral Analysis of Nigeria's Agricultural Policy in the Interconnected World..............621
46. The Human Rights Approach to Ensuring Food Security in Nigeria 645
47. Overview Of Fertilizer Demand And Investment Opportunities In Nigeria..............657
48. Emerging Issues for the Formulation of Policy on Agri-Input Delivery System in Nigeria..............675
49. Stocktaking Of The Soaring Food Prices In 2007/2008: Evidence From Nigeria 690
50. The Hungry and the Rest of Us 711

Index..............717

Foreword to Part 1

As correctly noted by Dr Ayoola in Chapter 12 of these essays, Nigeria has had over 100 years of organized public agricultural research dating back to the establishment of the Lagos Botanical Garden at Ebute Meta in 1893. More than a century later, Nigeria still remains food insecure, and the rural majority who depend mainly on agriculture for their livelihood remain in abject poverty. Nigeria continues to spend scarce foreign exchange on the importation of food items, the production of which Nigeria has a natural endowment. Why then is this paradox?

If organized public agricultural research holds the key to the technological and commercial transformation of Nigerian agriculture, two questions that immediately come to mind in trying to address the paradox of continuing agricultural underdevelopment in the face of a long history of organized public research are the following (1) Has the research been relevant, efficient, and problem solving? (2) Are there other factors or variables that are relevant which have been omitted from the equation?

Dr Ayoola has in the essays in this volume addressed some aspects of these questions, which particular stress on the answer to the second question. His frontal treatment of the policy framework and planning in the first set of essays serves to underscore the important point that even if the research is relevant, efficient and problem solving, but the policy environment remains hostile and disenabling, they labour in vain who seek to reduce food insecurity and mass poverty using new technologies generated by agricultural research. It is instructive that it took Nigeria almost 100 years after the commencement of organized public agricultural research from the first national agricultural policy statement to be articulated and published. Dr Ayoola in the different essays in Chapters 1 - 8 provides a rich discussion of food self-sufficient and self reliance, extending some of my earlier works on the analytical framework for examining national food self sufficiency within the framework of economic nationalism. Many years after this initial effort and some years after most of Dr Ayoola's essays on policy were written, the world has rapidly moved on. Some of these developments have been treated in the last set of papers in the volume, such as structural adjustment and privatization (Chapter 26-29)

But of equal importance and requiring the urgent attention of Nigerian Agricultural Economists are other developments such as the impact of trade liberation and the application of World Trade Organized (WTO) rules and regulations on Nigerian agriculture. The problem is that the new paradigm of international economic relations permits the massive influx of food items under WTO rules when Nigerian agriculture is suffering the worst handicap in rural infrastructural support services.

This underscores an important theme in these essays, and he second missing variable in the equation, which is the provision of rural infrastructural support services. The Rural Infrastructures Survey Team, of which Dr. Ayoola is a member, has under my coordination and on behalf of the Federal Ministry of Agriculture and Rural Development, generated massive amounts of data, for each Local Government Area in the country since 1979. The coverage, as he alludes to in the essays, includes physical infrastructure (rural roads, rural market, irrigation facilities); social infrastructure (primary schools, secondary schools and teacher training colleges, hospitals, maternity centres, dispensaries; and institutional infrastructure (rural institutions, farmer's co-operatives and self help groups, etc)

I can reveal here that the publication from the *surveys (Rural Infrastructures in Nigeria, University of Ibadan Press,* 1985) and the policy advocacy flowing there from, formed the bases for the establishment of Directorate of Food, Roads and Rural Infrastructure (DFRRI). It was on the basis of this work that I was able to initiate the creation of DFRRI in the Presidency in the 1986. (Unfortunately, DFRRI was poorly implemented, but that should be the subject matter of another volume in due course). Rural infrastructures are so fundamental that one could make the point that if only the government at all levels could saturate rural Nigeria with rural roads, rural markets, rural electricity and telecommunications, more than half of the country's food insecurity and rural poverty problems would have been solved.

The poor implementation of the DFRRI programme and its eventual atrophy highlight the other variable that has been omitted in the country's agricultural transformation efforts, that is the development of effective institutional and implementation capacity. This is a theme that is covered in several of the chapters in this volume, though with a slant towards extension (chapter 9 - 11). Much of the recorded failure to transform Nigerian agriculture can be attributed to the weak extension delivery systems and the weak linkages between research and extension. Several chapters in this volume are devoted to an assessment of the World Bank assisted Agricultural Development Projects (ADPs) that sought to provide an integrated package of new seed varieties, credit, improved agronomic practices, rural roads, small dams and virile extension services. The relative success of the first generation of the ADPs provided the impetus for the extension of the concept from the initial trial phase of enclaves to state-wide ADPs. I was involved with the country-wide coverage of the ADPs during my tenure as Head Federal Agricultural Coordinating Unit (FACU). Not only do policies matter, viable institutions also matter. Of particular relevance is the problem of institutional sustainability which Dr Ayoola discusses with revealing empirical evidence.

The documented instability in the political and professional leadership of federal and state ministries and agriculture is particularly worrisome especially in relation to the sustainability of agricultural policies and projects. Our concern is not just with the adequacy of institutional arrangements but with the instability and sustainability of institutions. This is a theme that has been of special interest to me over the years, and I am convinced that policy makers and analysts will profit from a careful reading of the analysis in some of the chapters in this volume.

The essays in this volume include rich discussions of innovations with practical examples of technical innovations such as the electrodyn sprayer and institutional innovations such as the national accelerated food production project (NAFPP) and the National Agricultural Land Development Authority (NALDA) for which I was the pioneer Chairman.

Farmers will openly adopt new innovations on the basis of proven profitability. This underscores the third omitted variable in the equation, that is, incentives. Dr Ayoola discusses fertiliser and pesticide subsidies, the agricultural credit guarantee scheme, and producer price support, and the problems created during implementation of well-intentioned policies.

Dr. Ayoola provides a rich dose of empirical analysis in this volume. I commend this to readers because I am convinced that the issues over which members of the policy community are likely to disagree are mostly empirical, not theoretical. The agricultural policy analyst has a duty to marshal empirical evidence, guided by sound analytical framework, as the foundation for policy advice and advocacy. This requires painstaking research in the field and ample amounts of data. In addition, it requires patience on the part of the policy researcher and analyst to be able to survive the slow motions of the public bureaucracy. Dr. Ayoola has imbibed this culture of working closely with the public bureaucrat, and his success on this score is to the benefit of policy makers, policy researchers and analysts and the food consuming public at large. I have found Dr Ayoola's volme very refreshing and rewarding.

...F.S Idachaba
International Service for National Agricultural Research (ISNAR)
The Hague, The Netherlands

Preface

These *Essays* are a decade-wise collection of non-experimental and non-econometric writings of mine over a period of 30 years, from the year of my doctoral graduation in 1988 to date in 2018. It was the late Professor Rufus Adegboye, the academic giant of agricultural economics department and also the so-called *baale* (imagine University of Ibadan as one big village), who planted it in my spirit, soon after defending my doctoral thesis, the need for me to philosophize continually; as he put it in felicitation with me, 'having earned yourself a degree of doctor of philosophy in this department, never should you be found wanting of what to say in any conceivable circumstances'. This registered in me as a charge for a holder of doctor of philosophy degree to constantly 'philosophize' on issues of the moment in his or her area of primary expertise; which, herein, is what the series of essays illustrate over time. Fired by this charge from outset, I have conceived it in my subconscious a series of professional writings, which I had compiled into a book form, in cycles of philosophical decades.

At the end of the first philosophical decade in 1998, the first book of essays was published as *Essays on Agricultural Economy – A Book of Readings on Agricultural Policy and Development Administration in Nigeria*, at which time I had matured to a Senior Lecturer at University of Agriculture Makurdi. The volume, which carries a foreword by the late Professor Francis Idachaba, another intellectual giant in the agricultural economics profession, comprised papers written on different aspects of the agricultural and rural economy of Nigeria, presented at various conferences and workshops or other speaking engagements in the country and abroad. The second philosophical decade (1998-2008) was equally an active period of paper writing for me, notwithstanding a major distraction I suffered from incessant court cases I filed against the university and prosecuted for about five years. Nonetheless, during this decade I had acquired considerable capacity in the use of computer, including the presentation of some of the papers in PowerPoint formats (Microsoft PowerPoint). The same situation prevailed during the third philosophical decade (2008-2018) whereby the use of PowerPoint was intensified. These developments led to availability of fewer number of full-length papers to select from for inclusion in the volume during the Second and Third decades. In any case, from the second to the third philosophical decade my professional attention to the agricultural economy had gradually shifted from that of a single-minded policy analyst to one of public spirited policy advocate, focusing more and more on outreach, with the view to making the benefits of my empirical policy research and brokerage activities available to public policy authorities. Thus my writings during the second and third periods obviously reflect this in increasing intensity as revealed in the titles of the essays.

Apart from the utility of the *Essays* as a repository of critical time series information of quantitative and qualitative types about agricultural policy process in Nigeria in relation to the rest of the world, thereby constituting a vital institutional memory in itself, the felt need for the present volume predicates so much upon the sustained demand for the earlier book by students, agricultural administrators, policy experts or practitioners, among others in the agricultural development community. However, in managing the content, while the first book was organized in thematic chapters, 29 of them, the present volume is organized in no particular thematic order.

Unfortunately, before completing the present volume of *Essays* Professor Francis Idachaba was no longer around to update his Foreword to the earlier volume, having passed away in August 2014. Nonetheless, as an honour he so well deserved, and in his much cherished memory, it is my decision to retain the same Foreword of his for the present volume as is.

Professor G. B. Ayoola
June 2018

Acknoweldegement:

I owe it a debt of cumulative gratitude to my technical colleagues and secretarial staff at Federal University of Agriculture Makurdi (FUAM) and also at The Farm & Infrastructure Foundation (FIF) who offered their assistance in proofreading the successive drafts of this volume, and in many other ways required at my instance at different stages of producing the book in its present form, namely, Victor Oboh and Orefi Abu (nee Okpe), John Ogbonaya, Tony Ochi; Omotola Ologbenla (nee Aweda) Abiola Bayode (nee Ojo). Busayo Kanjuni, Tope Davies, Ayodeji Taiwo, Lucky Otaru, Joel Olorunsola, Chidinma Anyanwu, to mention a few.

Moreover, many thanks go to my numerous colleagues in the universities and my past and present students at undergraduate and postgraduate levels, the experts in public service and private sector and other members of the agricultural development community at home and abroad, who have provided valuable feedbacks about the earlier volume of the book, and in whom I nurse the hope that the present volume would be found equally useful.

PART 1

FIRST PHILOSOPHICAL DECADE
(1988 – 1998)

The Policy Template For Food And Agricultural Strategies In Nigeria

In *Food and Agricultural Strategy Review*, Centre for Food and Agricultural Strategy, FASR No.1, October 1998. ISBN: 1114-3322 pp.1-10

The actions (and inactions) of government to achieve sustainable development of a sector are rooted deep in the policy context. Certain policies may be explicitly expressed in general or specific modes or articulated in definite statute documents while certain other policies might be implicitly subsumed in posterior actions only, without any conscious statement of them in public. By and large, the agricultural policy environment in Nigeria is a product of both the explicit and implicit policies from which the sequence of strategies for implementation has emerged.

This paper seeks to prepare the ground for the review of specific programmes and projects that are elements of the strategy. The objective is to illuminate the basic policy template for proper determination of the coordinates of each strategy therein, and for improved understanding of the issues associated with particular programmes and projects for agricultural development.

EVOLUTION OF AGRICULTURAL ADMINISTRATION AND POLICY

The discernible trend in agricultural policy and programme implementation in the post-independence period in Nigeria is the product of both pre- and postcolonial agricultural administration and the consequent policy environment. The full institutionalization of agricultural administration commenced in 1900 with the establishment of the Forestry Department.

Subsequently (1910), the first definitive Agricultural Department (Southern Nigeria) was established. This was quickly followed (1912) by the second Agricultural Department (Northern Nigeria). Both of them were merged to form one Agricultural Department for Nigeria in 1921. A Director of Agriculture who was responsible for the overall agricultural administration headed the successive department. The regional constitution was launched in 1954 with each of the three new regions (Eastern, Western, Northern) establishing its own Ministry of Agriculture and Natural Resources (MANR). An additional MANR was established (1963) when the Mid-Western Region was created. Since 1954 through 1967, there was no MANR at the federal level as a result of the inclusion of agricultural development on the residual legislative list. By a special clause of the constitution, therefore, agricultural administration and policy became the responsibility of the regions.

However, the military government created a full-fledged federal MANR in 1967, following expert advice (FAO 1966). The four regions were also restructured into twelve states, and later to nineteen states. This implies twelve and nineteen MANRs respectively. A similar exercise was carried out in 1986 when two new states (i.e. two MANRs) were created. The nomenclature of federal ministry has been modified variously to reflect its growing functions and size: Federal Ministry of Agriculture (1970), Federal Ministry of Agriculture and Water Resources (1980), Federal Ministry of Agriculture, Water Resources, and Rural Development (1984), and back to the starting point, Federal Ministry of Agriculture and Natural Resources (1990). The typical federal or state/regional ministry, like the colonial department, is responsible for the overall agricultural administration and policy management in its defined territory. The executive minister (federal) or commissioner (state) headed the ministry in the present structure.

In consonance with administration, agricultural policy evolved in overlapping phases. In the earliest phase, policies were not explicitly stated, but could be inferred from the programmes being implemented. Subsequently, statements of specific agricultural policy appeared in disparate documents, which further helped to confirm policy inferences. During the latter part of the colonial administration, conscious attempts were made to articulate specific agricultural policies. The highlights of these are as follows:

(i) Forest Policy 1937—This was based on the proposal of the Chief Conservator of Forests and a Forest Conference held to discuss same. It was directed to solve problems of depreciating forest capital as a result of unregulated exploitation.

(ii) Forest Policy 1945—This was a revision of the existing forest policy to reflect the new position of government that (a) agriculture must take priority over forestry, (b) the satisfaction of the need of the people at the lowest possible rates must take precedence over revenue, and (c) the production of greatest revenue compatible with a sustained yield.

(iii) Agricultural Policy 1946—This was the first all-embracing policy statement directed towards 'improvement in the general standard of living of the people by inducing proper use of the resources available to them'. For this purpose, Nigeria was subdivided into five agricultural areas: (a) Northern Province Pastoral or Livestock Production area, (b) Northern Province Export Crop (Groundnut and Cotton) Production area, (c) Middle Belt Food Production Area, (d) Export Crop (Palm Oil and Kernels) Production Area, and (e) South-West Food Production Export Crop (Cocoa and Palm Kernel) Areas.

(iv) Policy for the Marketing of Oils, Oilseeds, and Cotton 1948—Specific statement prepared to stabilize post-war prices.

(v) Forest Policy for Western Region 1952—This was declared on territorial basis for the trial period of regionalization.

(vi) Agricultural Policy for the Western Region 1952—This was also a territorial policy for Western Region agriculture;

(vii) Policy for Natural Resources of Eastern Region 1953—a territorial policy for Eastern Region agriculture;

(viii) Western Nigeria Policy of Agriculture and Natural Resources 1959—in which the farm settlement scheme was incorporated;

(ix) Nigeria Agricultural Policy (undated)—This was an attempt to unify the various policy directions of state governments, and was prepared by Federal Department of Agricultural Planning by assembling the bits and pieces of the state policies towards the formulation of a national policy for the agricultural sector.

(x) Agricultural Policy for Nigeria 1988—This is the most comprehensive of all the policies listed above. It is accompanied by detailed analysis of quantitative targets to make the country self-sufficient in food and agricultural raw materials, latest 2002.

Consequent upon this policy atmosphere, the post-independence development plans contained agricultural projects and programmes, which were executed to effect development. There are four such medium-term plans as follows: (i) First National Development Plan 1962-68, (ii) Second National Development Plan 1970-74, (iii) Third National Development Plan 1975-79, and (iv) Fourth National Development Plan 1980-85. The last plan period suffered severe balance-of-payments problems, which has culminated in the current Structural Adjustment Programme (SAP). Table 1 presents a typology of the different agricultural programmes and projects embarked upon in the post-independence period.

TRENDS IN POLICY ARTICULATION AND PHILOSOPHY

Historical account indicates that, by independence in 1960, the need for a well-articulated agricultural policy had been recognized. At that time, the existing agricultural policy documents, which were either region or commodity-specific, contained the objectives and philosophy of agricultural development. In the post-independence era, new documents have been produced, also containing objectives and underlined by philosophies that reflect the changing policy environment. Trends will be discerned from the thirty-year postcolonial experience in agricultural policy process.

First, the country has moved gradually away from the disparate sub-sectoral, regional, or commodity-based policy declarations towards the articulation of centralized and comprehensive agricultural policies. The period 1946–1967 featured each regional government following different policy documents in different directions. The absence of a federal agriculture ministry at that time meant that these policies lacked central coordination. In an aggregate sense, therefore, there was no internal consistency of agricultural development policies, which led to possible wastage of national resources. The separatist tendency was amplified by the First National Development Plan (1962–1968), which was merely an amalgam of regional policies that were not harmonized into a single national policy document for agricultural development in the country, which is to provide the overall policy direction for the states.

In this situation, the efficiency of agricultural resource use will be enhanced as a result of the possibility of internal comparative advantage arising from agroecological specialization of the country. In practical terms, this means the ability of the regions, which differ in climatic, edaphic, and socio-economic factors to use to the best advantage any particular benefits in skill and resource endowment. Through specialization of this kind, wasteful duplication of efforts will be avoided, which enhances allocative efficiency in agricultural resource use. Agroecological specialization also extends to the use of special environmental niches such as the temperate enclaves of Jos and Mambilla plateaus in special ways, which is only possible under the operation of a single policy document for the whole country. Harmonization of state agricultural policies has been achieved gradually through the second, third, and fourth development plan documents, which provide the experience incorporated into the preparation of the 1988 National Policy on Agriculture (FMAWRRD 1988).

The second trend to be discerned is the changing philosophy of agricultural development gradually from the need to support the metropolitan British economy during the colonial period to an inwards-looking posture. The tendency of colonial agricultural administration was towards the surplus extraction philosophy of development (Idachaba 1985). In this mode, immense quantities of forest products and other agricultural commodities were exported for the purpose of meeting the raw material requirements of British industries (Ayoola 1998). As a result, improved welfare of the people would be merely the residual outcome of the surplus extraction process. However, the situation had changed since independence. The country as a sovereign body pursues inwards-looking policies to the eventual benefit of its citizens. The deliberate attempt in this direction is reflected in the subsequent plan documents. The second plan, in particular, was quite explicit on this as the first document to provide for agriculture on a concurrent legislative list. The plan states, 'All government policies and actions should be guided solely by the best interests of the people of Nigeria' (Nigeria 1970).

The current policy atmosphere is dominated by the self-reliance philosophy, which is the terminal point of the nationalistic tendencies of post-independence development efforts. The philosophy of self-reliance as a strategy of economic development implies, in the main, a country's determination to use domestic resources for creating utility for its citizens. It represents a definite departure from the open-ended importation of goods and services, whose domestic resource costs of production are simply greater than their foreign exchange costs. For a determination of this kind to endure and be meaningful, however, a necessary, albeit not sufficient condition is the pursuit of self-sufficiency in the priority goods and services among which food items rank superior. The most common criticism of the self-sufficiency policy is based on the desire for the attainment of free market in the world economy. Antagonists hold the view that the policy limits the attainment of competitive equilibrium whereby the most efficient allocation of world resources can be achieved. Consequently, self-sufficiency policy is thought to be capable of imposing a low ceiling on the volume of global output attainable, thereby reducing the welfare status and standard of living of the entire world's people. On the other hand, the sponsors of self-sufficiency policy advance the arguments of social utility provision, infant industry argument, socio-economic discontinuities, and the theory of second best (Idachaba 1984; Ayoola 1998). In the first case, social utility provision results from the implicit gratification which people feel because they are able to feed themselves. This means that domestic availability of food is a public good which is capable of yielding social utility. In the second case, the infant industry argument rests on the need for certain industries to receive extra protection until they become securely established. Self-sufficiency policy helps to shield the infant industry well away from the outside influences of the imperfect world market until it develops adequate competitive competence. Nigerian agriculture—and food production in particular—fits into the infant industry classification because it is still characterized by technical and economic disequilibria resulting from its movement from old to new production surfaces. In the third case, the possibility of socio-economic discontinuities in the exporting countries makes high-level dependency, especially for food and other essential commodities, a high economic risk. The variety of these external shocks includes supply shortfall arising from prolonged drought, large-scale pest infestation, civil disturbances, and possibly war. The last case concerns the justification of self-sufficiency policy on the basis of the theory of second-best (Lipsey and Lancaster 1956). This states that once one or more of the conditions of pareto-optimality has been violated, there is no guarantee that the optimal solution is attainable through the pursuit of the remaining conditions. In the context of this discussion, the evidenced defects in structure conduct and performance of world commodity markets are such institutional restrictions, which make the maximum world output unattainable. Therefore, there appears to be no assurance that the strictest adherence of a developing country to the ideals of competitive market mechanism will make the maximum global output attainable.

The last trend to be discerned in the policy process is the gradual movement away from heavily supported agricultural production system towards a free-market agricultural economy. In this regard, the recent policy environment contains a number of market tendencies. First, government has embarked upon a programme of deregulation that has culminated in the abolition of the commodity boards. The commodity boards had earlier served for ten years (1976–1986) as the main instrument of agricultural price policy. Their establishment (Decree No. 29 of 1976) was informed by the need to protect the farmers against drastic falls in prices after harvest and pronounced fluctuations in prices in the world market. At the same time, it was envisaged consumers would be protected against abnormal food price increases during scarcity. These objectives could be achieved through the arrangements without provision for export restitution. In the case of food commodities, the boards served only as a buyer of last resort, also at fixed prices, and held strategic or buffer stocks until the time of scarcity when it resold to the consuming public. However, the market intervention policy was acting like a dog in the manger—it was not only failing to resolve the issues of market instability, insecurity of supply, unsuitable farm incomes, but was also preventing the solution by other means. Like the case of the earlier marketing boards, the inherent problems of the last generation of intervention boards were political, functional, and administrative in nature (Antonio 1984; Idachaba 1985; Ayoola 1989). Therefore, there appears to be substantial merit in the current agricultural market situation, which emphasizes decentralized price policies without the usual role of intervention bodies.

Another variant of the market tendency is the gradual movement away from a highly subsidized agricultural production system towards a relatively free input market. The prominent case in point is the reduction in fertilizer subsidy, which led to the rise in the official price from ₦10 per 25 kg bag in 1985 to ₦25 per bag 1989. The subsidy question is an important issue under the Nigerian agricultural production system. Originally, subsidies were introduced owing to the limited purchasing power of the small-scale farmers, which limited their demand for improved inputs and other technological innovations. However, the lessons of experience from the operation of direct subsidy programmes suggest that the policy became self-defeating in many ways. These include the following facts:

(a) A dependency mentality develops among farmers on government subsidy, thereby making it a permanent obligation of government and subsequently a huge fiscal burden;
(b) There is absence of proper allocative role of price following distortions created by subsidies, as production decisions of the farmers were not directed along the lines of efficient resource utilization;
(c) The level of morale is low among the private sector participants to build up substantial enterprise initiatives in the distribution of farm inputs and output of agricultural production, owing to price rigidities that are associated with subsidy, which ultimately produce a limiting effect on GDP;
(d) There is presence of substantial externalities whereby farm subsidy benefits illicitly flow to unintended channels within and outside the economy;
(e) The subsidy programme faces frequent exploitation to serve the selfish motives of civil servants, politicians, and other elite groups, to the detriment of the ordinary farmer.

As a result of the foregoing, the case is presently strong in favour of subsidy withdrawal, to be substituted with indirect subsidy modes such as provision of rural infrastructure and other facilitating functions of government. The case for infrastructural support is particularly supported on the additional basis that the problem of unintended beneficiaries will be minimized, because once installed, the farmers possess absolute, although not necessarily exclusive, right of access to the infrastructure.

Survey of Implementation Strategies

The employment of policy instruments for the development of agriculture at the initial stage was carried out at the various levels of authority, some loosely programmed and some not programmed at all.

A list of such policy instruments includes the generalized types such as agricultural extension, education/training, research, cooperatives, industries, infrastructure, legislation, credit input support, fiscal incentives, public enterprise, land reforms, social management services, international business assistance, settlement schemes, agricultural campaigns, and others. The implementation of a few agricultural development programmes will be briefly described and assessed for the purpose of discerning the implicit trend.

The first comprehensive agricultural programme was the farm settlement scheme of the old Western Region. Under the scheme, the graduates of the free primary education programme were established on farm sites together with credit, training, and infrastructural support. Only a few farm settlements exist to date, and the concept has petered out to an insignificant component of the Nigerian agricultural policy. Some of the reasons for failure of the farm settlement scheme include the following facts (FAO 1966):

(a) The philosophical motivation of young school leavers to run profitable farms under government assistance in training, credit, and infrastructure was an impracticable proposition 'to turn physically and mentally immature youths into serious-minded, hard-working farmers'; this is because young men can only be expected to settle down and devote themselves to productive work after they have passed the age of 20 and have married and thereby assumed responsibility for maintenance of other persons than themselves.
(b) The income-generating capabilities of the settlers were too low to meet the repayment schedules compared to the expectations of the prototype plans, which consequently prevented the fulfilment of the anticipated demonstration value of the scheme on surrounding farmers (also see Adegboye et al. 1969).
(c) The enormous capital outlay per settler renders the scheme incapable of making any significant contribution to the employment problem in the country given the rate of population growth.

The implementation experience of the farm settlement scheme is relevant to the employment component of the Structural Adjustment Programme (SAP). The two schemes are related in regard to the nature of participants. The Directorate of Employment, which was established to address the mounting unemployment problem associated with SAP, has provided immense funding and material assistance to new graduates at various levels of education. The failure of the farm settlement scheme to make desired impacts should inform the new directorate in the delivery of their service. In particular, it is instructive to note that apart from the limitations of the credit packages in terms of adequacy in timing of disbursements, it remains an empirical question to answer whether the agriculture graduates are technically and economically superior to the traditional farmers.

The integrated Agricultural Development Programme was mounted on a more extensive scale than the farm settlement scheme. It involves the establishment of command-area projects in sequence all over the country. The first generation of these was established at Funtua, Gusau, and Gombe in 1975. Newer enclave-type Agricultural Development Projects (ADPs) was established at Lafia, Ayangba, Ilorin, Bida, Ekiti Akoko, and Oyo North. At present, each state of the federation has one ADP, all at statewide levels. The system of ADPs is supported by two technical service agencies, which are Federal Agricultural Coordinating Unit (FACU), presently having its headquarters at Ibadan and Agricultural Projects Monitoring and Evaluating Unit (APMEU), which has its headquarters at Kaduna. Federal and respective state governments together with development loan assistance of the World Bank jointly fund the investment phase of a typical ADP. The original aim of the ADP is to establish an autonomous management system for: (i) input delivery through a network of farm service centres, (ii) construction of a massive rural road network for opening up new areas of cultivation and farm produce sourcing, and (iii) providing a revitalized, intensive, and systematic extension and training system backed up by timely input supply and adaptive research services.

The discernable trends in the implementation of ADPs are two:

First, there have been several reports of incessant political and other interferences, which create remarkable deviation from the designed paths in different aspects ranging from leadership posts, use of project facilities, location of a headquarters, to initial scale of project.

Second, despite the presence of built-in monitoring and evaluation, a number of projects have failed to make desired impact. Lastly, there is a strong tendency towards low funding commitments among the sponsors of the ADPs. Delayed or inadequate funding of agricultural projects could be responsible for the evidenced poor extension effectiveness, slow infrastructural build-up, and consequently, laggard technology uptake among the farmers.

The slow process of achievement among the ADPs led to the establishment of the Directorate of Food, Roads, and Rural Infrastructure (DFRRI) in 1986 as a means of accelerating the rate of agricultural and rural transformation. As a supra-ministerial body located within the presidency, DFRRI was expected to remove the usual administrative bottlenecks and bureaucratic snarls that characterized the mainstream ministries in the performance of agricultural services. However, the body had been criticized for its lack of proper role focus and programme accountability because of initial high attention it paid to low-priority projects as well as the failure of its organs to match their actions with the huge expenditure outlays (Idachaba 1988).

The River Basin Development Authorities (RBDAs) were established (Decree 25, 1976) with the aim to develop the economic potential of the massive water bodies in the country. In particular, they had specific mandates in irrigation services and fishery while hydroelectric power generation and domestic water supply are secondary functions. There were eleven authorities at the outset, which grew to eighteen in number during the second republic. Subsequently, the operation showed strong tendencies towards role confusion by performing non-statutory functions, particularly direct agricultural production. Many of them grew out of proportion while the operations of others suffered from intense political interference. As regards the latter, a good example is the litigation against one of the RBDAs (Ogun-Oshun) that paralyzed its functions for a reasonable length of time. The authority was taken to court jointly by six state governments (which were politically opposed to the federal government) to challenge its constitutional right to jurisdiction in their territory. It has taken substantial loss of public fund to reverse the trend shown by the RBDAs, through efforts to streamline their sizes, reduce their number to the original eleven and to dispose of their non-water assets.

SUMMARY AND CONCLUSION

At independence, the surplus extraction policy of colonial agricultural development was immediately replaced by inwards-looking policies. The latter have evolved gradually from region-based objectives to a single centrally coordinated policy document, which recognizes self-reliance as a strategy of agricultural development. The current emphasis on food self-sufficiency is premised on the need to create additional social utility protect the infant agricultural industry, and to limit the country's vulnerability to the effects of uncontrollable social climate and economic problems of other countries.

However, the great efforts made to mount agricultural programmes and projects have been largely frustrated by unsatisfactory implementation results. Based on available empirical evidence, it is necessary to take steps in reversing the trend towards persistent problems of political interference, role confusion, lack of accountability, and fluctuating commitments during the implementation of agricultural projects.

REFERENCES

Ayoola, G. B. (1998), 'The Dependency Problem in Food Grain and Rationale for Self-Sufficiency Policy in Nigeria', *Rural Development in Nigeria* 3 (1) (June 1998): 12–20.

Antonio, Q. B. O. (1984) 'Marketing Development in Nigeria: A Review of Relevant Government Policies' in J. P. Feldman and F. S. Idachaba (eds.), *Crop Marketing and Input Distribution in Nigeria* (Ibadan: Federal Agricultural Coordinating Unit), December 1984.

FAO (Food and Agriculture Organization) (1966), *Government Marketing Policies in Latin America*, Report of the FAO seminar, Bogotá, November/December 1966 (Rome: FAO).

Federal Government of Nigeria (1970), *Second National Development Plan (1970–1974)* (Lagos: Government Press).

FMAWRRI (Federal Ministry of Agriculture, Water Resources, and Rural Development (1989), *Agricultural Policy for Nigeria*, April 1989, (Lagos/Abuja).

Idachaba, F. S. (1985), 'Commodity Boards in Nigeria: A Crisis of Identity' in K. Archin, B. Aespen, and L. Van der Lann (eds.), *Marketing Boards in Tropical Africa* (London: KPI).

Lipsey, R. G. and K. Lancaster (1967), 'The General Theory of Second Best', *Review of Economic Studies* 34: 95–124.

Table 1. Typology of Agricultural Programmes and Projects in Nigeria, 1960–1989

Programme/Project	Description
Farm Settlement Scheme	Initiated in old Western Region, aimed at solving unemployment problem among primary school leavers. Policy instruments include agricultural extension, cooperative societies, and credit facilities.
National Accelerated Food Production Project (NAFPP)	Aimed at enhancing farmers' technical efficiency in the food production of selected crops (mostly grains). Policy instruments include subsidy, credit, adaptive research, and demonstration plots.
Operation Feed the Nation	A mass mobilization and mass awareness programme. Policy instruments include mass media, centralized input procurement, massive fertilizer subsidy, and imports.
River Basin Development Authorities (RBDA)	To tap the potentials of available water bodies, first eleven (11) were developed, then eighteen (18) and eleven (11) in number. Specific authorities' (RBDA) objectives are irrigation services, fishery development, and control of flood, water pollution, and erosion. Policy instruments include input distribution, credit services, infrastructure development, and manpower development.

Agricultural Development Projects (ADPs)	To enhance the technical and economic efficiency of small farmers in general. Policy instruments include rural infrastructure development (feeder road network, dams, etc.), revamped input delivery system, and revitalized agricultural extension system, autonomous project management, and domestic-cum-international capital.
Green Revolution Programme	To accelerate the achievement of the agricultural sector objectives. Policy instruments include food production plan, input supply and subsidy, special commodity development programme, review of Agricultural Credit Guarantee Scheme, increased resource allocation to RDBAs, etc.
Directorate of Food, Roads, and Rural Infrastructure	Established to facilitate programmes in food production, particularly through the provision of rural infrastructures

Source: Miscellaneous documents

Self-Sufficiency in Food Production: An Essential Foundation for Economic and Industrial Development in Nigeria

Commissioned paper presented at the 9th Annual Training Conference of the Industrial Training Fund, Zaranda Hotel, Bauchi, 9–10 November 1989.

The hardships of the past few years are the quintessence of economic recession. The prominent aspects of these hardships include massive unemployment, low capacity utilization, frustrated social programmes, substandard nutrition, and unaffordable prices. Besides, there is great strain on the foreign trade position and the financial system of the country, as well as pronounced friction in the dynamics of Nigeria's social organization. But it also provides a great many lessons for economic management by highlighting weaknesses in the structure of production exchange and distribution. The significant contribution of balance-of-payments disequilibria to the magnitude of the problem evidently makes the resurgence of self-reliance as a theme of economic and industrial development quite expedient.

The overall objective of this paper is to survey the entire agricultural economy of Nigeria with a view to examining some of the immediate and remote causes of the hardships. After establishing the rationale behind the self-reliance policy in food production, we shall (i) present the historical perspective and present outlook of the food self-sufficiency strategy, (ii) specify the future needs and prospects of the process, and (iii) highlight the appropriate institutional mechanism for attaining process efficiency in regard to the self-sufficient targets. The paper is subsequently concluded with broad recommendations.

RATIONALE FOR SELF-RELIANCE POLICY IN FOOD PRODUCTION

The self-reliance strategy of economic development implies, in the main, a country's determination to use its domestic resources in creating utility for its citizens. It represents a definite departure from the open importation of goods and services, whose domestic resource costs of production are simply greater than their foreign exchange costs. It therefore implies that for a determination of this kind to endure and be meaningful, a necessary, albeit not sufficient condition is the pursuit of self- sufficiency policy in the priority goods and services, among which food items should rank superior.

We broach the controversy over the self-sufficiency policy from the negative side. Then, the negativities are immediately neutralized, before we proceed in the positive direction to establish the rationale for the food self-sufficiency policy. The popular argument against self-sufficiency among economic theorists is that it prevents the attainment of equilibrium in the world market. In practical terms, it is taken as a deterrent to the free market competition, which therefore limits the efficiency of utilization of world resources and consequently reduces the level of global outputs attainable. Hence, the policy lowers the welfare status and standards of living of the entire world's people. This proposition follows directly from the theory of comparative advantage and it is apparently valid.[1]

How is the argument to be evaluated? First, we note that the proponents of free world market often say nothing about how the increased volume of world commodity output is to be distributed efficiently. Second, what makes us believe that with the strictest adherence of a developing country to the rules of competitive market the maximum attainable global output is still achievable?

As regards the first counterargument, ample empirical evidence exist to show that the problem is not so much inadequacy of world output as the absence of efficient distributive system to translate existing output into welfare parameters.[2]

---

[1] For good exposition on the theory of comparative advantage, see Kindleberger (1973).

[2] For instance, according to Timer et al. (1983), 'relative to an arbitrary average energy requirement of 2,500 calorie per day 1972 world grains production was 128% of requirement, while 1978 production was 143%'. Therefore, the caloric equivalents of world's annual output of basic grains are significantly greater,

To the question about the existence of other distortions in the world market than disparate pursuit of self-sufficiency policy among developing countries, the answer is provided in the theory of second best (Lipsey and Lancaster 1956). This states that once one or more of the conditions of pareto-optimality has been violated, there is no guarantee that the optimal solution is attainable through the pursuit of the remaining conditions. In the context of this discussion, the evidenced defects in structure, conduct, and performance of world commodity markets are such institutional restrictions that make the maximum output unattainable. It is quite irrelevant, therefore, to inquire whether a second best position can be attained by satisfying the remaining pareto conditions. Following this theory, the international criticism of the self-sufficiency policy in developing countries on the account of violation of the equilibrium conditions of the world market is spurious.[3]

On this note, we can proceed to establish the rationale for the pursuit of food self-sufficiency in Nigeria on three positive grounds:

1. Self-sufficiency can be perceived in the context of a welfare function. In this case, the domestic availability of food can be taken to be a public good, which is capable of yielding social utility. This utility is in the form of the implicit gratification that the citizens of the country feel from being able to feed themselves.
2. The infant industry argument applies. Certain industries should receive extra protection until they become securely established. Self-sufficiency policy helps to shield the infant industry well away from the outside influences of the imperfect world market until it develops adequate competitive competence. There is substantial merit in the notion that Nigerian agriculture and food production in particular, is an infant industry because it is still characterized by technical and domestic disequilibria resulting from its movement from old to new production surfaces. In this regard, however, the free market school warns that the artificial barriers mounted around the infant industries create inefficient resource utilization, the consequence of which is the domestic availability of inferior goods at higher prices. Even though impressionistic evidence confirms this point in the Nigerian case as regards the motor car and other industries, as well as rice and wheat, it is likely to be a short-run phenomenon only.

---

and are never lower, than the amount of nutrients that are needed for human survival. This happens in the presence of widespread hunger in Asia and Africa.

[3] For further discussions of issues in self-sufficiency policy, see Ayoola (1989) and Idachaba (1985).

3. The food security argument is also very important. It holds that high-level dependency, especially for food and other essential commodities, is an economic risk. This is because it opens the importing country to the effects of socio-economic shocks in the exporting country. The variety of these external shocks include supply shortfall arising from prolonged drought, large-scale infestation, civil disturbances, and possibly war. Table 1 provides a measure of Nigeria's vulnerability to possible socio-economic shocks in selected countries, which is high for UK and USA.

THE HISTORICAL PERSPECTIVE AND CURRENT OUTLOOK

The historical and contemporary events concerning food self-sufficiency can be highlighted in three segments: these are (i) policy and planning framework, (ii) strategies and programmes, and (iii) instrumentalities. These will be substantially discussed in regard to the post-independence period.

Food Self-Sufficiency: Policies and Plans

In all, Nigeria has experienced four official planning phases preceding the present structural adjustment phase since independence. The First National Development Plan 1962–1968 was launched as the first truly Nigerian attempt in national planning. In this plan, conscious efforts were made to quantify national objectives, but these were not comprehensive enough. For example, there were definite statements to attain 15% saving of GDP by 1967, 15% annual investment of the GDP, and minimum of 4% growth rate of GDP during the planning horizon. Conscious efforts were also made to ensure a common national planning framework such as that contained in the acceptance of general priorities by all governments in which the highest was accorded agriculture, industry, and training.

In regard to agriculture, the obvious aberration of the first plan was the absence of an agriculture ministry at the centre to coordinate regional development programmes. This feature is a spillover problem from the federal constitution (1954) that assigned agriculture a residual responsibility. Consequently, the first plan, like earlier plans,[4] only embodied separate

---

[4] The pre-independence planning efforts consist of two documents: 'The ten-year plan of Development and Welfare (1946–1954)' and 'Economic Development Plan (1955–1962)'.

programmes with projects in each region overlapping one another but without a central policy framework. At the individual regional policy levels, however, the first plan operated on dependency mentality, which provides for borrowing or importing management and funds.

The war economy occurred shortly before the official termination period of the plan (1967). The interruption of the civil war meant that all available resources were mobilized for its prosecution. The war years (1967–1970), therefore, constitute a vacuum in the planning process.

Subsequently, the Second National Development Plan 1970–79 was launched against the background of the need to remove the war effects, typified by the adoption of the three Rs—rehabilitation, reconciliation, and reconstruction. In general policy, it featured definite departures from the first plan, particularly with that of self-reliance and self-sufficiency. In terms of societal goals, the second plan aimed at the building of 'a united, strong and self-reliant nation', 'a great and dynamic economy', 'a just and egalitarian society', 'a free and democratic society', etc. In particular regard to agriculture, it is the first document to provide for the sector's development on the concurrent legislative list, which thereby made it a joint responsibility of the state and federal government.

The Third National Development Plan 1975–1980 came into being at the height of the oil boom, during which foreign exchange was not a serious constraint. This plan, therefore, relegated the notion of self-reliance and self-sufficiency to the lowest background, making sufficient provisions for the importation of the several inadequacies of the economy, including technology. Consequently, the plan period featured the greatest volumes of food imports (Table 2).[5]

---

[5] The historical antecedent of the federal agricultural ministry is relevant at this juncture: Following an FAO recommendation and in the presence of the limitation still imposed by the federal constitution, the Federal Ministry of Natural Resources and Research was created in 1965 (see Nigeria, *Official Gazette* vol. 52 N. 33 of 9 April 1965). Apparently, the word 'agriculture' was carefully avoided by the then civilian administration so as not to 'offend the political sensibilities' of the regions, which still possessed the constitutional right over agricultural development at that time (see Ayida, 1973). However, the military administration used an executive fiat to establish in 1966 the Federal Ministry of Agriculture and Natural Resources (see Nigeria, *Official Gazette* Vol. No. 10 of 7 February 1966). Since that time, the name of the ministry has been variously modified to become what is called Federal Ministry of Agriculture, Water Resources, and Rural Development.

The Fourth National Development Plan 1981–1985 followed immediately. Although a conscious re-emphasis on the notion of self-reliance was continued, agricultural imports also thrived during the plan period. But the later part of the plan period suffered from acute foreign exchange shortages, which led to widespread scarcity of essential commodities mostly consisting of food items. The frantic moves to resuscitate the plan include an abortive attempt to obtain a recovery loan from the International Monetary Fund (IMF). Another development plan has been set aside for four years under the indigenous Structural Adjustment Programme (SAP). Under the SAP, self-sufficiency in food production is accorded high priority. Consequently, food imports have been put under indefinite ban. Domestic policies to reactivate and sustain supply have been designed within the guidelines of market economy, direct government production is being discouraged, and emphasis is firmly on the small-scale farmers as the central focus of the food production process. And all these have been articulated in a national policy on agriculture (FMAWRRD 1988).

Food Self-Sufficiency: Strategies and Programme

We shall highlight a few prominent ones out of the numerous programmes of agricultural development that have been mounted since independence.

1. Farm Settlement Scheme (FSS)

This was initially launched by the old Western Region (Western Nigeria 1963). The concept of farm settlement was primarily necessitated by the need to stem an impending unemployment problem among the products of the free primary education scheme. The potential of agriculture was recognized as a possible means of engaging the graduating youths in gainful occupation. The essential elements of the farm settlement in the old Western Region are as follows:

- The establishment of farm sites to settle young school leavers together with credit assistance and infrastructural support
- The establishment of farm institutes for the training of prospective farmers and providing ad-hoc courses of a general or specialized nature for established farmers
- A school uniform scheme through a system of cooperative tailoring to produce school uniform.

The farm settlement still exists, to date, but has petered out to an insignificant component of the generalized agricultural development policy instrument. The buildings are scattered over the states with few or no settlers in place. A study conducted by FAO (1965) adduced some reasons for the woeful ineffectiveness of the scheme:

(i) The philosophical motivation of young school leavers to run profitable farms under government assistance in training, credit and infrastructure was an impracticable proposition 'to turn physically and mentally immature youths into serious-minded hard-working farmers'; the general opinion holds that young men can only be expected to settle down and devote themselves seriously to productive work after they have passed the age of 20 and have married and thereby assumed responsibility for maintenance of other persons than themselves.

(ii) The income-generating capabilities of the settlers were too low to meet the requirement schedules compared to the expectations of the prototype plans, which consequently prevented the fulfilment of the anticipated demonstration value of the scheme on surrounding farmers (also see Adegboye et al. 1969).

(iii) The enormous capital outlay per settler renders the scheme incapable of making any significant contribution to the employment problem in the country at the rate of population growth.

2. National Accelerated Food Production Project (NAFPP)

The earliest consolidated programming efforts of the new federal agriculture ministry were the National Accelerated Food Production Project (NAFPP) of its Federal Department of Agriculture (FDA). The project has its roots in the Accelerated Cereal Production programme earlier identified by a team of experts who surveyed the Nigerian food condition 'with a view to recommending means of implementing an integrated extension and research programme which could stimulate the masses of Nigerian farmers to dramatically increase food production' (FDA 1974). The initial concentration of the project was on rice, maize, sorghum, millet, wheat, and cassava.

The NAFPP was designed to employ 'green revolution' techniques, particularly including the utilization of high-yielding seeds, fertilizers, credit, and other inputs together with the intensification of specialized extension and adaptive research and training for staff. Technology transfer to the farm level involves pilot schemes, including 'mini-kit' and 'production kit' trials.

3. The Agricultural Development Project (ADPs)

The ADPs are integrated agencies in the sense that the same administrative system performs technological enhancement in many areas of agriculture and also provides a number of rural infrastructural support in the rural areas contemporaneously. In concept, an ADP is an area-based agricultural development strategy that is designed to yield fast neighbourhood effects among the small-farmers population. In terms of funding, it is originally financed under the tripartite arrangement by the federal government, respective state government, and the World Bank. The first generation of ADPs was the enclave types established at Funtua (April 1975), Gusau (April 1975), and Gombe (November 1975). Newer enclave-type projects were established at Lafia (1977), Ayangba (1979), Ilorin (1979), Bida (1980), Ekiti (1981), and Oyo North (1982). Presently, each state of the federation has one ADP, all at statewide level.

The core elements of the ADPs primarily include

- An input delivery and credit supply system through a network of farm service centres which ensures that no farmer travels more than 5–15 kilometres to purchase needed farm inputs
- A massive rural feeder road network to open up new areas for cultivation and facilitate the rapid evacuation of farm produce as well as the timely delivery of farm inputs
- A revitalized intensive and systematic extension and training system backed up by timely input supply and adaptive research services; and
- Effective project management together with built-in project monitoring and evaluation.

All ADPs are under the Federal Department of Agriculture and Rural Development (FDARD), which coordinates their activities through the technical support services of its two agencies (i) Federal Agricultural Coordinating Unit (FACU Ibadan—Planning) and (ii) Agricultural Projects Monitoring and Evaluation Unit (APMEU Kaduna—Monitoring and Evaluation).

4. River Basin Development Programme

The River Basin Development decree was promulgated in 1976 to establish eleven River Basin Development Authorities (Decree 25 of 1976). The original aim of these authorities was to develop the economic potential of the existing water bodies, particularly irrigation and fishery, with hydroelectric power generation and domestic water supply as secondary functions. However, the programme soon extended their spheres of activity to other areas, particularly production and rural infrastructure development. The number of authorities was also increased from eleven to eighteen. However, the situation has now been reversed, with the number of authorities reduced to the original eleven and their roles streamlined.

5. Agricultural Campaigns

The first prominent post-independence agricultural campaign was the Operation Feed the Nation (OFN) programme, launched in April 1976. Earlier, there had been a 'freedom from hunger' campaign, meant to sensitize the people towards the solution to the food problem. The second campaign was launched in 1980 as the Green Revolution programme. The aim of this was to intensify efforts of the existing agencies for accelerating the solution process to the food problem. A food production plan was drawn up for this purpose which aimed at attaining self-sufficiency in food by 1985 and turning the country to a net exporter of food by 1987 (Food Strategies Mission 1980). In addition, numerous food production campaigns have been launched at the state level, including (i) Food for All Programme (Kara), (ii) Back to Land Programme (prominent in Niger, Ondo, and other states), (iii) School to Land Programme (Rivers), (iv) Operation Grow More Food (Cross River), and (v) Wheat Round Up (Kano).

6. Agricultural Programme of the NDE

The immediate policy reaction of the present administration to the mounting unemployment problem was the establishment of the National Directorate of Employment (NDE). Not unexpectedly, a programme of agricultural employment was initiated for the engagement of the youthful population. The most popular aspect is the Graduate Farmers Scheme (GFS).

Under this scheme, the null hypothesis that is actually being tested is that the calibre of the person, which is farming, does not really matter. Government had therefore provided credit assistance to aid the graduate farmers. The relevant elements of the programme are as follows:

(i) Each graduate receives ₦110, 000 credits in cash and kind forms; the conditions are quite liberal, including two guarantors and the mortgaging of the university degree certificate of each beneficiary; interest rate is low with substantial moratorium.
(ii) Each graduate is allocated 10 hectares cleared land.
(iii) Participants are to enjoy the benefits of existing state facilitating functions, including the input delivery system and marketing services.

The preliminary data collected from participant observation in this scheme reveals new dangerous trends that bear resemblance to the experience of the farm settlement scheme. The first is about characteristics of the credit package. Inadequacy and form and the package are faulty. Adequacy concerns the ability of the credit amount to meet the cost of inputs, transport, processing of other farm expenses, let alone expenditure on housing and general upkeep of the graduates in their rural abodes. As regards form, the uniform loan amount presupposes that the cost of production and other variables are the same throughout the country; this premise is spurious. The second aspect is the characteristics of the borrowers. The relevant question is how one sees a university agriculture graduate in his professional and social status. Professionally, his business horizon seems far wider than the opportunities provided by the present package. In particular, the low level of capital provided in the package appears quite incongruous with the exposure of the graduate to efficient machinery and other farm technologies, which have been recently experienced in the university curriculum. Therefore, at the very best, the participants would only continue to feel that circumstances have only (probably temporarily) pushed them to partake in the scheme as a second best option. In corollary, the graduate farmers' scheme is at best a temporary solution to the unemployment problem. If as envisaged, and when the urban economy picks up, the rate of desertion to the preferred sectors is likely to be high as experienced under the farm settlement scheme. Besides, there is no evidence that the graduates are actually performing better in farming than the more experienced traditional farmers.

7. The Directorate of Food, Roads, and Rural Infrastructure (DFRRI)

The last agricultural development programme to be discussed is that executed under the Directorate of Food, Roads, and Rural Infrastructure (DFRRI). The establishment of DFRRI in 1986 was informed by the need to accelerate the process of enhancing the technical efficiency of the rural workforce, particularly farmers, through the provision of rural infrastructure. The distinguishing features of DFRRI are as follows:

- It is a supra-ministerial body with headquarters located within the presidency; this is done for the purpose of accelerating decision-making and eliminating the usual wasteful delays inherent in the ministries.
- It enjoys generous funding compared to other bodies; the sum of ₦450 million was initially set aside for its use out of the possible ₦900 million realizable from the newly introduced second-tier Foreign Exchange Market; the subsequent budget allocations are ₦360 million (1986), ₦400 million (1987), ₦500 million (1988), ₦300 million (1989).
- DFRRI is not modelled to implant its own bureaucracy for doing things, but to cause things to be done within the framework of existing machineries.

FUTURE NEEDS AND PROSPECTS

The realization of the available potentials depends on the provision of critical needs in the future, three of which have been identified.

Ecological Specialization

First is the pursuit of ecological specialization in food and agricultural production, as recently emphasized in the new agricultural policy. By ecological specialization is meant the permission and encouragement of the regions, which differ in climatic, edaphic, and socio-economic factors to use to the best advantage any particular benefits in skills and resources endowments. Through specialization of this kind, wasteful duplication of efforts is avoided, thereby creating high resources use, efficiency, and internal comparative advantage. Agroecological specialization also extends to the use of special environmental niches such as the temperate enclaves of Jos and others like Mambilla and Obudu Plateaus, in particular ways. For example, maize and other grains production efforts may require being concentrated in the derived savannah middle belt and certain parts of the upper north, so that the tree crop economy of the southern part would receive greater attention for output maximization.

The practice of agroecological specialization, however, has definite implication for marketing policy. The relevant issue in this context concerns some state governments that sometimes restrict commodity trade. There are reports that certain states of the federation sometimes impede the outflow of food items to other states, usually as a panic measure during shortages.[6] This practice in itself is capable of working counter to the policy of agroecological specialization. This happens, as every state will attempt to grow everything for the fear that specially endowed states might later obstruct free flow of the relevant commodities. In the end, the whole country loses out in terms of resource use efficiency, efforts wastage, and consequently low volume of national output.

Therefore, the practical implication of ecological specialization as a national strategy of food production is that one cannot speak of self-sufficiency in any one product on a state basis. Nigeria should operate, as one large, indivisible market in which the surplus output of specially endowed producing areas should move freely to deficient areas without hindrance. This requirement of the production and market structures is made highly necessary by the spatial distribution of industries in relation to raw material sources in Nigeria.[7] It is therefore necessary to zone the country according to commodity productivities along which development efforts will be applied, using efficiency criteria.

---

[6] A copious example is Kaduna State where the government issued instruction 'to prevent out-of-state traders from buying off and evacuation of this season's harvest of grains'. Consequently, the state-owned Farmers' Supply Company (FASCOM) instituted a programme 'against all those who will buy grains cash down and carry away from the state' (Anonymous).

[7] A copious example is Kaduna State where the government issued instruction 'to prevent out-of-state traders from buying off and evacuation of this season's harvest of grains'.

Technological Progress

The second need for future economic and industrial development is aggressive technological progress in agricultural production and marketing. We shall characterize agricultural technology as consisting of the nature and types of available inputs (e.g. seed, fertilizer, chemicals, tools, machines, farm power, etc.) and the ways in which they are combined (e.g. land-fertilizer ratio, labour-machine ratio, etc.). Technological progress is desirable because it reduces average unit costs of output, with input prices held constant. The relevant areas where this is needed in Nigerian agriculture include the following:

(i) Technology of land preparation
(ii) Technology of planting, seed and seedling
(iii) Technology of farm management practices
(iv) Technology of harvesting
(v) Technology of on-farm haulage, processing and storage.

These categories clearly imply that agricultural development is *sine qua non* for industrial development through which the technological drive can be provided and the momentum of progress maintained.

There are three phases through which technological progress in agriculture could be initiated. These are (i) agricultural research for technology creation, (ii) diffusion of agricultural technology, and (iii) development of technological intellect. For meaningful progress to take place, determined efforts are required in these areas, which specifically include the following:

- The adoption of a technology policy that is all-embracing and complete;
- The creation of effective agricultural technology through a virile national agricultural research system;
- The transfer of agricultural technology through the development of technological intellect;
- The adoption of appropriate agricultural technology on merit criteria, including ecological and sociocultural considerations, simplicity, relative availabilities of capital and labour, divisibility and riskiness for the particular application of small-scale farmers.

Rural Infrastructure Support

The view is widely held that the provision of rural infrastructure is the essential foundation for sustained increases in food production. The general notion underlying the rural infrastructure strategy is that it is difficult for the rural sector to contribute significantly to economic and industrial progress in the absence of basic facilities that also enhance their production activities, as well as their living standards. These facilities and their roles are generally described under three categories:

(i) Rural physical infrastructure, including

- Rural roads which cause accelerated delivery of farm inputs, reduce transportation costs, and enhance spatial agricultural production efficiency;
- Storage facilities which help to preserve foods in the forms that consumers need them and at the time they need them;
- Irrigation facilities, which assure farm water supply and stabilize food production by protecting the farm production system against uncontrollable and undesirable fluctuations in domestic food production.

(ii) Rural social infrastructure, including

- Clean water, decent housing, environmental sanitation, personal hygiene, and adequate nutrition, which help to improve the quality of rural life;
- Formal and informal education, which promote rural productivity by making the farmer able to decode agronomic and other information and carry out other desirable modern production practices; basic education also promotes feeding quality, dignity, self-respect, and sense of belonging as well as political integration of the rural people.

(iii) Rural institutional infrastructure, including

- Farmers' unions and cooperatives, which facilitate economies of scale and profitability of rural enterprises;
- Agricultural extension, which improves the technology status of the farm business.

However, the rural infrastructural development process of Nigeria contains a number of persistent problems and unresolved issues. Ayoola and Idachaba (1989b) highlight some of these, including the following:

(a) Poor commitments of the programme sponsors,
(b) Incessant perturbations in the institutional framework,
(c) Low level of in-house operational research and planning of projects,
(d) The resources required for meeting infrastructure requirements,
(e) The trade-off between rural infrastructure supply and other objectives.

These problems and issues border on the efficiency of the institutional mechanism within which rural infrastructure strategy operates. However, such institutional frictions are not limited to infrastructure development only, but also permeate through the complex fabric of the overall self-sufficiency strategy. Therefore, we now turn to discuss some of these as elements of appropriate institutional framework in broader perspectives.

APPROPRIATE INSTITUTIONAL FRAMEWORK FOR ECONOMIC AND INDUSTRIAL DEVELOPMENT

Commitment

The problem is not so much absence of projects for achieving self-sufficiency in food production as relatively ineffective projects. One reason for ineffectiveness of agricultural and rural development projects is the lack of sustained commitment of sponsors. Commitment can best be illustrated in terms of funding support. For instance, Table 3 shows that the actual fund allocation to ADPs is fairly lower than budgeted amounts (consider the total and average commitment values). In practical terms, dwindling funding commitment results in midstream abandonment of key project components as well as poor execution of some others.

Perturbation

Two levels of perturbation exist in the food production process. First, at the policy level, changing perception of the role of government has pronounced effects on the continuity of the overall rate of progress. As earlier highlighted, the successive plan documents contained fluctuating fortunes in respect of the self-sufficiency policy objectives. If the role perception of government continues to fluctuate in the future, the self-sufficiency attainments become uneven, and at each point of change, some part of the recorded achievements of the proceeding efforts will definitely be lost.

Second, at the organizational level, perturbations arise mostly from (a) frequent fusion and breakages of parts of the main ministry, and (b) frequent personnel changes in leadership positions (Table 4). The evidenced organizational instability and high rates of personnel turnover in the political, administrative and professional leadership positions create adverse effects. They end up in particular programmes being given more or less emphasis, redesigned, re-introduced or the implementation pace speeded up or slowed down, so as to reflect the new political, philosophical, ideological, and occupational biases of the new people involved. Again, the agricultural development pathway consequently becomes uneven, which in turn affects the economic and industrial development adversely.

Operational Research and Planning

There is the need to examine the inner mechanism of the agencies responsible for food production development. The important technical processes that need perfection include the following:

(i) The processes of budgeting and costing, screening, and evaluation;
(ii) The processes of tendering or contracting and force account implementation;
(iii) The process of cost control, quality control, and supervision.

In regard to these processes, the inherent problems are as follows:

(a) The existence of a great paucity of data that leads to the use of old sets of data and inadequate, inconsistent statistics in the planning and execution of agricultural projects
(b) Shortage of operational research activities to back up food production planning and implementation processes. This often results in the use of inappropriate generalizations of concepts and measurements.
(c) High implementation failures owing to lax supervision and maintenance outfits, as well as wasteful delays in the procurement and start-up of major projects

(d) Glaring role conflict and poor performance resulting from inappropriate harmonization of role expectation and role perception among the various agencies involved. For instance, the indiscrete expansion of the functions of RBDAs from the original location-specific, river-based objectives to include direct agricultural production activities had caused a thin spread of limited resources and the consequent inefficiency of programme performance. This informed the recent revision of the underlying policy concerning the authorities, which in itself is not without substantial social costs. In addition, DFRRI is on the verge of repeating the same experience by virtue of its reported direct agricultural production efforts in some states of the federation.

Balance of Rural and Urban Development Objectives

Although the attainment of self-sufficiency in food production is an important objective, it is not the only objective. Some other objectives which may sometimes compete with it are high income to farmers and low food prices for urban consumers at the same time. In addition, technological progress of a developing agricultural economy for self-sufficiency often implies increasing dependence on farm input imports, expatriate management, and foreign capital. This is so because the country lacks compensating domestic production capabilities in the necessary mechanical, chemical, and biological inputs required for the self-sufficiency production drives. Although the resultant effect is a change in the nature, rather than the degree of food dependency, the increased importation of goods and services often leads to increased urban income as a result of increased volume of distributive trade and other effects.

Even within the domestic economy itself, some internal objectives might be conflicting depending on the relative incidence on urban and rural populace. An illustration will suffice: the completion of a new feeder road network to enhance food production efficiency and rural living standard also opens up the rural communities concerned to outside markets. The inevitable increase in trade traffic and the sudden demand pressure will often lead to price increases and hence substantial reduction in the welfare of the generality of the people in the short run.

The above points imply that the self-sufficiency strategy must attempt to balance inconsistent objectives so as to achieve internal harmony that ensures positive incremental benefit to the society at large. The following issues, therefore, arise:

- What are the trade-offs, if any, between self-sufficiency policy objectives and alternative economic objectives and by what means can the trade-off curves be shifted in desired directions?
- What constitute the empirical relationships between economic growth and development on the one hand and the self-sufficiency policy on the other?

Public and Private Sector Roles

It has taken substantial efforts to persuade government that it cannot successfully do business alongside with governance. The argument, which defines the appropriate roles for the public and private sectors in agricultural development, can be presented from two angles. From the government angle, public obligations are based on four grounds as follows:

(i) Government is called to provide facilities that require lumpy expenditures, which are generally beyond the reach of ordinary individuals;
(ii) Government is called to provide facilities having substantial free-rider problem, which thereby discourages private individuals;
(iii) Government should provide facilities, which represent durable stocks of capital and also require regular maintenance costs, and hence have lifelong intergenerational consequences;
(iv) Government should provide facilities requiring diversified inputs from different ministries and disciplines, which it only controls.

Therefore, the role of government should be limited to facilitating functions only. On the other hand, the business enterprise initiatives in agriculture should be made to reside primarily in private sector participation. Or, for one thing, the superior sensitivity of private investors to signals and incentives means that the nation's scarce resource will be better allocated by them. For another thing, the small private farmers constitute the bulk of the population and the largest part of the labour force, which means that any unfair competition that reduces their income status will produce substantial welfare losses in the population at the same time.

Aside from the arguments in favour of the private sector as the main production energy for achieving food self-sufficiency, several other arguments exist in disfavour of government direct production efforts in agriculture. The latter set of arguments ranges from the general ineptitude of government officials to business opportunities, to their lack of sensitivity to the usual economic variables, to the wasteful bureaucratic delays caused by strict adherence to official protocol and finally to the widespread corruption usually as a result of reckless political intransigency and the absence of financial and programme accountability among government workers. Usually, government agricultural business ventures liquidate out of limelight because of the inordinate ambition, frivolities, and incompetence of the members of the management units. They consequently turn out to depend perpetually on annual budgetary support from public resources.

SUMMARY

The rationale for self-sufficiency strategy in food production as an essential foundation for economic and agricultural development is predicated upon social utility provision, infant industry argument, socio-economic discontinuities, and also the theory of second best. But the massive policy and programming efforts in Nigeria have achieved little in meeting pre-set food targets. The future prospects will definitely depend on how adequately certain needs are satisfied and some issues attended to. These include:

(a) The pursuit of ecological specialization,
(b) Significant progress in diversified agricultural technologies,
(c) Rural infrastructure support,
(d) Efficiency of inner processes of delivery agencies,
(e) Programme commitment and accountability, and
(f) The delicate balance of agricultural roles between the public and the private sectors.

REFERENCES

Adegboye, R. O., A. C. Basu, and D. Olatunbosun (1969), 'Impact of Western Nigeria Farm Settlement on Surrounding Farmers', *J. Econ. and Soc. Studies* 1 (2).

Anonymous, 'Format for 1984 grain purchase by FASCOM' Kaduna (undated).

Ayida, A. A. (1973), 'Business Issues in Financing the Nigerian Agriculture in the Seventies' in *Proceedings of the National Agricultural Development*

Seminar Federal Ministry of Agriculture and Natural Resources (Caxton Press).

Ayoola, G. B. (1988) 'The Dependency Problem in Food Grains and Rationale for Self-Sufficiency Policy in Nigeria' *J. Rural Dev. Nig.* 3 (1).

Ayoola, G. B. and F. S. Idachaba (1989a) Workshop paper, Society for International Development, Ibadan Chapter, 24–29 Sept. 1989.

Ayoola, G. B. and F. S. Idachaba (1989b) 'Technology and Nigerian Agricultural Development', International Conference, Federal University of Technology, Yola, 20–26 Aug. 1989.

FDA (Federal Department of Agriculture) (1974), *Annual Report*, Lagos.

FMAWRRD (Federal Ministry of Agriculture, Water Resources and Rural Development) (1988), *Agricultural Policy for Nigeria*, Lagos.

Federal Republic of Nigeria, National Development Plans, Lagos (four issues).

FAO (Food and Agriculture Organization) (1965), *Agricultural Development in Nigeria 1965–1980* (Rome).

Food Strategies Mission (1980), *The Green Revolution: A Food Production Plan for Nigeria* (2 volumes), Federal Ministry of Agriculture, Lagos, May 1980.

Idachaba, F. S. 'Self-Reliance as a Strategy for Nigerian Agriculture: Cornucopia or Pandora's box?' *Proceedings of 25th Anniversary Conference*, Nigerian Economic Society, 23–24 April 1984.

Kindleberger, C. P. (1973), *International Economics* (D. B. Taraporevala Sons & Co. PVT Ltd, India).

Lipsey, R. G. and K. Lancaster (1956), 'The General Theory of Second Best' *Review of Economic Studies* 24.

Western Nigeria White Paper on Integrated Rural Development, Official document No. 8, Ibadan, 1963.

Table 1: Indices of Nigeria's Vulnerability to External Shocks in Food Grains Production in the Major Sources of Import - 1979

| | Percentage Contribution to Total Imports – 1979 ||||||| Vulnerability index all grains %** |
	Wheat and spelt including meslin	Rice	Barley unmilled	Maize, corn unmilled	Other cereals unmilled	Meal flour wheat spelt including meslin	Meal and flour of other cereals than wheat	Cereal preparations	
USA	86.53 (1st)	18.68 (1st)	13.81 (3rd)	58.68 (2nd)	1.45 (5th)	40.22 (1st)	0.001 (5th)		26.72
Netherlands	0.04 (2nd)	1.80	-	-	10.50 (3rd)	-	1.24 (2nd)		1.14
France	5.31 (3rd)	-	-	1.98 (3rd)	0.71 (6th)	33.82 (2nd)	0.002 (4th)	12.21 (3rd)	5.27
UK	0.02 (4th)	1.35 (4th)	30.72 (1st)	-	46.35 (1st)	1.72 (5th)	98.75 (1st)	43.73 (1st)	27.60
Switzerland	-	0.75 (6th)	-	1.36 (6th)	-	-	0.01 (3rd)		0.04
Fed. Rep. of Germany	0.002 (5th)	2.34 (6th)	-	-	1.59 (4th)	11.61 (3rd)	-	13.58 (2nd)	5.76
Denmark	-	-	20.93 (2nd)	-	-	-	-		2.22
Belgium & Luxembourg	-	-	-	-	-	9.57 (4th)	-	1.76 (6th)	1.02
Ireland	-	-	-	-	-	-	-	6.78 (4th)	
China (Mainland)	-	1.06 (5th)	-	-	-	0.91 (6th)	-		0.02
Rep. of Benin	-	-	-	1.39 (5th)	-	-	-		0.04
Sweden	0.001 (6th)	-	-	-	-	-	-		0.06
Finland	-	-	-	1.46 (4th)	-	-	-		0.00
Austria	-	-	-	-	45.59 (2nd)	-	-		0.09
East Germany	-	-	-	-	-	-	-	5.31 (5th)	4.75
									0.22

Source: Underlying figures from Federal Office of Statistics, *Nigeria Trade Summary* (various issues). Culled from Ayoola (1988).

* Means that either the country concerned does not export the commodity to Nigeria at all, or that it does not fall into any position between the first and six among the class of external sources of the commodity in any one year within 1977–1979. Zero value implies that percentage contribution of the country total imports of the commodity is negligible (after round off to the nearest 100th).

** The vulnerability index (I) is constructed as the sum of the per cent contribution of a country to total import in year with respect to the ith commodity (x_{it}), weighted by certain arbitrary scores (Yit) which are allotted on the basis of the relative position of the country among a class of six topmost import sources (i.e. 1<Yi<6; e.g. the highest import source scores 6 while the lowest scores 1), normalized by the maximum score possible (i.e. for the eight commodity classes and for the three selected years, the maximum score possible is 6 x 8 x 3 = 144).

Table 2: Imports of Food by Commodity Class, 1975–1979 (1,000 MT)

Name	1975	1976	1977	1978	1979	Average (1975–1979)
Meat fresh, chilled, and frozen	3.71	13.76	22.33	31.78	21.24	18.56
Meat dried, salted, or smoked, not canned	0.14	0.21	1.00	0.50	0.03	0.38
Meat canned and meat preparation canned	17.9	3.96	6.12	4.66	1.86	3.68
Milk and cream	70.01	99.54	148.46	132.01	118.65	113.73
Butter	0.89	1.27	1.50	9.03	1.25	2.79
Cheese and curd	0.64	1.07	0.84	2.37	0.12	1.01
Fish fresh and simply preserved	17.41	3.91	31.26	46.57	75.08	34.85
Fish canned and fish preparations canned or not, including crustacean and molluscs	22.59	46.38	68.39	110.82	88.01	67.24
Wheat and spelt including meslin	407.31	733.13	719.66	878.88	80.49	563.89
Rice	6.65	45.38	413.2	564.65	567.90	319.57
Barley unmilled	0.04	0.02	0.09	0.17	1.45	0.35
Maize corn unmilled	2.21	9.86	36.81	66.26	40.48	31.12
Cereals unmilled other than wheat, rice, barley, and maize	0.94	0.84	0.13	3.35	9.71	2.99
Meal and flour of wheat, spelt including meslin	0.23	1.70	23.68	248.79	55.18	65.92
Meal and flour of cereals except wheat	0.11	0.96	1.17	6.78	40.25	9.85
Cereal preparation including preparation from flour and starch of fruits and vegetables	71.05	149.46	114.99	135.51	108.73	115.95
Fruits fresh and nuts not including oil nuts	11.12	5.40	1.07	6.49	2.90	5.40
Dried fruits including artificially dehydrated	0.10	0.07	0.03	0.19	0.32	0.20
Fruits preserved and fruits preparations	2.94	2.24	4.00	2.08	0.65	2.38
Vegetables, roots and tubers fresh and dry, not including artificially dehydrated	2.10	1.14	4.37	3.11	4.80	3.10
Vegetables, roots and tubers preserved or prepared, whether or not in airtight containers	7.60	18.36	23.98	13.32	2.52	13.16
Sugar	115.20	209.98	363.44	514.37	547.68	350.13
Sugar confectionary and other sugar	1.79	3.48	2.17	16.08	4.99	5.70
Coffee	1.19	24.88	2.09	3.34	2.96	6.89
Cocoa	0.10	0.02	0.05	0.07	0.09	0.05
Chocolate and chocolate preparations including chocolate confectionary	0.71	0.35	4.92	3.60	2.43	3.00
Tea and tea mate	1.15	2.20	3.12	97.17	4.43	21.61
Spices	0.70	0.89	0.72	1.35	0.14	0.90
Feeding stuff for animals, not including cereals	2.33	3.24	8.65	20.87	26.42	12.30
Margarine and shortening including card	-	-	-	-	0.0006	0.006
Total (1,000 MT)	770	1384	2017	2906	1811	1777
Total (value in millions of naira)	298	441	738	103	952	

Source: *Nigeria Trade Summary*, various issues.

Table 3: Funding Commitment in Nigerian Agricultural Development Projects Area 1987

Name of Project	Commencement Date	Federal	State	World Bank	Other Sources (c)	Total Commitment to project
Anambra State ADP	Jan. 1986	43.3	133.5	0	-	58.9
Bendel State ADP	Jan. 1986	88.3	93.9	9.7	30.9	55.7
Benue State ADP	Jan. 1986	78.6	63.7	0	34.2	4.1
C/River State ADP	Jan. 1986	83.7	81.3	0	x	55.0
Imo State ADP	Jan. 1986	75.1	90.2	13.8	73.7	63.2
Ogun State ADP	Feb. 1986	123.9	93.6	0	-	72.5
Plateau State ADP	Jan. 1986	101.8	53.3	0	112.0	67.0
Bida ADP	Jul. 1980	80.0	69.3	109.9	-	86.4
Gongola State ADP	Feb. 1982	99.0	50.1	-	150.7	99.9
Ilorin ADP	Mar. 1980	72.9	55.0	42.7	63.2	58.3
Ekiti Akoko ADP	Aug. 1981	78.1	56.2	15.7	67.8	62.0
Lagos State ADP	Jan. 1987	76.0	175.6	-	-	125.8
Oyo North ADP	Jun. 1982	155.4	72.2	105.9	64.7	99.6
Rivers State ADP	Jan. 1987	69.1	75.0	-	50.1	64.7
Bauchi State ADP	Jan. 1981	75.3	68.6	97.4	-	80.4
Borno State ADP	Feb. 1982	76.3	53.8	-	62.7	64.3
Kano State ADP	Dec. 1982	76.9	85.3	135.4	-	99.2
Sokoto State ADP	Apr. 1983	82.9	77.0	98.1	-	86.0
Southern Borno ADP	Jul. 1986	97.2	102.9	53.3	75.1	82.1
Average Commitment		86.0	81.7	47.5	71.4	75.0

Notes:

(a) - means not budgeted and no expenditure committed;
X means not budgeted but some expenditure is committed;
zero means budgeted but no expenditure is committed.

(b) Other sources of funding ADP mainly include internal revenue.

Source: Underlying data from APMEPU, *Project Status Report, 1988*

Table 4: Personnel Turnover in Agricultural Administration, Selected State Ministries

State Government	Political Leadership (Commissioners)			Administrative Leadership (Permanent Secretaries)		
	Approximate Period	Number of People	Average Length of Stay (Years)	Period	Numbers of Occupants	Average Length of Stay (Years)
Anambra	1976–1988	9	1.3	1974–1988	9	1.6
Bauchi	1976–1988	15	0.8	1976–1988	13	0.9
Benue	1976–1988	13	0.9	1976–1988	12	1.0
Borno	1976–1988	13	0.9	1967–1988	12	1.0
C/River	1968–1988	14	1.4	1976–1988	14	0.9
Gongola	1976–1988	9	1.3	-	-	-
Kaduna	1967–1988	11	1.9	1977–1988	13	1.6
Kano	1967–1988	12	1.8	1967–1988	20	1.0
Kwara	1967–1988	12	1.8	1967–1988	15	1.4
Oyo	1954–1988	17	2	-	-	-
Mean Length of Stay =			1.4	Mean Length of Stay =		1.2

Source: Underlying data from field survey, 1988

The Impact of Government Policies On Rural Life in Nigeria

*Invited paper presented at the symposium organized by the Professors World Peace Academy of Nigeria, University of Ibadan, 18 December 1989.

Rural life, in all its ramifications, is affected by actions and inactions of government. These ramifications include the nutritional status, availability of potable water, health conditions, and the level of social and political integration of the rural dwellers, among others. There is the need to assess the impact of government policies in all these areas. In doing so, it is clearly recognized that the problem is usually not with the policies *per se*, but with their implementation process. Nevertheless, it is also necessary to examine the *a priori* expectations of government policies.

Consequently, there are two choices in a study of the impact of government policies on living standards in rural areas. The first is to use the theoretical expectations of the policies as standards and then examine the normative impact of Nigerian policies on rural life. The second approach is to take pictures of rural life at two points in time and then examine the differences that have occurred as concrete impact. The paper employs both approaches not only because there is logic in comparing the two situations but also because both of them may not be possible in certain instances. Therefore, the theoretical foundation of government policies is first highlighted. Then, the relevant policies are briefly described, which are closely followed by their impact assessments. Finally, a number of guidelines are provided for enhancing the positive impact of government policies in rural areas.

THEORETICAL FOUNDATION OF GOVERNMENT POLICIES

The actions and inactions of government are rooted in select theories of economic growth and development. Eight of these can be briefly highlighted, following Essang (1975) as follows:[1]

(a) **The classical-neoclassical theory**—This posits that economic growth is a function of capital investments and employment of labour. Capital and labour are assumed to flow from sectors with low rates of return and marginal productivities to those with high rates of return and marginal productivities. This is the foundation of the demand for high-yielding enterprises in rural areas. Consequently, it is reasonable to measure the growth status of the rural lives by the magnitudes of rates of return to capital and labour and their marginal productivities.

(b) **Basic resource theory**—This states that economic growth depends on the presence, quality, and magnitude of basic resources in certain regions. These resources can be developed or exploited to create utilities. Consequently, one may assess the economic status of rural lives on account of the level of utilization of available resources.

(c) **Internal combustion theory**—This attributes economic growth and development to certain factors. These include technology, specialization, economies of scale, as well as the institutional, administrative, and political factors. On this basis, one may examine the situation with these factors to assess the impact of development policies in rural areas, using applicable socio-economic yardsticks.

(d) **Dual-economy model**—Two sectors of the economy are demarcated, namely rural and urban. The rural sector is assumed to possess surplus resources, particularly labour, which should be released to develop the urban sector. This helps us to evaluate the resource situations of rural areas as a way of evaluating government policies.

(e) **Export-led growth model**—This posits that policies designed to expand export markets will lead to greater utilization of idle resources, capable of enhancing incomes of producers, employment, and government revenues. This can be used as a basis for evaluating certain government policies that have been used to stimulate agricultural produce in rural areas.

(f) **Urban industrial impact theory**—This describes growth as a burning candle. This candle of economic growth is located in the industrial urban centre, and it illuminates the rural areas. Therefore, the intensity of this illumination is a decreasing function of distance from the urban centre. The logic of this theory is that nearness to urban centres determines the transportation cost of inputs and outputs and also the market for agricultural produce. This theory permits us to assess government policies on account of infrastructure available.

(g) **High input pay-off model**—This assumes farmers are efficient allocators of resources and also respond to economic stimuli, but operate under immense technical and economic inhibitions. Therefore, support is necessary in the forms of improved seeds and other technical inputs, as well as to output prices. It therefore provides basis for favourable price policies, which lower input prices relative for assessing agricultural research and price policies meant to enhance the productivities in rural areas.

(h) **Diffusion model**—This attributes productivity differences among farmers to the presence of different access to inputs and adoption capabilities. The need for agricultural extension policies, therefore, arises. Effective extension would improve the profitability of the farm business. Therefore, there is basis to evaluate the impact of such policies on the economic status of rural dwellers that are mostly farmers.

As earlier stated, the foregoing analysis helps us to evaluate the normative impact of policies, based on subjective theories. On the other hand, the temporal impact is concrete because it is based on two real lines—snapshot at an initial point in time and another snapshot at a later point in time. These pictures concern the value of certain variables that determine the quality of life in the rural areas. Such variables include the quantity and quality effects of policies as regards basic needs in the forms of food and nutritional status, health, housing, water supplies, sewage, drainage, and refuse disposal, as well as electricity and transport. We shall also evaluate the socio-economic status of rural dwellers in these dimensions, but first, a highlight of the relevant policies and programmes is necessary.

SOME RELEVANT POLICIES

The predominant theme of development in the colonial period was the surplus extraction philosophy, whereby immense products were generated from the rural areas to satisfy the demand for raw materials in metropolitan Britain. The early focus of extraction policies was placed on forest resources.

Agricultural exports followed suit, including cocoa, coffee, rubber, groundnut, and palm oil, to mention a few.

In the post-independence era, new policies were formulated to make for more balanced growth. In agriculture, the earlier surplus extraction policies quickly translated into the pursuit of an export-led growth. In this regard, the country was roughly demarcated into the cocoa west, groundnut north, and oil palm east. The stage following this is the import substitution era. The sponsors of import substitution policies think that industrialization is the best strategy by which to attain economic progress. Therefore, the most obvious route is through import substitution.

This entailed establishing domestic industries behind tariff and quota barriers. Manufacturing industries were considered the most appropriate with which to start the process, of which motor assembly plants were prominent examples in Nigeria. It was hoped that imports would be replaced and internal growth fostered, and that the costs of the strategy would be mostly borne by the advanced countries supplying the manufactured consumer goods (Pearce 1986).

Lately, the era of self-reliant policies has emerged. The self-reliance strategy of economic development implies, in the main, a country's determination to use its domestic resources in creating utility for its citizens. It represents a definite departure from the open-ended importation of goods and services whose domestic resource costs of production are simply greater than their foreign exchange costs. It therefore implies that for a determination of this kind to endure and be meaningful, a necessary, albeit not sufficient condition is the pursuit of self-sufficiency policy in the priority goods and services. The rationale for self-reliance strategy of economic development is predicated on social utility provision, infant industry argument, socio-economic discontinuities and the theory of second best (Idachaba 1984; Ayoola 1988).

In general, the rural areas are directly or indirectly affected by all policies to the extent that the agricultural enterprise and other rural industries are affected. Table 1 summarises the major agricultural programmes which have emanated within the context of surplus extraction, export generation, import substitutions, and self-reliance policies in Nigeria. These include the specific aspects of the production, market, incomes, and rural infrastructure policies.

IMPACT ASSESSMENT OF POLICIES

Normative Impact

The impact of policies formulated within the context of the growth theories will now be correspondingly discussed.

First, the classical-neoclassical doctrines have had limited impact in the rural areas. In general, government has undertaken more successful measures to raise the rates of return and marginal productivities of capital and labour in urban areas than in the rural areas. Consequently, capital and labour persistently move from the rural areas to urban areas. The effect of this resource movement would have been largely compensated for by improvements in the quality of labour, community services, and rural infrastructure, but these are relatively slow. The situation is also aggravated by the presence of bad institutional and organizational arrangements, which also deter investment and growth. The slow rate of investment growth in rural areas, however, perpetuates the vicious cycle of poverty: slow investment rate – low income – low saving – slow investment rate. The situation is further complicated by the absence of appropriate rural technologies necessary to break the vicious cycle of poverty as follows: technological breakthrough – higher profit – higher income – higher saving – higher investment – higher income.

Second, the basic resources theory is relevant to the Nigerian situation. Regions with basic resources have enjoyed higher incomes and grown faster than others. The striking examples are the cocoa-, groundnut-, and cotton-growing areas. In addition, petroleum resources presently explain part of the socio-economic status of the producing states. In regard to the rural areas, however, the impacts of these developments have been minimal. The case of marketing board operations will be cited to explain the limited impact of regional resource developments on rural lives. These boards served as the means of excessive and lopsided taxation of farmers' produce and also mismanaging farmers' money (Heleiner 1964; FAO 1966; Idachaba 1973).

Third, policies following the internal combustion theory have not produced meaningful impact in the rural areas. In general, the several internal factors—technology, specialization, economies of scale—are poorly developed, as also are the growth-stimulating institutional, political, and administrative arrangements. The process of technology creation in agriculture now involves the works of a multi-institutional research system comprised of specialized research institutes and agricultural universities, technology universities, as well as general universities (Table 2). Despite this, the rate of technology breakthrough is slow, thus limiting the profitability of farm business.

Another relevant internal factor, which is poorly developed, is specialization. In particular, ecological specialization in agricultural production is required in order to derive the benefits of internal comparative advantage. Agroecological specialization promises removal of wasteful duplication of efforts and enhanced resource use efficiency. This would translate to productivity improvement of the rural people, which could be enjoyed either as higher incomes or increased leisure. Both of these help to elevate the socio-economic status of rural people. In addition, the poor state of rural infrastructure and inadequate credit facilities limit the expansion of rural enterprises, and hence prevent the benefits of substantial economies of scale. At the same time, political instabilities and associated frequent changes in the administrative mechanisms create stop policies that are eventually ineffective in rural areas. Therefore, rural people do not feel the impact.

Fourth, the pursuit of policies in the context of the dual model of the economy has important welfare consequences in rural areas. The theory has particular relevance to labour, which is believed to be traditionally surplus in rural areas. As such, it needs to be released to develop the urban sector. The problem, actually, is not in the principle, but in the modalities of labour release. Truly, Nigerian agriculture has engaged too many people relative to the urban economy. But the release of the excess labour needs to be induced by technological progress, rather than non-profitability of farm business as it is actually the case. In the former situation, technology will replace labour when labour moves to urban areas. But in the latter case, the movement of labour will further dampen agriculture. Rural living standard, therefore, worsens as the youthful population drift to towns, leaving aged ones to perform farm and other operations. In recent times, when the absorptive capacities of urban industries fall, massive unemployment results and this further lowers the general standard of living.

Fifth, the implementation of export-led growth policies has raised important questions. It is correct that expanded export market yields higher income, government revenue and also stimulates growth. But how efficiently do these translate to better living standards in rural places? To begin with, the assumption of an elastic supply curve, on which it is based, does not hold true for rural products, particularly agricultural commodities. The expected income increase may also be reduced by countervailing measures of other countries of the world and other events in the international market. In the case of cocoa, for instance, the elasticity of supply is limited by the long gestation period, during which the market situation should have changed drastically. This is further complicated by the absence of intervention agencies in the market, as in the present case of Nigeria. Farmers are exposed to wide variations in prices, which destabilize their income expectations and plans, to produce immense negative welfare consequences in rural areas.

Sixth, the urban industrial impact model of economic growth provides the theoretical basis for rural infrastructure. Rural infrastructural development would essentially imply the burning of the candles at several points. The regions of illumination will obviously overlap and produce brighter locations. In practical terms, this means an upliftment of the living standard of rural people.

In realization of the advantages of rural infrastructure, government has mounted definite integrated programmes. The directly important ones among these include the Farm Settlement Schemes, Agricultural Development Projects (ADPs) and Directorate of Food, Roads, and Rural Infrastructure (DFRRI). But these agencies have produced limited impacts in rural areas as exemplified by the achievement indices of the ADPs (Table 3).

Seventh, the high-input pay-off model is well tested in the Nigerian case. The programmes in fertilizer, seed, and agrochemicals are important cases in point. In particular, government has undertaken massive support in subsidizing the prices of these inputs. However, the impact of subsidy in agriculture is limited for the following reasons:

- x The development of dependency mentality among farmers on government subsidy, thereby making it a permanent obligation of government and a huge fiscal burden. For example, budget allocations for fertilizer subsidy are as follows: ₦5 million (1976/77), ₦7.8 million (1977/78), ₦31 million (1979/80), ₦66 million (1980), ₦105 million (1981) (Idachaba 1981).
- x The low level of morale of the private sector participants to build up substantial enterprise initiatives in the distribution of inputs and outputs of agricultural production, owing to price rigidities with the consequent limiting effect on GDP
- x The absence of proper price relativities, following price distortions created by subsidies, which is needed to direct production decisions of farmers along the line of efficient resource utilization
- x The presence of substantial externalities and leakages, whereby farm subsidy benefits flow to unintended channels
- x The exploitation of subsidy programmes to serve the selfish motives of civil servants, politicians, and their friends, to the detriment of the ordinary farmer.

Eight, as earlier stated, agricultural extension is the main element of the diffusion model. In this connection, various methods of agricultural extension have been employed in the country. These include the following:

- Model and demonstration farms
- Agricultural shows and exhibitions
- Itinerant instructions, farm visits, and group meetings
- Agricultural competition
- Farmers' organizations
- Mass communications, including the use of posters, radio, television, film shows, etc.
- Home economics extension
- Training and visit extension.

These methods are usually applied as parts of definite extension projects and programmes, two of which are the National Accelerated Food Production Project (NAFPP) and Agricultural Development Projects (ADPs). In general, the impact of agricultural extension on the farmers' lives is seriously limited by the inadequate number and poor training of extension agents, lack of adequate transport facilities to penetrate the rural areas, inadequate backup facilities such as credit and inputs, among others.

Concrete Impact

This section provides some empirical evidence in evaluating government policies using selected concrete basic needs as yardsticks.

Starting with food as the most important basic need, the estimated per capita agricultural GDP, which fell from N109.8m in 1970 to N80m in 1981, indicates reduced availability of food over the period (Idachaba 1987), keeping importation level constant.

In the post-1983 era, the emergence of acute foreign exchange scarcity in the face of declining agricultural production should have further reduced food availability to very low levels. Following the ban on grain imports, these commodities have also caused diminution in the production capacities of feed industry, implying that animal protein is also in short supply. It is therefore suggestive that the observed downwards trend in the nutritional status of Nigerians is still in force. According to Olayide et al. (1972), the supply of calories per day was 2,198 kcal in 1969 (against an estimated requirement of 2,470 kcal), and according to the Food Strategies Mission (1981), it had decreased to 1,887 kcal in 1980. Similarly, per capita daily supply of protein decreased from 58.8 grams in 1969 (against requirement of 65 grams) to 45.1 grams in 1980.

The other basic needs will be collectively discussed as rural infrastructure. The types and roles of these include the following:

(i) **Rural Physical Infrastructure**

- Rural roads cause accelerated delivery of farm inputs, reduce transportation costs, and enhance special agricultural production and distribution efficiency.
- Storage facilities help to preserve foods in the forms that consumers need them and to the time they need them; on-farm storage also helps to stabilize inter-seasonal supplies.
- Irrigation facilities ensure farm water supply and stabilize food production by protecting the farm production system against uncontrollable and undesirable fluctuations in domestic food production.

(ii) **Rural Social Infrastructure**

- Clean water, decent housing, environmental sanitation, personal hygiene, and adequate nutrition help to improve the quality of rural life.
- Formal and informal education promote rural productivity by making the producers able to decode agronomic and other information, and carry out other desirable modern production practices; basic education also promotes feeding quality, dignity, self-respect, and sense of belonging as well as political integration of rural people.

(iii) **Rural Institutional Infrastructure**

- Rural groups and cooperative facilitate economies of scale and profitability of rural enterprise.
- Agricultural extension improves the technology status of the farm business.

At this stage, the important agencies of infrastructure provision consist of the ADPs and DFRRI. The ADPs are integrated rural development bodies, established in series as follows: Funtua (1975), Gusau (1975), Gombe (1975), Lafia (1977), Ayangba (1979), Ilorin (1979), Ekiti Akoko (1981), and Oyo

North (1982). All these are enclave types, while subsequent ones are statewide projects.

As regards DFRRI, it is a super-ministerial body established in 1986. The purpose is to complement the efforts of ADPs and other agencies so as to accelerate the rate of progress in providing rural infrastructure. The impact of DFRRI awaits a comprehensive evaluation study. However, the agency has been largely criticized in the past for the lack of proper focus and programme accountability (Idachaba 1988).

POLICY SUGGESTIONS FOR ENHANCING RURAL LIFE

The foregoing discussion reveals a number of shortcomings in the policy process meant to enhance the quality of life in rural areas. A few suggestions will be subsequently provided to correct the shortcomings in selected areas. These are briefly explained as follows:

1. Redistribution with growth

Most of the government economic policies produce lopsided benefits to favour the urban rather than rural sector. Some pertinent examples, which have been mentioned, are

- Industrialization policy, which produces items that are mostly consumed in urban areas such as motorcars and building materials whose prices are beyond the reach of the rural people.
- Trade policies that create emergency middlemen forwarding and clearing agents, among others, who enjoy enormous markups in the urban markets without creating compensating value added.
- The concentration of good things of life in urban centre, such as electricity, roads, potable water, etc. at the same time when the rural dwellers lack these basic needs.

Therefore, government needs to design a programme that redistributes the benefit of growth more evenly.

2. Rural Infrastructure to Improve Quality of Rural Life

The case has been established that the provision of basic infrastructure in rural areas will significantly improve the quality of life. Government is, therefore, required to pursue the rural infrastructure programmes with increased vigour. In particular, the operational efficiencies of the ADPs and DFRRI, as well as other agencies concerned with providing infrastructure to rural areas, need to be improved upon from time to time.

3. Appropriate Market and Incomes Policy

The efficiency of the market where the rural dweller buys and sells is very important to the quality of his life. There is need to ensure stability of supply and income. In regard to the agricultural market, close monitoring is necessary to safeguard the interest of the small farmers, especially to ensure a favourable input-output price ratio. In addition, the effect of urban wage policy has important implications for the rural wage rates and labour availability. Therefore, the rural-urban wage relations' needs to be properly studied. In particular, policies must be designed in such a way that the urban wage rate does not necessarily escalate the off-farm opportunity cost of farm employment.

CONCLUSIONS

Most government policies fail to have desired impact on the socio-economic status of rural people, not always because the theoretical efficiency of the policies are doubtful, but usually because of the great urban bias in the implementation process. In concrete terms, the quality of life of rural people centres mostly on the appropriate nutritional standard and availability of basic infrastructure. Consequently, policy efforts are required in the areas of redistribution with growth, provision of rural infrastructure, as well as appropriate market and income policies.

REFERENCES

Adegboye, R. O., A. C. Basu, D. Olatunbosun (1969). 'Impact of Western Nigeria Farm Settlement on Surrounding Farmers', *J. Econ. and Social Studies* 1 (2 & 9).

Antonio, Q. B. O. (1984), 'Marketing Development in Nigeria: A Review of Relevant Government Policies', in J. P. Feldman and F. S. Idachaba (eds.) *Crop Marketing and Impact Distribution in Nigeria*, Federal Agricultural Coordinating Unit, Ibadan.

Ayoola, G. B. (1988), 'The Dependency Problem in Food Grains and Rationale for Self-Sufficiency Policy in Nigeria', *J. of the Federal Department of Agriculture and Rural Development* 3 (1): 12–20.

Ayoola, G. B. (1989), 'Socio-Economic Factors of Ecological Disaster Management in Nigeria', *J. Forestry* 19 (1 & 2): 35–39.

Ayoola, G. B. (1989), 'Agricultural Price Policy under the Structural Adjustment Programme: Proposal for Alternative Mode of Market Intervention', Paper presented at the Silver Jubilee Conference of the Agricultural Society of Nigeria, Owerri, 3–6 September 1989.

FAO (Food and Agricultural Organization) (1966), *Agricultural Development in Nigeria 1965–1980*.

Federal Ministry of Agriculture, Water Resources, and Rural Development (1988), *Agricultural Policy for Nigeria*.

Idachaba, F. S. (1985), 'Integrated Rural Development in Nigeria: Lessons from Experiences', Paper presented at the Workshop on Designing Rural Development Strategies, Federal Agricultural Coordinating Unit, Ibadan, December 1985.

Idachaba, F. S. (1988). 'Commodity Boards in Nigeria: A Crisis of Identity' in K. Archin, B. Aespen, and L. Van der Lann (eds.), *Marketing Boards in Tropical Africa* (London: KPI Ltd) pp. 149–168.

Idachaba, F. S. (1988). 'Strategies for Achieving Food Self-sufficiency in Nigeria' Keynote Address, 1st National Congress of Science and Technology, University of Ibadan, 16 August 1989.

Idachaba, F. S. (1984). 'Self-Reliance as a Strategy for Nigerian Agriculture: A Cornucopia or Pandora's Box?' in *Self-reliance Strategies for National Development*, Nigerian Economic Society, 2 volumes.

Lipsey, R. G. and K. Lancaster (1956), 'The General Theory of Second Best', *Review of Econ. Studies* 24.

Nigeria (1970), *Second National Development Plan 1970–74*, Federal Ministry of Economic Development.

Table 1: Typology of Agricultural Policy in Nigeria, 1960–1989

Key Strategies	Relevant Details
Farm Settlement Scheme	Initiated in old Western Region; aimed at solving unemployment problem among primary school leavers. Policy instruments include agricultural extension, cooperative societies, and credit facilities.
National Accelerated Food Production Project (NAFPP)	Aimed at enhancing farmers' technical efficiency in the production of selected crops (mostly grains). Policy instruments include subsidy, credit, adaptive research, and demonstration plots.
Operation Feed the Nation (OFN)	A mass mobilization and mass awareness programme. Policy instruments include mass media, centralized input procurement, massive fertilizer subsidy, and imports.
River Basin Development Authorities (RBDAs)	To tap the potentials of available water bodies; first 11, then 18 and 11 in number. Specific objectives are irrigation services, fishery development, and control of flood, water pollution and erosion. Policy instruments include input distribution, credit services, infrastructure development, and manpower development.
Agricultural Development Projects (ADPs)	To enhance the technical and economic efficiency of small farmers in general. Policy instruments include rural infrastructure development (feeder road network, dams, etc.), revamped input delivery system and revitalized agricultural extension system, autonomous project management, and domestic-cum-international capital.
Green Revolution Programme	To accelerate the achievement of the agricultural sector objectives. Policy instruments include food production plan, input supply and subsidy, special commodity development programme, review of Agricultural Credit Guarantee Scheme, increased resource allocation to RBDAs, etc.
Directorate of Food, Roads, and Rural Infrastructure	Established to facilitate programmes in food production, particularly through the provision of rural infrastructure

Source: Miscellaneous documents

Table 2: Main Organs of Nigerian Agricultural Research System

	INSTITUTIONS	OWNERSHIP/ HEADQUARTERS
A.	Research Institutes	
1.	Agricultural Extension and Research Liaison Service	Federal / Kaduna (Kaduna State)
2.	Agricultural Extension Research Liaison and Training	Federal / Umudike (Imo State)
3.	Cocoa Research Institute of Nigeria (CRIN)	Federal / Ibadan (Oyo State)
4.	Federal Research Institute of Industrial Research (FIIRO)[a]	Federal / Oshodi (Lagos State)
5.	Forestry Research Institute of Nigeria	Federal / Ibadan (Oyo State)
6	Institute of Agricultural Research)	Federal / Zaria (Kaduna State)
7	Institute of Agricultural Research and Training (IAR&T)	Federal / Ibadan (Oyo State)
8	International Institute of Tropical Agriculture (IITA)	International Body / Ibadan (Oyo State)
9	Kainji Lake Research Institute (KLRI)	Federal / New Bussa (Kwara State)
10	Lake Chad Research Institute	Federal / Maiduguri (Borno State)
11	National Animal Production Institute (NAPRI)	Federal / Shika, Zaria (Kaduna State)
12	National Cereals Research Institute (NCRI)	Federal / Badeggi (Niger State)
13	National Horticultural Research Institute (NIHORT)	Federal / Ibadan (Oyo State)
14	National Root Crops Research Institute	Federal / Umudike, Umuahia (Imo State)
15	National Veterinary Research Institute	Federal / Vom (Plateau State)
16	Nigerian Institute for Oil Palm Research	Federal / Benin-City (Bendel State)
17	Nigerian Institute for Trypanosomiasis Research (NITR)	Federal / Kaduna (Kaduna State)
18	Nigerian Stored Products Research Institute	Federal / Yaba (Lagos State)

19.	Rubber Research Institute of Nigeria (RRIN)	Federal / Benin-City (Bendel State)
B.	**Specialized Agricultural Universities**	
20.	University of Agriculture, Abeokuta	Federal / Abeokuta (Ogun State)
21.	University of Agriculture, Makurdi	Federal / Makurdi (Benue State)
C.	**Specialized Technology Universities** [b]	
22.	Federal University of Technology	Federal / Akure (Ondo State)
23.	Federal University of Technology	Federal / Minna (Niger State)
24.	Federal University of Technology	Federal / Owerri (Imo State)
25.	Anambra State University of Technology	State / Anambra State
26.	River State University of Science & Technology	State / Port Harcourt (Rivers State)
27.	Tafawa Balewa University of Technology	Federal / Bauchi (Bauchi State)
D.	**General Universities** [c]	
28.	Ahmadu Bello University	Federal / Zaria (Kaduna State)
29.	University of Benin	Federal / Benin (Bendel State)
30.	University of Calabar	Federal / Calabar (Cross River State)
31.	University of Ibadan	Federal / Ibadan (Oyo State)
32.	Obafemi Awolowo University	Federal / Ile-Ife (Osun State)
33.	University of Ilorin	Federal / Ilorin (Kwara State)
34.	University of Maiduguri	Federal / Maiduguri (Borno State)
35.	University of Nigeria	Federal / Nsukka (Anambra State)
36.	Usman Dan Fodio University	Federal / Sokoto (Sokoto State)
37.	Ogun State University	State / Ago-Iwoye (Ogun State)
38.	Imo State University	State / Owerri (Imo State)

Notes:

(a) FIIRO conducts research into the industrial use of agricultural commodities.
(b) Technology universities conduct research into agricultural technology.
(c) General universities, which have faculties of agriculture, and research in all aspects of agriculture.

Sources:

(1) Agricultural Research Centres, Volume 2, pages 624–630
(2) Joint Admission and Matriculation Board Brochure 1988–89 Session, pages 7–227.

NAME OF ADP	PERIOD OF EXISTENCE	ROADS (KM)	BRIDGES (NO.)	BOREHOLES (NO.)	WELLS (NO.)	FARM SERVICE CENTRE
RECENT PROJECTS						
Anambra State ADP	1986	14.7	-	-	-	52.3
Bendel State ADP	1986	-	-	-	-	-
Benue State ADP	1986	-	0	-	-	-
Cross River State ADP	1986	-	-	-	-	-
Imo State ADP	1986	14.7	50	-	-	33.3
Plateau State ADP	1986	30.6	0.1	66.7	14	-
Ogun State ADP	1986	-	-	0	-	100
OLDER PROJECTS						
Bida ADP	1981–86	69.6	-	-	100	-
Borno State ADP	1983–86	23.7	51.5	100	-	126.5
Sokoto State ADP	1983–86	70.8	31.3	-	-	41.7
Gongola State ADP	1983–86	22.6	31.3	-	-	-
Ilorin ADP	1981–86	26.7	-	400	-	-
Oyo North ADP	1983–86	80.6	21.9	-	83.5	33.3
Kaduna State ADP	1985–86	34.3	-	-	-	-
Kano State ADP	1986–86	55.7	-	100	81.6	33.3

| Ekiti Akoko ADP | 1983–86 | 73.9 | - | - | - | - |
| Average | - | 43.2 | 25.8 | 133.3 | 88.6 | 60.1 |

- means not available.

Source: Underlying data from APMEPU, *Project Status* Report, 1987

4

Sectoral Planning of the National Economy in Perspective: The Case of Agriculture

In monograph, *Food and Agricultural Strategy Review*, Centre for Food and Agricultural Strategy, FASR No.2, December 1998. ISBN: 1114-3322

The functional defects of pure market mechanism constitute the basic premise on which to justify the widespread adoption of planned economy. A plan works as a super-instrument whereby economic agents, including government, make deliberate effort to design the best ways of achieving chosen goals. Such is the trademark of agricultural planning: an exercise in forethought to carefully specify objectives and systematically select alternative policy instruments towards finding the optimal means of meeting agricultural objectives. The argument against pure market mechanism rests on the inefficient utilization of resources and production of undesirable commodities resulting from lack of coordination of economic activity. Planning is particularly attractive as a means of addressing quality-of-life matters, such as employment, pollution, poverty, and inequality.

However, the aberrations of centralized planning system mostly include inefficient resource allocation arising from the need for coordination of several components of plans in action, as well as coordination of independent executors at each level of economic process. Rigidity and inertia may also result from the inherent bureaucracy, involving large organizations required to implement the plan. These features may manifest in persistent shortages of certain goods and services, at the same time when there are surpluses of others. Apart from these limitations, even with the best of coordination in place, practical experience indicates that planning may be insensitive to the wishes of the consumers

This paper puts the planning of Nigeria's agricultural economy in perspective with other sectors. The objective is mainly to examine the extent to which the merits and demerits of central planning have come to bear upon the agricultural development process. The paper is structured into an initial overview (II), highlight of special problems and features of agricultural planning (III), discussion of sectoral objectives and growth targets (IV), description of planned programmes and strategies (V), and explanation of linkage relations between agriculture and other sectors.

OVERVIEW OF AGRICULTURAL DEVELOPMENT PLANNING IN NIGERIA

An overview of agricultural development planning in Nigeria must include the aspects of institutional context, policy framework, and time frames. This section provides the historical perspective of these aspects to enhance the understanding of the issues relating to formulation, execution, and control of the agricultural sector plans in the country.

Institutional Context of Agricultural Planning

The evolution of institutions for general agricultural administration which commenced with the botanical garden at Ebute Metta (established 1893) rapidly stabilized within some ten years. The first Agricultural Department (Southern Provinces) was established at Ibadan in 1910, presumably carved out of an earlier Forests Department (established 1900). The Agricultural Department (Northern Provinces) was established at Samaru in 1912. Both departments were administratively merged into one Agricultural Department for Nigeria in 1921. This unified department was neither broken up nor merged with others for about the next twenty-five years. Between 1921 and 1954, the department served as the single institutional outfit for agricultural administration, including policy formation, plan preparation and implementation. There were only three directors of agriculture in that period: Mr. O. T. Faulkner (1921–1936), Captain J. R. Mackie (1936–1945), and Mr. A. G. Beatle (1945–54). Following the federal constitution in 1954, which established three regional governments, the era of unitary agricultural department for the whole country came to an end. Each region created its own Ministry of Agriculture and Natural Resources. At this time, agricultural development appeared on the residual legislative list of the constitution, meaning that the federal government had no responsibility in this regard.

After independence (1960), the regions mounted agricultural ministries with crops, livestock, and forestry as component departments or divisions. The Mid-West ministry came into being in 1963, leading to four-ministry stage. The successive state creation exercises led to twelve, nineteen, and twenty-one agricultural ministries at the state level in 1967, 1976, and 1986 respectively. Thirty states emerged from the state creation in 1991, excluding the Federal Capital Territory of Abuja, thereby leading to 30 or 31 agricultural ministries. The military administration (Gowon) disregarded the constitutional provision to implement the recommendation of FAO (1966) by creating a Federal Ministry of Agriculture and Natural Resources (FMANR) (Nigeria, official gazette vol. 55 No. 10 of 7 February 1966). The preceding civilian government developed cold feet to this, but could only establish the Federal Ministry of Natural Resources and Research (FMNRR) in which the name 'agriculture' was carefully avoided so as not to 'offend the political sensibilities of the regions' which were constitutionally responsible for agricultural development (Ayida 1973).

Between 1962 to date (29 years), the federal agricultural ministry has changed nomenclature several times, depicting merging and demerging at different stages: Federal Ministry of Agriculture and Natural Resources (1966), Federal Ministry of Agriculture (1979), Federal Ministry of Agriculture and Water Resources (1982), Federal Ministry of Agriculture and Rural Development (1984), Federal Ministry of Agriculture, Water Resources, and Rural Development (1985), Federal Ministry of Agriculture and Natural Resources (1990). Similarly, changes have taken place at the state level, as well as several organizations at sub-ministry levels involving divisions and departments that are frequently merged and demerged (Ayoola 1990). Within the same period of time, personnel changes have taken place several times in key leadership positions, including political (ministers/commissioners), administrative (permanent secretaries / directors general), and professional (directors/heads). Instances of reorganization of ministries in the states have also occurred through changes in nomenclature, as many as eight times, one each in Anambra State (1970–1988) and Kaduna State (1970–1985). The mean lengths of stay for political and administrative leaders in a sample of ten states of the federation are 1.4 years and 1.2 years respectively. These changes could have produced immense alterations in the content and duration of plans as well as the pace of their implementation so as to reflect the political, philosophical, and occupational biases of the new people involved.

Policy Context of Agricultural Planning

Table 1 describes the articulation of agricultural policy in Nigeria. Two aspects of the policy context are important for the assessment of agricultural planning, which are national cohesion and macroeconomic objectives. In terms of the former, we are concerned with degree of decentralization along geographical and commodity lines and how this affects plan formulation, implementation, and control. In this connection, the country started with disparate sub-sectoral, regional, or commodity-specific policies. The initial commodities of emphasis in the colonial period were forest products, oils, oilseeds, and cotton. In the 1960–67 period, each regional government followed different policy documents in different directions. The absence of a federal agriculture ministry at that time meant that these policies lacked central coordination. So also were the corresponding plans that were drawn to implement them.

In a sense, therefore, the initial policy environment lacked internal consistency of policies and plans, which led to possible wastage of national resources. The separatist tendency was implicit in the Economic Development Plan (1955–1962), which was merely an amalgam of four separate regional and one federal policy that were not harmonized into a consistent national document. Similarly, the initial commodity specificity of formulated policies implied that the structural relationships of various agricultural commodities were not explored for planning purposes. For example, when distinctive policy documents were prepared for oils, oilseeds, and cotton for the purpose of stabilizing the post-war prices in Europe, how could plans be drawn up to capture the effects of such policies on grains production and marketing which, though related, were outside the cotton policy?

Concerning macroeconomic policy objectives underlying agricultural planning, the concern is mainly to discern changes since the colonial period. Even though these objectives may either be written and expressed (i.e. explicit) or unwritten (implicit), they exist at any point in time (Idachaba 1984). At the outset, colonial agricultural administration operated a surplus extraction policy objective in general. In this mode, the goal of administration, including policy statements and planning, was mainly to support the British economy, whereby immense quantities of forest products and other agricultural commodities were exported towards meeting the raw material requirement of British industries (Ayoola 1989). Planning at that time was limited to budget allocation for pursuing this objective on an annual basis. The period was essentially devoid of concrete plan documents until the launching in 1946 of the colonial Ten-Year Plan of Development and Welfare as the earliest deliberate planning effort in Nigeria. The plan, which was prepared by a small

Central Development Board consisting of senior colonial government officers, was aimed at strengthening the colonial raw materials bases in groundnut, soybean, palm kernel, and cocoa, among others. The critical instrument of achieving this was development of physical infrastructure, particularly roadwork. The objectives, which were not quantified, were pursued through assorted projects. Before the plan came to a premature end in 1954, it was popularly criticized for its non-involvement of native Nigerians, and lack of necessary manpower to implement it.

A change in policy objective resulted from economic nationalism following independence, which invariably translated into new successive plan orientations. The initial orientation was akin to export-led growth objectives. In this regard, great policy attention was focused on cocoa, groundnut, cotton, oil palm, and other sources of foreign exchange earnings. The Western Nigeria policy on Agriculture and Natural Resources (1959) was a case in point. Huge trade surpluses emanated from the cocoa-based regional marketing board, which was used to finance urban projects such as roads, schools, water supply, and the University of Ife. The First National Development Plan (1962–1968) gave operational strength to the export-led development policy. This was achieved within the context of a common national planning framework and acceptance of general priorities by all governments, in which the highest was accorded to agriculture, industry, and training. However, the plan operated on dependency mentality, including importation of management, funds, and other inputs.

Another change in policy objective could be discerned and described as import-substitution in nature. In this regard, deliberate effort was made to develop the infant industries, possibly including agriculture, towards substituting imported final goods and services with local ones in the long run. This was consistent with the need to affect reparations after the civil war. The Second National Development Plan (1970–1974) was launched specifically to achieve this. Even though the theme of self-reliance was expressed in this plan, its operationalization was difficult, owing to the weak state of development of available resources. This plan provided for agriculture on the concurrent legislative list for the first time.

The objective of planning changed further with the advent of petrol-dollar windfall. Dependency became the underlying theme of economic development, whereby the huge foreign exchange reserves were used to sustain massive importation of goods and services over a period of time. The Third National Development Plan (1975–1980) took off at the height of the oil boom, in which the notion of self-reliance was relegated to the background. The statements of agricultural development were made within the context of importation of the inadequacy of the economy, including technology.

The objective of self-reliance reappeared in the Fourth Development Plan (1981–1985), as a result of threatening shortage of foreign exchange backup. This plan was abruptly terminated after the change in government (1983), giving way for special programmes in recession management, including the Structural Adjustment Programme (SAP). SAP is premised upon the themes of self-reliance and self-sufficiency as the philosophy of economic development.

Agricultural Planning Horizon Context

The term horizon of agricultural planning may be discussed in four frameworks: development plan, perspective plan, annual plan, and rolling plan frameworks. A special characteristic of planning, besides its coordinating role, is the introduction of time factor into the policy process. In the popular development plan framework, the planning horizon is medium-term in nature, lasting four or five years. As indicated earlier, four of such medium-term plans have been operated since independence in Nigeria (1962–1968, 1970–1974, 1975–1980, 1981–1985). The first plan was disrupted by the civil war which broke out in 1967. Planning was suspended at that time because all available resources were to be mobilized for the war prosecution, in disregard of the plan. Another perturbation occurred at the end of the fourth plan, which coincided with change in government. The drawing up of the fifth plan document, which was at advanced stage, was not to be completed in preference for the special programme of Structural Adjustment.

The perspective plan framework does not exist as an alternative to the medium-term plan. Rather, it 'places the medium-term intentions into the context of a statement about the desired pattern of economic development over the next twenty to thirty years' (Killick 1981). By this means, the perspective characteristics are generated from which the medium-term objectives will derive. The country has the experience of two perspectives plan documents. The first one was prepared by the FAO covering 1965–1980 (FAO 1965 *op. cit.*), and the second one was prepared under the aegis of a Joint Planning Committee, covering 1973–1985. A third exercise is under way for which the Federal Agricultural Co-coordinating Unit (FACU) has taken responsibility since 1988, and which is to cover the period 1990 to 2005.

The annual plan, commonly called annual budget, is a device for (a) specifying in detail the content of medium-term plan to be implemented in successive years and (b) modifying the medium-term plan in the light of changing circumstances, new information, and shifting priorities. The country is familiar with the budget rituals to mark the beginning of new years.

The financial budget is usually stated, including the expected allocations to specific projects at federal and state levels. Table 2 shows planned and budgeted allocations in respect of selected livestock projects in Plateau State. In general, cumulative budget allocation exceeds plan allocation (11%) but actual expenditure is far lower than both plan allocation (39%) and budget allocation (38%). The latter case may be due to one or the other two reasons: either that budgeted funds are not released in expected amounts, or the amounts released are not spent on the respective projects. In both instances, the plan fails in action.

The last context of time horizon to be highlighted is the concept of rolling plan. Like the annual plan, the rolling plan is aimed at introducing flexibility simultaneously with implementation. A rolling plan can be defined as a medium-term plan, which is rewritten every year. Using the example of Nigeria's three-year rolling plan (1990–1992) newly adopted, the plan is to be rolled forward every year (i.e. 1991–1993, 1992–1994, 1993–1995, etc.). In this mode, it is possible to incorporate new elements from annual experiences. Save that the demand on resources for preparation and monitoring is heavy, the concept of rolling plan is particularly justified in the Nigerian case, to be able to capture the influence of frequent changes in terms of trade, government, and other factors. The case of agriculture is the more relevant because of the vagaries of weather conditions, incidence of disease outbreak, and other uncontrollable factors.

SPECIAL PROBLEMS AND FEATURES OF AGRICULTURAL SECTOR PLANNING

The performance of agricultural development plans is greatly limited by several problems, mostly brought about by the peculiar features of agriculture itself. Two features of agriculture can describe most of the situation in the sector. The first is the absolute dependence of agricultural production on biological and ecological phenomena, both of which are, for all practical purposes, beyond the control of man as the critical agent of economic change. These biological and ecological processes do not obey definite laws, resulting in large variation of projection estimates in the plan. Examples are death of crops and livestock from massive pest infestation and disease outbreak. In one instance at the Benue State Cattle Ranch, all 180 beef cattle imported from USA died from skin disease, thereby causing failure of plan in that respect. In another example bordering on ecological disaster, both drought and flood occurred in one season of 1987, leading to massive crop failure and thereby failing the structural adjustment plan objective of food self-sufficiency in successive years. The list of other sources of biological and ecological disasters in the country includes infestation of locusts, quail birds, and rats, as well as fire outbreaks and oil spillage (Ayoola 1989).

The second feature of agricultural production which affects plans is the presence of small and numerous production units which the plan is intended to influence. It therefore becomes difficult to predict the behaviour of a great number of farmers, which has several non-business components, including sociological and cultural aspects (Ayoola and Ayoade 1991). In particular, smallholder farmers have substantial family character, which creates additional requirements on the plan to include and apply relevant instruments for effecting change.

Put together, the two features highlighted lead to the problem of uncertainty. This describes a situation in which the parameters of probability distributions of several variables are unknown but can only be estimated. In the aggregate, uncertainty leads to probabilistic production functions, whereby the technical relationships between inputs and corresponding outputs are not precisely known. At the micro level also, such as planning of specific investment projects, the implication of uncertainty is the possibility of cost overruns, yield variation, delay in implementation, and price fluctuations, all of which affect the standard decision criteria and rules for applying discounted measures of project worth. Consequently, the art of planning involves the additional exercise to capture a reasonable proportion of the uncertainty factor during formulation. The frequent treatments of uncertainty include sensitivity analysis and provision of contingency allowances, both of which are tantamount to intellectual guesswork.

Another critical problem of agriculture is the dearth of good-quality data for planning. In this connection, planning without facts often results (Stolper 1966), which is blind planning because the data are not available to capture the past events well enough for reliable indications for future economic activities. Consequently, plans have failed from false expectations and false hopes about the future of the economy. Poor-quality data themselves are the outcomes of apparent and real inconsistencies of concepts and methodology in cross-section and time-series terms. Table 3 illustrates poor correlation in production data obtained from different agencies.

SECTORAL OBJECTIVES AND GROWTH TARGETS

In section 2 above, we considered issues in macroeconomic objectives in terms of national cohesion and temporal consistency. Our concern here is to examine how agricultural sector objectives dovetail into the overall macroeconomic policy objectives and highlight the issues in growth targets. The discussion of the former takes its bearing from a theory of transformation of policy objectives in which the hierarchy of objectives involves societal objectives, macroeconomic objectives, sectoral objectives, and programme objectives. For instance, self-reliance as a societal objective translates into self-sufficiency in goods and services as a macroeconomic objective. At that level, agricultural sector objectives take off, such as self-sufficiency in food and agricultural raw materials from which the various programme objectives will originate (e.g. maize, fertilizer, and other programmes). The sequential transformation of policy objectives enables us to link lower-level objectives (e.g. agricultural sector objectives) to higher-level objectives. Hence, we can derive objectives of public agricultural planning from the established sequence of policy objectives.

In the foregoing context, it is possible to discern areas of conflicts or trade-offs among objectives using the vertical and horizontal relationships. The particular case in point is the existence of conflicts and trade-offs in the self-reliance philosophy of development. According to Ayoola (1988a), society desires self-reliance on grounds of social utility provision, infant industry argument, food security, and the theory of second best. However, implementation of past plans indicates that as the economy pursues objectives in self-reliance, it becomes more and more dependent on farm input imports, expatriate personnel, and foreign capital. For instance, the average annual growth rate of merchandise imports in the 1970–82 period was 17.2%, against the rate of only 1.5% in the 1960–70 periods. In contrast, merchandise exports grew at -1.6% and 6.6% in the corresponding periods (World Bank 1984). Suffice it to say that agricultural plans must be prepared in such a way that vertical and horizontal objectives are drawn to beneficially shift the trade-off curve between food import dependence and input import dependence. In this regard, it is critical for planning authority to understand the nature of this curve.

Growth Targets

The primary mode of exerting influence on the economy through a plan is the selection of growth targets. Subsequent stages of planning only clarify the implications of the rates for sectoral development and specify the contributions of the public and private sectors towards meeting the growth targets that are consequent upon this rate. Of course, controls are usually introduced to direct economic agents accordingly. For instance, the first plan (1962–8) was designed on the minimum growth rate of 4% leading to the quantitative estimates of commodity targets. The incremental demand, production, and supply to be achieved in the plan period are determined by difference between projected and targeted estimates. Evaluation (*ex ante*) is carried out at the same time, leading to the adjustment of the growth rate in cases of unacceptable economic impact of the successive plan proposals. Input budgets are residual activities, based on the resultant output projections.

The exogenous selection of growth rate for the agricultural sector is a primary point of theoretical departure of this paper. While the industrial sector can be driven at a predetermined rate because of relatively high exactitudes of production and supply of commodities, agriculture seems not to be so easily drivable, owing to the domination of the uncertainty factor. Therefore, the planning model of agricultural growth which is *rate-led* is likely bound to feature large forecast errors most of the time. A recent thesis following this thinking is for agricultural planners to adopt the concept of internal growth rates. In this thesis, commodity targets (not rates) are to be set over a given planning horizon which reflects the prevailing societal, macroeconomic, and sectoral objectives. Agriculture will only grow at an internal rate to meet the set commodity targets. In cases of unacceptable internal rates, either the set target or the planning horizon may be modified. Input budgets are still residual activities based on technical coefficients. We may term this new procedure *target-led* growth. The procedure avoids a subjective growth rate, which generally reduces the art of agricultural planning to mere iterative guesswork. The fundamental difference is that instead of targeting a *subjective* growth rate, which we only hope will meet the set agricultural objectives at the end of the plan, we target quantities, which we require to meet these objectives. The advantage of a target-led planning approach is that the commodity targets are real relative to the set objectives. But growth rates are exogenous to the system, which makes it unreal relative to the set objectives. The forecast error is likely to be lower, as we pursue real quantities rather than unreal rates for the development of agricultural sector.

In an extended target-led planning of Nigeria's maize economy as a test case (Ayoola 1988) employed a methodology which ensures an optimal internal growth rate. This was based on the realization that growth paths were usually not even. Therefore, the overall internal rate of growth cannot be distributed evenly among each successive year of the planning horizon. The methodology, which is essentially a 'calculus of variations' model of optimal control (Intrilligator 1971), explores the dynamic character of historical growth of the commodity. In the maize example, self-sufficient targets were set against alternative planning intervals, leading to optimal time path of production plan. The planning horizon which not only contained an acceptable internal growth rate but also met the set self-sufficient output targets was recommended. By implication, the derived input schedules were optimal. The method, which is perfectly general, can be used to design commodity programmes as attempted in the food production plan (Food Strategies Mission 1980) as well as preparation of food balance sheet as attempted in Olayide et al. (1976) and Olayemi et al. (1986). The method itself is consistent with the procedure suggested by Olayide and Heady (1974) towards improving and implementing agricultural planning machinery in Nigeria.

AGRICULTURAL PROGRAMMES AND STRATEGIES IN THE PLANS

Table 4 describes the key programmes and strategies in the post-independence planning era. Within the frameworks of these programmes and strategies, the instruments of agricultural development include generalized and specialized types, among which are agricultural extension, education/training, research, cooperatives, agricultural industries, infrastructure, legislation, credit, input support, fiscal incentives, public enterprise, land reforms, soil management services, international assistance, settlement projects, agricultural campaigns, and others. Some of the programmes will be highlighted critically.

Agricultural programming effort can be said to have commenced with the farm settlement scheme of the old Western Region. Under this scheme, it was planned that the graduates of the free primary education programme would be established on farm sites with government assistance in credit, training, and infrastructural support. The scheme failed to achieve the planned objectives because the plan did not capture the basic features of beneficiaries well enough. According to FAO (1966 op cit.), the philosophical basis was an 'impracticable proposition to turn physically and mentally immature youths into serious-minded, hard-working farmers'; this is because 'young men can only be expected to settle down and devote themselves to productive work after they have passed the age of 20 and have married and thereby assumed responsibility for maintenance of other persons than themselves'. Secondly, the income-generating capacity of the settlers was too low to meet the repayment schedules compared to the expectations of the prototype plans. This probably prevented the fulfilment of the anticipated demonstration value of the scheme on surrounding farmers (Adegboye et al. 1969). Suffice it to say that planners of the NDE agricultural programmes have important lessons to learn from the farm settlement scheme. In particular, it is instructive to note that the aspect of borrowers' characteristics is critical to the successful application of the credit packages.

The Integrated Agricultural Development Programme was mounted as a national strategy. Command-area projects were established in sequence, beginning with Funtua, Gusau, and Gombe projects in 1975. Newer enclave-type projects were Lafia (1978), Ayangba (1978), Ilorin (1979), Bida (1980), Ekiti Akoko (1981), and Oyo North (1982). At present, each state of the federation has one ADP, all at statewide levels. Federal and state governments, together with World Bank loan, jointly fund a typical ADP. The system of ADPs is serviced by two specialized units, which are Federal Agricultural Coordinating Unit (FACU, headquarters Ibadan), which is responsible for preparation, negotiation, take-off, and planning of projects, and Agricultural Projects Monitoring and Evaluation Unit (APMEU, headquarters Kaduna), which is responsible for monitoring and evaluation. The concept of ADP is to use autonomous management for the supply to smallholder farmers of (a) inputs through a dense network of farm service centres, (b) transport facility through the massive rural road network, and (c) new technologies through revitalized, intensive, and systematic extension and training backed up by timely input support and adaptive research services. Table 5 shows the performance of a number of ADP plans.

The River Basin Development Authorities (RBDAs) were established (Decree 25, 1976) with the aim to develop the economic potential of the massive water bodies available in the country. In particular, they have specific mandates in irrigation services and fishery. Eleven authorities were originally established; the figure grew to eighteen before it was reversed to the original number. A number of them grew out of proportion, and the operations of some others suffered from intensive political interference. A case in point was one of them (Ogun-Oshun RBDA) which had its functions disrupted for a reasonable length of time because six state governments (non-NPN) took the federal government to court to challenge the constitutional right of the government to agricultural jurisdiction in territories of the states. Substantial public funds were wasted to streamline sizes and functions of RBDAs, through the disposal of their non-water assets.

LINKAGE OF AGRICULTURE WITH OTHER SECTORS IN THE PLAN

The common bifurcation of the economy is usually along the lines of agricultural and industrial sectors. The least a good plan should do is to link up these sectors together very closely. For instance, it is generally accepted that it takes more than industry to industrialize. First, agriculture is the source of food for the increasing population, including people engaged in industry. Second, agriculture produces raw materials for industry to grow. These traditionally include cotton (textile industry), cocoa, coffee, tea (beverage industry), as well as palm oil, grains, rubber, to mention a few. In recent times, industrial growth has resulted from cassava, soya bean, yam, and others, which have found important uses as industrial raw materials. Lastly, agriculture serves as market for industrial output, including machinery, agrochemicals, and vehicles. In turn, industry makes technology grow in general and carries responsibility for developing, supplying, and maintaining infrastructure in the economy, as well as absorbing labour that is released as a result of productivity improvement in agriculture.

Agriculture-industry linkage is achieved in the plan through macroeconomic model. This involves the specification of structural relationships, which may be exogenously or endogenously determined. Solution of certain equations leads to the estimates of necessary coefficients for linking agricultural and industrial sectors. In practical terms, the problem to be solved can be stated as follows:

- The need to synchronize the rate of release of labour as agriculture develops and the absorptive capacity of industry to employ idle labour—if this cannot be achieved, then two things may happen, both of which are negative for the health of the economy: (a) in the case of excess growth of agriculture over industrial capacity, unemployment will result from idle agricultural labour, (b) in the case of excessive industrial capacity over agricultural growth, scarcity of agricultural labour will prevail.
- The need to pursue growth in agriculture that is consistent with industrial requirements for raw materials—proper linkage presupposes that there is equilibrium between raw materials requirements of industries and agricultural capacity to supply them; low-capacity utilization of agro-based industrial sector is the frequent result of inadequate supply of agricultural raw materials.
- The need to enhance productivity status of agriculture for the continued expansion of industrial plants—this borders on the performance of research and development (R & D) institutions to initiate and diffuse technologies among farmers; the market for several industrial products will be saturated at a rate depending on the degree to which the agricultural sector consumes them.

A 'Deathly Embrace' Explanation of Plan Ineffectiveness

The term 'deathly embrace' is used to describe a situation whereby two or more systems wait for inputs from one another. In such a situation as sometimes encountered in computing, all the systems will stagnate. Deathly embrace can also take place in the national economy in absolute or relative terms, depending on the degree to which some of the systems satisfy the input requirement of others. In the context of the inter-sectoral linkage discussion, it appears that both the agricultural and industrial sectors are involved in a deathly embrace through their inability to make necessary inputs for developing each other. Therefore, public planning has the fundamental role to play in specifying the expected inputs of each sector to the other in clear terms. Quantitative statements of these inputs will definitely enhance the operational content of the plan, including monitoring and evaluation of projects during implementation. Therefore, planning has a critical role to play in the process of disentangling the numerous instances of deathly embrace in the national economy.

CONCLUSIONS

The performance of agricultural sector plans hinges as much on the institutional framework and the policy environment within which the plans will be implemented as on the special problems and features of the sector, the mode of setting objectives and growth targets, as well as extent to which inter-sectoral relations are captured in the plan. In specific terms, the critical factors of agricultural plan performance include (a) degree of stability attained in the organization, personnel, and statement of policy objectives, (b) the appropriateness of planning machinery to the small farm production units, biological nature of production systems, and the problem of uncertainty, (c) reliability of data that are available for planning, (d) implementation problems such as political interference, inadequate funding, inefficient management capacity, among others, and (e) the degree to which the agriculture-industry linkage is facilitated.

REFERENCES

Adegboye, R. O., A. C. Basu, and D. Olatunbosun (1969), 'Impact of Western Nigeria Farm Settlement on Surrounding Farmers', *J. Econ. and Soc. Studies* 1 (2).

Ayida, A. A. (1973) 'Basic Issues in Financing the Nigerian Agriculture in the Seventies', in *Proceedings of the Natural Agriculture Development Seminar*, Federal Ministry of Agriculture and Natural Resources (Caxton Press).

Ayoola, G. B. (1988a), 'An Optimal Time Path for Self-Sufficiency in Maize: A Production Control Approach', PhD Thesis, Department of Agricultural Economics, University of Ibadan, Ibadan, Jan. 1988.

Ayoola, G. B. (1988b), 'The Dependency Problem in Food and Rationale for Self-Sufficiency Policy in Nigeria', *J. Rural Dev. Nigeria* 3 (1).

Ayoola, G. B. (1989), 'Socio-economic Factors of Ecological Disaster Management', *Nigerian J. Forestry* 1.

Ayoola, G. B. (1990), 'The Policy Aspect of Agricultural Production in Nigeria', *J. Productivity India* 30 (4).

Ayoola, G. B. and J. A. Ayoade (1991) 'Socio-economic and Policy Aspects of Utilizing Crop Residue and Agro-industrial By-products as Alternative Livestock Feed Resources in Nigeria', AFRRINET Workshop, Gaborone Sun Hotel, Botswana, 4–8 March 1991.

FAO (Food and Agricultural Organization) (1966), *Agricultural Development in Nigeria 1965–1980* (Rome).

Food Strategies Mission (1980), *The Green Revolution: a Food Production Plan for Nigeria* (2 volumes) Federal Ministry of Agriculture.

Idachaba, F. S. (1984), 'Self-reliance as a Strategy for Nigerian Agriculture: Cornucopia or Pandora's Box?' Proc. 25th Anniversary Conference, Nigerian Economic Society, 23–24 April 1984.

Idachaba, F. S. (1985), *Priorities for Nigerian Agriculture in the Fifth National Development Plan 1986–90*, FACU occasional paper No. 1, Ibadan.

Intrilligator, M. D. (1971), *Mathematical Optimization and Economic Theory* (Prentice-Hall Inc.).

Killick, Tony (1981), *Policy Economics, a Textbook of Applied Economics on Developing Countries* (London: Heinemann).

Olayemi, J. K., S. O. Titilola, and M. S. Igben (1975), *Nigerian Food Balance Sheet 1985–1995*, Nigeria Institute of Social and Economic Research Ibadan.

Olayide, S. O., D. Olatunbosun, E. O. Idusogie, and J. D. Abiagom (1972), *A Quantitative Analysis of Food Requirements, Supplies and Demand in Nigeria 1968–1985* (Ibadan, Nigeria: University of Ibadan Press), 113 pp.

Olayide, S. O. and E. O. Heady (1974), *Improving and Implementing Agricultural Planning to Achieve National Goals in Nigeria*, Rural Development Paper No. 13, Department of Agricultural Economics / Center for Agricultural and Rural Development, University of Ibadan.

Stolper, W. F. (1966), *Planning Without Facts: Lessons in Reconstruction and Economic Development in Nigeria* (Harvard University Press).

World Bank (1981), *World Development Report* (Washington DC).

Table 1. Nigeria, Articulation of Agricultural Policy, 1900–1989

Official Title	Brief Description
Forest policy 1937	Based on proposal of Chief Conservator of Forests after a Forest Conference. The problem of depreciating forest capital as a result of unregulated exploitation was addressed.
Forest policy 1945	Revision of 1937 policy; it incorporated the new position of government that (a) agriculture must take priority over forestry, (b) the satisfaction of the need of people at the lowest rates (prices) must take precedence over revenue, and (c) maximization of revenue must be compatible with sustained yield.
Agricultural policy 1946	First all-embracing policy statement in respect of agriculture; Nigeria was demarcated into five agricultural areas: (i) Northern Provinces Pastoral or Livestock Production Area, (ii) Northern Provinces Export Crop (Groundnut and Cotton) Production Area, (iii) Middle-belt Food Production Area, (iv) Southern Provinces Export Crop (Palm Oil and Kernels) Production Area, and (v) South-West Food Export (Cocoa and Palm Kernels) Area.
Policy for the Marketing of Oils, Oilseeds and Cotton, 1948	Commodity-specific policy, directed towards stabilizing post–Second World War prices in Britain

Forest policy for Western Region 1952	Territorial policy declared during the trial of the regionalization concept; focused on forest matters.
Agricultural policy 1952	Territorial policy focused on agricultural matters for the Western Region
Policy for Natural Resources Eastern Nigeria	Territorial; Eastern Region Resources of forest/agricultural matters
Western Nigeria policy the farm settlement scheme was Of Agricultural and the critical element. Natural Resources 1959.	
Nigeria, Agricultural Policy	Undated attempt of the Federal Department of Agricultural Planning to assemble numerous policies of federal government on agriculture.
Agricultural policy For Nigeria 1988	Latest policy statement; comprehensive; based on detailed analysis of quantitative targets; aims at self-sufficiency in food and agricultural raw materials latest 1002.[8]

---

[8] Ash down and carry away from the state" (Anonymous). Of Agricultural Department, (b) Sessional papers of legislative and government document.

Table 2: Planned allocation, cumulative appropriation and actual expenditure, in the Third Development Plan (1975–80), Plateau State, Nigeria, selected livestock projects

Project Title	Plan allocation	Cum Annual Appropriation Amount (₦000)	% of Plan	Actual Expenditure Amount (₦000)	% of Plan	% of app.
Cattle breeding and fattening ranch, Jos	1350	1000	74	566	57	42
Cattle breeding and fattening ranch, Nassarawa	500	630	126	227	36	45
Cattle breeding and fattening ranch, Wase	500	480	96	299	62	60
Improvement of Butura livestock farm	300	360	120	157	44	52
Establishment of piggeries, all LGAs and Kuru	1400	1800	129	799	44	57
Establishment of sheep and goat ranches in Jengre and Mangu	400	250	63	230	92	58
Establishment of rabbit breeding centre	9	15	167	2	13	22
Establishment of milk collection centres		1048				
	1000		148	442	42	44
Construction of meat market, Jos and Keffi	326	254	78	0	0	0
Purchase of equipment of 3 mobile Veterinary Clinics[13]	93	85	94	0	0	0
Mean	588	592	110	272	39	38

[3]**Source:** Field survey, 1990

Table 3: Comparison of data from different sources, production estimates, selected grain crops, 1980

			\multicolumn{4}{c}{1000 MT}
Crop	Source	Value	\multicolumn{4}{c}{Pair-wise difference (2-way)}
			FOS	CBN	FAO	USDA
Maize	FOS	653	0	92	97	1067
	CBN	745	92	0	805	975
	FAO	1550	897	805	0	170
	USDA	1720	1067	975	170	0
Rice	FOS	96	0	210	629	994
	CBN	306	210	0	119	784
	FAO	725	629	419	0	365
	USDA	1090	994	784	365	0

Source: Underlying data from Idachaba 1985

Table 4: Typology of Agricultural Programmes and Projects in Nigeria, 1960–1989

Programme/Project	Description
1. Farm Settlement Scheme	Initiated in old Western Region; aimed at solving unemployment problem among primary school leavers. Policy instruments include agricultural extension, cooperative societies, credit facilities.
2. National Accelerated food Production Programme (NAFPP)	Food Aimed at enhancing farmers' production project technical efficiency in the production of selected crops (mostly grains). Policy instruments include subsidy, credit, adaptive research and demonstration plots.
3. Operation Feed the Nation (OFN)	A mass mobilization and mass awareness programme. Policy instruments include mass media, centralized input procurement, massive fertilizer subsidy and imports.

4. River Basin Development Authorities (RBDAs)	To tap the potentials of available water bodies; first 11, then 18 and 11 in number. Specific objectives are irrigation services, fishery development, and control of flood, water pollution, and erosion. Policy instruments include input distribution, credit services, infrastructure development, and manpower development.
5. Agricultural Development Projects (ADPs)	To enhance the technical and economic efficiency of small farmers in general. Policy instruments include rural infrastructure development (feeder road network, dams, etc.), revamped input delivery system, revitalized agricultural extension system, autonomous project management domestic cum international capital.
6. Green Revolution	To accelerate the achievement of programmes for meeting the agricultural sector objectives. Policy instruments include Food Production Plan, input supply and subsidy, special commodity development programme, review of Agricultural Credit Guarantee Scheme, increased resource allocation to RBDAs, etc.
7. Directorate of Food, Roads, and Rural Infrastructure	Established to facilitate roads and rural infrastructure programmers

Source: Miscellaneous documents

Table 5: Performance of Financial Plans, Selected Agricultural Development Projects, Nigeria

ADP	Financing Period	Amount planned (a) ₦'000	Actual expenditure (b) ₦'000	Index of performance (b/a) %
Bauchi	1981–87	227,273.9	256,293.00	113
Borne	1982–87	133,451	55,571	42
Aduna	1982–87	1,045,533	215,873	21
Somoto	1983–87	282,761	181,186	64
Vida	1980–87	65,523	55,425	84
Gangola	1982–87	33,057	22,492	68
Ilorin	1981–87	79,384	42,029	53
Ekiti Akoko	1982–87	62,757	39,037.2	62
Oyo North	1983–87	51,628.9	52,544.3	102

Source: APMEU, *Project Status Report* 1988

From Food Deficiency to Food Sufficiency: Strategies to Overcome The Food Crisis in Nigeria

Invited paper presented at the National Agricultural Workshop and Exhibition organized by the Imo State Agricultural Development Project, Owerri (Imo Concorde Hotel), 13–15 September 1995.

The attainment of the food security objective requires, as a primary (but not sufficient) condition, that a nation be self-sufficient in its staple food items. This implies that the country does not and should not depend on imports of such items from other countries of the world in such a degree as to produce any perceptive diminution in the attained standard of living of the people, in case of shortfall in external supplies. Going by these specifications, we note that (a) the notion of food self-sufficiency is a subset of a generalized food security objective, (b) food self-sufficiency is conceived within the context of external (and *not* internal) trade in food, (c) absolute or 100 per cent food self-sufficiency is not necessarily desirable, and (d) food sufficiency is a dynamic concept, keeping pace with changes in consumer preference and other moving elements.

Against this conceptual background, we then proceed to discuss issues in food self-sufficiency vis-à-vis the strategies for achieving it in the Nigerian case. Throughout the paper it is assumed, as widely believed, that food deficiency really exists and that over the years, this has reached crisis proportions so that the series of policy actions to overcome it are unarguably warranted. A critical examination of these actions would lead to some programmatic recommendations thereafter.

Some Indicators of Food Deficiency

In Nigeria, the poor performance of the food sub-sector was concealed for long in the capacity of the country for massive imports of food and live animals using the windfall accumulation of petrodollar of the 1970s as purchasing power. However, the inadequacies of the overall agricultural sector were soon revealed by the emerging foreign exchange scarcity of the early 1980s, which manifested through pronounced shortfalls in the supply market for food commodities against the escalating population growth together with increasing demand. By and large, it became very evident that the growing dependency of Nigeria on food imports from outside countries was not sustainable in the long run. Thus, the first indicator of food deficiency status of the country is the quantity and diversity of its imports with respect to staple food and animals amounting to about ₦12.6 billion in 1992 as published by the Central Bank of Nigeria (1992).

Another indicator of food deficiency is the stress factor introduced to both official and private life in the management of the food supply situation. The critical source of this factor is supply shortfall large enough to be noticed or visible. In this connection, the latter part of 1970s witnessed an austerity measure necessitated by developments in the international market for crude oil. A campaign was launched by government tagged Operation Feed the Nation (OFN) to spontaneously stimulate domestic output of basic foods. Such panic measures would not beennecessary if deficiency was hitherto non-existent but had been superficially covered up by oil-backed imports of the items concerned. Later, a Green Revolution campaign was launched which was more systemic compared to the OFN campaign.

The ignominious rationing of essential commodities, including staple food commodities such as rice, in the early 1980s was also a clear indicator of food stress or deficiency. Considerable man-days were wasted in government and private offices when the junior and senior staff alike queued up, as refugees did, on working days to receive their shares of basic items. Their agencies or enterprises became the official channel for distributing food and other commodities to be purchased, a decision made necessary by gross inadequacy of domestic supply sources to meet demand at the same time that foreign exchange was too scarce to import the shortfalls. Subsequent to this event, a frantic effort was made to secure a lifeline facility from the International Monetary Fund (IMF) but was aborted in the face of growing public opposition. Finally, an economy-wide Structural Adjustment Programme (SAP) was launched in 1986 to the convenience of the World Bank - IMF and other agencies in international trade and finance. The latter programme was generally acceptable to the people as a means of upgrading the domestic production efforts rather than merely postponing the evil day, as the IMF loan suggested.

In connection with the IMF loan, the fear was expressed that another windfall of access to foreign exchange would more or less increase the imports capacity and hence reduce the food self-sufficiency position of the country, only to be revealed in later years again.

Thus, in the absence of visible evidence owing to lack of reliable data, to associate death with shortage of food or to determine the degree of hunger presenting as malnutrition, or to determine the number of people seen to be eating from dustbins (in the classical Umaru Dikko sense), certain indicators may be found in official and private events for use as proximate determinants of food deficiency. In the final analysis, regardless of the measures employed, the effect of food deficiency, whether in the temporary or permanent sense, is directly incident on the generality of the people, whose general standard of living collectively falls.

Parameters of Food Sufficiency

Let us examine the parameters of food self-sufficiency to predicate on them the huge policy attention to be highlighted subsequently. The first crucial parameter is that a food self-sufficiency policy should assure continuous social utility provision, derived from the implicit satisfaction of the general populace as a result of the ability of the country to feed itself. Social utility is a matter for national pride; quite naturally, beneficiaries of social utility feel proud among the community of nationals that they get what it takes to live (i.e. food—a critical biological necessity) from within and not from outside. Such utilities enjoyed by individuals may be added together for the whole population to become a great quantity in the aggregate. And this amounts to national pride that conditions certain nationals to feel good in relation to other nationals.

The second parameter of food self-sufficiency is the support needed for infant industries to grow fast. The infant industry argument is that food self-sufficiency, at least, shields the young enterprises from excessive competition until they grow enough to have adequate competitive competence to exist side by side with developed enterprise in the international market. Some of the special advantages not accruing to infant industries include international economies of scale and access to critical productive factors such as cheaper finance and benefits of economic cooperation. The food sub-sector deserves to be so pampered, within the context of the need to promote international market competitiveness, because it represents an infant industry at best and a major source of immense comparative advantage for developing economies.

The third parameter is what Idachaba (1983) refers to as socio-economic discontinuities. Every country is prone to socio-economic discontinuities such as weather hazards (flood, drought, etc). For instance, a section of the United States of America suffered from heat stress that adversely affected crop and livestock populations.

It is most undignifying that Nigeria should share the suffering should such events occur in other countries. From our earlier calculations based on 1979 imports data (Ayoola 1988), Nigeria depended critically on the United Kingdom as the first highest source of 'barley unmilled', 'other cereals, meals, and flours of other cereals than wheat', and 'cereal preparations'. At the same time, the UK was the fourth highest source of 'wheat and spelt including meslin unmilled', fourth highest source of rice, and the fifth highest source of 'meal and flour of wheat and spelt including meslin'. The nomenclature used follows the Standard International Trade Commodity Sections (SITCS). In the same year, the USA was the first (i.e. highest) source of 'wheat and spelt including meslin unmilled', rice, and 'meal and flour of wheat and spelt including meslin'; the USA was also the second highest source of 'maize corn unmilled', the third highest source of 'barley unmilled', and the fifth highest source of 'other cereals unmilled' as well as 'meal and flour of other cereals than wheat'.

Summarizing the roles of these countries as sources of food, the estimates for 1979 led to the inference that out of any 100 people that could possibly die in Nigeria from the shortage of imported food items, in all probability, supply shortfalls from the UK alone could account for about 28 while supply shortfalls from USA could account for another 26, totalling 54 people in respect of these two countries alone. When added to food imports from other countries such as the Netherlands, France, Germany, Belgium, and Luxembourg, Austria, to mention a few, the lives of about 71 people are exposed to the risk of possible socio-economic discontinuities occurring in other countries (Ayoola op. cit).

Thus, a food sufficiency policy should, as a matter of national survival, shield the citizens well away from the vagaries of supply arising from unexpected socio-economic discontinuities. If Nigeria has to sneeze in terms of food distress each time that certain countries catch a cold, then our sovereignty is obviously out of the question let alone national pride!

The last parameter of food sufficiency is within the context of the theory of second best. This theory states that once equilibrium suffers perturbation by way of violation of one of the required conditions, it cannot be necessarily re-established by merely pursuing one or another or even all of the remaining conditions. Translated into world trade affairs, it is a known fact that international market equilibrium is elusive because of frequent perturbation from undesirable practices of many countries. So, it cannot be re-established merely by the absence of self-sufficiency policies in food production by some of these countries where circumstances dictate such policies. Therefore, such countries cannot be held responsible for the failure of international market equilibrium to be established. For Nigeria, the food sufficiency policy is further justified in the present situation where the country is being constantly harassed with economic sanctions in certain parts of the world that see Nigeria as non-conformist to the tenets of current world political order. This event illustrates that we cannot wish away the expediency of a deliberate food sufficiency policy simply based on the idealized doctrine of world market equilibrium, which is constantly undermined in many other ways, anyway.

The Nigerian Food Crisis in Perspective

The most important perspective is the agricultural economy context within which the food crisis emanates directly. In this context, the food problem or crisis is seen as a result of two main forces, which are *technology* and *policy*. Fundamentally, the food crisis has emanated from the failure of agricultural technology to radically transform production enterprises from the traditional low level of efficiency to a higher level of efficiency, which is efficiency of utilization of productive inputs.

One aspect is process technology such as improved planting materials and mechanization for raising productivity of land and labour upwards. This type of technology is primarily responsible for any incremental output at given levels of land and labour resources and thus forms a critical source of food sufficiency for the country. Thus, the apparent failure of process technology to transform traditional agriculture has meant a low ceiling of food output in the first instance, so the constant filling of the food gap is not feasible since growth of food production lags behind growth of population.

The second aspect is product technology, which generates value added, for increasing the utility of raw products to the consumers. The post-harvest activities in storage, processing, and marketing fall into this category of technology. The greater food value of many products probably resides in these activities, so without them, only a small proportion of utility of food is available to the populace. Thus, the low ebb of product technology in Nigeria has blocked the needed multiplier effect of food utilities and consequently keeps the agricultural economy dull and low in performance.

Based on the foregoing and evidence from advanced agricultural economies, technology development is a primary condition for food sufficiency. The only viable source of technology generation is research, which should be appropriate in several respects as determined in previous work, including ecological considerations, socio-economic considerations, simplicity, capital versus labour considerations, as well as divisibility and risk factor (Ayoola 1990).

Next in the agricultural economy context of the Nigerian food crisis is the policy question. Policy intervention becomes necessary owing to the failure of the market system to be perfect as desired. Maximum allocative efficiency cannot be achieved through the market system only, because of several limitations, including:

- The inability of market forces to create social projects such as rural roads and bridges
- The adjustment time between the equilibrium positions, which can be too long
- The fact that market forces can be very wasteful from extensive market competition, implying low return to inputs
- The presence of several market failures which the market fails to correct on its own, e.g. (a) insufficient capital market for long-term (agricultural) investments, (b) unreliability of market forces to provide marketing facilities such as storage, grading, transportation, etc., and (c) the domination of developing economies by multinationals whose interests do not necessarily coincide with our developmental needs and aspirations
- The presence of externalities implying divergences between private objectives and social objectives
- The need to bring about economic equality, i.e. equal opportunities among classes and people
- The need to forestall dehumanizing of the employees by the employers.

Thus, planned intervention is logical and justified. The practical effect of policy, of course, is that it brings about new production and consumption patterns, which would not otherwise have been the case without intervention. This is true of the agricultural sector, where imported technologies may easily displace the traditional methods or 'superior' tastes may emerge for imported items. By and large, the policy environment might contribute to the food crisis through the faulty packaging of certain instruments to form strategies for closing the food gap, as well as the ways and means of implementing such strategies.

We have given perspectives to the discussion of food crisis in terms of technology and policy. The transition from food deficiency to food sufficiency represents an interface in the process of closing the food gap or overcoming the crisis. This interface also needs to be highlighted before examining the inner and outer mechanisms of the strategies required to overcome the food crisis.

The Interface from Food Deficiency to Food Sufficiency

This interface is important for what it is—an adjustment phase for food producers and consumers, and a precision management phase for policy authorities. All the participants have their critical roles to play in order to ensure a smooth transition from food deficiency to food sufficiency in the agricultural economy.

First, the producers must be mentally receptive to taking up the technologies required for generating large incremental food output. However, this is not the major problem area, given the fact that traditional farmers are rational and responsive to economic incentives. If these incentives are rightly provided to them, they will surely make the optimal decisions towards production efficiency, including optimal resource use in production, processing, and marketing enterprises. However, producers must be faithful in the way they undertake their ventures in response to this situation. I particularly wish they should resist the strong temptation to encourage smuggling and other sharp practices that undermine government policies to reduce imports during the interface. For instance, when importation of malted barley was banned, breweries in the country encouraged smuggling of it into the country by their refusal to modify their plants quickly to produce maize and sorghum lager beer as substitute. Therefore, the observed substitution effects noticed could not be sustained at the take-off rate (Ayoola and Mkpojiogu 1991).

Second, the consumers must do away with rigid tastes during this phase. It is quite obvious that for imports to be substituted with domestic production, the basket of goods and services available to consumers will change in form and quantity; this often necessitates change in taste to conform to changes in commodity space. Without the willingness of consumers to adjust their consumption patterns in this way, then the new production technologies cannot be sustained, as imports of foreign foods remain very competitive. A case in point is the ban imposed on wheat importation to permit substitution of bread in the diet of Nigerian consumers. But because Nigerian consumers refused to change their taste in this regard, to eat the newly emerging cassava or maize bread and biscuits and other products, it was difficult for the importation of wheat to stop, especially through the back door. Instead, the urban consumers merely adjusted to the high price of wheat bread to literally frustrate the new induced production initiatives towards self-sufficiency in food supply.

The last element of the interface is the role of government to support the smooth transition from food deficiency to food sufficiency. Government is called upon to promote private enterprise only; it is not expected to engage in direct food production under any guise, because it is not favourably disposed to do that. Strictly speaking, government has no business in business. Rather it should formulate and implement policies and strategies to create an enabling environment for private farmers and other enterprise owners to flourish and fulfil their objectives in production. In doing so, there are several policy instruments at government's disposal, such as agricultural extension, research, education, technical input support services, to mention a few. These instruments should be packaged into effective strategies to overcome the food crisis. The implementation of such strategies must be purposeful and effective. Government should be firm and decisive during implementation; it must be strongly willed and committed, as it must ensure programme accountability and financial probity in implementation. Also, stability of policy is an essential precondition for successful transition from food deficit status to food adequacy and then food surplus. As observed and subsequently shown in the next section, these virtues were greatly lacking in the present interface period to make the transition successful.

Among several policy failures during the structural adjustment, government failed to make the farmers organized. It was proposed at the University of Agriculture, Makurdi—what is now known as Makurdi Declaration—that the Nigerian farmers be organized into one body from the grass roots to the top? The nomenclature, Federation of Farmers Organizations in Nigeria (FOFAN) was coined for the body, which would have roles to participate in the policy process and help government to achieve its aim in agricultural extension and input supply, among others. We did all that at the instance of government itself, which sponsored two national workshops for the purpose; but the same government has refused to implement since 1993 till this paper presentation. Farmers want it and are eager about the national organization as an appropriate means to keep out the middlemen in the policy process. They gathered at Makurdi from all states of the federation with great enthusiasm, in readiness for the unified organization to take off, but it never did. Those who attended the last meeting of the NCA (National Council on Agriculture) would recall the speech of the spokesman of farmers in which the desire for a viable farmers' organization was strongly expressed. We need a visionary leader to break the long silence over the issue and, therefore, call on the Honourable Minister of Agriculture, Alhaji Mohammadu Gambo Jimeta, to do so by removing all obstacles to the take-off of FOFAN.

Strategies to Overcome the Food Crisis in Nigeria

Up to this point, we have alluded to some important aberrations of the agricultural policy process, which make the pace of development of the sector sluggish. At this stage, the discussion is shifted to the illumination of the key strategies on the ground in order to show the particular nature of these aberrations. In doing so, the detailed knowledge of these strategies is assumed to enable us focus squarely on the main issues affecting their successful implementation.

The main group of strategies to be considered is the integrated rural development types. First, let us revisit an issue of intellectual interest that arose at the outset of the World Bank–assisted Agricultural Development Projects (ADPs). The issue was raised as to the more appropriate mode of integration between simultaneity and administrative integration. The point here is that it is possible to integrate services without bringing them under the same administrative roof as is also possible to integrate administratively. Separate arguments have been advanced to justify each mode of integration under different circumstances. The argument that won in the case of ADPs in Nigeria was the one in favour of administrative integration of complementary agricultural services in a semi-autonomous mode. In that case, the ADPs do not depend on the mother agriculture ministry for their day-to-day operations, are self-accounting financially and self-determining in terms of their programmes. However, the ministry supervises their operations in a general sense with the Commissioner of Agriculture sitting on the board as chairman.

Now, after over two decades of operating this mode with the ADPs, how has it performed in the Nigerian circumstance? We have observations to the effect that the model made way for greater efficiency as a result of reduction of bureaucracy involved in traditional agricultural administration. It also made for higher staff productivity based on enhanced remuneration compared to the mainstream ministry. In particular, agricultural extension was more effective than before, largely because of shorter command structure of the staff involved.

But the ministry was quick to produce noticeable interferences through appointments, financial and other matters. Perhaps the most inimical influence of mother ministries on the ADPs was the encroachment on project facilities, including (a) project vehicles that may be permanently attached to the director general or commissioner at the expense of its assigned jobs on the field, (b) tractors and other farm equipment which spend more time on the farms of influential officers in the agricultural and even other ministries than in providing services for ordinary farmers, and (c) heavy equipment such as the rig, trucks, and road construction machines providing authorized and unauthorized services to people in power in their private estates. In fact, the issue of sustainability started on this basis. We questioned at midterm reviews (MTRs) and during redesigned exercises how to keep the influence of ministries low or totally out of project performance. The idea of capacity building of the ministry naturally emerged, and efforts are going on in this regard in many states. To the extent that interferences of the magnitudes witnessed represent definite leakages in the implementation process and to the extent that such leakages reduce project performance, to these extents, the ADPs have failed as integrated rural development projects for overcoming the food crisis.

The River Basin Development Authorities represent an excellent concept for upgrading the utility of water bodies for food production. In the original basin format (11 in number), the focus was on developing the surface and subsurface water resources for irrigation and associated purposes. However, that focus disappeared when, in 1984, they became one per state, totalling 19 at that time. Their disposition to venture into extensive non-water activities was only expected as a means of employing idle capacity, especially in those states where water resource development could not go too far. Before their number was streamlined to the original eleven, much resource wastage had occurred in terms of role confusion and conflicts with ADPs and other agencies, inefficient direct production activities, among other elements of unsatisfactory performance.

The decision to associate a RBDA with each state in 1984 could only be understood within the context of politics of the time. It would appear that the idea had earlier taken root during the Second Republic under President Shehu Shagari when the National Party of Nigeria (NPN) controlled Federal Government had become uncomfortable to overturn the non-NPN-controlled state governments in the control of the ADPs to score political points by way of populist projects such as fertilizer supplies, rural borehole supplies, rural roads constructions, and others. The ADPs would jealously prevent their programme from being used by the Federal Government against the political interests of their host state government. And this was quite possible with the administrative structure that gave the state government the upper hand in the actual implementation of the ADPs, although at the maximum risk of losing the federal subvention meant for the projects.

The situation could not be contained further as the election year approached (1983) and the Federal Government had to fall back on the RBDAs for the purpose. As this approach proved somehow effective in certain targeted states, legal actions could not be avoided. The six Unity Party of Nigeria (UPN) states finally dragged the Federal Government to court to challenge the activities of the RBDAs in their states, citing favourable clauses in the old constitution.

To the extent that their loss of sharp focus and role confusion were resource-wasteful, to the extent that their structural instability also retarded development progress, to the extent that their ventures in direct agricultural production were inefficient and to the extent that the political intransigence worsened their programme performance—to those extents, the RBDAs could not fulfil their mission and mandate to overcome the Nigerian food crisis.

When the Directorate of Food, Roads, and Rural Infrastructure came on board in 1986, the aim was to provide a major force for accelerating the pace of agriculture and rural development. As an agency in the highest political powerhouse (i.e. presidency) and under the leadership of a top military officer, it enjoyed all facilities required to transform the rural landscape. Few people would agree to understand why DFRRI could not have performed the wonders expected from it. A part of these reasons can be found in its decision to go to the cellular level of agricultural development instrumentation, including its activities in agricultural extension, seed production and distribution, and such highly technical functions not really meant for the agency. There were also a few sharp practices that were noteworthy, the role conflict between DFRRI and ADPs in many states being the most pronounced. Reports had it that DFRRI merely removed the signpost of some ADPs and replaced it with its own to lay claim to the construction of certain rural roads and boreholes. To the small extent that DFRRI's activities had increased the network of rural roads in quantity and quality together with the inherent regular and periodic maintenance problems, to the small extent that the DFRRI's borehole yield water across seasons, among other aspects, to these small extents DFRRI has helped in overcoming the food crisis in the country before it went underground.

The National Agricultural Land Development Authority (NALDA) as the latest in the series of integrated rural development strategies has a gap to fill in terms of focused area development. The agency addresses the land availability issue along with land capacity enhancement support and input supply, all directed at a limited area with particular attention on the special ecological specialization and other sources of comparative advantage of the area. But stress signals are emerging already as many of their enclave projects (one per state so far) have recorded low participation and deviation from recommended packages. Preliminary evaluation has shown that the agency has failed to provide the support services in the degrees desired to guarantee efficiency of production under the universal 4-ha.-per-farmer land-allocation model. In Ogun State for instance, the results of a farm survey has shown that the optimal farm size for arable production was only about 0.5 ha. under the prevailing cost and revenue structures. Therefore, the NALDA-imposed 4-ha. farms would need some efficiency enhancement multiplier of about 8 times at least, in terms of timely land preparation, supply of technical inputs, market facilitation, and other services. This is obviously unlikely to happen in the foreseeable future, given the general resource limitation faced by NALDA and other government agencies. Therefore, NALDA needs to undertake a constant review of the situation to ensure that progress is being made as expected towards overcoming the food crisis.

Summary and Conclusion

The proximate indicators of food deficiency include its large quantities and diversity of food and live animals imported over a long time and the need for occasional panic measures as official reactions to looming dangers such as the ignominious rationing of essential (food) commodities of the early 1980s. Few people would disagree that the food deficiency status of the country has reached crisis proportions, with the larger fraction of average income being spent on staple food items.

The paper posits that the main causes of the crisis can be traced to aberrant technology and policy. In this context, the transition from food deficiency to food sufficiency involves an interface that plays definite roles to the producers, consumers, and government. In the final analysis, the success of the strategies that are being implemented to overcome the Nigerian food crisis depends on the effectiveness of the assigned roles of these three actors in the policy process.

REFERENCES

Ayoola, G. B. (1988), 'The Dependency Problem in Food Grains and Rationale for Self-Sufficiency Policy in Nigeria', *Rural Development in Nigeria* 3 (1): 12–20.

Ayoola, G. B. (1990), 'Technological Progress in Agriculture: Some Issues' in *Capital Goods and Technological Development in Nigeria*, The Nigerian Economic Society Annual Conference, Minna, May 1990.

Ayoola, G. B. and D. O. Mkpojiogu (1991), 'Analysis of the Effects of Structural Adjustment Programme on Food Prices in Benue State, Nigeria', *J. Agric. Sci. Technol.* 1 (2): 81–85.

CBN (Central Bank of Nigeria) (1992), *Annual Report and Statement of Accounts for the year ended 31st December 1992* (Lagos).

6

The National Question in Nigerian Agriculture: A Proximate Analysis

'National question', as a debate on the political economy, has now graduated from informal media discussions to attract explicit intellectual attention. The issues involved span over wealth distribution, power concentration (political and economic power), class interests, including ethnic, geographical and minority group aspects. Within the development context, these issues are important in the sense that their optimal resolutions determine the degree of peace in the environment for achieving social and economic objectives. Therefore, the need arises to undertake the conscious analysis of these issues using informed opinion and empirical evidence covering different sectors of the economy. The current problem, however, is an apparent gap in knowledge about the forms and functions that the national question presents in the sectors.

The objective of this paper is to demonstrate that the agricultural sector is replete with concrete instances when national question has been, and is still being raised. Far from providing definite answers to the elements of the question, the analysis rests on a highlight of the crude sources of issues in the question to validate and discuss the hypothesis implicit in the debate generated. The data employed were sourced from service statistics of agricultural agencies and other miscellaneous materials. A proximate analysis implies the recognition of the presence of analytical inexactitudes, limiting the drawing of hard inferences and conclusions in certain instances.

CRUDE SOURCES OF ISSUES IN THE NATIONAL QUESTION

There are several sources of issues for raising the national question in the agricultural development process. The three among them that will be highlighted are federal presence, political leadership, and location/funding of projects.

The Federal Presence Question

The somewhat silent question about federal presence in agricultural development has arisen from a historical context. The main element is whether or not federal government should play a direct role in the development of agriculture, for example having a Federal Ministry of Agriculture, let alone its multifarious parastatal system and implementing a budget policy concerning the agricultural sector. To appreciate the rationale behind this question, let us go a little into the agricultural history of the country.

When the Richard Constitution (1946) was introduced into the country, agricultural administration was provided for as a regional responsibility. The situation survived the Macpherson Constitution (1954) through independence (1960). Agricultural development belonged to the residual legislative list, which, by a special clause in the constitution, was a regional matter. The team of agricultural experts from Food and Agricultural Organization (FAO 1966) that visited the country soon after independence later recommended the creation of a central organization for coordinating disparate regional activities. But this idea could not be implemented owing to what Ayida (1973) described as the need not to 'offend the political sensibilities' of the regions. The Federal Government made an initial attempt to go about the problem by creating a stylized organ, carefully named Ministry of Natural Resources and Research (Federal Government 1965). Later, using a military character, Gowon's regime boldly established the first truly federal ministry for agriculture by administrative fiat (Federal Government 1966). This was to later provide the basis for transferring agricultural development matters to the concurrent legislative list in the more recent constitutional reforms (i.e. as a joint federal and state affair).

The question then arose naturally from the more agriculturally prosperous regions concerning the implicit (unstated) objective of the policy to increase federal presence in the development process of the sector. These regions sponsored the idea that they would be worse off, given the inevitable redistributive consequences of the policy. The logic of this worry—an aberration of the national question—was based on the fact that the affected regions were enjoying a relatively greater boom from their export trade at the time; the export products then were cocoa (West), oil palm (East), rubber (Mid-West), and groundnut (North). These commodities yielded different degrees of foreign exchange earnings, which reflected clearly in the regional budget policies and manifested through different capacities to implement social and economic programmes.

It was actually imagined that this form of the national question had disappeared from people's memories over the lo..g time, until lately (Second Republic period) when it resurfaced in a political context: five state governments (all UPN-controlled)[1] took the Federal Government to court over the activities of the River Basin Development Authorities (RBDAs). In the case, they challenged the constitutional rights of the NPN-controlled Federal Government to undertake those (agricultural) activities in their respective territories. Observers, however, would note that the legal action was motivated, remotely and immediately, by the need for the party controlling the states to limit the ability of the other party controlling the Federal Government to score political points in the states where they had strong foothold, as election approached.

The Leadership Question

Although the instability of the political office holders in the federal agriculture ministry has been substantiated (Ayoola 1990), the real causes remain unclear. The changes in government, as was stated, appear inadequate to explain the very fast turnover rate of ministers or (in the past) commissioners or (presently) secretary. By and large, certain sections of the population got into thinking that they have not been favoured in the appointment of the political leaders of agriculture. The emerging question involves whether the greater frequency of members of other sections has not translated into greater progress of agriculture in the regions of the country that they represent. The sponsors of this question consider development similar to a burning candle which provides greater illumination as one gets closer. If this holds true, then the distribution of agricultural infrastructure and other forms of social capital bases for growth could have been overtly or covertly skewed region-wide in favour of persistent occupants of the leadership posts by people from a region.

Nevertheless, the strength of this argument may be gleaned from Table 1. Using the names of the sample of leaders obtained as correlates of their regional affiliations, combined with the duration of their tenure of office, one may or may not conclude in the same way that the leadership question has been silently put forward.

> [1]The two political parties in the forefront at that period were the Unity Party of Nigeria (UPN) and the National Party of Nigeria (NPN).

Location of Projects

By far, the loudest presentation of the national question in agriculture is by way of accusations of preferential location of projects. There could be several candidate sources of this question, but one instance would illustrate its justification using events in the World Bank–assisted Agricultural Development Projects (ADPs). An identification mission that visited the country in the early 1970s recommended certain arable crop projects to be fashioned as command-area development strategies. It happened that the first generation of three projects involving huge capital outlays was located in one region of the country: Funtua ADP (Kaduna State), Gombe (Bauchi State), and Gusau (Sokoto State). This even quickly served as confirmatory tests of the implicit objectives of the federal presence policy. The sponsors of the location question were worried, again naturally, how such an event could be simply attributable to chance. Keen observers of ADP process would notice the high intensity of the concern about this initial event. Certain sections of the country expressed the feeling that they could not wait till a future date while development was going on in other areas. These people were least impressed by the theory that provided for a learning curve. Thus, the original schedule of the ADP process had to be modified in two significant respects. First was the need to speed up the expansion of the ADP concept to other regions, ignoring the World Bank Fund that was associated with a slow pace (FMA 1981, p. 12). The result was the mounting of the so-called Accelerated Development Area Projects (ADAPs) at Anambra and Imo States. Second was the need to speed up the World Bank's process through joint negotiation of projects in single loan deals. The results of this were the contemporaneous approval of groups of projects termed multi-state (MSADPs, for short).

Other strategies whose activities had been incorporated into the national question are (i) Directorate of Food, Roads, and Rural Infrastructure (DFRRI) concerning its role in the provision of social capital base for development; (ii) the commodity boards, concerning their role in agricultural market intervention, (iii) the River Basin Development Authorities (RBDAs), concerning their role in the development of surface and subsurface water bodies for irrigation and other purposes. The issues raised about these strategies are not drastically different from the previous kinds; they are all tied together in the unified discussion following.

DISCUSSION

A careful discussion of the various forms of the national question as highlighted above is necessary on three intellectual grounds: (i) the need to inform participants in the debate about some unknown variables of the question; (ii) the need to articulate a balance of the arguments, albeit at the maximum risk of being judgemental; and (iii) the need to relate the proximate inferences that can be drawn to the sustainable development context.

By and large, the red thread running through the fabric of the several forms of the question as presented above is the untested hypothesis of a regional bias in federal development efforts in the agricultural sector. Certain rough-and-ready instances have been cited in an attempt to validate this hypothesis. The focus of this discussion, then, is to see the extent to which the federal policies in question can be justifiably rationalized. To achieve this, we set forth Table 2, which describes the key strategies in quick succession.

The NAFPP does not provide any evidence of regional development bias, direct or indirect, as the commodity focus appears region-neutral. The susceptibility of the ADP process to national questioning had been ascertained earlier. The redistribution effect of this would be achieved through productivity increases with the attendant increases in factor incomes; unless such incomes are mopped up by the tax instrument, the command-area projects would produce pace differentials in the development of the regions. In any case, the system soon adjusted to this as similar projects were mounted in all the states of the federation. The latter event has since neutralized the need for further questioning in regard to the ADPs.

[1] Regional price protection behavior was manifest in the commodity boards system. For instance, the governors of Ondo State and Kaduna State petitioned the Chief of General Staff about low prices of commodities predominantly produced in their areas (i.e. cocoa and groundnuts respectively). They demanded upward revision of the prices relative to others.

[2] For instance, the Kaduna State Government once decided to prevent out-of-state traders from buying grains in the state (see FASCOM 1984).

The fact that the commodity boards were active in export trade to different degrees posed the possibility of taxing agriculture in one area to finance development in other areas. Given that budget policy is region-neutral while the boards existed, this proposition might hold true, even in the absence of rigorous proofs.[2] Table 3 shows the buoyancy status of the boards as at 1986 when they were liquidated as part of the structural adjustment of the economy. It appears that the Nigerian Cocoa Board had the greater capacity to have contributed to development financing in the country than others. There were further indications that the non-exporting boards, especially the Nigerian Grains Board, actually depended on budget allocations for most of the period of existence. This might strengthen the earlier argument about federal presence as an avenue to use budget policy for achieving regional transfer of agricultural incomes.

The region-sensitivity of RBDAs arises from the irrigation services as their main instrument for developing the agricultural sector. The fact that the felt need for such services is region-specific makes the activities of RBDAs prone to inclusion in the regional bias argument. Irrigation services help to lower the risk of crop failure at the same time when the chances of multiple cropping increase. Thus, the incidence of irrigation is on profitability of agricultural enterprises. In fact, provision of the services constitutes indirect subsidy policy in the production of vegetables and grains during the dry season. Nevertheless, the merit of questioning these services in terms of regional bias undermines the more national thinking about developing ecological advantages for the overall growth of the economy. The argument also lacks merit on account of food security. What is merely required to complement the success where it exists is a single national market that permits the free flow of goods and services. In this connection, one would discourage the kind of hollow policy victory sought by certain state governments occasionally, wherein a wedge is put in the flow of food items in the country.[3] Such a policy is not consistent with the effort to establish one single national food market for Nigeria.

Another question having the vertically opposite effect to the ones about RBDAs is the one that could be raised about the activities of NALDA. Some 30,000–50,000 hectares of arable land would be developed in each state of the federation. As revealed in the formulation of some of these projects, a major thrust of government would come as assistance in land clearing (Table 4). Incidentally, land clearing poses a great problem in areas where irrigation services are less required. Thus, the direction of net benefit flow to implement RBDA and NALDA is opposite, though not necessarily equal, thereby significantly reducing the strength of the question about income effect of budget subsidy for implementing these strategies across the regions.

CONCLUSIONS

The national question in agriculture presents mainly in the form of accusation and counter-accusation about the possibility of regional bias, implicit in the role of Federal Government. Some of the arguments are quite robust but can be balanced against the policy objectives of sustainable development and food security. Efforts to strengthen the agricultural economy with a single free national market for food and fibre would remove much of the basis for questioning.

REFERENCES

Ayida, A. A. (1973), 'Basic Issues in Financing the Nigerian Agriculture in the Seventies' in *Proceedings of the National Agricultural Development Seminar*, Federal Ministry of Agriculture and Natural Resources, Lagos.

Ayoola, G. B. (1990), 'Agricultural Productivity in Nigeria: Some Policy Aspects', *Productivity* 30 (4): 394–401.

FAO (Food and Agriculture Organization) (1986), *Agricultural Development in Nigeria 1965–1980* (Rome).

Federal Ministry of Agriculture (1981), *Annual Report*.

FASCOM (Farmers Supply Company) (1984), 'Format for 1984 Grain Purchase by FASCOM'.

Federal Government (Nigeria), *Official Gazette* 52 (33) (9 April 1965).

Federal Government (Nigeria), *Official Gazette* 53 (10) (7 February 1966).

NALDA (National Agricultural Land Development Authority) (1992), *Design and Formulation of Agu-Ukehe Smallholder Agricultural Project in Enugu State*.

—— (1992), *Design and Formulation of Mbagbe-Ugambe Smallholder Agricultural Project in Benue State*.

Table 1: Inventory of political leaders in agricultural administration (including water resources), federal level, 1966–1993

Name[1]	Official Title of Office [2]	Approximate Tenure [3] Period	Duration (Yrs)
C. O. Komolafe (W)	Minister of State+	1966–1969	4
Yahaya Gasau (N)	Fed. Commissioner++	1969–1972	4
J. O. (Y) Okezie (E)	Fed. Commissioner++	1972–1975	4
J. E. Adetoro (W)	Fed. Commissioner++	1975–1976	2
E. O. Ekpo (E)	Fed. Commissioner+++	1975	1
(T) U. W. Osisiogu (E)	Minister++++	1975–1976	2
B. O. W. Mafeni (W)	Fed. Commissioner+++	1976–1979	4
Ibrahim El Yakubu (N)	Minister++++	1976–1979	4
Adamu Ciroma (N)	Minister++	1979–1983	5
Ibrahim Gusau (N)	Minister+++	1979–1982	4
Emmanuel Aguma (N)	Minister of State+++	1979–1983	5
Mdagi Mamudu (N)	Minister++++	1979–1982	4
Usman Sani (N)	Minister of State+++++	1982	1
Olu Awotesu (W)	Minister of State	1979–1982	4
Ken Green (E)	Minister of State	1982	1
Emmanuel Atanu (N)	Minister++++	1983	1
Etong Okoi-Obuli (E)	Minister+++++	1983	1
Buka Shaib (N)	Minister++++++	1983–1985	3
Ismaila Isa Funtua (N)	Minister++++	1983	1
Bala Sokoto (N)	Minister+++++++	1983	1
Alani Akinrinade (W)	Minister++++++	1985–1986	2
M. Gado Nasko (N)	Minister++++++	1986–1989	4
Samaila Mamman (N)	Minister++++++	1986–1989	4
Shetima Mustapha (N)	Minister+++	1990–1992	3
Abubarkar Hashidu (N)	Minister++++	1990–1992	3
Abubarkar Hashidu (N)	Minister++++++	1992	1
Garba Abukadir (N)	Minister++++++	1993	1
Isa Mohammed (N)	Secretary++++++	1993	1

1. Plausible direction of implicit regional bias is indicated in parenthesis as follows, using the colonial administrative boundaries: N = North, E = East, W = West.
2. The nomenclature of the federal ministry of affiliation is stated as follows:

+	Federal Ministry of Natural Resources and Research (FMNRR)
++	Federal Ministry of Agriculture and Natural Resources (FMANR)
+++	Federal Ministry of Agriculture and Rural Development (FMARD)
++++	Federal Ministry of Water Resources (FMWR)
+++++	Federal Ministry of Agriculture, Water Resources, and Rural Development (FMAWRRD)
++++++	Federal Ministry of Animal and Forest Resources.

3. As actual dates are not available, periods are indicated by years only; duration is roughly estimated by the number of years over which the tenure spreads. Owing to presence of parallel ministries, multiple tenures, and concurrent ministers, the total of duration column exceeds the period of analysis (i.e. 1966–1993).

Table 2: Typology of selected projects/programmes in agriculture and rural development, Nigeria

Project/Programmes	Objectives	Relevant details
National Accelerated Food Production Project (NAFPP)	A project to enhance the technical efficiency of farmers in the production of selected crops	Subsidy, credit, adaptive research, demonstration plots.
Agricultural Development Projects (ADP)	Generalized enhancement of farm production efficiency	Expatriate (World Bank) funding, technology adaptation and transfer, technical input supply (fertilizer, improved planting materials, etc), rural infrastructure support (rural roads, boreholes).
River Basin Development Authorities (RBDAs)	Development of available water bodies (surface and subsurface aquifers) for agricultural production and other uses	Decree No. 25 of 1976 (as subsequently amended). Irrigation services. Currently, eleven authorities exist at strategic locations on the major rivers (Sokoto-Rima, Hadejia-Jama're, Chad Basin, Benue, Niger, Ogun-Oshun, Benin-Owena).
Commodity Boards	To undertake the marketing and development of major commodities	Decree No. 29 of 1977. Intervention buying. Seven boards were established initially for cocoa, groundnuts, cotton, rubber, grains, tuber and root crops. Technical Committee on Produce Prices existed to recommend prices for the various commodities on annual basis. The system was scrapped in 1985 as part of Structural Adjustment.
Directorate of Food, Roads, and Rural Infrastructure	To raise the quality of life of rural people	Decree No. 4 of 1987. Rural infrastructure (rural roads, water supply, electrification) etc.
National Agricultural Land Development Authority	To assist farmers in the development of land and labour resources thereby enhancing the development of special ecological advantages of different areas in the country	Mounting of command-area smallholder projects towards developing 30,000–50,000 hectares of arable and permanent crops in each state of the federation.

Source: Ayoola (1988)

Table 3: Indebtedness of commodity boards to Central Bank of Nigeria as at 30th November 1986

Board	Total Loan Granted 1976–1986	Total Loan Plus Accrued Interest	Payments i.e. sales proceeds plus refund	Outstanding Balance
Cocoa Board	2233.9	2275.9	2275.9	-
Palm Produce Board	662.2	853.9	325.1	528.8
Cotton Board	359.7	449.7	289.3	160.4
Rubber Board	257.5	309.4	154.5	154.9
Grains Board	184.0	212.5	58.1	154.4
Groundnut Board	42.2	53.4	38.2	15.2
Total	3739.5	4154.9	3141.1	1013.7

Table 4: Capital Cost Profile in Selected Smallholder NALDA Projects

	Agu-Ukehe Project (Enugu State)	Mbagbe-Uganbe Project (Benue State)
Land Development		
Land acquisition	5.00 (100)	5.00 (100)
Land demarcation	960.00 (100)	900.00 (100)
Layout/parcelling	375.00 (10)	375.00 (100)
Land clearing	2160.00 (70)	1000.00 (60)
Other Investment Costs	**3,500.00**	**2,280.00**
Civil works	5300.00 (100)	11373.75 (100)
Plant and equipment	10352.05 (100)	20632.50 (100)
Livestock (parent stock)	450.00 (100)	600.00 (100)
Planting material	6750.00 (50)	3000.00 (5)
	19,602.00	**37,885.75**

Source: NALDA (1992a, b)

*Figures in parenthesis are the percentage contribution of government.

The Unknown Variables in Nigeria's Food Self-Sufficiency Equation

Paper presented at the First National Congress of Science and Technology, University of Ibadan, 14–20 August 1988.

The renewed policy recognition of self-reliance as a theme of Nigeria's economic development began in the early 1980s when it became apparent that the oil-boom era would soon come to an end. At that time, the balance of payment problems accentuated the emergent scarcity of essential commodities, consisting mostly of food items, when petrol-dollar earnings slumped substantially. Among the lessons of this general scarcity, it had been clearly demonstrated that the use of massive importation of goods and services as a means of raising the welfare standards of citizens could have only a short-run advantage. In the light of this experience, the strategy of self-reliance easily suggests itself as the only testable alternative on which sustainable economic growth and development can be based.

The self-reliance strategy of economic development implies, in the main, a country's determination to use its domestic resources in creating utility for its citizens. It represents a definite departure from the open-ended importation of goods and services whose domestic resource costs of production are simply greater than their foreign exchange costs. It therefore implies that for a determination of this kind to endure and be meaningful, a necessary, albeit not sufficient condition is the pursuit of self-sufficiency policy in the priority goods and services, among which food items should rank superior. The economic rationale for doing this can be explained briefly on four grounds as follows:

(i) Self-sufficiency can be discussed in the context of a welfare function, in which case the domestic availability of food can be conceived as a public good that is capable of yielding social utility. In the present context, the utility is in the form of the implicit gratification which the citizens, Nigerians, will feel as if they are able to feed themselves.

(ii) There is the infant industry argument for food self-sufficiency policy. This considers that certain industries should receive extra protection until they become securely established. Self-sufficiency policy helps to shade the infant industry well away from the outside influences of the imperfect world market until it develops adequate competition competence. There is merit in the idea that Nigerian agriculture, and food production in particular, is an infant industry because it is still characterized by technical and economic disequilibria resulting from its movement from old to new production surfaces.

(iii) The food security argument also holds to justify self-sufficiency policy. On this basis, it is presumed that high-level dependency, especially for essential commodities, is economic risk because it opens the importing country to the effects of socio-economic shocks in the exporting country.[1] These external shocks range from prolonged drought to large-scale pest infestation to civil disturbances and possibly war.

(iv) The theory of second best justifies self-sufficiency policy. Succinctly put, this theory, first put forward by Lipsey and Lancaster (1956) establishes that once one or more of the conditions of pareto-optimality has been violated, there is no guarantee that we will ever attain another optimality by pursing the remaining conditions. In the context of the current discussion, the evidenced defects in structure, conduct, and performance of world commodity markets are such institutional restrictions which make the best welfare positions unattainable; and it is quite irrelevant to inquire whether a second best position can be attained by satisfying the remaining pareto conditions. Consequently, any international criticism of self-sufficiency policy in developing countries cannot be acceptably based on account of violation of the equilibrium conditions for the attainment of pareto-optimality on a global scale.

Given this suitable rationale, the intent of this paper is to examine the salient analytical issues involved in the food self-sufficiency policy scientifically by presenting the problem in an equation form. The second section conceptualizes a self-sufficiency function as the basis for the subsequent discussions. The third section examines the left-hand side of the equation, and the fourth section examines the right-hand side. The fifth section presents the inferences made from an empirical effort. The sixth section highlights on the residual variables of the equation. The paper is concluded in the seventh section.

The Food Self-Sufficiency Function

The first issue to address in the food self-sufficiency function is how to define and determine an acceptable measure of self-sufficiency. This must be done first before thinking about the acceptable level of it to desire. The usual practice is to choose an index, often called the self-sufficiency index or ratio. This is defined as the ratio of domestic output of a commodity to the total demand of the same commodity, i.e.

$$I(t) = P(t) / D(t)$$

Where   I   self-sufficiency index
        P   domestic output of the commodity
        D   total demand of the commodity
        T   the time.

Next is to determine the variables that influence the level of I that can be employed as policy instruments. In other words, we require the determinants of P and D, which can be used to manipulate "I". If these variables are represented by an instruments vector X, then we can write:

$$I(t) = f(X(t)).$$

This means that the level of self-sufficiency in a given food item depends on the values of certain variables collectively called X. This is the self-sufficiency function that will form the basis of the subsequent discussions. It contains all information pertaining to the way in which aggregate production and demand determine the value of self-sufficiency index in a given commodity. The two sides of the equation are to be examined with the objective of revealing the analytical problems intrinsically embedded in the food self-sufficiency policy.

First, as it were, what are the characteristics of such a function? The first difficulty facing the analyst is to determine whether this function is naturally single-valued or stochastic, discrete or continuous, linear or non-linear, etc. In essence, the form of the function itself is the beginning of the analytical problem, and this has implications for policy just as much as the ones to be exposed shortly. In effect, these preliminary problems have no general or specific solutions. What the economists do, therefore, is to settle for a process of compromise and batter, which does not necessarily ensure that the resultant public programming efforts are actually moving the nation further towards or away from the desired level of self-sufficiency. Such a compromise is revealed in the assumption that this function is single valued, and continuous with nth order derivatives. It is to be defined for non-negative values of the index and input levels. Naturally, the domain of the function does not include the entire non-negative quadrant, being restricted to the range between zero and unity for any one commodity; and it is concave when I is maximized. These types of assumption essentially make the results of analysis imprecise.

The Dependent Variable

In this section the left-hand side of the equation is to be closely examined. To begin with, that variable $I(t)$, defined in terms of output and demand, seems structurally deficient in actual practice. The proxy in common use, because of the inherent problems of evaluating the denominator D, is put as follows:

$$I(t) = \frac{P(t)}{P(t) + M(t)}$$

Where $M(t)$ = the net imports of the food item.

Obviously this identity ignores one important aspect, which is the pent-up demand. Unsatisfied demand is a legitimate component of the self-sufficient demand, but its level is not certainly known at any particular time. Even if an attempt is made to estimate it, the results will still be based on sample data rather than the population and are conditional on the goodness of the demand model used. Consequently, the measure on hand is one that does not reflect total sufficiency. Suffice it to say that all such quantitative analyses that are based on the index will produce lesser results than desired.

One reason that can be put forward in support of neglecting pent-up demand is the presence of close substitutes, which exist among food items, particularly grains, which suggests that the magnitudes of demand with regard to individual food items will be rather low. The low magnitudes of pent-up demand should have been almost wholly absorbed by imports. Again, this is like begging the issue, and the argument now exposes the index to another shortcoming. If availability of foreign exchange is a precondition for the index, then the measure on hand cannot apply at all phases in the economic cycle, which further makes it short of what is required.

Lastly, two categories of food items make econometric nonsense of the index measure of self-sufficiency. One of these includes sorghum, yam, and others, which are non-traded commodities. The formula then contains $M(t) = 0$ for all values of t, thereby collapsing $I(t)$ to unity in the range. The other category includes wheat, uniquely being a commodity not produced for all practical purposes. Similarly, $P(t)$ is zero for all values of t, thereby collapsing $I(t)$ to zero in the range. In both cases, there is no variation to explain in the dependent variable, to be able to determine how the policy instruments will influence the level of self-sufficiency. Thus, the index measure on hand lacks universal applicability across all the food commodities. This poses severe limitation on its usefulness as a performance variable of policy.

The Independent Variables

In examining the right-hand side of the equation, we are concerned with the issues associated with the policy instruments. 'Policy instruments' means a macroeconomic variable that can be used by government for changing the behaviour of the economy. In particular we are presently interested in the variables which determine the level of the self-sufficiency index. One important source of problems of policy management is the unending inter-relationships within the economy. Just as the self-sufficiency index has instruments, the specified instruments also depend on another set of variables, which also have instruments and so on. In practice, this chain is abruptly shortened in order to fix one idea at a time. It is therefore not expected that the controllability of the economy using the chosen performance index could be complete.

The numerous other issues associated with the usual instruments of self-sufficiency policy can be summarized into pertinent questions. These include whether or not the instruments are commodity neutral; can they be subjected to policy management with relative ease? How large will the response of the target variable be to a given change in the instrument act upon the causes of the problem which it is directed at? Is it cost effective more than the others? Are the effects of the instruments confined to the furtherance of the objective it is intended to promote (i.e. is it strictly selective)? How flexible is a particular instrument (i.e. whether it allows for stoppage when unintended results are being obtained)? It is quite difficult to obtain precise answers at once with respect to the instrument variables of food self-sufficiency policy.

In the usual compromise situation, economists have suggested a number of instruments with little regard to their ability to fulfil the intended standards, which are implicit in those questions. In a study by Olufokunbi (1984), it was made conclusive that overall price index is an important explanatory variable of food self-sufficiency. This supports the positive significance of the role of the general price level in self-sufficiency relations as reported by Sverberg (1978). In addition, the central arguments in the future growth plans for self-sufficiency in China are seed, fertilizer, irrigation, land, and tractorization, as reported by Wiens (1978). Sarma (1978) added the importance of credit, research, and extension in food self-sufficiency in India. Given that these are in the stylized self-sufficiency function earlier defined, the extents to which they possess the desirable qualities in practice are largely unsatisfactory.

An Empirical Illustration[9]

In an original work (Ayoola 1988), a commodity specific self-sufficiency equation was fitted for the whole country. The choice of commodity was maize because it enjoyed substantial political visibility in the economy. The explanatory variables of the self-sufficiency index used include land area in hectares, market price in kobo per kg, government expenditure on agriculture in naira of budget allocations, rainfall, and time as proxy for technology. The hypothesized logic under the available data is the simple quadratic equation with neither additive nor multiplicative interaction terms.

The empirical results can be summarized into two major aspects. First, the statistically significant explanatory variables of self-sufficiency index are land area, government expenditure on agriculture, and agricultural GDP. Further analysis indicates that government spending is the most effective in terms of average and marginal effect coefficient, while land is the most effective in terms of elasticity coefficients. The supremacy of government spending over the other factors is probably due to the great multiplier potential it possesses. That of land is because it serves in providing nutrients

---

[9] This section summarizes the results in part of a study being considered for publication in a scientific journal in greater depth.

for plant growth as well as anchorage and it is the common basis for the application of the other factors of maize production. Secondly, evidence abounds to suggest the hypothesis of critical minima in factor utilization for self-sufficiency. This hypothesis is based on the feature of land, which is not statistically significant at low level, but it is at the high level, but it is at the high level as an explanatory variable of self-sufficiency. Another basis is that of government expenditure on agriculture, which is significant at both levels but it carried wrong sign at the low level. This suggests that there is probably a critical minimum or threshold point below which increases in the use of an instrument will not influence self-sufficiency level in the crop. In the context of land use for maize, this minimum level corresponds to the subsistence phase in agricultural production where most Nigerian farmers are still trapped. Even though the present study does not determine this minimum level, the suggestion for policy is that additional levels of a factor should be applied beyond respective threshold level.

Residual Issues in Food Self-Sufficiency Policy

The aim of this section is to examine some qualitative arguments in the food self-sufficiency equation. The first is occurrence of conflicts and trade-offs. Nigeria's experience lends credence to the fact that achievement of, or attempt to achieve self-sufficiency in food often necessitates increasing dependence on farm input imports, expatriate management and foreign capital. The massive importation of fertilizer in the Operation Feed the Nation (OFN) years and the initially high import tolerant management expertise in the World Bank–assisted Agricultural Development Projects (ADPs) are relevant examples in this regard (see Fig. 1). Suffice it to say that if a self-reliance programme does not feature a leftwards and downwards shift in the trade-off curve between food import dependence and input import dependence, the country will only succeed in changing the forms, rather than the degree of self-sufficiency.

Next is the issue of sustainability, which concerns the ability of the economic systems to keep production efforts going continuously without failing. In this regard, the major worry is how to prevent the usual discontinuities from occurring so that an attained level of self-sufficiency will, at least, be sustained. Idachaba (1987) identifies four major groups of these discontinuities. These are the unstable institutional arrangements, including instabilities in the political, administrative and probably professional leaderships, ever-changing or reversible macroeconomic policies, irregular natural hazards of the agricultural environment, and the highly distorted and uncertain international economy. One addition is implementation indiscipline.

Another issue is the place of international trade theory in self-sufficiency policies. There seem to be conflicting ideals; while self-sufficiency policy represents a clear departure from open trade which can generate comparative advantages, it is still generally recognized that no one country is completely endowed in the economic sense. Although the argument for self-sufficiency is at clear variance with the principles of comparative advantage in the global sense, it makes a lot of political economy sense. For one thing, the case for comparative advantage theory is often not persuasively advanced, as determinants often used to divide the national commodity basket into tradable and non-tradable categories are usually not conclusive enough. This weakens the argument in favour of self-sufficiency in a way. For instance, calculations of domestic resource costs (DRCs) in a study by World Bank (1986) indicate that Nigerian maize is internationally competitive. But this depends on the foreign exchange rate used, which is five naira to one dollar, but agreeably unstable. The same crop lost competitiveness at stronger-naira selected rates to which the results were sensitized (these being 1:1 and 3:1). The competitive position also shifts with changes in yield, reference price, other international prices, transport costs, and mechanization. However, the economic results are quite neutral to scale.

Even if the possibility of these changes is ignored, some protection is still needed to put Nigerian maize on competitive setting. This protection often takes the form of tariff restructuring, subsidies, or outright bans. In such a situation, self-sufficiency is called for so that the burden of protection is not borne for too long, and the commodity in question is given the chance to improve in trade position when conditions change. Besides this fact, it is also possible that the social cost of protection is greater than the social benefits of self-sufficiency after all. What can therefore be suggested, in the final analysis, is that there is the need for a country to periodically review its commodity baskets in order to delineate certain types for self-sufficiency programmes and the others for open trade. In any case, the eligibility criteria need not be economic only; socio-political factors such as survival implications and national security are also important.

The twin issues of risk and uncertainty are also important to the discussion of the food self-sufficiency policy. Despite the New Nigerian Agricultural Insurance Agency, there is considerable belief that research attention is generally inadequate on discovering optimal farm input use under risk and uncertainty, and predicting how farmers will choose inputs in these circumstances. Some unsettled empirical matters in this regard are how to measure risk, risk preferences, and testing of alternative descriptive models. Since no other aspect of economic life is more fraught with unknown consequences than agriculture, it is high time that substantial attention was placed on risk analysis. The last issue to be discussed is the limitation posed by data in the analysis of food self-sufficiency policy. All along we have examined the two sides of the equation as if the Federal Office of Statistics is behaving normally. To say the least, Nigeria's agricultural database is largely untidy. Two bottlenecks are frequently encountered. The first is that there are discrepancies sometimes in the same set of data, from the same source, but available in several publications. This has been discovered several times without any useful footnote as to whether or not a revision has taken place over time. Policy analysts then resort to arranging direct interviews with data officials, and subsequent exposure to massive piles of raw figures sometimes eventually leads to apparently better estimates but these are often different significantly from the existing published forms. The second type of data problem is the existence of different figures for the same variable and time by different agencies. The correlation between any pair of data sets by source is never perfect. The implication of this is that supplementing data of one source with that of another source may not be totally error-free. Finally, the greatest degree of frustration is usually experienced with government establishments when unpublished data must be obtained directly from them. Apart from the fact that such statistics are not usually available in any processed form, the official procedures are too tedious and wasteful in terms of research time and money. All put together, the aberrations of data in Nigeria are a major restriction in the development of policy analysis for food sufficiency.

Summary and Conclusions

Nigeria's food self-sufficiency policy is shrouded in a great deal of terminological and definitional inexactitudes that make its quantitative analysis difficult as an economic problem. If the problem is stated in an equation form, a scientific examination of relevant issues involved then becomes possible. The independent variable lacks conceptual, definitional and empirical clarity and number and values. The eventual outcome is the lack of a consistent and reliable framework of food policy analysis towards self-sufficiency. This generally leads to compromise and batter in the design and implementation of public programmes meant to achieve the preset self-sufficient targets.

The role of qualitative arguments is also significant in explaining the inability of policy efforts to yield desired results. The number of unsolved issues includes those of conflicts and trade-offs sustainability questions, the balance between self-sufficiency and international trade policies, the risk and uncertainty elements, and the deficiencies of data. However, an empirical illustration confirms that some instruments currently being used can be quite effective.

Far from providing concise answers to the questions raised, this paper brings the several dimensions of the issues involved into focus.

A multidisciplinary approach can then be worked out to evolve a new culture of systematic and dependable framework of analyzing food self-sufficiency policy in Nigeria.

REFERENCES

Ayoola, G. B. (1988), 'An Optimal Time Path for Self-Sufficiency in Maize: A Production Control Approach', PhD thesis, Department of Agricultural Economics, University of Ibadan.

Idachaba, F. S. (1987), 'Sustainability Issues in Agricultural Development', Invited symposium lecture, Agricultural and Rural Development Department, World Bank, Washington DC, 1987.

Lipsey, R. G. and K. Lancaster (1956), 'The General Theory of Second Best', *Review of Economic Studies* 24.

Olufukunbi, B. (1984), 'Food Prices and Inflation in Nigeria: Implications for Self-Reliance in Agricultural Production' in *Self-Reliant Strategies for National Development*, Nigeria Economic Society, Annual Proceedings.

Sarama, J. S. (1978), 'India: A Drive toward Self-Sufficiency in Food Grains', *Am. J. of Agric. Econ.* 6 (5).

Svealberg, P. (1978), 'World Food Sufficiency and Meat Consumption', *Am. J. of Agric. Econ.* 6 (5).

Wiens, T. B. (1978), 'Agriculture: Continued Self-Reliance', *Am. J. of Agric. Econ.* 1 (3).

World Bank (1986), *Nigeria Agriculture Sector Memorandum: Collection of Papers in the Domestic Resources Cost Analysis (DRCs)* (Washington DC, USA).

8

Food Self-Sufficiency and Mass Production: What Approach?

Food self-sufficiency as a theme of agricultural development predates the current Structural Adjustment Programme in Nigeria. Since the publication of the second national development plan (1970–74), the theme has attracted widespread attention both among academics and policymakers, not least because it appears to provide some form of social utility, protection of infant industry, and food security (see Idachaba 1983; Ayoola 1988). The fact that it also negates the arguments now fashionable in World Bank / IMF circles (deregulation, liberalization, market orientation, comparative advantage, etc.) has naturally reinforced its popularity and significance.

This brief paper will examine the concept of food self-sufficiency against the desire for mass food production in Nigeria and specify the proper approach for making progress. It will argue that although several projects have been planned and implemented in this direction, development *per se* (and agricultural development for that matter) is not only a technical problem, but also a social and political one.

SELF-SUFFICIENCY AS A THEME OF AGRICULTURAL DEVELOPMENT

Understanding the Concepts

The impression one gets from the debate is that the conceptual basis is not so clear to the generality of participants. Therefore, there is need to delineate the key concepts involved by meaning and functions. First, food self-sufficiency may be defined as the situation wherein food production is a large enough quantity to satisfy the requirements of consumers in the country. Thus absolute sufficiency denotes no importation of any form of food. The sponsors of this strategy operate on the philosophy that food is crucial for the survival of a population, and as such, nature must have provided enough resources for producing food items in any environment to sustain a carrying capacity of lives therein. The problem then is the ability of the population to exploit the available resources to feed itself. It is not likely or even necessary that a country attains absolute food sufficiency, so that most developing economies talk in relative terms only—that is relative to what the key staples are, and compared to other countries.

The distinction between self-sufficiency and self-reliance is also important. To rely on oneself does not imply having enough: it only implies that one should be content with what one has, whether the available quantity is adequate or not. The import of the distinction drawn is that while self-reliance policy is applicable to the overall economy, food is so crucial that a nation cannot rely on less than the quantity required by its people. Therefore a country that operates on the notion of self-reliance must carry such to self-sufficiency extent for food.

The pair of concepts is dynamic, not static, in the sense that the physical target for achieving them are ever moving. This results from two sources, which are population increases and changing nature of the structure of human wants. Therefore, to make progress in achieving food self-sufficiency, the targets must not only account for effects of demographic transition, but also anticipate future changes in the structure of food demand.

Justifying the Policy

The justifiability of food self-sufficiency rests on issues affecting both the domestic and international economies. The first merit of the strategy is that social utility is produced through the ability of a nation to feed itself, resulting in the implicit gratification of its people. This can be translated into national pride in the people and a pleasant sense of independence of their lives. Given the diversified food imports, their volumes and values, as well as increasing trends, this type of social utility eludes Nigerians to date.

Secondly, food self-sufficiency touches on the infant-industry argument. In the naive form, this can be put thus: if a country does not import processed food, then the inevitable excess demand pressure will stimulate growth of the infant food-processing industry. Some observations during the early period of structural adjustments attest to this. For instance, a significant consequence of the ban on wheat import has not only stimulated domestic production efforts, but in addition, emergence of non-wheat bread and other flour products, e.g. maize - cassava flour products. Similarly, the breweries perfected and sold sorghum and maize beer to replace imported malted barley as raw material of alcohol industry (Ayoola and Idachaba 1990).

By far, the most important argument for the adoption of self-sufficiency policy is food security. As Idachaba puts it (ibid.), the possibility of socio-economic discontinuities predisposes the country that imports the bulk of its food items to incessant shocks arising from weather problems, war, and shift in the balance of international politics as they affect the other countries where the food items are being imported from. This vulnerability status is dangerous for the survival of any nation, especially in an uncertain world as is currently the case. An analysis (Ayoola 1988) indicated that Nigeria was vulnerable to such external shocks in great degrees with respect to food and live animals. The interesting conclusion was that were 100 Nigerians to die from lack of calorie food (grains) in 1979, the numbers that could be attributed to blockage of imports by country of origin are 27 (USA) and 28 (Great Britain).

Nevertheless, orthodox international economists speak of several negative effects of the commercial policy necessary to implement self-sufficiency programme as negation of market orientation; it is established that such a policy limits the attainable level of world food output and efficiency of utilization of the world's economic resources. These come about through distortions and inappropriate price relativities in commodity market space. True, these points are not in dispute based on theory, but countries are still encouraged to pursue food self-sufficiency, owing to the presence of numerous other sources of distortions, including tariff and non-tariff barriers that reduce the positive analysis to more wishful thinking.

Resource and Technology Base for Self-Sufficiency

The resource and technology requirement for achieving food self-sufficiency cannot be taken for granted. Given that land area is not a constraint to food production in Nigeria, sustained attention on human capital development and entrepreneurship are the critical elements of a disciplined mass food production effort. Within the domestic economy, the resource allocation should reflect the sense of priority given to the food production sector.

The technology question is paramount. As idle resources disappear in the domestic economy, increase in productivity appears as a more promising solution to large incremental food output than opening of more land for cultivation. The problem is the notion of technology that developers have, which is narrow and restricted to Western or modern methods of production. This condition worsens the food self-sufficiency status of the country as it only leads to changes in the form, rather than the degree of insufficiency. As the country strives to reduce food imports through increased domestic production, it simultaneously increases its magnitude of agricultural inputs. While this might be excusable for being a transition phenomenon in the early period, it is not tenable after about a quarter of a century of self-sufficiency campaign.

Technology is crucial, without which the green revolution is an absolute farce. The country should own its own technology, either by importing it or even 'stealing' it but well adapted or by enhancing its own indigenous stock. Getting the technology basis right for mass food production is not limited to tractorization and such large-scale technologies, but involves coalescing the indigenous knowledge for upgrading.

PRACTICAL APPROACHES TO MASS FOOD PRODUCTION

Small-Scale Farmers

As exigencies of land tenure, abject poverty and other factors make the small-scale farmers in greater numbers than large-scale farm enterprises, the importance of maintaining special policy focus on the former is self-evident. Thus it is obvious in the Nigerian situation that economic benefits accruing to the class of small-scale farmers will mean a high degree of policy success owing to their wide geographical spread.

The current approaches for stimulating small-scale farmers towards mass food production are highlighted as follows:

1. **Agricultural Development Projects**—through massive infusion of World Bank funds, the ADPs were established to provide extension services, technical input support, and rural infrastructure (rural road network and water supply). The ADP revolution commenced in 1974 as command area projects at Funtua, Gusau, and Gombe. Table 1 described the ADPs as at 1979. The Federal Agricultural Coordinating Unit (FACU) and the Agricultural Projects Monitoring

and Evaluation Unit (APMEU) exist to provide technical support services for the proper functioning of the ADPs.

2. **River Basin Development Authorities**—as water is the critical element in the context of agriculture, efforts to sustain food production in dry areas and periods through irrigation services are very important. The RBDAs were established to develop the potentials of available water bodies in the country for agriculture, fishing, and other purposes. From an initial number of eleven authorities, the RBDAs expanded to eighteen during the Second Republic when their activities were largely politicized and diffused. Subsequently, they are streamlined to the original number and size following the disposal of non-water assets and role clarifications.

3. **Directorate of Food, Roads, and Rural Infrastructure**—DFRRI was established in late 1986 to accelerate the rate of infrastructure development in the rural areas. It was originally designed as a supra-ministerial body for channelling the proceeds of the liberalized foreign exchange market to rural development. DFRRI is involved in the provision of rural roads, water supply, electricity, and community development services.

4. **National Agricultural Land Development Authority**—The motivation for establishing NALDA was based on the need to (i) reduce energy wasted by small-scale farmers in sourcing agricultural inputs by facilitating their supply directly from the relevant sources, (ii) eliminate the problem of lack of access to land, (iii) reduce the cost of land development to farmers, among others. NALDA is already mounting command-area projects of the order of 1,200 ha. in several states. The statute provides that 30,000–50,000 ha. be developed in each state of the federation. NALDA would capture the benefits of special environmental niches in the course of its activities nationwide.

5. **Special credit and insurance services**—The Nigerian Agricultural and Cooperative Bank (NACB) provides specialized credit services to the small-scale farmers. The Agricultural Credit Guarantee Scheme (ACGS) facilitates loans from commercial banks through a special fund operated by Central Bank of Nigeria to give indemnity against non-repayment. Recently the Nigerian Agricultural Insurance Company (NAIC) was established to reduce the risk burden on food producers. Table 2 shows the number and value for loans granted under ACGS over the years. The concept of community banking holds great promises as a means of mobilizing rural savings but its role in food production is too early to appraise. Less clear is the role

of the People's Bank in motivating enterprise initiative in the food production sector.

The foregoing strategies constitute the facilitating approach, which contrasts with the participatory approach previously practiced by government. The current policy environment permits that farmers only be facilitated through these strategies to generate incremental food production. The existing government farms could be demonstration or adaptive research farms of the ADPs.

Government used to be directly involved in food production, especially as panic measures at periods of short supply. The consensus of academic and policy opinion is now that such a practice is discontinued. As experience showed, government was decidedly a bad farmer. Its farm could not pass any known test of economic, financial, or technical efficiency; moreover, it constituted unfair competition to genuine enterprise initiatives without justifying any of the cost components with commensurate output. In the Ogun State example, government established large farms in the early 1980s as a commercial outfit. Management failure led to unorganized use of senior civil servants as labour to harvest maize. The outcome was so low an output as to be less than the value of seed sown.

Even when government floated a food production enterprise solely owned or in joint venture arrangements, factors like boardroom politics, infighting, seniority tussle, and gross absence of accountability would limit performance. The list of such ventures that failed is long, including crop, livestock, and fisheries enterprises at state and federal government levels.

Organized Private Sector

Some of the initial positive responses of the economy to SAP instruments were the active involvement of the organized private sector in large-scale farming. This was stimulated by the backward integration campaign, where firms extended into agricultural production for sourcing of their raw materials. Notable among such firms were breweries that established large hectares of maize and sorghum and the flour industry that grew cassava. A number of petrochemical firms also diversified to growing and processing agricultural products (e.g. AGIP, Total, etc.). This development raised hopes that the successful management of the affected firms would be brought to bear on the organization and running of farms. However, no sooner they entered farming than they pulled out, probably owing to unremunerative returns and a host of other frustrations.

An effort of the federal agricultural ministry to revitalize the organized large-scale farming was aborted after a productive seminar at University of Agriculture, Abeokuta in late 1990. It appeared government enthusiasm lapsed, as there was no follow-up action to actualize the useful suggestions made. Generally, private bodies would require adequate incentives to sustain their investment in agriculture. The forms these could take in addition or as replacement of existing incentives need to be properly investigated in relation to the opportunity cost of private sector capital.

Special Ecological Potential

The 1988 agricultural policy (FMAWRRD 1988) emphasizes the need for ecological specialization a great deal. This pertains to the cultivation of crops and raising of livestock in regions to which they are well adapted by nature and market characteristics. The country must shift gradually away from the practice of growing everything everywhere towards delineating special belts for producing certain items. The ensuing specialization will lead to increased output of all the items.

A logical extension of this proposal is the tapping of special environmental niches for the production of suitable items. Examples of such a potential includes the cold weather of Jos and Mambilla Plateaus. While all the existing strategies may adopt the ecological specialization principle in their operations, NALDA appears best suited to illustrate the content through the carefully planned land use formations in the states. Towards this end, it is pleasant to observe the trend involving special emphases on (i) oil palm in the Agu Ukehe (Enugu State) project, (ii) citrus in the Mbagbe-Ugambe (Benue State) project, (iii) cashew in the Elebu (Kwara State) project, among other NALDA projects.

However, the scale of ecological specialization possible will depend upon the degree of liberalization of the national food market. Unless there is a guarantee of only one national food market, individual states may be discouraged from carrying the principle too far. Suffice it to say that the actions of some states to occasionally partition the food market through instituting wedges on the path of free flow of food items is inimical to the implementation of the ecological specialization policy. An old instance was when the Kaduna State Government directed in 1984 that 'out of state' traders be prevented from buying grains in the state (FASCOM 1984). This could prevent states where maize is less endowed or less comparatively advantageous to forcefully cultivate the crop. In a more recent situation, the injudicious actions of every state to declare itself as a wheat-producing area, for the purpose of benefiting from the special government funding support could be informed by such market obstructions.

Market Intervention Policy

In the past, government had intervened in price formation for the purpose of stabilizing the produce market and farm incomes. The aim of doing so was to prevent the dampening of farmers' enthusiasm arising from seasonal price variation. In this connection, government operated three generations of marketing or commodity boards, all failing in meeting the set objectives (Helleiner 1964; Antonio 1984). Theoretically, the boards should mop up excess food supply at the harvest period, thereby avoiding a glut that would make prices to fall below a tolerable floor. The stock would be sold at scarcity time so as to prevent price from rising beyond a tolerable ceiling. In the process, both farmers and consumers would be better off, and the former encouraged producing in the following season.

We are currently more interested in the last set of six or seven commodity boards, for its active involvement in food items, especially the Nigerian Grains Board (NGB). As Ayoola and Idachaba (1991) ascertained, price formation failed to reflect the need to protect real producer prices in the face of inflation. While nominal price of maize over the ten-year span of NGB (1977/78–1986/87) increased at 18.8%, the real price increased only by 7.6%. The Technical Committee on Produce Prices that was responsible for the annual fixing of prices based its decisions on geopolitical considerations rather than technical market factors. The conclusion was that the commodity boards failed to achieve the desired objectives of market stabilization and maintenance of farm income, which largely informed their abolition in 1986. The current approach is to leave the food market relatively free of open government intervention. Only a strategic reserve is being operated to alleviate the suffering of the people when the need arises, but not to produce any perceptible influence on farmers' incomes and decisions.

Price Policy

Price policy concerns the actions of government to put a wedge between international price and domestic price, usually of tradable commodity. Though the tariff and non-tariff barriers on inputs and output belong to this regime, the most widely debated now is the subsidy issue. The purpose of subsidy on food production inputs is technically to influence behaviour of demanders (farmers) via the mechanism of elasticity of demand. This mechanism worked so well in the case of fertilizer that farmers have become mentally dependent on fertilizer subsidy. Between 1972 (75,675 tons) and 1985 (1.16 million), consumption had risen fourteenfold. Therefore, government faces severe public opposition in the effort to withdraw fertilizer subsidy.

In participating in the active debate on correct pricing of fertilizer, there is need to examine both sides of the argument intellectually. So far, those who maintain hard-line views that fertilizer subsidy be withdrawn have been as equally misleading as those who maintain hard-line views that it be retained. It appears unhelpful to draw conclusions based solely on the merits of hard-line views, as both sides of the argument exhibit merits, limitations, and demerits alike.

Olufokunbi and Titilola (1992) represent the proponents of fertilizer subsidy withdrawal. In their argument for withdrawal, a list of benefits accruing to society from fertilizer subsidy over the past years were presented, including (a) widespread adoption, (b) income redistribution, (c) shield against inflation, (d) decision variable towards social optimization of inputs, (e) incentive to new entrants. The drawbacks of subsidy are (a) creation of dependency mentality, (b) buffering of marginal and inefficient farmers, (c) discouragement of private sector initiatives, (d) huge fiscal burden. This line of argument conforms to the positive economics context. Applying subjective weights to the factors and justifying the hypothesis of unintended beneficiaries, the conclusion was reached that the subsidy on fertilizer, like that of rice and salt, be withdrawn, while suggesting the alternative ways of judicious use of the resulting saving.

On the other hand, Okoye (1992) represents the opponents of withdrawal. The argument (concerning petroleum subsidy) centred largely on the political economy kind of issues. Extended to agriculture, the presence of several reverse subsidies made subsidy withdrawal untenable. It appears, in net terms, those farmers, and taxpayers at large, are subsidizing government in the forms of widespread corruption and overpriced contracts. Therefore, subsidy should not be withdrawn for the fear that farmers would suffer.

While not faulting the logic of placing the subsidy question within the larger political economy context, the positive economics context is that which touches on the underlying issues directly. Though superficial somewhat, it is difficult to dismiss the conclusions emanating from the latter summarily, particularly the presence of unintended channels which defeats the very purpose of fertilizer subsidy. Nevertheless, the comparison of fertilizer with rice and salt appears misleading; while fertilizer is a production input, rice and salt are consumer items. This distinction is fundamental to the issue raised because consumers act in a dense commodity space where substitution is easier than the situation of farmers, who do not operate on such leverage. This suggests that the outcome of withdrawing subsidy on their prices might not necessarily be the same. Even at that, one does not clearly see the basis for expecting that farmers would suffer so much following subsidy withdrawal. Having ascertained the possibility of cost effect on food price variation empirically (Ayoola and Mkpojiogu 1991), there is basis to expect that farmers will be able to pass on the incremental cost of production to consumers substantially. As to the inflationary consequences on the larger society, it is hoped that if the saving is judiciously ploughed back into the agricultural production system, negative impact might not be phenomenal as popularly envisaged.

Thus, the balance of the arguments is that fertilizer subsidy is gradually withdrawn to be compensated by wise ploughing back of the saving therefrom into the agricultural system as other forms of production incentive. The main problem lies with government, whose programme accountability and fiscal discipline has been called to question several times. Therefore, the last section of this discussion highlights some of the issues on the social and political side.

SOCIO-POLITICAL CONSIDERATIONS

Indigenization of Consumer Preferences

A hard fact in the pursuit of the food self-sufficiency objective is that foreign food items that cannot be locally produced must be substituted with domestic food items. This implies for society a determination to modify current tastes and preferences to suit the new situation. If this does not happen then, domestic pressure on the cheaper imported food items will frustrate the policy instruments for attaining food self-sufficiency as the competitive competence of the home food industry will be permanently suppressed.

As can be observed, the pressure of the flour industry, which was derived from the pressure of the bread-consuming public, led to the reversal of the ban on wheat importation. The situation could have been different if consumers quickly adjusted their consumption pattern in favour of composite flour products such as maize-cassava bread, sorghum-based biscuits, among others tested, following the ban. Therefore, society has a critical role to play in keeping the morale of policy authorities steady in this direction.

Political Stability and Will

Food self-sufficiency has a political character in that agricultural administration, the way it is organized, falls in a political arrangement. In particular, as one regime succeeds another regime, there is a strong tendency to change the leadership structure and move professionals around. Since new people in office will naturally emphasize policies reflecting the new political, philosophical, and occupational biases of themselves or the pressure group and people they represent, food policies undergo frequent perturbations. Such perturbations manifest through structural changes in the organization of the ministry and high rate of personnel turnover. So also do specific functions become perturbed, such as the high tempo of the federal agriculture ministry under Shetima Mustapha as minister to form a viable apex association of Nigerian farmers formulated at the University of Agriculture, Makurdi, which became frustrated soon afterwards when another minister was appointed. When the large number of political leaders for the ministry since inception is considered (Table 3), the high magnitude of functional perturbation in the agricultural policy environment becomes self-evident.

Given the great affinity of Nigerian government to undergo incessant changes at the political level, the need arises to devise ways and means of protecting current policies from the effect of such changes. Otherwise, the rate of progress to achieve food self-sufficiency and mass domestic production will be too slow to justify the huge policy effort being committed.

CONCLUSIONS

The current approach to the attainment of food self-sufficiency and mass production rests on the ability of government to facilitate rather than participate in direct production efforts. The crucial lines of action include technology development, continued emphasis on the small-scale farmers, sustained participation of the organized private sector, ecological specialization, conducive market, and price policy environment. In addition, the consuming public should be willing to make their taste and preferences flexible to accommodate the changes in the basket of food commodities newly emerging. Finally, government needs to ensure that agricultural policies and programmes are regime-neutral.

REFERENCES

Antonio, Q. B. O. (1984), 'Marketing Development: A Review of Relevant Government Policies', in J. P. Feldman and F. S. Idachaba (eds.) *Crop Marketing and Input Distribution in Nigeria* (Ibadan: Federal Agricultural Coordinating Unit).

Ayoola, G. B. (1988), 'The Dependency Problem in Food and Rationale for Self-Sufficiency Policy in Nigeria', *J Rural Development in Nigeria* 3 (1).

Ayoola, G. B. and D. O. Mkpojiogu (1991), 'Analysis of the Effects of Structural Adjustment Programme on Food Prices in Benue State, Nigeria', *Journal of Agriculture, Science and Technology* 1 (2).

FMAWRRD (Federal Ministry of Agriculture Water Resources and Rural Development) (1988), *The Agricultural Policy of Nigeria*.

Helleiner, G. K. (1964), 'The Fiscal Role of the Marketing Boards in Nigeria's Economic Development, 1947–61', *The Economic Journal* LXXIV (September): 56–610.

Idachaba, F. S. (1984), 'Self-Reliance as a Strategy for Nigerian Agriculture', *Proceedings of the Nigerian Economic Society*.

Idachaba, F. S. and G. B. Ayoola (1991), 'Market Intervention Policy in Nigerian Agriculture: An Ex-post Performance Evaluation of the Technical Committee on Produce Prices', *Development Policy Review* September 1991.

Okoye, Mokwugo (1993), Article on Subsidy in *African Guardian* 8 (20) (June 3, 1993).

Olufokunbi, B. and Titilola, T. (1992) 'Fertilizer Pricing and Subsidy in Nigeria: Issues and Implications'.

Table 1 Typology of Agricultural Development Project in Nigeria

Name of Project [1]	Area Covered (km)	Proposed Commencement Date	Proposed Termination Date	Proposed Project Cost (₦-million)	Remark[2]
Akwa-Ibom State ADP*	7276	1988	1991	47.25	-
Anambra State ADP*	467571	1986	1991	38.10	-
Bauchi State IRDA*	66000	1981	1988	192.80	Gombe ADP
Bendel State ADP*	36900	1986	1991	29.10	-
Benue State ARDA*	46300	1986	1991	37.70	Ayangba ADP
Borno State ADP*	117000	1988	1992	133.00	-
Cross River State ADP*	21374	1986	1991	27.60	-
Ekiti Akoko ADP**	500	1981	1987	48.20	-
Gongola State ADP*	90000	1987	1992	220.09	-
Imo State ADP*	12800	1986	1991	34.20	
Kaduna State ADP*	43000	1984	1993	672.80	Funtua ADP
Kano ARDA*	43000	1982	1989	265.20	-
Kastina State ADP*	23662	1989	1993	696.75	-
Ilorin ADP**	11775	1980	1988	41.80	
Lagos State ADP*	3600	1987	1991	56.50	-
Niger State ADP*	58500	1987	1992	146.00	Bida ADP
Ogun State ADP*	16400	1982	1987	41.60	-
Plateau State ADP*	54000	1986	1991	64.00	
River State ADP*	19420	1987	1991	74.04	-
Sokoto ARDA*	102000	1985	1989	274.30	Gusau ADP	Gusau ADP

1. * = Statewide ADP, ** = enclave ADP.
2. Preceded by enclave projects as indicated; - = preceded by enclave project.

Source: APMEU Digests on Agricultural Development Projects, Kaduna, 1989

Table 2 Operations of Agricultural Credit Guarantee Scheme Fund (ACGSF)

Year	Number of loans guaranteed	Value of loans guaranteed (₦, 000)
1978	341	11,284.4
1979	1105	33,596.7
1980	945	30,945.0
1981	1295	35,642.4
1982	1076	31,763.9
1983	1333	36,307.5
1984	1642	25,154.9
1985	3337	44,242.1
1986	5203	68,417.4
1987	16209	102,152.7
1988	24538	118,611.0
1989	34518	129,300.3
1990	30704	98,493.4
1991	22014	-

Source: Central Bank of Nigeria *Statistical Bulletin* Vol. 2, No. 2, December 1991.

Table 3: Inventory of political leaders in agricultural administration (including water resources), federal level

NAME	TITLE OF OFFICE AND NAME OF MINISTRY [1]	APPROXIMATE TENURE [2] PERIOD	ESTIMATED DURATION (YRS)
C. O. Komolafe	Minister of State, FMMRR	1966–1969	4
Yahaya Gusau	Federal Commissioner, FMANR	1969–1972	4
J. O. Okezie	Federal Commissioner, FMANR	1972–1975	4
J. E. Adetoro	Federal Commissioner, FMANR	1975–1976	2
E. O. Ekpo	Federal Commissioner FMARD	1975	1
J. U. W. Osisiogu	Minister, FMARD	1975–1976	2
B.O.W. Mafeni	Federal Commissioner, FMARD	1976–1979	4
Ibrahim I. Yakub	Minister, FMWR	1976–1979	4
Adamu Ciroma	Minister, FMANR	1979–1983	5
Ibrahim Gusau	Minister, FMARD	1979–1982	4
Emmanuel Aguma	Minister of State, FMARD	1979–1983	5
Mdagi Madudu	Minister, FMWR	1979–1982	1
Usman Sani	Minister of State, FMA	1982	1
Olu Awotesu	Minister of State,	1979–1982	4
Ken Green	Minister of State	1982	1
Emmanuel Atanu	Minister, FMWR	1983	1

Eteng Okoi-Obuli	Minister, FMA	1983	1
Buka Shaib	Minister, FMAWRRD	1983–1985	3
Ismaila Isa Funtua	Minister, FMAFR	1983	1
Bala Sokoto	Minister, FMAFR	1983	1
Alani Akinrinade	Minister, FMAWRRD	1985–1986	2
M. Gado Nasko	Minister, FMAWRRD	1986–1989	4
Samaila Mamman	Minister, FMAWRRD	1986–1989	4
Shetima Mustapha	Minister, FMARD	1990–1992	3
Abubarkar Hashidu	Minister, FMWR	1990–1992	3
Abubarkar Hashidu	Minister, FMAWRRD	1992	1
Garba Abukadir	Secretary, FMAWRRD	1993	1
Isa Mohammed	Secretary of State, MAWRRD	1993	1

The Institutional Dimensions of Unifying Agricultural Extension System in Nigeria

*Paper presented at the National Workshop on Unified Extension System and Women in Agriculture, organized by Federal Agricultural Coordinating Unit, Gateway Hotel, and Ijebu Ode, 11–17 February 1991

The felt needs for homogeneity of messages and removal of parallel costs have culminated in the proposal to unify the agricultural extension delivery process in Nigeria. In operational terms, a unified extension system uses a single line of command to reach farmers through multidisciplinary individual extension agents. These agents will carry pre-tested integrated extension messages that recognize the inherent interdependence among agricultural enterprises at the household level. The proponents of this concept have the aim to use the World Bank-assisted projects (ADPs) as the basic institutional umbrella to effect unification, apparently owing to the relative superiority of the projects as effective agricultural extension agencies in the country.

The objective of this paper is to critically examine the overall existing institutional framework in which the unified extension system will work. First, a historical perspective is provided (Section II). Second, the institutional dimensions are illuminated (Section III). Third, and lastly, a number of sustainability issues are raised and highlighted.

II. HISTORICAL PERSPECTIVE

The degree of success attained in future actions is deeply rooted in the past. Therefore, this section seeks to briefly survey the past of extension work in Nigeria towards identifying relevant lessons of experience for designing and implementing a unified extension system in the country.

The modern training and visit (T&V) extension mode is closely similar to the early activities in itinerant instructions, farm visits, and group meetings. In the colonial outfit for agricultural administration (Nigeria, Agriculture Department, various issues) the European staff spent 762 man-days and 387 man-days in 1915 and 1916 respectively on tour, giving instructions to farmers on their farms, lecturing, holding organized meetings, and advising them: .Even though these people (called travelling teachers of agriculture, or simply travelling instructors) were not restrictive in the nature of information they carried, the area of primary attention was crop agriculture. The activities which supported itinerant instruction were mainly (i) model/demonstration farms established at several places (e.g. Agege, Warri, Benin, Sapele, Kwale, Abeokuta, Oyo, Ikot-Ekpene); (ii) agricultural shows and exhibition (e.g. Lagos exhibition of 8–10 December 1910 and others in 1949/50 at Dadawa, Daura, Kafinsoli, Gombe, Bauchi, Vom, Anchau, etc.; a Farmers Week was held at Moor Plantation in 1949). Other activities of the colonial extension system were staging of agricultural competition and organization of young farmers' club.

In recent times, except for sectoral specializations that have taken place, the extension activities are fundamentally similar. The central activity is the T&V system, characterized by systematic visits of farmers by the village extension agents (VEAs) and the delivery of messages through training of tested technologies. This activity is supported with demonstration farms, mostly in crop agriculture. The specializations that have taken place include other aspects of agricultural extension such as livestock, fisheries, forestry, and home economics. Each discipline has developed specialized delivery modes for carrying their messages to farmers and setting up demonstration or model units. Demonstration farms are hard to come by in livestock extension, but fisheries extension contains some examples of model units. In the latter case, extension work concerns the improvement of fishermen's knowledge in the use of available fishing inputs and new techniques of fish farming and processing. For instance, there are several demonstration ponds all over the country, such as (a) Bida and Suleja (Niger State), where fish culture techniques are demonstrated on lands not suitable for crop cultivation; (b) boatbuilding centres at Yelwa, Jebba, Kabba, etc.; (c) fisheries extension units at Uta-Ewa (Cross River State), Oyorokoto (Rivers), Koko (Bendel), Orioka (Ondo), Igbekki (Ogun), and Epe (Lagos).

The historical development in the organization aspect of extension administration is also important for this discussion. In the colonial setting, the evolution of institutions for agricultural extension activities followed the following sequence:

(a) Botanical Garden at Ebute Metta (Colony of Lagos) established in 1893
(b) Forest Department at Olokemeji in 1900
(c) Agricultural Department (Southern Provinces) established in 1910
(d) Agricultural Department (Northern Provinces) established in 1912
(e) Agricultural Department (Nigeria) established in 1921
(f) Regional Agricultural Departments (North, East, West), experimented since 1946, established in 1954.

In the postcolonial period, the regional agriculture ministries have multiplied in number from three in 1960 to four in 1963 (plus Mid-West) to twelve in 1967 (following creation of twelve states), to nineteen in 1976 (following another state creation exercise) to twenty-one (following another state creation exercise in 1986). The Federal Government had no constitutional backing for extension works until 1966 (see Official Gazette 1 1966[10]) when the Federal Ministry of Agriculture and Natural Resources (FMANR) was created. This ministry has transformed variously in nomenclature through (a) Federal Ministry of Agriculture (FMA), (b) Federal Ministry of Agriculture and Water Resources (FMAWR), (c) Federal Ministry of Agriculture, Water Resources, and Rural Development (FMAWRRD), and lastly coming to the initial point (d) Federal Ministry of Agriculture and Natural Resources (FMANR). Traditionally, an agricultural ministry is divided into discipline-specific departments, including crops, livestock, fisheries, and forestry.

The typical agriculture ministry (state and federal) in the postcolonial period was characterized by the delivery of extension services, not only through direct efforts, but also through a varied number of parastatal agencies, programmes, projects, and schemes. Prominent among these are (i) Farm Settlement Scheme (old Western Region), (ii) National Accelerated Food Production Project (NAFPP), (iii) National Accelerated Fish Production Project, (iv) River Basin Development Authorities (RBDAs), and (v) Agricultural Development Projects (ADPs). In recent times, a number of other bodies were involved in substantial activities in agricultural extension, including Directorate of Food, Roads, and Rural Infrastructure (DFRRI) and certain non-governmental organizations.

The foregoing historical perspective reveals the inherent features of the Nigerian agricultural extension system, which we can elicit to include (a) the different rates of growth and development of extension activities among the various agricultural disciplines; crop extension is most developed while livestock extension is at rudimentary level of development; (b) there had

---

[10] Nigeria, *Official Gazette* Vol. 55 No. 10 of 7th February 1966.

been little or no change in the type and number of agricultural extension methods over time and across disciplines; (c) the institutional arrangements for administration of extension activities are ever-changing. The implications of these and other concomitant features for the current effort in extension unification constitute the basic institutional dimensions and sustainability issues to be discussed subsequently.

III. BASIC INSTITUTIONAL DIMENSIONS

Unification of extension will entail equalization of activities such as demonstration units, model units, adaptive research, infrastructure and input services. These facilities have been established to cater for crops agriculture, and they have accumulated over a long period of time. The multidisciplinary extension agents also need equivalent services in livestock, fisheries, forestry, and home economics. Unification will therefore require immediate expenditures in these aspects. In particular, the livestock sub-sector will have to evolve the modes of demonstrating smallholder enterprises similar to the system already existing for crops. Unless activities are quickly equalized for all sectors, individual agents will continue to carry extension messages containing substantial disciplinary biases.

Adaptation involves the making of necessary modifications in the existing techniques of extension to suit the nature of new components. The ADPs will require new institutional framework for (a) identifying technologies in livestock, fisheries, and forestry, (b) subjecting the technologies to trials on-station and on-farm, (c) communicating tested technologies to farmers, and (d) providing feedback to researchers in these areas for subsequent cycles of activities.

Equalization and adaptation have important implications for the structure and functions of existing institutions. These are briefly explained as follows:

1. Interdepartmental relationship: The issues in this regard include (a) the need to harmonize the instruments and objectives of sectoral development at the policy level among the different departments; (b) the need to bring all the subject-matter specialists from all departments together under the ADP in order to enhance integration of extension messages under a single institution in each state of the federation; (c) the need for all departments to jointly embark on in-depth assemblage and modification of existing extension techniques to facilitate their use in specific disciplines through a single agent;

2. Input-support services: In this regard, the ADPs must consider the effect of diversified inputs on their present level of management capacity. As modes of procurement, handling, and distribution of inputs will differ from discipline to discipline, unification of extension under the ADP will invariably necessitate multidisciplinary approach to input delivery and improved management of the input services section.
3. Training and retraining: The existing staff of the ADPs and those transferred from ministries need a reorientation to avoid possible role conflict and confusion under the proposed unified extension system. This is different from the usual classroom training. Work-study needs to be conducted for the purpose of mounting systematic training and retraining of the people on job.

IV. SUSTAINABILITY ISSUES

The sustainability of an institutional reform depends on a set of factors, primarily including financial commitment of funding authority, stability of the evolving organization, personnel turnover and intergenerational continuity. The role of these factors in the overall agricultural performance of Nigeria has been substantiated (see Ayoola 1990).

Funding commitment is particularly important in the case of ADPs. The historical performance of federal and state governments in this aspect leaves much to be desired (Table 1). The World Bank closely ties its disbursement schedules to the expressed commitment of the local financiers. Consequently, many projects have suffered very low drawdowns and payment of high commitment charges on approved loans. In a way, extension unification is expected to enhance funding because of greater cost effectiveness and savings. On the other hand, a free-rider effect may result, whereby a dominant sector provides funds for most of or all extension works. Therefore, evolution of appropriate funding system should normally be a consideration in the process of unification.

The second sustainability issue is the endurance of the new institutional outfit resulting from unification. Persistent shuttles of functional units characterize the country's agricultural institutions. The farthest way in which this can set the system back will be from outright reversal of the policy, once put in place. Consider Table 2, which shows the movement of the Home Economics units from agency to agency in selected states. Such instability will cause adverse effects on the whole extension system in the event of withdrawal of some or all of the sectors from the proposed unified system. The possible reorganizations of departments in the main ministries will also produce similar negative effects.

The third issue in sustainability of unified extension system is the traditional high rate of turnover of critical personnel. The performance of ADP depends largely on the presence of core professionals over the life of the projects. The main ministries are characterized by frequent changes in political, administrative, and professional leadership in key positions which consequently lead to alterations in the content and duration of the programmes, as also in the pace of their implementation so as to reflect the new political, philosophical, and occupational biases of the new people involved. Hence the need for great efforts to stabilize the policy environment within which the unified extension system will work merits consideration.

Lastly, the sustainability of the proposed unified extension system will depend largely on the factor of intergenerational continuity as regards the production of extension intellect and technicians. The system of education, which produces extension personnel, includes agricultural faculties in seventeen general universities and a host of middle-level colleges, schools, and polytechnics. Essentially, while retraining of existing stock of extension workers proceeds, the new intakes and graduating students from the educational system require curriculum reorientation to reflect the unification principle. The curricula of the institutions need to be strengthened with the various disciplines involved in the provision of integrated extension message. The existing sub-sector biases in the programmes of these institutions need to be rationalized (Table 3) within the concept of agroecological specialization, consistent with the provisions of the national policy on agriculture (FMAWRRD 1988).

CONCLUSIONS

The proposed unification of agricultural extension services has substantial merit mainly as a means of creating homogeneity in the messages delivered to farmers, as well as achieving cost savings in the extension delivery process. However, it faces severe institutional constraints in the context of existing sub-sectoral differences, organizational and functional instability, funding commitment, as well as sustainability of the emerging composite agricultural extension intellect.

Firstly, each sub-sector demands special skills in the extension agent, whose capability is limited to develop new ones. Secondly, unification will need lumpy expenditure outlay to facilitate contacts with the targeted clientele, which the past experience has not indicated to be guaranteed. Lastly the middle-level agricultural colleges producing the agents need substantial enhancement efforts. Ways of correcting those aberrations are suggested, including the need to provide an inbuilt mechanism for operating the concept of ecological specialization as an important objective of the current agricultural development policy for Nigeria.

REFERENCES

Ayoola, G. B. (1990), 'Agricultural Productivity of Nigeria: Some Policy Aspects' *Indian J. Productivity* 30 (4): 394–401.

Nigeria, *Annual Reports*, Agricultural Department (various issues).

Nigeria (1988) *Agricultural Policy for Nigeria*, Federal Ministry of Agriculture, Water Resources and Rural Development, Lagos.

Table 1: Funding Commitment in Nigerian Agricultural Development Project Selected Projects, circa 1987

Project	commencement date	\multicolumn{5}{c}{Actual as per cent of budgeted expenditure}
		Fed.	State	World Bank resources [b]	Other	All sponsors
Anambra	Jan. 1986	43.0	133.5	0.0	-	58.9
Bendel	Jan. 1986	88.3	93.9	9.7	30.9	55.7
Benue	Jan. 1986	76.6	63.7	0.0	34.2	44.1
Cross River	Jan. 1986	83.7	81.3	0.0	X	55.0
Imo	Jan. 1986	75.1	90.1	13.8	73.7	63.2
Ilorin	Mar. 1980	72.9	55.0	42.7	63.2	58.2

Notes

(a) - means not budgeted and no expenditure committed;
X means not budgeted but some expenditure is committed;
zero means budgeted but no expenditure is committed.
(b) Other sources of funding ADP mainly include internal revenue.
Source: Underlying data from APMEU, *Project Status Report 1987*

Table 2: Breakage and Fusions in the Agriculture Ministry, Selected Functional Units and Selected States

Functional Unit	State	Institutional Shuttles
Home Economics	Bauchi	i. Ministry of Social Development
		ii. Ministry of Rural Development and Cooperatives
		iii. Ministry of Agriculture and Natural Resources (MANR)
		iv. Ministry of Education (Adult and Non-formal Education Agency
		v. MAMSER* (under Women's Programme
	Kwara	i. Agricultural Services Division of Agriculture Ministry
		ii. Ministry of Local Government and Rural Development
		iii. Ministry of Agricultural and Natural Resources
	Niger	i. Agricultural Services of MANR
		ii. Ministry of Local Government
		iii. Ministry of Rural Development
		iv. Ministry of Local Government (again)
		v. MANR (again)
Produce	Bendel	i. MANR
		ii. Ministry of Trade and Cooperative
Agricultural Cooperatives	Kano	i. MANR
		ii. Ministry of Trade

*MAMSER (Mass Mobilization for Social and Economic Recovery Programme)

Source: Culled from Ayoola 1990

Table 3: Sub-sectoral Emphasis of Selected Middle-level Agricultural Institutions Producing Extension Agents

Name of Institution	Academic Programmes	
	Ordinary Diploma	Higher Diploma
1. College of Agriculture, Samaru, Zaria	i. General Agriculture (o) ii. Agric. Mech.	i. Crop Production ii. Agricultural Mechanization
2. College of Agric. and	iii. Home Economics	iii. Home Economics iv. Farm Management
3. College of Animal Husbandry, Kaduna	i. Animal Health Husbandry ii. Poultry Prod. iii. Range Management	i. Animal Health and Agric. and Husbandry ii. Poultry Prod. iii. Range Management
4. College of Agric., Kabba	i. Horticulture	i. Horticulture ii. General Agriculture
5. College of Agric., Bakura	i. Irrigation Agronomy	i. Irrigation Agronomy
6. School of Agric., Akure	i. General Agriculture	-
7. School of Animal Health and Husbandry, Ibadan	i. Animal Health and Husbandry	i. Animal Health and Husbandry

8.	School of Agriculture, Ibadan	-	i. General Agriculture ii. Horticulture and Landscape Design iii. Agricultural Mechanization

Source: Field survey 1990

Effective Communication with Women for Agricultural Development in Nigeria

Keynote paper presented at the National Conference on Communication and Women in Agriculture, organized by the African Council for Communication Education (ACCE) Nigerian Chapter; Conference Center, University of Ibadan, Nigeria. 20–22 November 1991.

The premise of new policy emphasis on women is the realization that they possess immense potential for accelerating economic growth and development. The concept of women in agriculture consists in harnessing this potential for creating new value added in the agricultural sector. Communication theory will play an important role in this process as the means of analyzing the complex medium of reaching women as well as understanding their implicit and explicit responses to information stimulus. Such knowledge, in turn, is a necessary condition for enhancing the effectiveness of the agricultural policy.

This paper analyzes the channel of information flow to and from women in agriculture. The goal is to provide guidelines for designing new methodologies and enhancing the effectiveness of agricultural schemes in communicating with the female gender.

There are three main structural components of the paper: role description, channel decomposition, and impact assessment. In each case, the implications for policy are drawn towards specifying an optimal communications support for the Women in Agriculture Programme.

ROLE OF WOMEN IN AGRICULTURE

Women play important roles in agricultural production and consumption systems, which should be highlighted before analyzing the underlying communication issues. These roles could be gleaned from the aspects of crop and livestock production, food processing and marketing, and the home-based functions.

Women as Producers

Women are active participants in food production in Nigeria, which involves the keeping of poultry, goats, sheep and tilling of the soil to grow energy food crops (yam, maize, cassava, cowpea, etc.). Their specific activities in crop production include land clearing, weeding, fertilizer application, and harvesting, which vary in the degree of women's involvement among the ethnic groups. Among the Tiv people, for example, women engage mostly in lighter farm works such as weeding and leave the heavy works for men. This line of division of labour is not so glaring in the case of Igbo farmers, where women also make major decisions as men do on all aspects of production of crops. Igbo women are known to do most works, both on their farms and also their husbands' farms.

---

*

The roles of women in livestock production, though concentrated on small animals, are substantial. The keeping of poultry, goats, and sheep is a traditional activity of rural women. Vabi et al. (1988) revealed that about a quarter of sampled Fulani women owned ruminant animals that grazed with those of their husbands. Evidence from Unaji (1991) also indicated that most animals (76%) are kept by women above 30 years of age, most of which were raised through the purchase of foundation stock (53%) and by free-range system (73%).

The role of women in crop and livestock agriculture becomes enormous when production is defined in the larger sense than on-farm activities. The off-farm aspects include processing and marketing of produce to drive production to its logical extent. Therefore, women create temporal and spatial utilities through several activities, including dehusking of legumes, threshing of cereal grains, milling, grating, heat treatment and fermentation of cassava, and others, as well as undertaking investments in the preservation and distribution of the end products (Sama 1991). The techniques of performing these tasks are handed down from one generation to the next in the traditional environment.

Women as Consumers

The consumption role of women in food production is consequent upon their multiple roles as housewives, home keepers, and child bearers or rearers. In these roles, their creativity largely determines what to produce, how much of it, and where. The diversity of the menu for husbands and children available relates directly to the basket of foods to be produced on farms. Therefore, the efforts to increase the utilization of certain food items such as soyabean products, when directed at women as target, could yield large results.

The emphasis on the 'kitchen value' of a woman as a basic criterion for assessing her quality as a wife is popular among the various ethnic groups in Nigeria. In many of the Nigerian cultures, marrying a woman is synonymous to getting a hand to feed the man well. This invariably implies a good knowledge of food for different age groups, on different occasions, and for different events, to sustain the overall health of the family. In this process, women become the most important factor for generating demand for agricultural commodities.

In furtherance of this role, women determine the degree of demand pressure placed on the food and fibre system through their fertility behaviour. As child bearers, their decisions to produce new children explain the most part of the population issue. Overpopulation, which causes excess demand for agricultural produce, is a direct consequence of such decisions. Therefore, the quantity of farm produce available for exportation to generate foreign exchange and thereby keeping the dynamics of world trade steady can be traced to actions of women. Of course, the fertility behaviour also explains the degree of availability of labour to execute farm works. The need to balance these advantages and disadvantages of women in the future economy shows the complexity of the policy environment to deal with women matters.

An important issue in the economic role of women arises from the environmental angle, which concerns the available sources of energy for home consumption activities. The use of firewood, which is very popular in rural-urban areas, is cheap though it does not only reduce the available time for women to do their farm and non-farm income-generating activities, but also accelerates the process of deforestation and desert encroachment. This raises fundamental questions about the sustainability of the country's agricultural system, including the need to satisfy demand for forest products such as wood for furniture and pulp and paper production.

THE COMMUNICATION CONTENT OF WOMEN-ORIENTED PROGRAMMES

The highlight of women's role helps us to appreciate the demand for appropriate policy intervention in certain aspects. The success of this hinges delicately on the free flow of information from the relevant authorities to women and vice versa. The aim of this section is to establish a systematic basis for analyzing the issues in communication with women for the purpose of agricultural development.

Framework of Analysis

We may analyze the communication channel into four interconnected components. The first is the message. The message is a piece of information to be communicated to effect changes in the agricultural system. It constitutes the real essence of communication. Without a message, no issue of communication will arise. Therefore, the illumination of the features of an agricultural message becomes very important. In particular, we shall examine the nature and adequacy of agricultural messages being sent to women.

The second component of the communication channel is the source of this message. The source where a message originates is responsible for the adequacy or otherwise of the message. This makes it impractical to separate the discussion of message from that of its source without creating overlaps in thought. Therefore, we shall highlight the agricultural messages and their sources simultaneously to exploit their intricate relationship.

The third component of the channel is the medium of communication. The medium is the system in which the message flows. Many problems of agricultural communication lie in the range. Therefore, the features of the medium of communication with women will be surveyed to identify the important areas for policy intervention.

The last component of the communication chain is the target. A target is the end point for receiving and implementing the message. The best of message passing through an efficient medium may miss the target partially or completely, depending on the latter's disposition to receive. Therefore, the proper description of the target is critical for effective transmission of messages. We shall examine the disposition of women as targets in the chain of agricultural communication.

The foregoing framework of analysis helps to organize our thought along the critical areas of the communication system. It is first necessary to briefly characterize the various components before applying them to assess the impact of women-oriented agricultural schemes.

The Message for Enhancing Women's Productivity in Agriculture

The adequacy of technological messages for improving agricultural productivity is a reflection of the existing research system, which is the ultimate source of the messages. The system comprises a network of commodity-based agricultural research institutes, several agricultural faculties of federal- and state-owned general universities and two specialized agricultural universities.

Certain features of the agricultural research system affect the messages for improving women's productivity significantly. A prominent one among them is the lack of proper consideration of the women's aspect of designing and transmitting the messages. A case in point is the hydraulic oil press, which was introduced to the old Western Region farmers to enhance the efficiency of oil extraction from palm fruits. Reports indicated that this machine was initially adopted on a large scale but later rejected based on inadequacy of the relevant message. As the palm oil extraction enterprise was a traditional activity of women, the machine led to substantial release of women's labour. Therefore, the eventual rejection of this innovation was largely due to the idleness of farmers' wives, which their husbands detested.

The example of higher-nutrient yellow maize also produced similar consequences to the oil press technology. In the same region of the country, the yellow maize innovation, which was successfully introduced, was finally rejected in the rural communities because of the unusual colour of pap made by women from the product. These examples indicate the importance of gender considerations in generating innovations and packaging the messages to the clientele. The agricultural research systems must therefore make conscious efforts to stimulate the reaction of women to the innovations before sending them as messages to the consuming populace. The time and money to achieve this will be justified by the huge losses associated with failure of the innovations at the mass adoption stages.

The Medium of Communication with Women in Agriculture

The important issue in a medium of communication is the magnitude of noise. The term noise describes any element present in the medium which may impede flow of the message, distort the message, or at least, impair its reception by the target. Therefore, the presence of noise elements essentially discounts the value of information passing through the medium in the particular case of women in agriculture. Effectiveness of communication would be limited by noises in the agricultural extension system, which is the most important medium for transmitting messages about productivity enhancement innovations. Other media for transmitting agricultural messages to women include government agencies, community associations, the audiovisual medium, and market agents.

The following noise elements characterize the alternative media for transmitting messages to and from women in agriculture.

(i) Agricultural extension staff, who carries the messages, face serious problems of mobility in reaching its clientele and low level of up-to-date technical knowledge about the innovations being communicated.

(ii) Inadequacy of basic infrastructure of physical, social, and institutional types that are necessary for unimpeded flow of agricultural information to and from women exists.

(iii) The system of supplying farm inputs that are needed for actualizing the contents of messages being communicated to women in agriculture is very inefficient.

(iv) There is confusion in the messages sent by various agencies concerned with transmission of innovation because of deficient definition and perception of individual roles in the communication chain.

(v) There is lack of significant mobilization effort to stimulate easy transmission of agricultural messages to women through their community associations, cooperative societies, and other forms of grass-roots bodies.

(vi) Selective exposure exists through the print and electronic media, whereby direction of information flow is tangential to the rural communities and mostly towards urban communities.

(vii) The markets for farm produce are ill-developed and hinder the generation of income for women from produce in excess of immediate household consumption.

There is no doubt that effective communication with women for agricultural development is greatly limited by the presence of these and other noise elements. The severity of these elements in affecting the rate and quantity of information flow differs noticeably, as also the degree of impairment they cause to the receptivity of the clientele. However, each aspect contributes immensely to the observed low level of agricultural productivity among women.

Women as Targets in Agricultural Communication

The environment is the main issue about the target of communication. Environment here means the sociocultural and economic settings in which women live, which influence their ability as targets to implement the messages contained in agricultural innovations and also their proficiency as change agents in the development process.

A few of these features will be discussed in order to highlight the underlying issues involved. First are the multiple roles of women in the society as producers, mothers, and housekeepers. The various ways in which women perform these roles have already been cited in Section Two. This time we are concerned with how the multiple nature of women's role affects the effectiveness of communication. One way is the time dimension of multiple role performance. As the many roles consume their time greatly, little time is left for leisure to women. This means that most messages sent to them might miss the target in the desired position. It is logical to say that the felt need to solve a given problem will arise when women are engaged in the particular action in which the problem emanates. That is to say that an agricultural message may not be appreciated fully by women where and when they are primarily engaged in non-agricultural functions. Therefore, technological messages, which seek to reduce the time of performing the different tasks, such as time-saving sources of energy for cooking or easier method of washing clothes, stand a good chance of being effectively communicated.

Another way in which the multiple-role situation can affect communication effectiveness among women is the need for them to acquire multiple skills to deal with the situation. Therefore, techniques which improve the efficiency of carrying the many tasks simultaneously will be easy for women to appreciate when the messages reach them. This creates the need for training facilities for impacting different skills to women. The other way is the effect that women are involved in the balancing of these roles. This means that the innovation being communicated must not worsen the conflict situation that exists in the process of performing the multiple roles. Domestic harmony entails that technological information enhance the performance of tasks with minimum trade-offs. Therefore, such messages which promote this harmony will be more effectively communicated than others.

The second feature of women in the overall environment in which they dwell which affects effective communication with them is their social and economic status in this environment (Ijere 1991). The obvious elements of this status include, for example, the leadership structure of gender relations in which men are regarded as physically and mentally superior to women. Therefore, most decisions are vested in the hands of the male or husbands on behalf of the female or wives. Interestingly, some of these decisions concern the issue of who meets with wives, which reduces the exposure of women to agricultural information from male extension agents. The situation is particularly serious with women in purdah, which is very rampant in the northern part of the country. Similarly, women lack direct access to land and other aspects of family wealth, which inhibits their ability to implement an agricultural message quickly. A related issue is the economic power structure in which women do not control the use of their income as freely as possible. Their husbands have a say in the way and manner the proceeds of their labour are expended. Even though we will not raise issues of fairness to women here, the rate and level of implementation of economic information are substantially reduced by a system which perpetually vests the right to spend on the person that did not earn the money. It is still common in rural areas to observe that husbands are engaged in quiet implicit monitoring of flow of income of their wives, so that the latter fail to exercise efficient choices among the investment portfolio facing them. This reduces effective communication with women because of the limited ability to acquire critical inputs of agricultural production as recommended in extension messages.

The last feature of the environment to be highlighted is the presence of taboos and folklore. Different stories exist in the traditional system, which demarcates certain no-go areas for women in the enterprise system. Some of these border on religion and other cultural beliefs which forbid one or both sexes in the rearing of pigs or raising other certain kinds of livestock and cultivation of certain types of crops. In this connection, much of the communication efforts are wasted in the situation that the target audiences are inhibited from implementing the messages. Therefore, effective communication of such messages should include the aspects that, first and foremost, debunk unfounded rumours and unfavourable belief.

COMMUNICATION SUPPORT FOR WOMEN IN AGRICULTURAL SCHEMES

In this section, the various agricultural schemes are examined to assess their effectiveness in communicating with women and identify areas of necessary support. The aim is to suggest ways of strengthening the communication contents of these schemes, which are discussed with respect to adequacy of the messages, legitimacy of sources of the messages, and the level of noise in the medium.

Agricultural Development Projects (ADPs)

These projects have evolved from the first-generation enclaves at Funtua (1974), Gombe (1975), and Gusau (1975) through the second-generation enclaves (Oyo North, Ilorin, Ekiti Akoko, Bida, Ayangba, and Lafia) now to statewide project, with all states covered including the Federal Capital Territory, Abuja. They all seek to enhance the productivity of agriculture through a comprehensive programme of extension services, rural infrastructure, and input support which are well described in literature (Ayoola 1990; Idachaba 1990; Ayoola and Idachaba 1990). The Federal Government jointly funds the ADPs' respective states together with World Bank assistance.

The medium of communication in these projects is the agricultural extension component. The extension messages originate through research collaboration among the projects, research institutes, and universities. In this relationship, the result of basic and applied research are made available to the projects, which undertake further research of adaptive nature before packaging them as agricultural messages for women and men alike. Therefore, the messages originate from dependable sources and are quite adequate in terms of ecological and other necessary considerations.

However, the mediums of transmitting the messages are deficient in certain aspects. One of these is the ineffective coverage as revealed by generally low extension worker-farmer ratio (APMEU 1988) and the lack of enough transport facilities to mobilize the extension agents (see Ayoola and Ochai). There is a new effort among the projects to employ the services of women as extension workers, which, although it addresses the question of accessibility and information bias, might introduce new noise elements to the medium because of inadequate training of the women involved and the observed wholesale approach of involving them. Most of the women extension agents were originally home economics staff of the agriculture ministry who lacked the necessary skills to transmit technical messages in agriculture. According to APMEU (1990), the extension contact in the Ilorin ADP was low, being only 9% coverage of the farming households, mainly because of low staff quality and immobility of the staff. The ADPs are characterized by weak development communication supports in terms of video and audio production, graphics and printing, as well as information centre. It was established that the media or information services of Bauchi State ADP suffered adversely from the ineffective use of educational methods, non-essential use of the video equipment, among other reasons (Wake 1985).

Directorate of Food, Roads, and Rural Infrastructure (DFRRI)

This directorate was established in 1986 to fill gaps in rural road density and water supply, as well as provide necessary support for accelerating food production in the country. The communication content of DFFRI's programme involves the mobilization of rural communities to establish the basis for generating initiatives for community projects and building capacity for maintaining them. Towards this end, the Federal Department of Agricultural Cooperatives was merged with DFRRI as a technical arm for organizing community associations. This effort gives credibility to the medium, though it does not necessarily ensure the adequacy of the messages, unless they are derived directly from proven research results, which are desegregated by gender, ecology, and other factors. An observed noise element present in the DFRRI's mode of communicating with rural people now is the low morale of the development officers, arising from poor facilities and incentives. Casual field observations show that DFRRI has large idle capacity among the cooperatives experts seconded from the Federal Ministry of Agriculture and Natural Resources. This explains why the number of viable community associations is insignificant.

DFRRI also communicates with women for the supply of certain farm inputs such as seedlings, fish fingerlings, among others. These messages generally originate from the services of consultants to the agencies. Their adequacy, therefore, cannot be guaranteed on a sustained basis. In this connection, DFRRI requires formal collaboration with the relevant research institutions for the continuous flow of technological messages to be communicated to women. This is more so for the need to facilitate the medium of transmitting the messages. Since DFRRI cannot create its own parallel agricultural extension system, the need arises for the agency to work hand in hand with the ADPs which have the viable structure for delivering extension messages to women.

Better Life for Rural Women

This programme, which enjoys tremendous political support throughout the nation, is distinguished by its specific focus on women as the target group. Extensive network of communication links has been established to send messages about techniques and requirements for undertaking investments in cottage industry. The main issues arising are the sustainability of these channels of communication and that concerning whether or not the intended targets are actually being reached. On the first issue, it is feared that the direct association of the First Lady with the programme gives an identity of high political content to the programme. As such, the continuity of the programme as a means of transmitting agricultural messages to women is subject to the emerging political environment.

The issue of intended versus unintended targets of the Better Life programme is an empirical one as it depends on the appropriate definition of rural women and the results of an impact assessment study. Casual field observations, however, indicate that the problems of domination by the urbanized rural women exist as well as the low impact of the programme at the grass-roots level compared to the huge financial commitment to the programme. Therefore, there is need for support, particularly in streamlining the medium and sharpening the focus on rural women as the real target.

MAMSER Programme

The Mass Mobilization for Social and Economic Recovery (MAMSER) programme is to be the central source of all mobilization energy during the structural adjustment period. Lopsided emphasis is placed on the political aspect of mobilization under this programme. MAMSER would not have faced the crisis of identity in the post-party formation period if only it had balanced coverage of the economic and social system in its development activities. In the present situation, support is needed to strengthen the communication system necessary for migrating towards new emphasis on economic mobilization. Given the present atmosphere, focus on women matters promises a source of large incremental impact of MAMSER. The message should be properly sourced and transmitted through an efficient medium for achieving an effective communication with women. Therefore, MAMSER needs to develop a strong working relationship with the accredited bodies for generating agricultural information, such as the universities and other organs of the research system, and those established for providing the proper medium for transmitting the information to reach women most effectively.

National Commission for Women

Given the permanent nature of its tenure as a special bureau for handling women matters in all spheres of development, the commission holds the key to the sustainability of the various sectoral efforts in this direction. The degree of effectiveness attained in commission with the relevant agencies and also directly with women by the committee depends on two critical aspects: research and documentation. The intention is not for the commission to establish its own outfit of researching into women matters, but to promote research activities among the institutes and universities necessary for desegregating results along gender lines.

Secondly, the commission should be the ultimate repository of information on the economic and social activities of women in the country. Therefore, it needs to be strengthened with a functional documentation system, involving effective generation, storage, and retrieval of relevant data. To achieve this objective in the agricultural sector, the commission requires strong relationship with the universities and ADPs where capacity exists for communicating effectively with women farmers, processors, and marketers of farm produce.

The University of Agriculture

The Universities of Agriculture are strategically placed to facilitate the system of communicating with women towards enhancing their productivity in agricultural and other rural enterprises. Their specific role, as conceived within the framework of their mission and mandate, includes the steady conduct of research into all aspects of the subject and the establishment of pilot projects to test the effectiveness of various modes of communication with women in different social, economic and ecological settings. Towards this end, the University of Agriculture, Makurdi has launched its special Women in Agriculture programme which involves not only the steady conduct of research to illuminate the current status of activities, but also the establishment of a permanent structure of testing alternative communication methodologies.

SUMMARY AND CONCLUSIONS

In realizing the potential of women in agricultural development, a proper understating of the channel of information flow to and from them is important, in order to provide guidelines for designing new methodologies and enhancing the effectiveness of agricultural schemes in communicating with the female gender; this in terms of their roles and impact for policy implementation, with particular reference to programmes of government with high content for poverty reduction, such as the Women in Agriculture Programme; the Better Life for Rural Women, and the National Commission for Women, among others.

REFERENCES

APMEU (Agricultural Projects Monitoring and Evaluation Unit) (1989), *Completion Report* of Bauchi State Agricultural Development Project (Vol. 1).

Ayoola, G. B. and F. S. Idachaba (1990), 'Self-sufficiency in Food Production as an Essential Foundation for Economic and Industrial Development in Nigeria' in ITF ed, 9th Annual National Training Conference of the Industrial Training Fund.

Ijere, M. O. (ed.) (1990), *Women in Nigerian Economy* (Enugu: Acena Publishers).

Sama, N. J. 'The Latent Socio-economic Role of the Rural Women in the Kom Female Farming System in Cameroon', *Cameroon Journal of Agricultural Economics and Rural Sociology* (in press).

Unaji, P. U. (1991) 'The Role of Women in Livestock Production and Marketing in Benue State, Nigeria: A Case Study of Okpokwu L.G.A' B.Sc. Thesis, Department of Animal Production, University of Agriculture, Makurdi, Nigeria.

Vabi, M. B., C. E. Williams, and P. A. Francis (1988), *Involving Fulani Women in Livestock Extension Strategies: A Case Study of South Central Nigeria* (in press).

Wake, Robert (1985), Bauchi State Agricultural Development Project and Federal Coordinating Unit (FACU), *Report on a visit to Nigeria*, AERDC, Reading University, August 1985.

Agricultural Education In Nigeria: Problems And Prospects

Convocation lecture, presented at the 13th Convocation Ceremony of the Kashim Ibrahim College of Education, Maiduguri, Borno State, Nigeria, 18 September 1993.

Education for agricultural development is a universal theme for regenerating and sustaining the knowledge base required to guarantee the continued supply of food and fibre to the population. As a traditional approach to sustained development, the successive breeds of agricultural intellect, as products of the education process, apply themselves to the real-life situations involving problems of teaching, research, and extension of agriculture, as well as commit resources into the enterprises therefrom. This cycle of activities has existed in Nigeria to be recognized as a veritable strategy for specifically meeting the green revolution objectives and generally uplifting the social and economic status of the people.

Yet after nearly one century of operating the theme, it is a puzzle that the country remains largely food-insecure at the domestic front, let alone generating considerable inflow of foreign exchange at the external front. Apparently, something is wrong with an educational system that prides itself on as many educational institutions at the primary, secondary, and tertiary levels as in Nigeria for meeting the objectives of a particular sector of the economy, but these persist as a poor performer for so long a time. Among the several challenges facing Nigerian agriculture, the low manpower retention capacity in occupations that are directly contributing to the growth of the sector appears as the most intractable. This can be traced to the problems inherent in the system for creating and sustaining the manpower, in the first instance, before identifying contributing factors in the social and economic environment attracting people away from the primary occupations.

This paper examines the agricultural education system for (a) the relevant historical events (Section II), (b) the problems inherent in the system (Section III), and (c) the prospects for enhancing the effectiveness of agricultural education in Nigeria (Section IV).

HISTORICAL OUTLINE OF AGRICULTURAL EDUCATION

Colonial Period

From the onset, the colonial agricultural administration placed emphasis on agricultural education and training. The training of agricultural cadets, as they were christened at that time commenced at the historical Ebute Metta Botanical Station in 1900 when four pupils, namely Samuel Adewusi, Nathan Philips, Josiah Thomas, and Cornelius George, were admitted into the garden to be taught in the practice of tropical horticulture and general cultivation, including garden works and the use of tools. The account in the colonial Annual Reports of Forests Department indicated that these cadets took lessons from Oliver Indian's *Botany*, Nicholl's *Tropical Agriculture*, and Silver's *Elementary Botany* as the basic textbooks. We can correctly identify this event as the step board for formally launching agricultural education as a conscious strategy in Nigeria. Another form of agricultural education existed involving the placement of agricultural pupils under the superintendent of forests and agriculture for sometime before they were qualified for positions in government or other services.

Following the creation of the first Agricultural Development (Southern Provinces) in 1910, a number of agricultural pupils were included on the staff to be trained on the job at the Experiment Station. In 1915, vacation courses of instructions were given to schoolteachers at the Onitsha and Calabar Agricultural Stations, involving theoretical and practical agriculture. Agricultural officers also organized lectures frequently for the members of the agricultural societies. The scheme for the instruction of agricultural pupils was revised in 1922 when two courses (Senior and Junior), each of four years duration, were introduced. At this juncture, the country witnessed the first type of student crises when, in 1923, the first class of students to undertake the new curriculum protested to object to the use of their own manual labour in cultivating their plots. In a swift reaction to this situation, the authorities disbanded that pioneer class for unruly behaviour. Then a new class of 'distinctly better standard of education' was selected from a very large number of applicants. The 1924 Annual Report revealed that these new intakes were assigned smaller plots than the previous class for their practical work; but they soon started to request for more in excess of the size assigned to their predecessors. Henceforth, students were to be recruited biennially through a public examination, subsequently laying the foundation for the establishment of the now popular school of Agriculture, Moor Plantation in Ibadan. Consequent upon a proposal in 1925, the school building was opened in May 1927 with a European schoolmaster appointed, in the person of M. R. J. Newberry.

The school curriculum was reorganized with greater practical bias and outdoor activities like fieldwork and farm mapping among others. The class immediately undertook a tour of Northern provinces, while in 1927, nationwide scholarship entrance examination was held at three centres (Lagos, Ibadan, and Calabar) leading to 135 awards of forty-eight pounds per annum. A similar exercise was carried out in 1929 when twelve new entrants were offered scholarship valued at thirty-six pounds per annum.

According to the official records, low literacy level prevented the development of agricultural education in the old Northern provinces by 1932; however, a departmental school had existed in Samaru for training the agricultural assistants. The first class consisted of three pupils that were college products from Katsina. A makeshift class was later created, consisting of a dozen boys from middle schools who were trained to serve as junior assistants. But the products were later found unfit for extension work, based on poor communications skills and young ages, leading to the discontinuation of the junior class.

Further development during the latter part of the colonial period in creating agricultural education institutions was quite involved but can be summarized as follows:

1. The School of Agriculture, Moor Plantation, Ibadan, was upgraded to an agricultural college in 1935, undertaking two/three-year courses. In 1936, a special course was mounted at the college for the elementary rural schoolteachers by the Education Department.
2. At the Farm Centre in Kafin Soli, a course of instruction for training the sons of illiterate farmers in practical mixed farming later commenced at the instance of the Emir of Katsina.
3. An Elementary Training Centre was established at Bauchi, where students received training in mixed farming.

4. Following an earlier proposal, a training school for African forest staff was opened at Ibadan on 1 May 1941 by Sir Bernard Bourdillon supported by the Colonial Development Fund.
5. The Samaru school became a joint forestry and agricultural school. A Forest Guards School existed at Zaria and another one at Naraguta, Jos.
6. For the first time, in 1938, two Nigerians were sent overseas to undergo refresher courses at the Imperial College of Tropical Agriculture in Trinidad, followed by another set of seven in 1944. Later, in 1948, two Nigerian junior service officials of the Forest Department were sent to the Imperial Forest Institute, Oxford, to pursue an honours degree course in botany. Another Nigerian was admitted into the new University College, Ibadan to read a science degree.
7. Farm schools were opened at Oyo (1942) and Ogbomosho.
8. A number of training centres for agricultural education were opened at Riyom (1946), and later at Kano, Adamawa, Ilorin, Omu Aran, Bamenda, Enugu, and Nkwelle.
9. The Veterinary School was built in 1943 at Vom, which was fully operational and residential in 1946, using a fund under the Colonial and Welfare Acts. The school ran a junior course (two years) and a senior course until 1950 when the latter was discontinued in preference to referring students to the Ibadan University College, which commenced a degree course in veterinary science.

The Postcolonial Period

In large measure, the postcolonial institutions for agricultural education took their roots from the colonial outfit. After independence, the further operationalization of the regional autonomy concept led to the strengthening of the inherited institutional matrix and creation of new schools, colleges, and institutes either wholly or partially meant for agricultural education. A significant development was the establishment of two specialized universities of agriculture in 1988 and another one in 1992.

Thus the pedigree of postcolonial agricultural education institutions now features three layers—low, middle, and high levels. At the low level, we have the numerous primary and secondary schools that teach elementary-cum-introductory aspects of agricultural science. They collectively breed the foundation stock of pupils among who would be selected to pursue further programmes in agricultural education. The middle-level institutions comprise schools and colleges of agriculture awarding National Diploma and Higher

Diploma in different programmes, ranging from General Agriculture to Agricultural Extension/Farm Management, Crop Production Technology, Animal Health Production, Home Economics, to mention a few. Prominent examples of middle-level agricultural education institutions include the ones at Ibadan, Samaru, Kabba, Lafia, Bakura, Umudike, Akure, Ikorodu, and Yandev. Also in the middle category are the polytechnics and colleges of education. The polytechnics run courses leading to the award of diploma certificates in agricultural engineering/technology, especially agricultural engineering and food technology. On the other hand, the colleges of education are concerned with producing expertise for effective teaching of agricultural science in the primary and secondary schools.

The conventional universities predominate at the higher level of agricultural education. There are presently sixteen of such federal universities with faculties of agriculture, except three (Abuja, Jos, and Lagos). In addition, we have five federal Universities of Technology (Bauchi, Akure, Owerri, Minna, and Yola) also running programmes in agricultural technology and related courses. There are also three federal Universities of Agriculture (Abeokuta, Makurdi, and Umudike), whose only foci are on agriculture in all dimensions conceivable. The state-owned universities are eleven in number, comprising nine conventional and two technology types. These high-level institutions offer both first, second, and terminal degrees in diversified aspects of agriculture in the country.

The Problems with Agricultural Education in Nigeria

The problems with agricultural education in the country must be discussed against the background of the two-period evolutionary process of institution building while also reflecting the factors underlying the pace and direction of the process in each period. In this connection, the development of agricultural education during the colonial period contained a number of crucial features as can be gleaned from the foregoing historical account. First, institution building for agricultural education was merely a product of circumstances. Rather than being undertaken within the context of any long-range plan for social capital accumulation, the decision to mount courses in specific places, at specific time, was usually need-driven. As such, the products of the haphazard arrangements did not attain a reasonable degree of uniformity in their ability to apply themselves to alternative, especially spatial, circumstances. Also in the absence of stated plan objective, the process, in itself, was diffuse and directionless. Nevertheless, the efforts represented a framework upon which to base the post-independence arrangements, using the stock of indigenes that although were mostly partially educated in agriculture, had acquired vast experiences in their various jobs.

Second, the disparate establishment of agricultural schools and training centres lacked some form of professional coordination. Although the colonial agricultural administrators were mindful of the standard of admission requirements, in the absence of a central coordinating body, it was difficult to harmonize these requirements and the curriculum of instruction in different parts of the country. It was not surprising, therefore, that a whole set of agricultural students was found incompetent to utilize the education provided to them, as the historical account contained. It was significant to note, however, that the task of administering the process of Agriculture Department which obviously ensured that the education was geared towards the actual need which it was meant to meet, in the first instance. It also offered a degree of lateral integration, albeit low, to ensure uniformity and harmonization of standards also crucial but lacking, in collaboration with higher-level institutions. The repercussion of this is the absence of avenues for demonstrating superior skills to low-level students. Added to the dearth of laboratories, libraries, and other relevant facilities, the lack of vertical integration could only lead to rudimentary education at best.

Third, an education system in agriculture that was dominated by expatriate staff could possibly benefit the indigenous population minimally. As the knowledge about mixed farming system and other cultural practices had not grown enough to form the contents of books used for instruction, the impact of agricultural education available at that time on the agricultural economy was insignificant.

Fourth, the process of colonial agricultural education was very low in response to the felt need for agricultural manpower. It is inexcusable that since 1900 when the institutionalization of agricultural development began through the establishment of the first Forests Department, through 1910 when a separate Agricultural Department was established in the North (both merged together as one Agricultural Department in 1921), up to 1948, the colonial government could only boast of offering only two Nigerians admission leading to first degrees. If the rate of upgrading the educational status of Nigerian staff in the Agricultural Department was two graduates in about fifty years, the insensitivity of colonial administration to the agricultural manpower needs of the country is self-evident.

Fifth, agricultural education in the colonial period at the higher level, which started late, was also not imparting agricultural knowledge per se. As the historic account revealed, the first two candidates sent to the Imperial Forest Institute at Oxford pursued an honours degree course in botany, while the one at the University College, Ibadan, was offered admission for a science degree. Therefore, up till the decade preceding independence, no Nigerian could be said to possess a degree certificate in agriculture in the true meaning of the discipline. Although the advanced knowledge of science is directly relevant to agriculture in theory, it does not manifest the proper traits of an educated agricultural expert in practice. Thus, the colonial agricultural education process in Nigeria only succeeded in breeding a foundation stock of pseudo-professionals on which to base the ensuing post-independence development activities.

Lastly, and related to the last point, is the narrow scope of agricultural activities in which Nigerians were educated to perform. This concerns (a) the researchers for the nation, (b) the absence of facilities to build indigenous capacity for teaching subjects in agriculture, and (c) the deliberate omission of elements in the curriculum for creating enterprise initiatives among Nigerians who received the limited education provided. By and large, the main purpose of the colonial agricultural education was the production of people to provide assistance to the expatriate staff of the department in performing their assigned duties in the field and offices. In particular, the need to reach the farming population more effectively necessitated the disproportionate emphasis on agricultural extension work as the primary objective of education. Therefore, we cannot say that the full potential of the agricultural knowledge acquired through the colonial period was realized before independence in 1960.

Following independence, the ills in the inherited system for agricultural education have been gradually attended to. A more systematic, planned process has evolved over time that is strengthened with vertical and lateral integration efforts. Moreover, a home-grown curriculum has come into being, which is enriched by locally produced books and better understanding of the indigenous farming system. The accelerated pace of education witnessed was consistent with the felt need for agricultural manpower soon after independence. The present system has drawn clear lines between pure science and agriculture so that the inherited role confusion is hoped to peter out totally in future. There is also greater degree of professionalization of agriculture, involving higher degrees in several subject matter aspects and enhanced basis for products of schools, colleges, polytechnics, and universities to practicalize their training through private enterprise efforts. In addition, the tripod of teaching, research, and extension has assumed an integrated outlook in the current system.

Nevertheless, the post-independence process of agricultural education is also plagued with a legion of problems limiting its effectiveness; a few among these are discussed as follows:

POOR FUNDING

The inadequate and erratic funding of agricultural education in the country is inconsistent with the developmental role ascribed to the sector. The result of this can best be illustrated in selected middle-level agricultural institutions. A situation analysis carried out in 1991 revealed substandard conditions of educational infrastructure that determine the quality of graduates in selected middle-level institutions awarding diploma in General Agriculture (Table 1). When evaluated against minimum standards as recommended by the National Boards for Technical Education, some critical instances of inadequacy of facilities were (a) Ikorodu with respect to farm facilities, staffing situation, and laboratory/workshop facilities, (b) Ibadan with respect to laboratory/workshop facilities, and (c) Lafia with respect to laboratory/workshop facilities. Generally, the situation is desperate for staff requirement to teach ND students, and the post of farm manager as also for laboratories and workshops.

POOR EDUCATIONAL MANAGEMENT

Educational management in agriculture at all levels is characterized by aberrant perception, distribution, and performance. First is the situation whereby some institutions are owned by the federal government and others by state governments. Some middle-level institutions are integrated into a university's organization framework and some are not; some universities, together with their agriculture faculties, have undergone rapid institutional reorganization by way of creation as full-fledged institutions to be subsequently merged with older universities as campuses. Some universities' orientations have changed from conventional to specialist (e.g. technology, agriculture) or vice versa; some universities (of agriculture) are responsible to the Agriculture Ministry while others are responsible to the Education Ministry. Lastly, we understand the Universities of Agriculture, probably with related lower-level institutions, will be coordinated laterally through another autonomous government agency at the federal level, while the conventional universities will remain under the coordination of the National Universities Commission. These disparities in the education system have translated into great variability among these institutions with regard to the quality of educational services provided. Of course, there is no basis to suggest harmonization along the lines of demarcation identified, but their presence obviously exerts additional pressure on the management capacity available.

The major ways in which inadequate management manifests in the educational systems are (a) application of non-merit criteria to head institutions, (b) confused priorities and poor sequencing of subject matter areas, (c) wasteful duplication of courses in several institutions, among others.

MANPOWER EROSION

It is common knowledge that very few of the products of agricultural education at all levels want to be employed in the core areas of agriculture, including the teaching of agriculture in the available institutions. Table 2 reveals that, regardless of growth in the number of academic staff in form of instability of individual staff. Using the University of Agriculture, Makurdi, as an example, the different departments have exhibited considerable degrees of academic staff instability based on calculated indices. This ranges, in the extremes, between the most stable (Department of Mechanical Engineering), currently having twenty academic staff, but in which only 5.6% of the eighteen pioneer staff had left, to the most unstable (Department of Agricultural Education) currently having five academic staff but in which 50% of the eight pioneer staff had left.

The practice sets such an institution into a process of attrition based on several problems created. First, the education offered will contain several stopgaps that affect the quality of graduates produced. Second, substantial resource leakage results through the trained staff on which huge funds had been committed, only to negotiate the terms of his bond to effect desertion.

By nature, the decisions of individual staff, creating this problem for the educational system, are mostly economic. Based on the high degree of mobility of the labour of academic staff and given the significantly more remunerative jobs outside the institutions, we contend that manpower erosion is inevitable in the education system in the present circumstances.

STUDENT CRISIS

The historical account pointed our attention to what could be taken as the first student crisis on record. The first set of students to undergo a new curriculum of instruction protested against the use of manual labour on their assigned plots in 1923. Therefore, students' crisis is not a new phenomenon in the country, as such. What is new is the way management treated the case by summarily dismissing a whole class of students, a situation never witnessed in the post-independence history of education in Nigeria. Yet, the spate of student crisis in the country has implicitly reduced the credibility and value of the educational system to low ebb.

The Prospects for Enhancing the Effectiveness of Agricultural Education in Nigeria

The three main sources of prospects for resolving the problems facing agricultural education in the country are (i) the potential of the three universities of agriculture, (ii) the proposal for the establishment of the National Board of Agricultural Education, and (iii) the recent placement of universities of agriculture under the agriculture ministry. The universities of agriculture have the potential to change the orientation of agricultural education towards the production of graduates with adequate practical exposure in farm and extension works for undertaking ventures in agricultural enterprises after training. The simultaneity and administrative integration of teaching, research, and extension aspects in these universities will also lead to more functional educational services, needed to foster sustainable agricultural development through the graduates. The close association of the agricultural universities with the agriculture ministry will make the former directly relevant to the origin of agricultural problems and ensure high degree of harmony between fiscal priorities and the projects in the universities. This should translate to greater efficiency of the educational duties at the high level.

The National Board of Agricultural Education (NBAE) will provide the backup support for achieving lateral and vertical relationships among the relevant institutions. Critical aspects of this involve setting priorities for funding and programming for the optimal distribution of courses in the system. The design and implementation of minimum standards will also be an important aspect of the NBAE as an approach to stemming the continuous diminution in the quality of instructions available.

The prospects for improving the pattern and level of funding depend upon the combined fiscal commitment of government and greater role of the private sector supporting educational programmes in the institutions, the issue of brain drain should be addressed through a conscious, planned equalization process with particular focus on the salary and non-salary components of the condition of service prevailing in the educational system. Any solution short of a system of remuneration that reflects the off-teaching opportunity cost of teachers' labour would be palliative at best and non-lasting in nature. Also, given the dynamic nature of economic systems, the solution proffered should contain appropriate built-in stabilizers to reflect possible changes overtime. The logic underlying this suggestion is that teachers at various levels of the educational system will not be satisfied in perpetuity to settle for a lower quality of life than those openly exhibited by graduates produced by them.

Among the many suggestions for curbing students' crisis in the educational system, those that rest the final authority on the respective institutions and encourage active students–management forum appear most appealing. Far from being an apologist of high-handed approach to resolving students' crisis, we feel that management requires strengthening with relevant instruments to deal with specific cases in finality.

SUMMARY AND CONCLUSIONS

The Nigerian agricultural education system has come a long way through the seemingly disparate arrangements during the colonial period to the present matrix of more systematic institutions. The main problems revolve mainly around the poor and erratic funding, poor micro- and macro-management performance, manpower erosion, and students' crisis. The prospects for a more effective agricultural education system in Nigeria depend on government's commitment and on the strength of private sector morale to support the system. Moreover, the problems of students' crisis must be eliminated to create a conducive environment for successive generation of agricultural intellect in the country.

REFERENCES

Ayoola, G. B. (1990), 'Agricultural Productivity in Nigeria: Some Policy Aspects', *Productivity* 130 (4): 294–401.

Idachaba, F. S. (1981), 'Agricultural Research Staff Instability—The Nigerian Experience', *Nigerian Journal of Agricultural Sciences* 3(1): 71–72.

Nigeria (Colonial Government), Annual Report of Agricultural Department (various issues).

Table 1: Infrastructural facilities in selected middle-level agricultural institutions (1991)

	Schools/Colleges of Agriculture*					
	Ikorodu (400)	Ibadan** (320)	Akure (100)	Kabba** (480)	Lafia	Samaru (407)
Farm facilities: Meteorological station	-	n.a	+	+	n.a	+
Survey equipment	-	n.a	+	+	n.a	+
Pest control equipment	-	n.a	+	+	n.a	+
Farm machinery/tool	-	n.a	+	+	n.a	+
Irrigation equipment	-	n.a	+	-	n.a	+
Nursery facility	-	n.a	+	-	n.a	+
Green house	-	n.a	-	-	n.a	+
Crop farm	-	n.a	+	+	n.a	+
Animal farm	-	n.a	+	+	n.a	+
Fish pond	-	n.a	-	+	n.a	=
Library facilities: No. of seats	70	92	120	256	27	85
Space: Classroom (No.)	4	2	7	6		
Lecture theatre (No.)	0	0	0	0		
Office Rooms	1	0	20	16		
Others	5	4	1	11		
Staffing: Principal Lecturer***	0	3	0	0	1	0
Senior Lecturer	1	2	3	1	1	2
Lecturer	0	4	4	4	4	4
Asst. Lecturer	0	1	0	0	5	6
Senior Technical	0	5	8	8	2	8
Farm Manager	0	0	0	0	0	0
Laboratory/Workshop: Biological Science	-	+	+	+	+	+
Chemical Science	-	+	+	-	+	+
Physical Science	-	-	+	-	+	+
Soil Science	-	-	-	+	-	+
Metal/Wood Maint.	-	-	+	-	-	+
Feed mill	-	-	+	-	-	-
Drawing rooms	-	-	+	-	-	+

*: Figures in parentheses are the student populations

Note + = available; - = not available; ** = awards HND in addition to ND certificates, while others award ND only; *** = additional staff required for HND.

Table 2: University of Agriculture, Makurdi strength of academic staff and calculated indices of instability, 1990/1993

Department	Academic staff strength (Number) Pioneer (1990)	Current (1993)	Instability index (%) (1990/1993)
Crop Production	9	14	11.1
Soil Science	7	8	28.6
Forestry & Wildlife	2	6	50.0
Animal Science	12	11	16.7
Fisheries	2	7	50.0
Agric. Econs	7	6	42.9
Agric. Extension	2	5	50.0
Food Sci. and Tech.	12	10	16.7
Agric. Engineering	6	13	16.7
Civil Engineering	4	5	25.0
Electrical Eng.	7	6	14.3
Mechanical Eng.	18	20	5.6
Agric. Education	8	5	50.0
Biological Sci.	10	10	10.0
Chemistry Dept.	8	6	25.0
Physics Dept.	3	5	33.3
Maths/Computer/Statistics	6	6	16.7

*Instability index (I), as used by Idachaba (1982) is defined for the ith individual member of staff whose value is set equal to 1 if he is on the job; 0 if he has left at the point in time, among k number of academic staff in the department at the initial time t1 (i.e. January 1990) and current time t2 (i.e. September 1993).

Technology and Nigerian Agricultural Development

Invited lead paper presented at the International Conference on 3rd World Strategies for Technological Development, Federal University of Technology, Yola, 20–26 August 1989.

In four years time, Nigeria shall have consciously pursued organized agricultural development for precisely one century[11]. Yet the current generations of farmers still grow their crops and tend their livestock with methods essentially unchanged since the outset. The evidenced lack of technological progress has constrained the expansion of agricultural output directly. Indirectly, it contributes immensely to rural poverty and inefficient labour utilization. Ultimately, it worsens the balance-of-payments problem through laggard export performance and food import dependence. It limits the capacity of the rural market to assist industrial growth, and also aggravates inflation as well as unemployment.

Therefore, the problem of technological backwardness in agriculture is worth examining critically. This paper does this within the context of economic development objectives. We shall first briefly highlight the meaning and salient economic principles underlying the drive towards technological progress (section one). Afterwards, the dimensions of agricultural technology shall be specified and development efforts in the Nigerian case shall be outlined (section three). The fourth section characterizes the appropriateness of agricultural technology. The fifth section discusses a number of sustainability issues in agricultural technology. The concern of the sixth section is to identify

---

[11] The event that is being referred to is the establishment of the Botanical Station at Ebute Metta, (1893) where initial development works took place including meteorology, herbarium, flower and vegetable garden, walks and paths as well as planting of economic trees such as coffee, cocoa, kola, and rubber in the gardens. For details, see Dawodu (1900).

the critical elements of virile agricultural technology policy. The paper is concluded in section seven.

MEANING OF TECHNOLOGY AND SALIENT ECONOMICS OF TECHNOLOGICAL PROGRESS

The apparent anarchy in the use of the term 'technology' permits us to define it in our own way. In general literature, technology connotes mechanical, electrical, electronic, and other such scientific impressions. For the present purpose, however, the meaning of technology shall be extended to include the totality of how society performs particular activities. Specifically therefore, the technology of society consists of the nature and types of available inputs and the various ways in which these inputs are combined to carry out particular activities. To fix ideas, agricultural technology consist of the nature and farm types of available inputs (e.g. seed, fertilizer, chemicals, tools, machines, farm power, etc.) and the ways in which these are combined (e.g. land-fertilizer ratio, labour-machine ratio, etc.). Defined in this perfectly general way, we are able to view technology far beyond the existence of modern scientific inventions, to incorporation of the aspects of natural inputs such as seed or seedlings, and also the influence of society's sociocultural values on the level of agricultural technology attained.

The highlights of the demand side of technological progress are important in explaining the rationale for great efforts in this direction. Society desires technological progress because it reduces average unit costs of output, with input price held constant. In corollary technological progress results in greater output from constant amount of inputs; alternatively smaller amount of inputs produce the same quantity of output.[12] In essence, technological progress leads to increased productivity, which is generally desirable and expedient. In turn, high productivity increases the rate of revenue inflow as well as return to labour and management. Moreover, if the entrepreneur wishes to maintain a given level of output, then higher productivity creates more time for leisure, thereby enhancing a consumer's utility level and hence general welfare. One qualification is necessary at this point. It is not all-technological change that is progress. For progress to occur, first, there should

---

[12] In economic language the conditions resulting from technological progress lead to shifts in the average costs (AV) curve (downward and leftward), production function (upward and leftward) and isoquant (downward and leftward) over a passage of time under standard assumptions. Samuelson (1976 is a good text on the subjects).

be increases in productivities of all or some of the inputs. In addition, the unit cost must have either also fallen, remained the same, or at worst, increased less proportionately than productivity.

But if higher costs result from change, then the original technology status, to more than offset the value of incremental output, then the 'improvement' has not led to progress.

Another point must be clarified in the discussion of economics of technological progress. The desirability of progress is not universal. This statement concerns the effect of technological progress on marginal productivities of labour and capital.[13] Certain changes may be capital saving (labour-using), others might be labour-saving (capital using), while others might be neutral.[14] Therefore, capital-saving technological progress creates new jobs, which may be a great desire of developing economies operating under relatively scarce capital and surplus labour. On the other hand, in the advanced economies where labour has been marginalized and capital is relatively more abundant, labour-saving technological progress might be more worthwhile.

Lastly, we should mention that the concept of technological progress means more than an advance in indigenous knowledge; it also includes outside knowledge. It also means more than mere existences of innovation; it includes diffusion of this innovation. Finally, technological progress is not confined to industrial and mechanical aspects only; it includes improvement in economic organization of enterprises as well.

DIMENSIONS OF AGRICULTURAL TECHNOLOGY

Two main types of technological progress can be distinguished. One is process innovation, which results in greater output of a specified product from a given quantity of inputs. The other is product innovation, which results in either a totally new product or an improvement in the quality of the same product. Most of the technological progress in agriculture takes

---

[13] Given fixed level of prices of labour and capital, the amounts employed of each is determined by their marginal productivities. A technological advance shifts the marginal productivity curves upwards and to the right, thereby changing the factor proportions.

[14] A capital-saving change raises productivity of labour relative to that of capital by increasing the demand for labour relative to capital. A labour-saving change does the reverse. A neutral change leaves the relative marginal productivities unchanged. See Samuelson (op. cit.).

the form of process innovation, in which new methods cause reductions in the cost of producing similar crops and livestock products. The bulk of product innovations in the economy take place in the industrial sector. The stock of process innovations will now be surveyed to characterize agricultural technology. The various dimensions will be briefly described by activity.

(i) Technology of land preparation

This concerns the means of carrying out clearing and tillage operations. The traditional methods use simple tools based on muscle power—cutlasses, axes, among others. Using bulldozers carries out presently extensive clearing of land. Tillage operations may be carried out by use of hand hoes of different shapes to suit the soil types and other local conditions. However, tractor-power-based methods are now popular, including ploughing, harrowing, and ridging. These are more successful in the northern part of the country, where total clearing is possible, than in the southern part, where heavy economic trees and thick forests necessitate partial clearing only. The same complement of tools—ploughs, harrows, ridges—can also be used based on animal power. Animal traction technology is popular in the trypano-tolerant zones of the country.

(ii) Technology of planting, seed and seedling

Planting is traditionally done by hand. Presently the planting of grains is sufficiently mechanized, through the tractor-mounted planters for maize and beans. A host of small-scale manual types also exists, such as rolling injection planters and manual jab planters. However, the technology of planting tree crop seedlings and yam is still at the low ebbs. Seedling planting by mechanical means may not be hopeful because of the fragile nature of living plants. As to yam planting, the problem is mainly that of lack of uniformity in shape of planting materials. In the case of cassava, most planting is still by hand, but recently a machine is being introduced which successfully plants the cuttings on flat soil. The technology for planting vegetables is still generally hand-based.

The potentials of planting materials are improved upon by genetic hybridization, selection, and other means of plant breeding. New crops have been bred for disease resistance, uniform ripening, palatability, product quality, market acceptability, and yield increases. Maize seed technology is very advanced. The yield of hybrid maize, for instance, is generally not less than five multiples of the open-pollinated varieties. Barring hereosis and the consequent high susceptibility to environmental hazards (drought, disease attack, etc.), hybrid maize technology is revolutionary. High-yielding varieties of cowpea seed are also locally available, following technological improvements in the traits of colour, insect resistance, padding uniformity, among others. In addition, tree crop seedlings have been bred for various traits such as low height, fleshiness of pericarp, oil quality and quantity (e.g. oil palm); low height, pericarp thickness, seed content, sugar content (e.g. orange); and so on. Some may be dressed to reduce attack by rodent, and soil-borne pathogens.

(iii) Technology of selected management practices

Soil replenishment is popularly by fertilizer application, particularly inorganic fertilizers. Traditional organic fertilizers are used which are prepared as compost or other forms. Fertilizers may be applied by hand (band or ring) or by tractor-mounted applicators and broadcasters. Weed control can be by hand or by the use of herbicides. The spectrum of available herbicides includes pre-emergence and post-emergence types, selective types, and others. The methods of herbicide application include assorted types of sprayers—knapsack, ultra-low volume, boom sprayers—which may be general or crop specific. In addition, some mechanical methods are now in use on small-scale farms, such as the row weeders. Insect and disease controls are generally by spraying chemicals. Sometimes aerial spraying is performed at low altitudes during epidemics.

(iv) Technology of harvesting

The general method of harvesting produce is by hand, except for grains (particularly maize and rice), where advanced technology exists. In the latter case, a combine harvester may be used, which also performs activities simultaneously such as dehusking, shelling, cleaning, and others. The harvesting tools are usually the simple types—knives (maize), harvesting hooks, otherwise called go-to-hell (cocoa, orange). Climbing may be necessary in tree crops such as kola and oil palm. Climbing of oil palm generally involves the use of ropes, except on experimental plots of the Nigerian Institute for Oil Palm Research (NIFOR), where ladders are used on limited scale.

(v) Technologies of on-farm haulage, processing, and storage

The transport of materials within the farm is generally done on human head. In some cases, bicycles are used. Other available methods are animal-driven carts, tractor-mounted trailers, and motor vans. Most of on-farm processing is carried out manually, including dehusking, shelling, threshing, drying, and fermenting (cocoa). In respect of grains, however, mechanical methods are available in the forms of maize shellers, cowpea; rice threshers, and artificial dryers. The storage technology concerns the type of housing, chemicals, and management techniques of keeping produce during storage. The traditional storage systems still prevail. These include the use of barns, pits, pots, and rhumb. Largely, wholesalers, transporters, millers and seaports operate warehouses and stores. Maize in the south is stored in cribs, and yams are stacked on vertical poles. The use of silos is mainly by government and a few large-scale farmers.

MODES AND ACCOMPLISHMENTS OF AGRICULTRUAL TECHNOLOGY DEVELOPMENT IN NIGERIA

Technological progress for the development of agricultural sector has been pursued in three interwoven directions. These are research for new technology creation, adaptation, and transfer of available technologies, and development of the technological intellect.

Agricultural Research for Technology Creation

The colonial Agricultural Department commenced research for discovering new technologies as an in-house activity. Ad hoc experiments were indiscriminately conducted to introduce new crop varieties of better performance than the indigenous ones. By 1942, some specialist staff (agriculturists, utilization assistants, forest engineers) had been styled research officers. The Oil Palm Research Station was established in 1937, which has now metamorphosed variously to become the Nigerian Institute for Oil Palm Research (NIFOR), Benin. A series of West African research institutes established substations in Nigeria before independence, which later became nationalized with new ones added over time. The present agricultural research system is comprised of several specialized institutes, specialized agriculture and technology universities, and a host of agricultural faculties (Table 1).

Diffusion of Agricultural Technology

Diffusion of available technology takes place through transfer and adaptation. Transfer includes transfer from abroad and subsequent transfer to the farming households who are the end users. Adaptation involves establishment of home-made and foreign technologies to suit specific local conditions, usually through adaptive research. The principal means of technology transfer is importation of foreign technology and extension of this and homemade ones within the country. The components of the two forms are materials, design, and capacity. The imported technology materials transfer includes fertilizer, chemicals, and tractors. Table 2 shows values of selected agricultural input imports for 1986. Sometimes, blueprints of technologies may be imported as well as prototype designs. Capacity transfer takes the form of foreign experts who are in the country to establish proven technologies and transfer the same to local experts through bilateral and multilateral arrangements in trade or research. Many expatriate personnel now serve in the Agricultural Development Projects (ADPs) and research institutes, particularly the International Institute for Tropical Agriculture (IITA).

The bulk of conscious transfer efforts are through agricultural extension. Some of the extension methods in practice are model and demonstration farms, group meetings, agricultural competition, farmers' organizations, clubs and unions, training and visit (T&V) method, mass communication, and home economics extension. The traditional agencies for carrying out extension activities are the State Ministries of Agriculture and Natural Resources. These have gradually been replaced with the ADPs, which are small-farmer-based and supported by the World Bank. Table 3 presents some facts about the ADP system in the country.

Development of Technological Intellect

The technological intellect comprises individuals having formal education in the different areas of agricultural technology. They form the basis for perpetuating the existing level of technology. Specifically they include graduates at the technical, vocational, and professional levels.

In regard to the creation of agricultural intellect, huge efforts have been made in the past. The training of agricultural cadets commenced at the historical Ebute Metta Botanical Station in the 1900s. Subsequently, educational institutions have grown immensely in the country in the forms of farm institutes, secondary schools, schools of agriculture, colleges of education, technical schools, polytechnics, and universities.

Selected Accomplishments

As earlier established, poor state of agricultural technology will lead to low productivities and poverty. The deferential technology statuses could therefore explain the substantial productivity gaps between developed and developing countries as illustrated in Table 4. For the selected years, the productivity of agriculture in Nigeria is only 3.8%, 4.1%, and 6.1% of productivity of agriculture in Canada, Australia, and the UK respectively. It is not by mere coincidence that the per capita incomes of Nigeria in relation to these countries follow the same pattern: 7.6% (Canada), 7.7% (Australia), and 8.9% (UK).[15]

Within the country itself, the product of agriculture is inferior to those in other sectors (Table 5). An estimate of output per man in agriculture is averagely lower than in mining for 1975–80 period (27%). This manifestation is quite in spite of increasing employment of agricultural inputs. This suggests that taken by themselves, increased levels of capital and also labour could only contribute modestly to the growth of output. The most important cause of increased volume of total output is productivity of resources; that is increased efficiency of their utilization through technological progress.

THE APPROPRIATENESS OF AGRICULTURAL TECHNOLOGY

An appropriate agricultural technology should relate first to the development objectives of the sector, and to the production problems of the farmers. The acid test of the existing technologies can be based on six criteria: ecological considerations, sociocultural considerations, simplicity, labour intensity, divisibility, and riskiness. These will be briefly applied in turn to selected technologies as follows:

---

[15] At this time four pupils were initially admitted into the garden: Samuel Adewusi, Nathan Philips, Josiah Thomas, and Cornelius George, who were trained in the practice of tropical horticulture and general cultivation such as garden works and the use of tools. For details, see Dawodu (1901).

Ecological Considerations

The first case to cite in this regard is fertilizer use. Presently, there is practically no system in use which gives proper regard to the different ecological conditions in the application of fertilizers. The soil scientists tell us that the type and quantity of fertilizer to apply in a given situation varies with the soil type, weather condition, nutrient status of soil, and vegetation. Yet farmers continue to practice blanket applications with no considerations to these factors. The inevitable result is low return of output to fertilizer. The farming public possesses only a nebulous idea about the necessary factors to consider and how to perform necessary measurements.

The optimum fertilizer-use policy involves the determination of the properties of soils, its hunger status, and crop requirements for the purpose of making nutrient budgets on a regular basis. Complications arise in fertilizer use through inefficient distribution channels, inadequate availabilities, and sharp practices of public officers and other people involved. The existing ecological inefficiency in the use of fertilizer, coupled with evidenced institutional wastage could have probably offset the apparent output responses to the fertilizer technology.

Another technology worth mentioning under ecological considerations is mechanization. It earns credit on grounds of ecological versatility to the extent that tractor ploughing and harrowing are technically feasible in most parts of the country. Also, a single model of tractor could be used under varying ecological conditions. However, the case of land clearing machines is different.[16]

More often than not, adequate attention is not given to the suitability of bulldozers as regards the engineering properties of agricultural soils. In effect, the valuable but fragile topsoils are destroyed or scraped off completely, leaving the nutrient-deficient subsoil for cultivation. The consequence is low productivity return to land clearing technology, and frequently, new problems or erosion.

---

[16] Only recently, reports had it that fertilizer racketeering was widespread in the country. In Benue State, for example, fertilizer bags were opened up and retailed using unorthodox measures (ten naira per measure) that could yield about ₦100 per bag instead of the statutorily fixed price of ₦15. The same state investigation revealed fake fertilizer sales receipts, as a result of which some officers lost their jobs. Desperate women buyers in the state have also performed a feat by preventing the carting away of trailer loads of fertilizer at night into unapproved channels.

Sociocultural Considerations

For a technology to be appropriate, it must fit into the existing sociocultural values system. Take the case of the high-lysine yellow maize technology, which was introduced for incorporation into baby meal to supplement infant nutrition in rural households. It was unacceptable to both the farmers, because of poor keeping quality, and the mothers. because of the high shaft content, which limits the quantity of pap (*ogi* or *akamu*) per unit weight of corn. In another example, one reason to explain the apparent rejection of small yam tubers (seed yam) for home consumption despite its virtues of better store ability, slow deterioration wastage, mechanizability, and others is the fact that they are shamefully too small for use in yam competitions at festivals. Therefore, the package of technological recommendations needs to include statements about either how to modify the sociocultural system to suit proven technologies or how to make the technological innovations fit well into the sociocultural system. In particular, most agricultural technologies have failed to satisfy the tradition of mixed-cropping and mixed-farming practices on small farms.

Simplicity

Most of the available technologies are not adapted to the special characteristics of the agricultural sector. Some of these characteristics are low levels of education and modern skills. This position limits the extent to which farmers can decode complex information accompanying some technologies. Some of these include the tractor and its accessories (i.e. maintenance difficulties), herbicides (walking speed, timeliness of spraying, dilution ratio, weed specificity, etc.), fertilizer application (nutrient contents, crop-specific requirements, application rates, nutrient status of soil, etc.).[17]

Therefore, a technology is appropriate only if it is quite simple for end users to understand. Otherwise, additional extension efforts must be applied for successful transfer to farmers.

---

[17] This is not to say that traditional farmers are unskilled. A number of other people such as teachers, civil servants, and similar urban elites who have ventured into farming under the mistaken belief that they are more skilled than the traditional farmers have learned the contrary to their costs. Therefore, the qualifications of 'modern' for skill are important.

Capital and Labour Considerations

It has been mentioned earlier that new technology has definite implications for factor proportions: it is either labour saving (capital using) or capital saving (labour using) and usually not neutral. On this basis, the inappropriateness of certain technologies in the developing economy can be explained at two levels. One, the small farmers are generally financially poor and therefore cannot access capital-intensive technologies requiring lumpy investible resources. This is why the tractor technology places heavy expenditure burden on government in terms of credit and subsidy requirements to ensure successful technology transfer. For the same reason of poverty, which makes small farms yield small revenues, most of the tractor loan schemes have failed in the country. We therefore need to shift attention from technologies generated in the capital-rich countries, to those that can be afforded more easily at the small-scale level.

Secondly, a developing country like ours, which often relies on agriculture for solving emergency unemployment problems, cannot pursue labour-saving technologies too far. Alternatively, capital-intensive technologies in agriculture must be backed up by effort to expand the absorptive capacity of the industrial sector for labour. However, each instance needs to be treated differently, as some exceptions exist. For instance, farm machinery and irrigation, which may offer new opportunities for multiple cropping result in the employment of idle labour during the slack season. In the Japanese example (Boserup 1946) the technology that leads to the harvesting of tree crops in the regions with relatively long cold seasons, also leads to an increase in the yearly demand for labour.

Divisibility and Riskiness

Certain technologies need to be divided for efficient utilization by the small-scale farmers. Fertilizer and seed satisfy this necessity, further than tractor. This is because mechanization is lumpy, thereby requiring a minimum level of utilization before it could become profitable, while seed and fertilizer can be divided into small components for application at different scales. The type of technological progress that is required is the small-scale neutral ones shown in Figure 1. In this case, the progress leads to complete downwards shift of the average cost curve (AC: from AC1 to AC2).[18]

---

[18] The shape of the AC curve is concave towards the origin owing to the presence of decreasing returns to sale and increasing returns to scale at low and high levels of

Indivisibility is related to riskiness of a technology. The well-known fact that the small farmer is generally risk-averse implies that he carried out an implicit discounting of the future value of an innovation before adoption is completed. The hybrid maize technology is a suitable example in this regard. In this case, the hybrid is practically a freak of nature; it is as disadvantageous as it is advantageous. In particular, it is claimed that the yield superiority of hybrid seed over the open-pollinated varieties, which is in the multiples of between five and ten, could be totally wiped out if environmental hazard occurs. The farmer interpreted this feature as high risk, because he also knows that the ordinary maize could turn out to perform better under the same insufficient conditions. Consequently, he is conscious to adopt hybrid maize fully together with the associated greater costs of establishment and management: heavy fertilizer dosage, chemicals for disease and insect control, ample rainfall, among others.

The measures against risk in agricultural development process could be of two types. One measure is to emphasize those technologies having little risk element or, if possible, no risk at all. However, this is not always possible. The second measure is to mount substantial efforts to reduce the magnitude of risks involved in selected technologies. In this regard, the new National Agricultural Insurance Scheme has a definite role to play. Thus, there is the urgent need to expand the scope of functions under this scheme by extending beyond crop or livestock covers to the facilitation of adoption of high-risk farm technologies.

SUSTAINABILITY ISSUES IN AGRICULTURAL TECHNOLOGY

Agricultural technology must not only be appropriate, but it must also be sustainable within the context of the existing institutional arrangements. Since government serves as the prime mover of technological progress, including its creation through research, and transfer primarily through agricultural extension, then its organizational actions and inactions mostly determine the sustainability of the level of technology attained (Idachaba 1987).

---

output respectively. Both small-scale farmers and large-scale farmers will enjoy lower costs from technological progress. However, progress in mechanization fits into the features of Figure 2. Costs are higher (AC1 less than AC2) if the derived from average costs by big farms proportionately increase at higher levels of output than q (AC2 less than AC1).

The general issues in sustainability are the institutional arrangements, the policy environment, and the international environment. These are policy changes, ministerial instabilities, and funding commitment.

Policy Changes and Sustainability of Technological Progress

The perceptions of the role of government in agriculture change too incessantly to put the country on a steady path of technological progress. The initial phase of agricultural development directed efforts towards the surplus extraction of export produce to satisfy the raw material needs of industries in metropolitan Britain. The main policy emphasis at that time therefore generated technological progress in the production of a selected few among agricultural commodities. Notable instances include the introduction of new varieties of cocoa and other products as well as the hydraulic oil press. The first tractors were imported for use on large plantations, and the technology of small farming was poorly developed. Later, marketing boards were established for cocoa, oil palm, groundnut, and cotton, which advanced the technology of producing these crops towards increased volume of their exports. Also, the first research organizations were mainly concerned with technologies in oil palm and cocoa production.

In the independence era, the changes in government policy and their effects on non-sustainability of agricultural technology can be outlined as follows:[19]

(i) The heavy taxation of agriculture through the marketing boards to raise revenue was tantamount to reductions of produce prices, which led to marked reductions in the output of the affected commodities (Idachaba 1987 op. cit.). Farms were deserted for a long period, which led to pronounced discontinuities in the growth of production technology.

(ii) The indiscriminate involvement of government in direct agricultural production, which frequently failed, created bad demonstration effect on private sector initiatives and also set unfair competition against them. This limited concentration of efforts on the rapid uptake of production technologies among the private farmers.

(iii) A number of macroeconomic policies have been working in the opposite direction of the path of technological progress

---

[19] According to Idachaba (1987), sustainability in general agricultural development.

in agriculture. Firstly, the over-valued exchange rate, which cheapened food and agricultural input imports relative to local products, created substantial dependency mentality, which dampened the initiatives in production and indigenous technology development. It also led to inefficient utilization of imported technology, such as inorganic fertilizers, chemicals, and tractors, among others. Secondly, artificial trade barriers were counterproductive in the sense that the protective walls around local industries caused inefficiency. Sustainable technological progress could not result in this circumstance because produce prices did not reflect the true costs of production. Thirdly, the minimum public-sector wage policy distorted the rural-urban labour market by affecting the value of the factor for the implicit discounting off-farm opportunity cost of farm employment. The resulting drifting of youth to towns led to discontinuities in technological progress in agriculture.

(iv) The poor status of rural infrastructure in the physical, social, and institutional forms means that the packages of available technologies are not complete. Bad rural roads network, for example, means limited accessibility of many technologies to the end users, as may also be caused by few farm service centres and storage support facilities. Poor educational standards limit the ability of the rural people to decode complex technological information and appreciate the effort of extension agents. Poorly developed rural groups also reduce the availability of capital to acquire new technologies.

Ministerial Instabilities and Sustainability of Technological Progress

There are two main ways by which the organization for agricultural administration frequently changes to produce discontinuities in the technology process. One of these is perturbation through merging and demerging of the ministry or a section of it, as well as creation of new sections and the scrapping of old sections. This happens at both the federal and state levels. The second is by high rates of personnel turnover in key leadership positions. In particular, the high rates of turnover in the political, administrative, and professional leadership positions are a cause for immediate concern. Table 6 provides selected evidence of these.

The end result of organizational instabilities in the forms of breakage fusions and personnel turnover in key leadership positions is adverse. They end up in programmes being given more or less emphasis, redesigned, re-introduced, or the implementation pace speeded up or slowed down so as to reflect the new political, philosophical, ideological, and occupational biases of the new people involved. Consequently, the technology development pathway becomes uneven.

Funding Commitment and Sustainability of Technological Progress

Usually, the sponsors of new technology packages fail to translate their original intentions into purposeful funding commitments during execution of the project. The case with the ADPs substantiates this point (Table 7). The effect of poor funding in midstream is adornment of key project components and poor performance of the others. In the end, the agricultural technologies meant to be developed and transferred are truncated. The same is the situation with agricultural research and several other aspects that could lead to progress in technology development.

CRITICAL ELEMENTS OF EFFECTIVE AGRICULTURAL TECHNOLOGY POLICY

The foregoing discussions enable us to coalesce the critical elements of a virile agricultural technology development policy. These are outlined in the following postulates.

(1) Effective agricultural technology must be all-embracing and completed. The components that must be included are mechanical, electrical, electronic, management, and sociocultural aspects. In addition, all the activities of agricultural production, transportation, processing, and storage should be considered together.
(2) Effective agricultural technology must be based on creation through virile national agricultural research system, transfer through active agricultural extensions system, and perpetuation through the developmental technological intellect.
(3) Effective agricultural technology must be appropriate on grounds of ecological and sociocultural considerations, simplicity relative availabilities of capital and labour, divisibility and riskiness.

(4) Effective agricultural technology must be sustained by government through the endurance of agricultural and macroeconomic policies, stabilization of the agricultural administration, and disciplined funding commitment.

CONCLUSION

On their own, increased application of capital and labour only make modest contributions to increase in output. The most significant explanatory variable of the variation in the volume of output is steady technological progress. In Nigeria, agricultural technological progress has been seriously hampered by the lack of a completely integrated approach, laggard research performance, and ineffective extension system. Practical accomplishments of technological development and policy efforts have been minimal, owing to the absence of stable sectoral and macroeconomic policy environment, complementary rural infrastructure, and sustained funding. Therefore, sustained policy efforts are required to correct the inadequacies of the Nigerian agricultural technology process.

REFERENCES

Boserup, E. (1965), *Conditions of Agricultural Growth* (London: Allen Unwin).

Dawodu, T. R. (1900, 1901), *Annual Report on the Botania Station* (Government Printer).

'Botanic' C. Punch (1901), *Report on Lagos Forest for Year to March 1901* (Government Printer).

Idachaba, F. S. (1987), 'Sustainability Issues in Agricultural Development' invited symposium lecture, Agriculture and Rural Development Department, The World Bank, Washington DC 8–9 Jan. 1987.

Samuelson, P. (1976), *Economics* (New York: McGraw-Hill).

Table 1: Main Organs of Nigerian Agricultural Research Systems

S/N	INSTITUTIONS	OWNERSHIP/ HEADQUARTERS
A. Research Institutes		
1.	Agricultural Ext. and Research Liaison Service	Federal / Kaduna (Kaduna State)
2.	Agricultural Ext. Research Liaison & Training	Federal /; Umudike (Imo State)
3.	Cocoa Research Institute of Nigeria (CRIN)[a]	Federal / Ibadan (Oyo State)
4.	Federal Institute of Industrial Research (FIIRO)[a]	Federal / Oshodi (Lagos State)
5.	Forestry Research Institute of Nigeria (FRIN)	Federal / Ibadan (Oyo State)
6.	Institute of Agricultural Research (IAR)	Federal / Zaria (Kaduna State)
7.	Institute of Agricultural Research and Training	Federal / Ibadan (Oyo State)
8.	International Institute of Tropical Agriculture	International Body / Ibadan (Oyo State)
9.	Kainji Lake Research Institute (KLRI)	Federal / New Bussa (Kwara State)
10.	Lake Chad Research Institute	Federal / Maiduguri (Borno State)
11.	National Animal Production Institute (NAPRI)	Federal / Shika, Zaria (Kaduna State)
12.	National Cereals Research Institute (NCRI)	Federal / Badeji (Niger State)
13.	National Horticultural Research Institute	Federal / Ibadan (Oyo State)
14.	National Root Crops Research Institute	Federal / Umudike, Umuahia (Imo State)
15.	National Veterinary Research Institute	Federal / Vom (Plateau State)
16.	Nigerian Institute for Oil Palm Research	Federal / Benin-City (Bendel State)
17.	Nigerian Institute for Trypanosomiasis	Federal / Kaduna (Kaduna State)
18.	Nigerian Stored Products Research Institute	Federal / Yaba (Lagos State)
19.	Rubber Research Institute of Nigeria (RRIN)	Federal / Benin-City (Edo State)

B. Specialized Agricultural Universities	
20.	University of Agriculture, Abeokuta	Federal / Abeokuta (Ogun State)
21.	University of Agriculture, Makurdi	Federal / Makurdi (Benue State)
C. Specialized Technology Universities[b]	
22.	Federal University of Technology	Federal / Akure (Ondo State)
23.	Federal University of Technology	Federal / Minna (Niger State)
24.	Federal University of Technology	Federal / Owerri (Imo State)
25.	Anambra State University of Technology	State / Anambra State
26.	Rivers State University of Science and Technology	State / Port Harcourt (Rivers State)
27.	Abubakar Tafawa Balewa University	Federal / Bauchi (Bauchi State)
D. General Universities[c]	
28.	Ahamadu Bello University	Federal / Zaria (Kaduna State)
29.	University of Benin	Federal / Benin (Edo State)
30.	University of Calabar	Federal / Calabar (Cross River State)
31.	University of Ibadan	Federal / Ibadan (Oyo State)
32.	Obafemi Awolowo University	Federal / Ile-Ife (Oyo State)
33.	University of Ilorin	Federal / Ilorin (Kwara State)
34.	University of Maiduguri	Federal / Maiduguri (Borno State)
35.	University of Nigeria	Federal / Nsukka (Anambra State)
36.	Usmanu Dan Fodiyo University	Federal / Sokoto (Sokoto State)
37.	Ogun State University	State / Ago-Iwoye (Ogun State)
38.	Imo State University	State / Owerri (Imo State)

Notes:

(a) FIIRO conducts research into the industrial use of agricultural commodities.
(b) Technology universities conduct research into agricultural technology.
(c) General universities have faculties of agriculture, and research into all aspects of agriculture.

SOURCE: (1) *Agricultural Research Centres*, Volume 2, pages 624–630

Table 2: Imports of Selected Agricultural Inputs, 1986

S/NO.	ITEMS	QUANTITY		VALUE (₦)
		UNIT	AMOUNT	
1.	Fertilizer, Crude	Metric ton	168,458	32,270,016
2.	Agric. Machinery for Soil Cultivation	Kilogram	4,654,085	40,007,574
3.	Agric. Machinery for Harvesting	Kilogram	2,909,432	38,358,587
4.	Milking Machines and Dairy Equipment	Kilogram	148,137	3,098,053
5.	Agric. Tractors, Tracked	Kilogram	503,898	11,921,561
6.	Agric. Tractors, Wheeled, not Exceeding 40 BHP	Kilogram	468,680	14,054,032
7.	Agric. Tractors, Wheeled, Exceeding 40 BHP	Kilogram	45,283	751,497
8.	Agric. Machinery NEC	Kilogram	3,941,205	44,396,269

SOURCE: *Nigeria Trade Summary*, December 1986 (Federal Office of Statistics)

Fig. 1

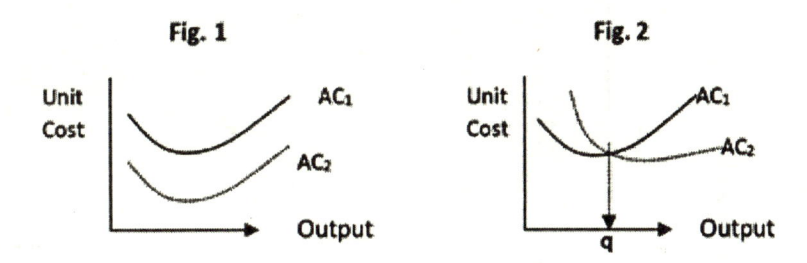

Unit Cost — AC_1, AC_2 — Output

Fig. 2

Unit Cost — AC_1, AC_2 — Output, q

Both small-scale farmers and large-scale farmers will enjoy lower costs from technological progress. However, progress in mechanization fits into the features of Figure 2. Costs are higher (AC_1 less than AC_2) if the technology is adopted at output levels less than q. And the benefits derived from average costs by big farms proportionately increase at higher levels of output than q (AC_2 less than AC_1).

Indivisibility is related to riskiness of a technology. The well-known fact that the small farmer is generally risk-averse implies that he carries out an implicit discounting of the future value of an innovation before adoption is completed. The hybrid maize technology is a sustainable example in this regard. In this case, the hybrid is practically a freak of nature; it is as disadvantageous as it is advantageous. In particular, it is claimed that the yield superiority of hybrid seed over the open-pollinated varieties could be wiped out if environmental hazard occurs. The farmer interprets this feature as high risk, because he also knows that the ordinary maize could turn out to perform better under the same insufficient conditions. Consequently, he is conscious to adopt hybrid maize fully together with the associated greater costs of establishment and management: heavy fertilizer dosage, chemicals for disease and insect control, ample rainfall, among others.

The measures against risk in agricultural technology development process could be of two types. One measure is to emphasize those technologies having little risk element or, if possible, no risk at all. However, this is not always possible. The second measure is to mount substantial efforts to reduce the magnitude of risks involved in selected technologies. In this regard, the National Agricultural Insurance Scheme has a definite role to play. Thus is the urgent need to expand the scope of functions under this scheme by extending beyond crop or livestock covers to the facilitation of adoption of high-risk farm technologies.

Table 3: Selected Features of Agricultural Development Projects

S/N	ADPs	Area Cov. (km²)	Commencement Date	Termination Date	No. of Farm Families	Total Cost (₦) Million
1.	Anambra State ADP	Statewide (16,727)	January 1986	December 1989	382,000	38.90
2.	Bendel State ADP	Statewide (36,300)	January 1986	December 1989	336,000	39.1
3.	Benue State ADP	Statewide (46,300)	January 1986	December 1989	385,000	37.7
4.	Cross River State ADP	Statewide (28,600)	January 1986	December 1989	540,000	27.6
5.	Imo State ADP	12,700 km²	January 1986	December 1989	-	34.2
6.	Ogun State ADP	Statewide (16,400)	January 1986	December 1989	150,000	8.16
7.	Plateau State ADP	Statewide (54,000)	January 1986	December 1989	311,000	64.0
8.	Bida Agric. Dev. Proj. (Niger State)	Enclave (17,000)	July 1980	June, 1986 (Extended to June 1987)	60,000	-
9.	Gongola State ADP	Statewide (11,775)	February 1982	January 1982	470,000	131.2
10.	Ilorin ADP (Kwara State)	Enclave (11,775)	1979	Dec. 1984 (Ext. to June 1987)	90,000	-

11.	Ekiti Akoko ADP (Ondo State)	5,000	28 August 1981	March 1986	70,000 (1981)	48.2
12.	Lagos State ADP	Statewide (3,600)	January 1987	December 1990	-	56.5
13.	Oyo-North ADP	Enclave (12,310)	June 1982	May 1987 (Ext. to Sept. 1988)	55,000	-
14.	Rivers State ADP	Statewide (19,420)	January 1987	December 1990	-	38.38 (Reprogramme cost is ₦74.04m)
15.	Bauchi State ADP	Statewide (66,000)	January 1981	Dec. 1988 (2 years extension)	425,000	-
16.	Borno State ADP	Statewide (79,292)	Feb., 1982	January 1988	262,000 (Excl. South Borno ADP area)	135.15 (Reprogramme cost ₦85.8)
17.	Kaduna State ADP	Statewide (68,00)	July, 1985	December 1990	430,000 (1981)	127.8
18.	Kano State ADP	Statewide (43,000)	January, 1982	Dec. 1988 (Reprogramme date)	430,000	265.0
19.	Sokoto State ADP	Statewide	January, 1983	June 1983 (Ext. to June 1989)	616,000	274.3
20.	Southern Borno ADP	Enclave (33,000)	July, 1986	1991	200,000	39.3

SOURCE: Agricultural Projects Monitoring and Evaluation Unit, Kaduna, Nigeria (APMEU) Project Status Report Volumes I–IV

Table 4: Agricultural Productivity and Per Capita Income Statuses of Nigeria and Other Selected Countries

COUNTRY	AGRICULTURAL GDP IN DOLLARS PER CAPITA AGRICULTURAL POPULATION*	GNP PER CAPITA** (DOLLARS)
Nigeria	369	860
OTHER COUNTRIES		
Libya	1416	8510
Canada	9771	11320
Costa Rica	1037	1430
Trinidad and Tobago	2296	6840
Argentina	2365	2520
Australia	8945	11140
Singapore	4999	5910
Austria	4869	9880
Norway	6803	14280
United Kingdom	6005	9660
Average of other countries	4851	7129

SOURCE: *Figures are for 1984 and have been obtained from Food and Agriculture Organization, *The State of Food and Agriculture*, Rome, 1986.

**Figures are for 1982, and have been obtained from *World Development Reports*, 1984.

Table 5: Productivities of Agriculture and Mining, Nigeria, 1975–80

	Average over 1975–80
Employment in agriculture (no.)	16,102,423[a]
Employment in mining (no.)[b]	43,768
Agricultural GDP at constant 1977 factor costs (₦m.)	7,535
Mining GDP at constant 1977 factor costs (₦m.)	7,508
GDP per person employed in agriculture	668
GDP per person employed in mining	1,720

Footnotes

(a) Agricultural employment is based on 1979 population estimate (84,496,105), and average values of labour force participation (37.15%), and agricultural labour force participation (56.05%) over 1975–80.
(b) Mining includes metaliferous mining and Nigerian Coal Corporation.

SOURCE: Underlying data from

(i) Idachaba, F. S. (1985), *Rural Infrastructures in Nigeria*, Ibadan University Press
(ii) Federal Office of Statistics, *Annual Abstract of Statistics*, 1982.
(iii) Central Bank of Nigeria, *Annual Report and Statement of Accounts*, 1984.
(iv) World Bank, *World Table* (various issues).
(v) Food and Agriculture Organization, *The State of Food and Agriculture*, Rome, 1986.

Table 6: Personnel Turnover in Agricultural Administration, Selected Areas

STATE GOVERNMENT	POLITICAL LEADERSHIP (COMMISSIONERS)			ADMINISTRATIVE LEADERSHIP (PERMANAENT SECRETARIES)		
	APPROXIMATE PERIOD	NUMBER OF OCCUPANTS	AVERAGE LENGTH OF STAY (YEARS)	PERIOD	NUMBER OF OCCUPANTS	AVERAGE LENGTH OF STAY (YEARS)
Anambra	1976–1988	9	1.3	1974–1988	9	1.6
Bauchi	1976–1988	15	0.8	1976–1976	13	0.9
Benue	1976–1988	13	0.9	1976–1988	12	1.0
Borno	1976–1988	13	0.9	1967–1988	12	1.0
Cross River	1976–1988	14	1.4	1976–1988	14	0.9
Gongola	1976–1988	9	1.3	-	-	-
Kaduna	1967–1988	11	1.9	1977–1988	13	1.6
Kano	1967–1988	12	1.8	1967–1988	20	1.0
Kwara	1967–1988	12	1.8	1967–1988	15	1.4
Oyo	1954–1988	17	2	-	-	-
Mean length of stay =			1.4 years	Mean length of stay		1.2 years

SOURCE: Underlying data from field survey, 1988.

Table 7: Funding Commitment in Nigerian Agricultural Development Projects

Name of project	Period of existence	Actual as percentage of budget expenditures [a]			
Anambra State ADP		46.6	133.5	0	-
Bendel State ADP		88.3	93.9	9.7	30.9
Benue State ADP		78.6	63.7	0	34.2
Cross River State ADP		83.7	81.3	0	X
Imo State ADP		75.1	90.2	13.8	73.7
Ogun State ADP		123.9	93.6	0	-
Plateau State ADP		101.8	54.3	0	112.0
Bida ADP		80.0	69.3	109.9	-
Gongola State ADP		99.0	50.1	-	150.7
Ilorin ADP		72.9	55.0	42.7	63.2
Ekiti Akoko ADP		78.1	56.2	45.7	67.8
Lagos State ADP		76.0	175.6	-	-
Oyo North ADP		155.4	72.2	105.9	64.7
Rivers State ADP		69.1	75.0	-	50.1
Bauchi State ADP		75.3	68.6	97.4	-
Borno State ADP		76.3	53.8	-	62.7
Kano State ADP		76.9	85.3	135.4	-
Sokoto State ADP		82.9	77.0	98.1	-
Southern Borno ADP		97.2	102.9	53.3	75.1

Notes:

(a) - means not budgeted and no expenditure committed, X means not budgeted but some expenditure is committed, zero means budgeted but no expenditure is committed.

(b) Other sources of funding ADP mainly include internal revenue.

Socio-Economic Problems in Commercializing Agricultural Research Findings and Technological Breakthrough in Nigeria

Lecture for the Senior Executive Course No. 21, National Institute for Policy and Strategy Studies, Kuru, Jos, 16 May 1990.

The main purpose of establishing research and development (R&D) institutions is to build up the science and technology capabilities that will generate innovations for national development. However, in a developing-economy setting, like Nigeria, these institutions have achieved a limited degree of success unlike the case with developed economies. In the course of this lecture, the differential performance of R&D institutions in the developed and developing economies will be ascribed to their dissimilar economic conditions. In particular, the paper will describe the role of the factors of these conditions in commercializing agricultural innovations in Nigeria.

Firstly, the outline of R&D process in Nigerian agriculture will be presented as background information and also a source of the usual commercialization problems. Secondly, selected cases will be described as a means of elucidating innovation specific issues. Thirdly, the socio-economic factors influencing the commercialization of agricultural research findings and technological breakthrough will be highlighted, with a view to identifying the associated problems. Lastly, the paper will be concluded together with recommendations.

AGRICULTURAL RESEARCH AND DEVELOPMENT PROCESS IN NIGERIA

The R&D process in Nigeria will be discussed in three phases: mechanisms for agricultural administration, mechanism for agricultural research, and mechanism for agricultural extension. Agricultural administration provides the initiative for broad policies, research generates technological innovations, and extension diffuses the innovations. This continuum of R, D & E determines the technologies available for agricultural development, their successful adoption, and hence the degree of success attained in their commercialization.

Institutional Mechanism for Agricultural Administration

The evolution of institution for agricultural administration can be described in three stages, including (a) the pre-1954 era, (b) 1954–1967 era, and (c) the post-1967 era. In the earliest era, agricultural administration commenced with the establishment of the Forests Department in 1900. This succeeded the Botanical Station, earlier established at Ebute Metta in Lagos in 1893. The first Agricultural Department (Southern Nigeria) was established in 1910. The second Agricultural Department (Northern Nigeria) was established in 1912. Both of these were combined to form the unified Agricultural Department for Nigeria in 1921. The department administered all development activities in agriculture, particularly research, extension and input support services.

In the 1954–1967 era, each regional government established its own Ministry of Agriculture and Natural Resources, and there was none at the federal level. Consequently, there was an interregnum in the central coordination of agricultural development activities at the national level. Each region was solely responsible for the conduct of agricultural input supply, extension, and research in its jurisdiction.

By an executive fiat, and following expert recommendations (FAO 1967), the military administration created the Federal MANR in 1968/69. This replaced an earlier institution by the name Ministry of Natural Resources and Research (MNRR), which could carry out tangential agricultural development functions only.[20]

---

[20] The regional constitution was signed into law in October 1954. By a special clause of this constitution, agricultural development was included on the residual legislative list, which made it a regional responsibility only. Therefore,

Since creation, the Federal MANR has undergone modifications in name, size and function over the time.[21] Presently, there is a system of agricultural administration comprising one federal and twenty-one state ministries in Nigeria.[22] In addition, there is autonomous management of parastatal bodies which carry out specific agricultural development functions.

Institutional Mechanism for Agricultural Research

The institutional mechanism for agricultural research has evolved parallel to the administrative arrangements. In the earliest stage of this evolution, research was carried out on ad hoc basis by specialist staff in the agricultural departments. Institutionalization of research commenced in the colonial period with the establishment of satellite stations of the pan-territorial West African research institutes in Nigeria.

At a later stage, these stations were nationalized after independence (Act No. 33 of 1964) and became parastatals of old Federal Ministry of Economic Development. This ministry had mandates to conduct research in 'soils and crop nutrition, improvement of food crops and the control of plant diseases and pests' (FDAR 1964). The Agricultural Research Council of Nigeria was subsequently established (Decree No. 25 of 1971) with power to establish agricultural research institutes in the country and transfer the existing ones to itself for general arrangements as also permitted by the enabling law (Decree No. 35 of 1973). These institutions are now administratively under the Federal Ministry of Science and Technology (FMST). Table 1 describes the agricultural research institutes in Nigeria.

In addition to research institutes, there are several agriculture faculties in the general university which are under the National Universities Commission of the Federal Ministry of Education (FME). Lately, two Universities of

---

the federal agriculture ministry could not be created so as not to 'offend the political sensibilities' of the regional governments. However, Gowon's military administration broke the trend by creating the Federal Ministry of Agriculture and Natural Resources (Nigeria, Official Gazette Vol. 55 No. 10 of 7th February).

[21] At the federal level, the nomenclature of the agriculture ministry has now changed in a full circle: Federal Ministry of Agriculture (FMA 1970); Federal Ministry of Agriculture and Water Resources (FMAWR 1979); Federal Ministry of Agriculture, Water Resources, and Rural Development (FMAWRRD 1985); Federal Ministry of Agriculture and Natural Resources (FMANR 1990).

[22] Another state creation exercise took place in 1991, leading to a total of 30 states excluding the Federal Capital Territory, Abuja.

Agriculture have been established in 1988 as parastatals of the FMANR. Table 2 describes the university-based agricultural research system in the country.

Institutional Mechanism for Agricultural Extension and Input Services

The evolution of the extension system is concurrent to that of agricultural administration and research. Traditionally, the extension works were assigned to staff in the agricultural department or ministry without an autonomous body being in charge. In addition, the research and development parastatals organized their individual extension works. The same is true of Input Supply Services. Subsequently, special programmes were mounted to intensify the extension and input services. One of these is the National Accelerated Food Production Programme (NAFPP), which was aimed at providing an integrated extension and input support services to farmers within the framework of the mainstream ministry.

The most important experience of separately institutionalized extension services is provided by the Agricultural Development Projects (ADPS). A typical ADP is an area-based development agency jointly funded by the federal and state governments, together with loan assistance from the World Bank. The ADPs have been generated in sequence, which are described in Table 3. In the original setting, the core elements of the ADP concept include:

(a) an input delivery and credit supply system through a network of farm service centres;
(b) a massive rural feeder road network to open up new areas for cultivation and facilitate rapid evacuation of farm produce as well as the timely delivery of farm inputs;
(c) a revitalized, intensive, and systematic extension and training system backed up by timely input supply and adaptive research services; and
(d) effective project management together with built-in monitoring and evaluation.

The entire R, D & E continuum, briefly described above, has the responsibility to

(a) generate useful agricultural innovations through a virile research system,

(b) perfect innovations and develop them into technological packages through the technology development institutions, and
(c) diffuse the technological innovations through effective extension system.

The last function, extension, is an essential precondition for commercialization of the agricultural technologies. Only the successfully adopted innovations can be profitably commercialized. This is because commercialization is primarily a private-sector activity, which has profitability as the main driving force. It will be explained through the subsequent case studies how the inherited and existing institutional matrix for R, D & E has guaranteed only a limited degree of success as a foundation for commercializing agricultural research findings and technological breakthroughs in Nigeria.

SELECTED CASES IN COMMERCIALIZING AGRICULTURAL INVENTIONS

In this section, we shall conduct a number of case studies for the purpose of illuminating the socio-economic factors affecting the commercialization process of agricultural inventions. Three broad categories of these inventions shall be delineated, which are:

(a) innovations in genetic engineering and biocontrol,
(b) innovations in process technology, and
(c) innovations in product technology.

In each case, selected inventions will be considered in some detail.

Innovations in Genetic Engineering and Biocontrol: The Case of Hybrid Maize

There is consensus among geneticists that the performance of an individual organism depends wholly on two attributes: genotype and phenotype. 'Genotype' means genetic constitution of the individual. 'Phenotype', on the other hand, refers to the totality of the environmental effects on the individual. A hybrid, therefore, represents an organism whose genetic constitution is a combination of the elements of others in a way that makes the hybrid to perform far better than the parent lines. This superior performance of the hybrid over its parent stock is termed heterosis, or hybrid vigour. Breeding in general is the science and art of manipulating the gene constitution of the organism to excel in its performance over the traditional stock. The resulting high-yielding or improved variety may be a hybrid or simply any one variety with certain desirable attributes that condition its superior performance over the unimproved varieties. Some of such desirable attributes include disease resistance, insect resistance, nematodes resistance, and early maturity. It is also possible to breed for palatability, height, colour and other desirable traits.

The hybrid maize variety exhibits an extremely outstanding performance, with its yield exceeding those of traditional varieties in multiples. For example, on-farm demonstrations record shows an average yield of 6 tonnes per hectare for hybrid varieties, as compared to improved but open pollinated varieties (4.8 tonnes) and local varieties (2.3 tonnes) (IITA 1983).

The breeding process leading to the release of a hybrid maize variety is long. The development process involves:

(a) the 'combining ability' test for the purpose of identifying promising materials
(b) 'inbred line' extraction and advancement
(c) inbred line testing for combining ability, and
(d) mating of hybrids and testing them for possible release.

The process requires highly skilled professional and technical personnel, in addition to the long time required. It also requires heavy funding for field assistance and establishments of a good seed storage system.

A number of hybrid seed varieties have been released to the farming population in Nigeria. The National Seed Service (NSS) at Ibadan has promoted their development and distribution in the country. Apart from the numerous contract growers engaged in the multiplication of the seed, some private business initiatives have been made. Two prominent hybrid maize producers in the country are Obasanjo Farms and Leventis Farms. These two farms have been supplying substantial quantities of hybrid maize to farmers in the past few years. It is generally claimed that the available hybrid maize seeds in the market can yield about fivefold compared to local varieties.

However, the hybrid maize is a freak of nature; its advantages are also its disadvantages. The production and utilization of the hybrid maize seed is conditional on certain rules and regulations. Firstly, it is said that the seed of last year's harvest cannot be used for planting this year—a rule against a traditional practice of the small farmers. The reason is often that segregation must have occurred which has broken down the combined genetic ability. Therefore, if segregated seeds are used for planting this year, the high performance should not be expected next year. Consequently, the farmer cannot save seeds till next year, but has to buy new seeds in each planting season. The gene combination of the hybrid maize is frequently more susceptible to disease attacks and other environmental stress factors than traditional varieties. As a result, the hybrid maize, at such a time of stress, may practically perform worse than the usual varieties.

Thirdly, the hybrid maize must be pampered for it to express its full potentials. It requires optimum conditions, which are often beyond the reach of small farmers—high dosage of fertilizer, optimum rainfall, effective weed control, and possibly soil inoculation or seed freezing. Lastly, the farmer who wishes to produce his own hybrid seed requires specialized knowledge and skill. He needs to obtain the parent stock, which will be planted on isolated plots to prevent crossing, which means that the pollen grain of the traditional varieties should not be allowed to be blown by wind to hybrid lines. There is also the need to demarcate male and female lines. The tassels of the male plants produce pollen grain to be used in pollinating the female plants. Selfing must also be prevented. Sometimes, there is the need to collect pollen grains in bags to hand-pollinate the female plants. Otherwise, detasseling or male sterility techniques may be employed to exclude the possibility of selfing.

The length of time involved in development and high cost of management required have translated to high selling price of the hybrid maize seed (₦5/kg in 1987), compared to local varieties (70 kobo per kg). The farmer only adopts because of the expectation that the incremental cost of production involved in planting hybrid seed will be more than offset by the incremental revenue. This is because the expected yield increment is a multiple of expected cost increment. However, the potential yield of hybrid maize has not been realized in general as a result of the limitations posed to farmers by unavailability of capital, inaccessibility to inputs, and the lack of necessary knowledge. This situation affects its commercialization adversely.

Innovations in Process Technology: The Cases of Electrodyn Sprayer and the Yam-Pounding Machine

The electrodyn spraying machine represents a breakthrough as a means of effective control of insects of cowpea. These insects, particularly the legume pod borer (Maruca), account for the most of the high yield loss in cowpea. It appears that the improved varieties of cowpea are more susceptible to insect damage than the traditional varieties, in which case the expression of the higher-yielding potential or the other desirable traits of the modern varieties are obscured by the infestation of insects. According to Ogunwolu (1990),

Maruca impairs reproductive development of the cowpea plant by attacking the flowers very early in the reproductive phase.

The insect population increases in the growing season, thereby possibly preventing pod formation completely. Breeding for insect resistance has not recorded appreciable positive results.

The general practice on farms is to control by using insecticides such as Cymbush, Decis, and others. The usual problem, however, is that the traditional spraying machines are not very effective in functional and cost terms. The pneumatic knapsack sprayer has a fluid chamber apparatus to be mounted on the back and a spraying gun which draws the insecticide solution directly with a trigger to fire the droplets through a graduated nozzle. Farmers do not find it easy to use because (a) it is heavy to carry, owing to the weight of water in the solution chamber, (b) the wastage rate is very high as droplets are carried by wind away from the targeted plants on to the bare soil, (c) there is danger of unrestrained insecticide use on environment and farmers' health, (d) there is problem of speed adjustments to spray the required concentration, (e) pumping causes early tiredness.

The electrodyn-spraying machine has been invented to correct for the deficiency of the knapsack sprayer. It uses four batteries in a long container, to which a disposable bozzle containing the insecticide is firmly attached. The machine is an ultra-low-volume sprayer using electrical energy to charge the very fine droplets that are sprayed on the surface of the leaves. The special advantage of the electrodyn sprayer over the knapsack sprayer arises from the fact that (a) it is more cost-effective as the electrically charged droplets are attracted by plants and achieve wider coverage of the target, (b) it is lighter to carry because water is not required, (c) there is no need for pumping, which saves the farmers energy or consequently labour costs per unit area, (d) environmental hazard is minimal, including danger to farmers' health. Consequently, there is widespread report of successful adoption of the electrodyn innovation by cowpea farmers, particularly in the ADPs. However, there are complaints that (a) the cost of replacing the batteries is too high, (b) there is no maintenance capability, (c) the machine is expensive and scarce, (d) the special bottle (bozzle) of insecticide which is disposable is not easy to find and denies the farmer the flexibility in the choice of insecticide, (e) there is need for training the farmer for effective use of the machine as regards speed of movement, angle of orientation, and other necessities. Besides, the electrodyn innovation is an imported technology, which makes its commercialization problems not only of adoption but also marketing in nature.

Among the other examples in process technology described in Table 4, the yam-pounding machine is an outstanding indigenous case in point that is useful in examining socio-economic factors of commercialization. The machine was the product of research at the University of Ife (now Obafemi Awolowo University). It was produced in 1976, following two years of research work, in response to the problems of the laborious nature of pounding yam and hygiene. The traditional process of pounding yam consumes great energy of women who sweat profusely. Part of this sweat drops in the mortar and becomes part of the sticky mash as product. Wastage of the yam can be high, and the product may contain lumps.

The yam-pounding machine uses a mechanized mortar and pestle of varying sizes. The home type is portable, and it can pound boiled yam that is enough for feeding five adults, in one minute. There is no wastage of yam, and the machine is more hygienic than the traditional process. There is practically no problem of lumps in the product.

The prototype was demonstrated to the public, and three initial machines were sold as pilot. Subsequently, a Japanese manufacturer copied the prototype and produced a more compact machine that is more attractive than the model. The patent was later sold to Addis Engineering, a local manufacturer, and the machine was distributed through Leventis Nigeria Limited as corporate entrepreneur. The user feedback led to re-innovations in the form of installing an all-purpose grinder as attachment. As a result, the Nigerian machine became more competitive in the market.

However, the survey report by Adjbeng-Asem (1988) indicates limited success in adoption and commercialization of the machine. The first reason is its expensive nature relative to the value system of the consumers of pounded yam. The original Nigerian model sold at 150 naira. The Japanese imitator sold at between 250 and 300. The grinder-attached Nigerian model sold at 1,050 naira. Consequently, the generality of the people could not afford the machine, which made it unable to displace the traditional method of pounding yam among the low-income people. In addition, the machine creates more utility to owners as status symbol than it does as a process innovation.

There are complaints that the product of the machine is different from the traditional product, despite the scientific confirmations that there are no significant differences. It therefore seems that culture plays a role. To corroborate this point, interviews with consuming households indicate that strong preference is still shown for the traditional process of pounding yam. An interesting reason obtained is that the husbands attach value to the sound of the mortar and the presence of the sweat of their wives in the pounded yam. According to some respondents, the sound of the mortar makes the consumers salivate and improves the appetite. On the other hand, husbands in rural areas say that the sweat makes the pounded yam sweeter.

Innovations in Product Technology: The Cases of Pandoyam, Babeena, Soyabean- and Sorghum-based Products

Table 5 describes a number of commercializable product innovations that have been invented in Nigeria. The Pandoyam is a yam powder. It is produced and marketed by Cadbury Nigeria Limited as another substitute to the traditional pounded yam. As a dry product, it has the added advantage of storability. Unlike the machine alternative, the pounding process is eliminated, which is substituted for by the less energy-consuming stirring process. Pandoyam contains nutritional additives, which enhance the food value of the resulting product. However, utilization is limited by the fact that (a) the product is finer and lighter than traditional pounded yam, (b) it is not cultural, (c) it is more expensive than equivalent weights of traditional pounded yam, (d) it is only available in supermarkets, which makes it an urban consumer item only.

Glaxo Nigeria Limited manufactures the Babeena baby milk cereal. The special advantages over similar products are (a) it has high local content, (b) it is a competitive substitute for imported baby foods. The release of the innovation at the height of scarcity of baby foods has led to its successful commercialization.

Another class of important product innovation consists of soyabean-based products. The wave of soyabean utilization caught on in many areas of the country a few years ago. A lot of it was due to the effort of the International Institute of Tropical Agriculture (IITA), where the protein content of the crop had been enhanced to 40% level. This makes the commodity an appropriate substitute to animal protein (fish, eggs, meat, milk). The improvement was carried out on the existing varieties, which were reported to have been introduced to Benue and Plateau areas of Nigeria by Christian missionaries many years ago. The process of improvement at IITA, which took about twelve years, involved the incorporation of certain traits such as free nodulation supplementary inoculation with the rhizobia bacteria. In the process of improvement, the yield rose from mere 200 kg per hectare to about 1.5 tonnes per hectare.

Subsequently, a soyabean utilization programme was mounted, and many recipes were prepared to reduce malnutrition problems among the low-income people. The products include soymilk, soy bread, soy garri, soy snacks (cakes etc.), soy cooking oil, and others. The release of these products coincided with the period of foreign exchange scarcity, which caused reduction in importation of animal products such as milk, meat, and fish. Therefore, it has generated immense substitution effect for the imported protein that is very expensive. However, the industrial sources of these products have not generated adequate supplies relative to demands, which has translated into high prices. Consequently, the products have not reached the vast majority of the people, who cannot afford the high prices. In turn, the industrial concerns complain that the technology of manufacture has to be imported, which only changes the nature of dependency from products importation to process importation. Besides, there are widespread complaints of limited supply of soyabean to the industries. Taraku Mills, for example, operates only at 30% of installed capacity despite the fact that it is located in Benue State, where production of soyabean is concentrated (Ojo 1989).

Home processing of the crop also faces certain bottlenecks. First, the long cooking that is always involved in soyabean process to inactivate the anti-nutritional factor wastes heat energy more than the other beans. Secondly, hardly do we have a whole-soyabean product; it goes with cowpea in moi-moi and akara, with maize in pap, and with wheat flour in cake and other snacks. The composite nature of soyabean products appears unsatisfactory at homes in which the other components cannot be afforded. In the case of soymilk, the problems are (a) the short shelf life which is at most two days in refrigerators before the milk product becomes slimy and (b) the presence of leafy odour. The poor storability means that large quantities cannot be prepared at a time. Consumers find it difficult to undergo the time-consuming process of preparing soymilk (boiling, cooking, grinding, sieving) only to produce one or two cups needed for one breakfast. As regards the unpleasant odour of soymilk, treatment may be effected with flavour additives, particularly vanilla. But such materials are only available in supermarkets, which are located in the major towns and cities. Therefore, the rural population has little access to the products. In this situation, commercialization can only achieve limited degree of success.

The last class of product innovations to be described includes the sorghum-based products. Sorghum, also called guinea corn, has attracted immense attention as a substitute raw material for wheat. Unlike wheat, sorghum thrives in Nigeria and is widely utilized as food and feed crop, especially in the vast northern part of the country. As a result of foreign exchange scarcity, the potential of sorghum is being explored to produce whole and composite flour used in the bakery and to produce malt for brewing beer or baby food manufacture, as well as energy source in livestock feed production.

However, a number of problems have limited the degree of success in commercializing these products. Some of these problems were identified at the ICRISAT (1989) conference as follows:

(a) Sorghum beer is not readily acceptable in the market because the consumers have developed rigid tastes for barley-based beer. This has

been used to explain the apparent failure of Mayor, a sorghum-based lager beer recently introduced to the market, but which is not popular.
(b) The efforts made towards backward integration to increase the supply base of raw sorghum failed among the processing industries. This is due to heavy capital outlay necessary to undertake farming, coupled with the small quantities of output relative to installed capacity of industries, as well as the low return of agriculture. This explains the reports that a number of early adopters in the backward-integration process plan to withdraw their productive assets from grains production, including sorghum.
(c) Biscuits and cakes made from whole-sorghum flours are hard and very brittle, which consumers do not prefer. On the other hand, composite flours (sorghum-maize mixtures) lead to increased costs, while the properties of hardness and brittleness are still imparted to the products in degrees depending on the ratio of mixtures.
(d) In breadmaking, whole-sorghum bread is not feasible; the main problems with composite (wheat-sorghum) bread include higher production cost as a result of greater quantity of sugar, and yeast requirements. There is also the need for higher temperature and greater amount of water. In addition, the shelf life of composite sorghum-wheat bread is too short (3–4 days) unless it is specially preserved up to seven days. Furthermore, the volume of the composite bread per unit quantity of flour is smaller than the wholewheat bread, which translates into low revenue per unit quantity of flour in bakeries.

SOCIO-ECONOMIC PROBLEMS IN COMMERCIALIZATION

The foregoing case studies provide the background information necessary for identifying the important factors influencing the commercialization of agricultural innovations. This section highlights these factors and discusses the attendant problems.

The Institutional Factor

The problems with the institutional arrangements for agricultural R, D & E can be stated as follows:

(i) **Instability**—The existence of pronounced instability in the organization and policy has been documented by Ayoola and Idachaba (1989). There has been incessant merging and demerging of constituents divisions or departments in the agricultural ministry. The political, administrative, and professional leadership positions also change frequently. As a result, agricultural policy changes correspondingly to reflect the philosophical, ideological, political, and professional biases of the new people involved. The emergence of stop-go policies ultimately produces substantial discontinuities in the efforts towards generation of research findings that can be developed into useful technologies and extended to farmers. This situation is not conducive to the development of commercializable breakthroughs in agriculture.

(ii) **Funding**—The issue of underfunding of the R, D & E in agriculture is a key factor that limits commercialization of the research findings and breakthrough. The problem is attributable to poor funding commitments of government as the main sponsor of R, D & E activities, as well as the absence of significant private-sector participation. For example, the budgetary allocations to R & D are generally less than 0.2% of the GDP in Nigeria compared to Korea's 2% (Mohammed 1989). The absence of private-sector funds prevents the agricultural development process from deriving the benefits of contract research for specific commercializable innovations. The industrial firms rely on the R & D results of their parent companies abroad, a practice which, although it creates positive externalities in Nigeria, suppresses fast creation of indigenous technology.

(iii) **Implementation failures**—Most policies that are aimed at inducing technological development are poorly managed. In particular, the presence of substantial leakage elements limits the degree of implementation success, which can be illustrated with the cases of the flour mills and breweries. In these industries, the ban on wheat and barley malt had initially stimulated the creation of new product technologies, particularly the use of sorghum as substitute raw material. Large costs have been borne in terms of adaptation of existing machines for the milling and malting of sorghum. Successes have been recorded in the form of market acceptability of the products, albeit at initially higher prices. However, the continued presence of wheat bread and barley-malt beers in the market has prevented the flow of the potential benefits of the innovations to the labour and management involved.

The two causes of implementation failure that faced the local sourcing of sorghum raw materials policy are (a) inability to control smuggling in of wheat and barley malt, whose importation had been officially banned; the reason advanced is not appealing (that there still exist stockpiles of barley malt after over four years of banning its importation); it is known as that the shelf life of barley malt is shorter than one year; (b) inability to establish complementary regulations to prevent circumventing the policy; for instance, while wheat importation was banned, importation of bread itself was not banned, which encouraged capital flight whereby bakeries closed up in Nigeria and reopened in the neighbouring countries, where wheat bread can be and is now being imported from. As a result of continued smuggling and the possibility of capital flight, the existing breakthroughs in agricultural technology could not be successfully commercialized as the infant innovator-industries face excessive competition with superior products.

Sociocultural Considerations

The importance of sociocultural factors in the commercialization of agricultural innovations is revealed in the cases of the yam-pounding machine, yellow maize, the yam miniset, and the hydraulic oil press. The commercialization of these innovations was not completely successful, owing to one or another of the following problems: (a) yam-pounding machine does not arouse the sensation of hunger because it does not make noise with the mortar and pestle; also the pounded yam does not contain the 'sweat of wives'; (b) people are not accustomed to the colour of pap resulting from yellow maize; (c) the hydraulic oil press deprives wives of their traditional business of selling the by-products of oil palm production; (d) the yam miniset are too small to be presented at festivals and as dowry; (e) sorghum-based lager beer does not satisfy the taste of consumers. These examples provide evidence of the significance of sociocultural issues in the diffusion of agricultural innovations, which need to be based on the social and cultural values of the clientele. There is a strong need to (a) involve the users from the beginning, (b) follow up the technologies for necessary modifications and re-innovations based on user responses and feedback.

In a more general sense, the sociocultural factors mentioned above merely underlay the issues in appropriate technology. An examination of the cases studied reveal that certain innovations might not be appropriate on other grounds, some of which Ayoola (1990) identifies to include (a) ecological considerations, (b) simplicity, (c) relative availability of capital and labour, (d) divisibility, and (e) risk factor. As only the appropriate technology can be adopted and hence successfully commercialized, a number of efforts in this direction have failed because the innovations concerned do not fit well into the requirements of some combination of these factors.

The Question of Entrepreneurship

The entrepreneurship ability hinges on the presence of a number of elements which Adjerebeng-Asem (1989) has empirically established to include the following: (a) venture capital, (b) infrastructure, (c) technical knowledge and capability, (d) performance of basic research, (e) patent laws, (f) market opportunities, (g) motivation. Access to capital and patent laws were found to be particularly important. Innovativeness is determined by the composite value of these variables in an individual, which makes or does not make him an entrepreneur. However, judging from the great number of commercializable research findings and technological breakthroughs that have been adopted but not significantly commercialized, it appears that innovativeness is quite low, and consequently, entrepreneurial ability is poor in Nigerian agriculture. The practical manifestation of this feature is the presence of many innovators that are not translated to entrepreneurs.

The problem of venture capital can be mostly sourced to the available capital market. Ndiomu (1989) puts it, 'banks tend to be conservative with regard to strange or new innovations'. The patenting system also fails to give the required protection to the inventor, who fears the loss of ownership of the invention and therefore feels reluctant to disseminate the secret for commercialization purposes.

CONCLUSION AND RECOMMENDATIONS

The analysis of agricultural R, D & E has shown that innovation is one thing, adoption is another, and entrepreneurship is another. Even though the presence of entrepreneurial ability appears to be the predominant factor of commercialization, the contributions of the available institutional mechanism, sociocultural considerations, and appropriateness of technology are also important factors.

The following recommendations have basis in the context of the analysis:

(1) There is an urgent need for institutional integration whereby all the relevant agencies and corporate bodies come together to facilitate the process of generating commercializable research findings and

technological breakthroughs. The elements of this are (a) prevention of policy and organizational instability in the process, (b) removal of the causes of failures in implementation of relevant policies, and (c) injection of substantial funds into the R & D – commercialization continuum.

(2) Mechanisms need to be set up to monitor the appropriateness of the existing and newly emerging innovations so that problems can be nipped in the bud before the commercialization stage is reached.

(3) The problems facing the development of entrepreneurship in agricultural technology should be addressed through (a) motivation of individuals to develop innovativeness, (b) assistance in venture capital generation, and (c) the provision of protective patent laws.

REFERENCES

Adjebeng-Asem, S. (1988), 'Social Factors Influencing the Translation of Innovation into Entrepreneurship: Three Successful and Three Unsuccessful Cases', IDRC Canada, Manuscript Report 189e.

Ayoola, G. B. (1990), 'Technological Progress in Agriculture: Some Issues', a paper presented at the 9th Annual Training Conference of Nigerian Economics Society, Minna, May 1990 (forthcoming).

Ayoola, G. B. and Idachaba, F. S. (1989), 'Self-sufficiency in Food Production as an Essential Foundation for Economic Development in Nigeria', a paper presented at the 9th Annual Training Conference of the Industrial Training Fund, Zaranda Hotel, Bauchi (9–10 November 1989).

FAO (Food and Agricultural Organization) (1966), *Agricultural Development in Nigeria 1965–1980* (Rome).

Federal Department of Agricultural Research (1964), *Guide to the Federal Department of Agricultural Research Nigeria*.

ICRISAT (International Crops Research Institute for the Semi-Arid Tropics) (1989), Proceedings of the Symposium on the Current Status and Potential of Industrial Uses of Sorghum in Nigeria, Kano, 4–6 December 1989.

International Institute of Tropical Agriculture (1983), *Annual Report*.

Ndiomu, C. (1989), 'Commercialization of Research and Development Results in Nigeria', a paper presented at the International Workshop on Commercialization and Evaluation of Research and Development Results, 24–28 April 1989.

Ogunwolu, E. O. (1990), 'Damage to cowpea by the legume borer *Maruca testulatis Geyer*, as influenced by infestation density in Nigeria', *Tropical Pest Management* (in press).

Ojo, A. A. (1989), 'Growing Soyabean in Nigeria', Paper presented at the Conference of Kaduna Chambers of Commerce.

Table 1: Agricultural Research Institutes in Nigeria

NAME OF INSTITUTE	OWNERSHIP/HEADQUARTERS
Agricultural Extension and Research Liaison Service	Federal / Kaduna (Kaduna State)
Cocoa Research Institute of Nigeria (CRIN)	Federal / Ibadan (Oyo State)
Federal Institute of Industrial Research (FIIR)	Federal / Oshodi (Lagos State)
Forestry Research Institute of Nigeria	Federal / Ibadan (Oyo State)
Institute for Agricultural Research	Federal / Zaria (Kaduna State)
Institute of Agricultural Research and Training (IART)	Federal / Ibadan (Oyo State)
International Institute of Tropical Agriculture (IITA)	International Body / Ibadan (Oyo State)
Kainji Lake Research Institute (KLRI)	Federal / New Bussa (Kwara State)
Lake Chad Research Institute	Federal / Maiduguri (Borno State)
National Animal Production Institute (NAPRI)	Federal / Shika, Zaria (Kaduna State)
National Cereals Research Institute (NCRI)	Federal / Badeggi (Niger State)
National Horticultural Research Institute (NIHORT)	Federal / Ibadan (Oyo State)
National Root Crops Research Institute	Federal / Umudike, Umuahia (Imo State)
National Veterinary Research Institute (NVRI)	Federal / Vom (Plateau State)
Nigerian Institute for Oil Palm Research (NIFOR)	Federal / Benin-City (Bendel State)
Nigerian Institute for Trypanosomiasis Research (NITR)	Federal / Kaduna (Kaduna State)
Nigerian Stored Products Research Institute (NSPRI)	Federal / Ilorin (Kwara State)
Rubber Research Institute of Nigeria (RRIN)	Federal / Benin-City (Bendel State)

Source: Field survey 1988

Table 2: Location and mandate crops of research institutes in Nigeria

Crop Research Institute	Location	Mandate Crops
Institute for Agricultural Research (IAR)	Zaria	Sorghum, maize, cotton, sunflower, groundnut, and cowpea
Institute for Agricultural Research and Training (IAR&T)	Ibadan	Maize and kenaf
National Institute for Horticultural Research (NIHORT)	Ibadan	Vegetable, horticultural, and ornamentals
National Crop Research Institute (NCRI)	Badeggi	Rice, benniseed and sugar cane, and soybean
Lake Chad Research Institute (LCRI)	Maiduguri	Wheat, barley, and millet
National Root Crop Research Institute (NRCRI)	Umudike	Cassava, yam, Irish potato, cocoyam, ginger, and sweet potato
International Institute of Tropical Agriculture (IITA)	Ibadan	Cowpea, yam, maize, soybean, cassava, banana/plantains
International Crop Research Institute for the Semi-Arid Tropics (ICRISAT)	Kano	Sorghum, millet, pigeon pea, and groundnut
West African Rice Development Association (WARDA)	Ibadan	Rice

Table 2: University-Based Agricultural Research System

Name of Institution	Ownership/Headquarters
A. Specialized Agricultural Universities	
University of Agriculture, Abeokuta	Federal / Abeokuta (Ogun State)
University of Agriculture, Makurdi	Federal / Makurdi, (Benue State)
B. Specialized Technology Universities	
Federal University of Technology	Federal / Akure (Ondo State)
Federal University of Technology	Federal / Minna (Niger State)
Federal University of Technology	Federal / Owerri (Imo State)
Anambra State University of Technology	State / Anambra State
Rivers State University of Science and Technology	State / Port Harcourt (River State)
Tafawa Balewa University of Technology	Federal / Bauchi (Bauchi State)
C. General Universities	
Ahmadu Bello University	Federal / Zaria (Kaduna State)
University of Benin	Federal / Benin (Bendel State)
University of Calabar	Federal / Calabar (Cross River State)
University of Ibadan	Federal / Ibadan (Oyo State)
Obafemi Awolowo University	Federal / Ile-Ife (Oyo State)
University of Ilorin	Federal / Ilorin (Kwara State)
University of Maiduguri	Federal / Maiduguri (Borno State)
University of Nigeria	Federal / Nsukka (Anambra State)
Usman Dan Fodio University	Federal / Sokoto (Sokoto State)
Ogun State University	State / Ago-Iwoye (Ogun State)
Imo State University	State / Owerri (Imo State)

Source: Miscellaneous reports

Table 3 Generation of Agricultural Development Projects in Nigeria

Name	State Located	Commencement Date	Termination Date	Total Cost (₦ Million)
A. FIRST GENERATION ADPs				
Gombe ADP	Bauchi	Nov. 1975	Nov. 1980	NA
Gusau ADP	Sokoto	April 1975	April 1980	NA
Funtua ADP	Kaduna	April 1975	April 1980	41.8
B. NEWER ENCLAVE ADPs				
Bida ADP	Niger	June 1980	June 1987 (incl. 2 yrs' ext.)	41.8
Ekiti Akoko ADP	Ondo	Aug. 1981	March 1987	48.2
Ilorin ADP	Kwara	March 1980	June 1987 (incl. 2 yrs' ext.)	41.8
Oyo North ADP	Oyo	June 1982	Sept. '87 (ext. to Sept. '88)	41.6
South Borno ADP	Borno	July 1986 (Eff. Mar '87)	June 1991	36.8
C. STATEWIDE ADPs				
Anambra State ADP	Anambra	Jan. 1986 (Effect. 1987)	Dec. 1990	38.9
Bauchi State ADP	Bauchi	June 1981	Dec. 1981 (incl. 2 yrs' ext.)	192.8
Bendel State ADP	Bendel	Jan. 1986 (Effect. 1987)	Dec. 1990	29.1
Benue State ADP	Benue	Jan. 1986 (Effect. 1987)	Dec. 1989	37.7
Cross River ADP	Cross River	Jan. 1986 (Effect. 1987)	Dec. 1990	27.6
Gongola ADP	Gongola	Feb. 1982	Feb. 1987	131.2
Imo State ADP	Imo	Jan. 1986 (Effect. 1987)	Dec. 1987	34.2
Kaduna State ADP	Kaduna	Jan. 1984 (Eff, July '85)	1990	127.8
Kano State ADP	Kano	Dec. 1982	Dec. 1988 (incl. 2 yrs' ext.)	265.2
Ogun State ADP	Ogun	Feb. 1986 (Effect. 1987)	Jan. 1990	8.16
Plateau State ADP	Plateau	Jan. 1986 (Effect. 1987)	June 1990	34.0
Rivers State ADP	Rivers	Jan. 1987 (Effect. 1987)	Dec. 1991	74.04
Sokoto State ADP	Sokoto	April 1983	June 1989 (Revised)	274.3
Borno State ADP	Borno	Feb. 1982	Jan. 1988 (Previously 1986)	135.15
Lagos State ADP	Lagos	Jan. 19987	Dec. 1990	56.5

Source: APMEPU, *Project Status Report*, 1988

Table 4: Selection Agricultural-process Innovation in Nigeria

Process	Comments/Features
Kiln	Improved smoking technique developed by FIIRO and NIOMR. It saves time, labour, and fuel consumption.
Pounded-yam machine	Developed by the University of Ife (now OAU) in 1974, it uses mechanized mortar and pestle to pound yam.
Palm wine bottling	Developed by FIIRO between 1971 and 1986. Avoids wastage
Mechanized fufu making	Developed by FIIRO between 1971 and 1986
Mechanized garri making	Developed by FIIRO between 1971 and 1986
Moi-moi production factory	Developed by PRODA to reduce labour and saves time.
Mechanical peeling and grading of cassava	It was developed by FIIRO between 1971 and 1986. It saves time.
Maize sheller	It was developed by PRODA between 1971 and 1986. It is 98% efficient. Saves time and labour.
Maniola machine	It was developed in 1976. This machine washes slice, salt and fry chips.
Cowpea thresher	Developed in 1976, this machine threshes, and the chaff is sucked out by a fan. Loss of only 0.1%
Insulated transportation containers for fish seeds	Developed by NIOMR. Reduces mortality to about 1% and carries 20,000 fish fry by road to any part of Nigeria.
Canning of tuna and the Bonga fish	Developed by NIOMR in 1988. Improves the consumption quality and reduces losses.
Sprouting the cassava ministem	It was developed by IITA in 1989 for rapid multiplication of cassava. It uses perforated polythene bags.
Yam miniset technique	For seed yam multiplication. Many seed yams are produced from single tuber.
Electrodyn sprayer	Effective control method for pest. Ease of operation and reduced risk.

Legend

FIIRO - Federal Institute of Industrial Research
PRODA - Projects Development Institute
NIOMR - Nigerian Institute for Oceanography and Marine Research
IITA - International Institute of Tropical Agriculture

Source: Miscellaneous sources including Adeboye T.

Table 5: Selected Agricultural Product Innovation in Nigeria

Products	Comments/Features
Cassava flour	Developed by FIIRO between 1971 and 1986
Cassava starch	Developed by FIIRO between 1971 and 1986, used in furniture making, etc.
Garri flour	Developed by FIIRO between 1971 and 1986, is an energy diet food
Instant fufu	Developed by FIIRO between 1971 and 1986. Consists of cassava, plantain, and yam
Maize flour	It was developed by FIIRO between 1971 and 1986
Soyogi baby food	Developed by FIIRO between 1971 and 1986, high-protein diet to replace pap.
Composite flour	Developed by FIIRO between 1971 and 1986, wheat-sorghum, wheat-cassava, wheat-maize composites replace wholewheat in bread, cake, etc.
Sorghum flour	Developed by FIIRO between 1971 and 1986. High-carbohydrate diet.
Potato alcohol	FIIRO product developed between 1971 and 1986
Bottled palm wine	FIIRO product developed between 1971 and 1986
Pito (local beer)	FIIRO product developed between 1971 and 1986
Table vinegar	FIIRO product developed between 1971 and 1986
Tomato puree, ketchup, and powder	FIIRO product developed between 1971 and 1986
Salad cream and mayonnaise	FIIRO product developed between 1971 and 1986
Full-fat soy grits and oil	FIIRO product developed between 1971 and 1986
Smoked fish	FIIRO product developed between 1971 and 1986
Sorghum malt	Developed by FIIRO between 1971 and 1986. Substitutes for barley malt.
Peanut butter	Developed by FIIRO between 1971 and 1986

Legend

FIIRO - Federal Institute of Industrial Research

Source: Miscellaneous sources including

(1) Adeboye, T., 'Diffusion of pre-commercial invention developed in government-funded research institute in Nigeria' (unpublished), West African Technology Studies Network Kaduna - Nigeria.

The Policy and Rural Infrastructure Factors in Stimulating Maize and Sorghum Production

Invited paper presented at the Joint National Workshop of the Nigerian Institute of Management and the Raw Materials Development Council, July 1989.

The grains economy of Nigeria has demonstrated strikingly contrasting features in its response to development efforts. The country, which exported substantial quantities of maize in the early period of development, later became a gross importer in recent time.[23] On the other hand, sorghum, which is a close substitute of maize, is generally a non-trade commodity. Given the excess supply in the international market (Timer et al. 1983) and the policy of subsidizing exports by the developed countries, the import of maize have been relatively cheap, thereby dampens domestic production initiatives and consequently creates acute dependency relations.

However, the emergent balance-of-payment problem, which set in the early 1980s, led to sudden scarcity of essential commodities. This condition

---

[23] Historical records reveal an average annual export of maize to the tune of 8,515 tonnes over nine successive years (1906–1914) (see Annual Reports of the Colonial Agriculture Department, various issues). Recent evidence to the contrary reveals the average annual import of maize to the tune of 22,543 tonnes over seven consecutive years (1973–1979) (see FOS, Nigeria Trade Summary, various issues).

This paper views the problem of maize and sorghum shortages within the context of technological backwardness in agriculture. The problem is stated in Section I. Section II discusses the policy responses. Section III is concerned with the practical efforts for providing rural infrastructure. Section IV concludes the paper.

made imperative the renewal and growing policy recognition of the theme of self-reliance as a strategy of general economic development. The remedial Structural Adjustment Programme (SAP) considers increased domestic production of maize and sorghum as well as other grains and agricultural commodities of value as food and raw materials to be of prime importance. Their importation is to be discouraged, initially by simple quantity restriction and later by putting an outright ban in force. This situation consequently poses great challenges to local production capabilities with clear implications for raw materials sourcing and also food supply to the teeming population. Yet, the rate of progress appears slow, and this is accentuated by deficient policy and poor state of rural infrastructure.

STATEMENT OF THE SHORTAGE PROBLEM IN MAIZE AND SORGHUM

The shortage problem in maize and sorghum will be highlighted in terms of fluctuations of productions and availability against the existing prospects. Then the causes of the problem are briefly characterized.

Fluctuations in Production and Availability

The net production and net availability of maize and sorghum in Nigeria show considerable annual fluctuations. As regards maize fluctuations, net production and net availability are 34% and 28% respectively (see coefficient of production on Table 1); they are 19% and 18% respectively of sorghum. However, there is apparently no convincing upwards or downwards trend in net production and net availability of these commodities over the 1965–1985 period. Net production is initially closely associated with availability in maize until about 1975 when they take different paths with widening divergences. These divergences, which indicate the magnitudes of sufficiency shortfall in maize imply the ever-increasing rate of dependency on maize imports.

The reason for talking in net terms is to make allowances for seed and wastage (25%), which do not generally enter the markets of these commodities as food and raw materials. Further, the data of production and availability (total supply) are placed side by side for the purpose of determining the degree of divergence between the growths in supply and demand over time. Imports complement domestic production to make up the total availabilities of the commodities (see Ayoola 1988a).

In the case of sorghum, the small divergences between net production and net availability which existed up to the beginning of the 1970s soon disappeared onwards. This does not necessarily mean the attainment of self-sufficiency in sorghum, but rather suggests that excess demands on sorghum had been wholly absorbed by maize imports which heightened in the period. The high substitution effect probably resulted from evidenced relatively low price of imported maize compared to the price of sorghum as a non-traded item.

Prospects in Maize and Sorghum Utilization

The economic importance of maize and sorghum rests on their advantages as food and feed crops. This is especially because their grains have high calorie value and are also highly digestible. There are several other instances of utilization of maize in Nigeria, each of which could have a parallel substitute in sorghum.

The identification of thirty-four different forms of direct local consumption of maize in Nigeria is very dated (Akinwumi 1970). These include breakfast and other meals as well as alcoholic drinks. The industrial processing of maize leads to products such as flour, meal, sugar, starch, alcohol and spirits. Possible areas of intermediate investments in these regards include milling of corn, making of corn whisky, popping and canning of corn. A particular type of maize that has not been exploited at all is the sweet corn, which the advanced countries have canned variously using addictives to preserve the freshness.

Further, the whole plant of maize and sorghum has immense utilities. Apart from the grains used in livestock rations, the stovers are also feedstuff to cattle, poultry, horses, and sheep, residues of the cannery such as husks, cobs, shanks, and silks have been used in making silage and soilage. Some other uses of the stems of these crops include the making of corncob pipes and cellulose for home fuel.

Suffice it to say that Nigeria has untapped potentials for increasing GDP in the utilization of maize and sorghum. In the event of the recent ban on grains importation, there is no alternative means of stemming the current excess demand problem other than the pursuit of aggressive domestic production drive.

Probable Causes of Persistent Maize and Sorghum Shortage Problem

Even though the influences of tradition may remain strong on the production performance of maize and sorghum as well as other agricultural commodities in general, the argument of the backward-sloping demand curve will not appeal to us much as an explanation for the slow technological progess.[24] To expatiate upon this, the tradition-bound school propounds that the peasant farmer seeks to market only just enough of his products to bring him in some target amount of income. Therefore, if price of maize or sorghum rises for instance, the farmer reduces his output because he does not need to sell as much output as before to achieve his target income. Conversely, he responds to falling price of output by increasing his output. The conclusion drawn from this theory is that a supply response results from the traditional farmers' behaviour, which is opposite to the conventional types of economic behaviour. That is to say that the farmer does not respond to economic incentives approximately. This is presumed to be a tradition among peasant farmers as also are social taboos, extended family obligations, costly funeral traditions, and the sheer inertia to change the cultivation methods for generations.

However, the idea of backward-sloping demand curve is now considered old-fashioned in the face of overwhelming evidence to the contrary. In place of this, economic reasons are to be offered for the technological backwardness of agriculture in general. In this regard, the low productivity of maize and sorghum will find explanations in the poor state of technology development and transfer, shortage of complementary inputs, and the inadequacies of the available rural infrastructure. The critical ones among these economic reasons are enunciated as follows:

i. The unsuitability of traditional agriculture to evolve adequate research capacities—this results from poor funding of agricultural research, and the absence of priority rating of research problems, among others.
ii. The existence of inefficient channel of communication between farmers and researchers—this is the outcome of poor extension-research linkages and defective relationships between farmers and extension agents.

---

[24] For a discussion of the theory of backward-sloping supply and transfer peasant farmers, see Killick (1983).

iii. The inaccessibility of farmers to new physical inputs that accompany new technologies, e.g. seed, fertilizer, insecticide, spraying machines, tractors, etc.
iv. The lack of adequate training facilities needed for the farmers to acquire new skills in order to take full advantage of the new technology, e.g. irrigation methods
v. The poor state of rural infrastructure, including physical (e.g. rural road network, storage facilities, irrigation facilities) social infrastructure (e.g. formal and informal education), and institutional infrastructures (e.g. farmers' unions, cooperative societies)
vi. The shortage of farm labour at the peak of planting and harvesting seasons—this problem is not reduced by the use of capital-intensive equipment, owing to the unavailability of credit facilities in time and volume dimensions.
vii. The land tenure system which puts a ceiling on the size of farm and hence efficiency of utilization of available resources

POLICY RESPONSE TO STIMULATE MAIZE AND SORGHUM PRODUCTION

The impact of macroeconomic policies on maize and sorghum production will be discussed in correspondence with the series of national development planning efforts. The first plan period (1962–1968) relied basically on the individual efforts of the regional governments for the purpose of agricultural development in general. This is the consequence of the lack of adequate constitutional jurisdiction for federal presence. There was a visible absence of an agricultural ministry at the federal level to coordinate and complement the regional activities. A relevant policy effort in relation to the present issue is the Rice/Maize Programme of the old Western Region. This and subsequent programmes with particular relevance to maize and sorghum development are summarized in Table 2.

The second plan period (1970–1974) was the first to provide for agricultural development on the concurrent legislative list, which meant that agricultural development was now a joint responsibility of both the federal and state governments. A full-fledged agriculture ministry was finally established at the federal level to accelerate development works. The immediate development strategy to be adopted is the National Accelerated Food Production Project (NAFPP). Table 2 also describes the features of NAFPP, including its relevance to maize production development.

The underlying policy of the third plan period (1975–1980), which was launched in the height of the oil boom, was the reliance on imports. Consequently, domestic production initiatives became dampened in the face of the cheaper imports of maize and other grains. A strategy adopted to reinvigorate production was the launching of the Operation Feed the Nation (OFN) programme. This was to be followed by a series of other development strategies with direct relevance to maize and sorghum, among other food items. These are also described in Table 2 to include (i) River Basin Development Authorities (RBDAs), (ii) Nigerian Grains Board, (iii) Green Revolution Programme, (iv) National Cereal Research Institute (NCRI), (v) Nigerian Grains Production Company (NGPC), and (vi) Agricultural Credit Guarantee Scheme (AGGS).

The fourth plan period (1981–1985) renewed the policy of self-sufficiency in agricultural development. However, the implementation of the plan suffered acute foreign exchange shortages, which in turn led to the mass shortage of essential commodities, particularly food items. The frantic efforts to resuscitate the plan include an abortive attempt to obtain a loan from the IMF and the implementation of a homemade Structural Adjustment Programme (SAP). The current policy environment which emerged under SAP, and following the suspension of the fourth plan, is characterized by free market mechanism, export promotion, import restriction (including the outright ban on maize among grain imports), and other features.

Let us now raise and discuss the salient issues within the broad policy framework already outlined.

The Absence of Ecological Specialization Policy

Ecological specialization means the permission of each region to use the best advantage of any particular benefits in skills and resource endowments. Through specialization of this kind, wasteful duplication efforts are avoided, thereby creating high resource use efficiency and internal comparative advantage. Agroecological specialization also extends to the use of special environmental niches such as the temperate enclaves of Jos, Mambilla, and Obudu areas in particular ways. In regard to maize and sorghum, efforts need to be concentrated on the derived savannah middle belt and certain parts of the upper north, so that the tree crop economy of the southern part would receive more attention.

The issue of agroecological specialization, however, has implication for market availability of the commodities on a nationwide scale. In this connection, it is worth mentioning that as a matter of deliberate policy, the practice of interstate restriction of trade in the commodities must be discouraged. Certain states of the federation sometimes impede the outflow of food items to other states, usually as a panic measure during shortages. This practice in itself is capable of working counter to the policy of agroecology specialization. This happens, as every state will attempt to grow everything, for fear that specially endowed states might later obstruct free flow of the relevant commodities. In the end, the whole country loses out in terms of resource use inefficiency, effort wastage, and low volume of national output.

Therefore, the practical implication of the absence of ecological specialization as a national strategy of agricultural development is that one cannot speak of self-sufficiency in maize or sorghum, or any other farm product for that matter, on a state-by-state basis. Nigeria should operate as one large indivisible market in which the surplus output of specially endowed producing areas should move freely to deficient areas without hindrance. It is therefore necessary now to zone the country according to commodity productivities along which development efforts will be applied, using efficiency criteria.[25]

Deficiencies in the Technology Policy

There are two relevant issues in respect of technology as far as maize and sorghum are concerned. First is the rate of technological development. In this regard, productivity-including technology must continue to evolve at a faster rate than it is now. This implies that new seed varieties and types must be bred fast to satisfy the different environmental conditions and market preferences. Also, new and better farming systems that solve agronomic problems must be introduced continuously. New machines and farm power sources that do not only remove drudgery but are cost-effective and fit into the edaphic and climatic features of the farming environment are also required.

There is presently the need to assess the available stock of technology as regards their suitability and sustainability. In addition, a collaborative effort of the research system and private sector is required in the technology development. The research system is to be responsible for ceaseless development of the available technologies as a matter of policy. We are yet to see the breweries, bakeries, and feed manufacturers, for instance, providing funds

---

[25] For instance, Kaduna State Government once issued instructions 'to prevent out-of-3/4 state traders from buying off and evacuation of this season's harvest of grains'. Consequently the state-owned Farmers' Supply Company (FASCOM) instituted a programme 'against all those who will buy grains cash down and carry away from the state'.

for research into breeding investigations in maize and sorghum to make these commodities more suitable for their different uses.

Input Policies that Distort Agricultural Markets

The relevant issue in this regard is the subsidy question. Originally, farmers are considered to deserve being pampered with subsidy support, owing to their limited purchasing power to demand improved technologies. However, the lessons of experience from the operation of direct subsidy programmes suggest that the policy might be self-defeating in many ways. This experience can be summarized in the Nigerian case as follows:

i. There has developed a dependency mentality among farmers on government subsidy, thereby making it a permanent obligation of government and a huge fiscal burden. There is the absence of proper price relativities, following price distortion created by subsidies, which is needed to direct production decisions of the farmers along the lines of efficient resource utilization.
ii. There is a low level of morale of the private sector participants to build up substantial enterprise initiatives in the distribution of inputs and outputs of agricultural production, owing to price rigidities with the consequent limiting effect on GDP. There is the presence of substantial externalities whereby farm subsidy benefits illicitly flow to unintended channels within and without the economy.
iii. The subsidy programmes are frequently exploited to serve the selfish motives of civil servants, politicians, among other elite groups, to the detriment of the ordinary farmers.[26]

The case is therefore presently strong in favour of indirect subsidy as the means of government support in agriculture. Consequently, the provision of rural infrastructure easily suggests itself as the viable alternative to heavy direct input subsidies. The argument for this is that once the infrastructure is installed (e.g. rural road network), the farmer possesses absolute, albeit not necessarily exclusive right of access to them (more on rural infrastructure shortly).

---

[26] The fertilizer subsidy programme is quite relevant to maize and sorghum. Selected budget allocations are as follows: ₦5 mill. (1976/77), ₦7.8 mill. (1977/78), ₦321 mill. (*199/90), ₦66 mill. (1980), ₦105 mill. (1981). See Idachaba (1981).

Pricing and Industrial Policies that Reduce the Profitability of Agriculture

We cannot overstress the importance of profitability of the maize and sorghum enterprises: that they pay is a precondition for speedy uptake of innovations. But, willingly or not, the government has often pursued policies which depress profitability of agriculture as a business. This limits the flow of investment funds there and also discourages quick adoption of innovations. Pricing policies are a case in point. Through the Nigerian Grains Board for instance, government had determined the price floors and ceilings for the maize and sorghum farmers. Essentially the board turned out to be a method of taxing the farmer to subsidize the consumer. Even though the stabilization of producer prices is justifiable and worthwhile, the activities of the agency have discouraged farmers from taking new market opportunities.

Policies of industrial protection produce similar effects. The incentives in agriculture are reduced by the worsening terms of trade of farmers through the rising prices of industrial inputs and consumer goods, which they buy relative to the prices of their produce. The manufacturing industries erected behind high tariff burdens and other forms of protection are not efficient, which makes farmers to pay high prices for often inferior goods. The pro-urban bias in the supply of basic infrastructure also affects profitability of agriculture. This happens by way of the relatively high agricultural costs, which result from the poor quality of education, health services, water supply, and electricity.

Discriminatory Policies in Favour of Direct Production and Capital Intensity

Government has erroneously felt that it could be a good farmer in several instances. Each time this happens, however, it eventually gets disappointed. Examples of these instances abound in the activities of agriculture ministries and the agencies like RBDAs.[27]

---

[27] The government factor in the Nigerian Grains Production Company is also inimical to profitability motives. With its high level of capital intensity and the extensive land areas under maize and other grains situated all over the country, the company has reported impressive performances, yet the underlying fact is that agricultures is a business. The present sluggish bureaucracy and other negative characteristics of government are not congruent with profitability motives. Therefore, the role of government in agricultural development in general should be limited to facilitation only.

Ineffective Grains Storage Policy

Storage is essential for the creation of time and place utilities from the marketable surpluses of maize and sorghum. These commodities particularly require an efficient storage system for the following reasons:

i. The bulk of their marketed quantities are generated from small-scale farmers who are widely dispersed geographically. This means that the multi-stage and multi-channel assembly points must be equipped with storage facilities to preserve quality over the required time.
ii. In general, the key consumers are located far away from the key producers. For example, Kaduna, Plateau, Benue, and Gongola States that are the leading suppliers of maize, and located in the northern part of Nigeria, the poultry industry that consumes the commodity as livestock feed is concentrated in the southern part of the country.
iii. Maize requires drying before storage, particularly first crop in the south, which is often harvested at high moisture levels.[28]

The initial government support in storage was the temporary buffer and reserve stocks of the old regional governments, which were operated to stabilize supplies and prices, and also as contingencies in case of production shortfalls. Traditionally, the bulk of the products are stored on farms, especially in cribs. The need for efficient storage system arose after the 1973/74 droughts, which made it necessary to build up a national reserve stock for food in general.

The Joint Federal Grain Storage Consultative Group and the FAO food security mission designed a Federal Grain Storage Scheme for Nigeria in 1975. Some of the recommendations of the body were the build-up of strategic federal reserved stocks and the operation of buffer stocks by states, through:

---

[28] For example in Ogun State (1984/85), government cultivated extensive areas to crops and established large livestock operations (including 300 heads of cattle imported from America) in its effort to demonstrate the profitability of agriculture 'if practiced in the modern way'. The general staff of the agriculture ministry, including top administrators, were drafted to perform farm operations like harvesting of maize. The effort failed on both output and profitability grounds, owing to long process of decision-making, heavy overheads, inadequate planning, and strict adherence to protocol. The product is in Lagos, Oyo, Ogun, and Bendel States. Similarly, sorghum, a predominantly northern production affair, serves the breweries that are concentrated in the south. Therefore, the distribution network of the commodities requires long time in storage.

i. The stores were empty most of the time because of the failure of the complementary grain price policy to attract purchase away from the open market.
ii. The states were ineffective in their buffer stock operations.

Concrete efforts should be made to implement the new agricultural storage strategies (FMAWRRD 1988), which includes the following:

i. The federal government should limit its role to the design of a suitable package of incentives to encourage the private sector in the erection and operation of on-farm storage.
ii. The local conditions should be considered in the development of storage facilities.
iii. The state governments should operate 10% of total grains produced in the country as buffer stock, and the federal government should operate 5% as strategic reserve.
iv. Research into storage technology should be accelerated to cope with the new demand pressures on maize and sorghum.

RURAL INFRFASTRUCLTURAL SUPPORT FOR STIMULATING MAIZE AND SORGHUM PRODUCTION

The provision of rural infrastructure is generally a public obligation for the following reasons:

- the lumpy expenditures on rural infrastructure are beyond the reach of ordinary individuals;
- there is marked divergence between infrastructural benefits that accrue to individual sponsors and those that accrue to the rest of the society, which leads to a free rider problem;
- rural infrastructure have intergenerational effects in regard to the benefits and maintenance costs;
- the provision of rural infrastructure cuts across many ministries and disciplines, thereby necessitating central coordination by government.

Two institutional arrangements for the provision of rural infrastructure can be distinguished. The first is the arrangement whereby each ministry performs its own separate functions in an unintegrated manner. Under this arrangement, the Ministry of Education establishes schools in rural markets. The Agriculture Ministry supplies electricity, etc. The major problem often encountered in this arrangement is the poor maintenance outfit. The furtherance of the separate activities is often left to the Local Government Authority to carry out, which happens to be the poorest tier of government, and which has low technical staff and limited financial capabilities.

The second institutional arrangement involves the provision of rural infrastructural facilities through special agencies in an integrated approach. In this case, specific infrastructural facilities towards definite functions are satisfied. The key agencies in this regard are (i) Agricultural Development Projects (ADPs), (ii) River Basin Development Authorities (RBDAs), and (iii) Directorate of Food, Roads, and Rural Infrastructure (DFRRI). The main features of these bodies have been summarized earlier in Table 2. Their poor performances in the provision of rural infrastructure are exemplified with available data on ADPs in Table 3.

As regards maize and sorghum development, a few issues on the rural infrastructure factor will now be examined.

There are three phases through which technological progress in agriculture could be initiated. These are (i) agricultural research for technology creation, (ii) diffusion of agricultural technology, and (iii) development of technological intellect. For meaningful progress to take place, determined efforts are required in these areas, which specifically include the following:

- the adoption of a technology policy that is all-embracing and complete,
- the creation of effective agricultural technology through a virile national agricultural research system,
- the transfer of agricultural technology through effective agricultural extension delivery,
- the perpetuation of existing technology level through the development of technological intellect,
- the adoption of appropriate agricultural technology on merit criteria, including ecological and sociocultural considerations, simplicity, relative availabilities of capital and labour, divisibility and riskiness for the particular application of small-scale farmers.

RURAL INFRASTRUCTURE SUPPORT

The view is widely held that provision of rural infrastructure is the essential foundation for sustained increases in food production. The general notion underlying the rural infrastructure strategy is that it is difficult for the rural sector to contribute significantly to economic and industrial progress in the absence of basic facilities that also enhance their production activities as well as their living standards. These facilities and their roles are generally described under three categories:

(i) Rural physical infrastructure, including

- rural roads which cause accelerated delivery of farm inputs, reduce transportation costs, and enhance spatial agricultural production efficiency;
- storage facilities which help to preserve foods in the forms that consumers need them and to the time they need them;
- irrigation facilities which assure farm water supply and stabilize food production by protecting the farm production system against uncontrollable and undesirable fluctuations in domestic food production.

(ii) Rural social infrastructure, including

- clean water, decent housing, environmental sanitation, personal hygiene, and adequate nutrition which help to improve the quality of rural life;
- formal and informal education, which promote rural productivity by making the farmer able to decide agronomic and other information and carry out other desirable modern production practices; basic education also promotes feeding quality, dignity, self-respect, and sense of belonging as well as political integration of the rural people.

(iii) Rural institutional infrastructure, including

- farmers' unions and cooperatives which facilitate economics of scale and profitability of rural enterprise,
- agricultural extension which improves the technology status of the farm business.

However, the rural infrastructural development process of Nigeria contains a number of persistent problems and unresolved issues. Ayoola and Idachaba (1989b) have highlighted some of these, including the following:

(a) Poor commitments of the programme sponsors,
(b) Incessant perturbation in the institutional framework,

(c) Low level of in-house operational research and planning of projects,
(d) The resource required for meeting infrastructure requirements,
(e) The trade-off between rural infrastructure supply and other objectives.

SUMMARY

The economic importance of maize and sorghum rests on their advantages as food and feed crops. Their shortage problem was described in terms of fluctuations of productions and availability against the existing prospects. The causes of this problem include: inadequate research capacity; inefficient channel of communication between farmers, extension agents and researchers; inaccessibility of farmers to new physical inputs that accompany new technologies; the lack of adequate training facilities needed for the farmers to acquire new skills in order to take full advantage of the new technology; poor state of rural infrastructure; shortage of farm labour at the peak of planting and harvesting seasons; and, a land tenure system which puts a ceiling on the size of farm and hence efficiency of utilization of available resources.

These give rise to salient issues within the broad policy framework; namely the absence of ecological specialization policy; deficiencies of technology policy; aberrant input policies that distort agricultural market; the pricing and Industrial policies that reduce the profitability of agriculture; discriminatory policies in favor of direct production and capital intensity; and ineffective grains storage policy. These coupled with the rural infrastructural development process of Nigeria that contains a number of persistent problems and unresolved issues as well, make the policy solutions to maize and sorghum shortage a nightmare to be addressed.

REFERENCES

Adegboye, R. O., A. C. Basu, D. Olatunbosun (1969), 'Impact of Western Nigeria Farm Settlement on Surrounding Farmers', *J. Econ. and Social Studies* I (2).

Abdullahi, Ango (1989), Keynote address, 25th Annual Conference of Agricultural Society of Nigeria, Owerri, August 1989.

Anonymous (undated), 'Format for 1984 grain purchase by FASCOM', Kaduna.

Ashwe, Chi Chi (ed.) (1988), *The Nigerian Economist*, 23 June 1988 edition.

Ayoola, G. B. (1988), 'The dependency problem and rationale for self-sufficiency in food grains', *J. Rural Dev. Nigeria* Vol. 3 (1).

Ayoola, G. B. (1989), 'Agricultural Produce Price Policy under the Structural Adjustment Programme: Proposal for Alternative Mode of Market Intervention', Proc. 25th Annual Conference, Agricultural Society of Nigeria, Owerri, August 1989.

Ayoola, G. B. and F. S. Idachaba (1989), 'Nigeria's Rural Infrastructure: Past and Present', Proceedings of Workshop on Social Mobilization, Poverty Alleviation and Collective Survival in Nigerian Society for International Development, Ibadan Chapter, University of Ibadan, 24–29 September 1989.

Blinder, Alan S. (1988), 'The Challenge of High Unemployment', Richard T. Ely Lecture, AEA papers and Proceedings, *Am. J. Agricultural Economics* 78 (2) (May 1988).

Faux, Jeff, 'A Cheaper Dollar is not Enough', *Challenge*, May–June 1988.

FMAWRRD (Federal Ministry of Agriculture, Water Resources and Rural Development (1988), Agricultural Policy for Nigeria 1988.

Galbraith, James K. (1988), 'Let's Try Export-led Growth', *Challenge*, May–June 1988.

Idachaba, F. S. (1989), 'Strategies for Achieving Food Self-sufficiency in Nigeria', Keynote address, 1st National Congress of Science and Technology, University of Ibadan, 16 August 1989.

Idachaba, F. S. (1981), 'A Farm Input Subsidy Policy for the Green Revolution Programme in Nigeria', Paper presented at the meeting of the Ministerial Committee on Farm Input Subsidies, Jos, 8–9 December 1981.

Lipsey, R. C., and K. Lancaster (1956), 'The General Theory of Second Best', *Rev. Econ. Studies* 24.

Martin, Ricardo and Marcelo Selowsky (1988), 'External Shock and the Demand for Adjustment Finance', *The World Bank Economic Review* 2 (1) (Jan. 1988).

Western Nigeria (1963), White Paper on Integrated Rural Development, Official Document No. 8, Ibadan, 1963.

World Bank (1983), *World Development Report*, Washington DC, USA.

World Bank (1986), *Poverty and Hunger: Issues and Options for Food Security in Developing Countries*, Washington DC, USA.

Table 1: Net Production and Net Availability of Major Grains, Nigeria, 1961–1984

YEAR	MAIZE Net Produc.	MAIZE Net Avail.	RICE Net Produc.	RICE Net Avail.	WHEAT Net Produc.	WHEAT Net Avail.	SORGHUM Net Produc.	SORGHUM Net Avail.
1965	870.0	870.2	173.3	174.7	11.3	47.3	3176.3	3216.3
1966	731.3	731.4	88.5	89.9	9.8	186.5	2370.0	2410.0
1967	562.5	562.6	227.3	228.5	7.5	129.3	2541.8	2581.8
1968	844.5	844.9	206.3	207.7	7.5	113.2	2427.0	2467.0
1969	941.3	941.4	211.5	211.8	7.5	182.9	3039.0	3111.8
1970	1082.3	1091.1	209.3	209.9	5.3	264.0	3039.0	3079.0
1971	952.5	956.4	210.0	202.8	5.3	364.2	2845.5	2885.5
1972	478.5	480.9	333.8	334.0	5.3	301.9	1723.5	1732.5
1973	920.3	922.3	364.5	370.4	3.0	311.0	2343.0	2345.0
1974	396.0	398.0	401.3	402.9	4.5	322.5	3549.0	3550.0
1975	1022.3	1024.3	482.3	486.3	13.5	330.5	2194.5	2195.0
1976	806.3	823.3	163.5	170.2	15.0	750.5	2212.5	2212.5*
1977	582.8	619.6	306.0	351.0	15.8	735.4	2494.5	2494.5*
1978	368.5	425.5	127.5	540.5	15.8	894.6	1797.0	1799.0
1979	368.3	408.8	117.0	681.7	15.8	820.6	2097.3	2478.3
1980	483.0	651.0	75.0	642.9	18.0	1201.0	2477.3	2478.3
1981	479.3	624.0	135.0	1151.0	18.8	443.8	2651.3	2651.3*
1982	469.5	814.5	152.3	445.1	18.8	1624.0	3060.8	3060.8*
1983	770.0	870.0	74.3	774.3	24.0	1522.3	3173.3	3173.3*
**CV	33.7	28.4	52.6	63.6	50.5	82.9	18.5	18.0

Source: Underlying data from Federal Office of Statistics, Lagos

* Quantities of sorghum imports are not available for these years. However, they are not expected to make significant differences in the net production figure because this crop is generally a non-traded commodity.

** CV means Coefficient of Variation = $\dfrac{\text{Standard deviation}}{\text{Mean 1}} \times 100$

Table 2: Selected Strategies of Agricultural Development with Substantial Relevance to Sorghum and Other Grains, 1960–1988.

Strategy	Relevant details
1. Rice/Maize Project	A programme of the old Western Region.
2. National Accelerated Food Production Project (NAFPP)	Ongoing, designed to enhance the technical efficiency of the farmers in respect of selected crops, including sorghum policy—instruments include subsidy, credit, and adaptive research as well as demonstration plots.
3. River Basin Development Authorities (RBDAs)	Originally established (first 11, later 18, now 11 in number) to undertake development activities in their respective areas. They are not crop-specific.
4. Operation Feed the Nation (OFN) programme	A mass mobilization campaign towards increased agricultural production (launched in 1977). Grain production received encouragement under OFN.
5. National Fertilizer Board	To undertake bulk purchase of fertilizer and distribute them to states (established 1977). It benefits grain production immensely.
6. Nigerian Grains Board	One of the seven commodity boards (established 1977) to undertake the marketing and development functions of maize, sorghum and other grains.
7. Agricultural Credit Guarantee Scheme (ACGS)	A scheme to facilitate credit availability to the farmers, administered by Central Bank of Nigeria (commenced 1977). It is not crop-specific.
8. Green Revolution Programme	Launched in 1977, to accelerate the attainment of self-sufficiency in food, particularly grain production.
9. National Grains Production Company (NGPC)	Established in 1979, to expand food grain production, provide efficient storage and handling facilities, develop food grain processing as well as an efficient marketing system. Enter into joint venture production operation.
10. National Cereal Research Institute	To conduct research into grains.
11. Nigerian Agricultural Insurance Agency	To offer protection to farmer from the effects of natural disasters, established 1988. It is not crop-specific.

Source: Miscellaneous documents

Table 3: Performance of ADPs in the Provision of Rural Roads, Selected Cases, 1986

ADP/Period	% Target achieved in rural road targets		
	Constructed	Maintained	Rehabilitated
Recent Projects			
Anambra State ADP (1986)	14.7	0	0
Bendel State ADP (1986)	NA	0	0
Benue State ADP (1986)	NA	9.1	0
Cross River State ADP (1986)	NA	0	NA
Imo State ADP (1986)	14.7	47.8	73
Plateau State ADP (1986)	30.6	5.3	-
Ogun State ADP (1986)	NA	NA	NA
Older Project			
Bida ADP (1981–86)	69.6	114.6	329.7
Borno ADP (1983–86)	23.7	NA	NA
Sokoto ADP (1983–86)	70.8	NA	NA
Gongola ADP (1983–86)	22.6	NA	NA
Ilorin ADP (1981–86)	26.7	91.3	NA
Oyo North ADP (1983–86)	80.6	-	48.4
Bauchi ADP (1983–86)	NA	NA	NA
Kaduna ADP (1985–86)	34.3	25	NA
Kano State ADP	55.7	58.3	40
Ekiti Akoko ADP (1983–86)	73.9	41.6	47.5

Source: Underlying data in km length of road from APMEPU (1987)

The Nigerian Grain Market and Industrial Utilization of Sorghum: A Perspective

Invited paper presented at the Joint ICRISAT, WASIP, and IAR symposium on the current status and potentials of Industrial Uses of Sorghum in Nigeria, Conference Room, Central Hotel Kano, 4–6 December 1989.

The grain economy of Nigeria was characterized by persistent excess demand for the most part of the 1970s. This was particularly related to the developments in poultry, bakery, and brewery, which were stimulated by the petroleum-based increases in per capita income during the period. Massive grain imports were necessitated by the failure of the available production system to create compensating supplies that could mop up the surplus demand in the grain market. However, the increasing trend of grain imports has now been stemmed under the ongoing Structural Adjustment Programme. As a result, the grain-based industries have had to embark on aggressive local sourcing of raw materials. This in turn has led to substantial structural changes in the affected industries, most of which have mounted programmes in backward integration. The prominent grains involved are maize, wheat, and sorghum.

This paper presents an economic perspective of the general grain market and the increasing industrial utilization of sorghum in Nigeria. The demand and supply sides of the grain market are first explored. This is followed by a survey of the main intervention instruments in the market. Finally, the implications for industrial utilization of sorghum are discussed before the paper is concluded.

FEATURES OF THE GRAIN MARKET

The important aspects of the supply side of the grain market are discernible trends in net production and net availability of major grains, the supply sources, output elasticities, as well as the supply price. On the demand side, the issues to consider are the utilization of grains, substitutability in grain commodity space, and demand price. Therefore, the market for food grains, in which the increasing utilization of sorghum will be undertaken, is the result of linear and non-linear combinations of the interacting supply and demand factors.

Supply-side issues

The net production and net availability of grains show considerable annual fluctuations.[29] Table 1 presents this picture for the major grains. As regards maize, variability is 34% for net production, and 28% for net availability. The data reveal widening gaps between net production and net availability from 1975 to 1983, which indicate the high magnitudes of sufficiency shortfall in maize and also imply an upwards trend dependency ratio. Similar features hold for rice. The case of wheat is unique because the volumes of domestic production are insignificant compared to those of imports over the entire period illustrated. In the case of sorghum, the small divergences between net production and net availability which existed up to the beginning of the 1970s soon disappeared. This does not necessarily mean the attainment of self-sufficiency in sorghum but rather suggests that the pent-up excess demand on sorghum has been wholly absorbed by maize imports, which heightened during the period. The high substitution effect probably resulted from the relatively low price of imported maize compared to the price of sorghum as a non-traded commodity.

The bulk of the domestic supplies of grains come from the small-scale farmers. These producers operate under limitations imposed mostly by poverty and inadequate knowledge. Therefore, they require substantial assistance in funding, innovations, and infrastructure. To the extent that the shortage of these support facilities exist, production has been grossly

---

[29] The reason for talking in net terms is to make allowances for seed and wastage, which do not generally enter the markets of these commodities as food and raw materials. Imports complement domestic production to make up availability (total supply) resulted from the relatively low price of imported maize compared to the price of sorghum as a non-traded commodity.

inefficient, and marketing, including storage and processing activities, has been greatly constrained to create the desired form and place utilities. Consequently, supply of grain continuously lags behind demand, thereby making the massive importation inevitable following improvements in the average income position.

The generally low output elasticities of grains with respect to the available economic incentives are a composite outcome of several factors. These include, among others,

- ineffective agricultural extension system that fails to enhance farm productivity through improved access of farmers to productive resources,
- failure of agricultural research to meet the challenge of fast technology creation and improvements necessary for efficient farm production,
- slow rate of production and poor management of the agricultural intellect necessary to replace the ageing farm population and also meet the demand for agricultural expertise of the urban economy,
- production policies which feature incessant perturbations and hence stop-go execution of important projects.

The supply price is directly related to the cost of production. In the case of grain, programming efforts to keep production cost low are substantial as shown in Table 2. These programmes contain elements of heavy input subsidies, concessional interest rates, and other support services. However, from the experience of perennial support to grain production, it is conclusive that the sustainable means of keeping the supply price is low through productivity improvements. Productivity increases are desirable in this regard because they bring about more output for given input levels, thereby reducing average cost of production. Therefore, supply price of grains is too high relative to normative standards because productivities of land, labour, and capital factors remain too low.

Demand-side issues

The demand for cereals primarily stems from their advantages as food and feed crops. This results from the fact that their grains have high caloric value and are so highly digestible. Consequently for example, they serve as light and heavy meals as well as alcoholic drinks, among other uses. For example, thirty-four different forms of utilizing maize have been identified (Akinwumi 1970). In addition, the industrial utilization of maize includes the production of flour, sugar, starch, alcohol, and spirit. Other value added resulted from milling of corn. In more advanced countries, sweet corn is canned variously, using additives to preserve the freshness and sugar content. Furthermore, the whole plant of maize, sorghum, and other grains has immense utilities in livestock rations, including the feeding of the stovers to hogs, cattle, poultry, horses, and sheep. Residues of the cannery such as husks, cobs, shanks, and silks have been used in making silage and soilage. Other uses of grain plants also include making of corncob pipes and cellulose for home fuel.

The potential utilization of sorghum, as a particular example, derives from the presence of high degree of substitutability in the grain commodity space. This means that all the major grains can be perfectly substituted for one another in the domestic and industrial uses. In effect, grain prices are likely to sympathize with one another through substitution effects in the long run. The industrial users of grains thus, have a great advantage to explore here. Efficient input demand management in the industries will depend on the availability of a flexible production system which can shift from a grain in short supply to that in high supply at particular times. In domestic uses, this flexibility is already present following the observation that grain-based foods, and tubers, have largely substituted for wheat bread.

This level of demand price relates to the issue of flexibility in consumption behaviour among grain users. The upwards effect of competition among users will be minimized if the numbers of grains that are being used for particular products increase. In the absence of this, it is normal to expect that competition of limited supply of particular grains will raise the demand price of grains. This, when combined with the factor of high production cost earlier discussed, creates another upwards pressure on the market price of grains, apart from that already created by supply price.

MARKET INTERVENTION POLICY

The Nigerian grain market contains other forces than demand and supply. Government also intervenes for the purpose of moderating the effects of the natural forces. The most pertinent issue in this regard centres on the defunct Nigerian Grains Board (NGB), which was established in 1976 as part of the third-generation intervention boards system that fixed prices for agricultural products under the activities of the Technical Committee on Produce Price (TCPP) and the Price Fixing Authority (PFA). In regard to grains, guaranteed minimum prices (GMPs) were fixed on an annual basis. GMP was a floor price at which the Grains Board was obliged to buy directly from farmers. This arrangement was aimed at mopping up the excess output immediately after harvest to disallow prices from falling below desired limits. The purchases would be held as buffer or strategic stocks, which would be resold at hitherto unstated prices at excess demand periods, in order to prevent prices from rising too high. The entire intervention arrangement was expected to be capable of maintaining farmers' incomes at desired level, and to prevent price increases from exceeding affordable limits. However, owing to the several problems encountered (Ayoola 1989 *op, cit.*), commodity boards system was scrapped at the onset of the Structural Adjustment Programme, as part of a deregulation exercise.

Therefore, the new challenge facing the industries towards increased utilization of sorghum is to be met in a relatively free-enterprise grain market. Even though this policy of non-intervention itself has not been stabilized, the new agriculture will be limited to facilitating function in production only.

IMPLICATIONS FOR INDUSTRIAL UTILIZATION OF SORGHUM

This section briefly highlights four implications which the foregoing discussions have for the increasing utilization of sorghum.

Backwards Integration—This is a type of vertical integration whereby a firm extends itself into a previous stage of the production process. Therefore, the current effort of breweries and industries to establish large-scale grain farms is an attempt in backwards integration. These attempts are made in response to the need for local sourcing of raw materials, following the import restrictions of the Structural Adjustment Programme. These efforts need to be encouraged because backward integration promises reduction in working capital requirements, elimination of prohibitory transaction cost, and greater price competitiveness as a result of avoiding successive profit markups. Therefore, the industries should intensify efforts in the overall production system.

However, this proposition places heavy demand on the industries' capital structure. Heavy capital outlays are required to be incurred by the industries to achieve efficient farm organization, including the purchase of machinery and other capital items. In order to avoid this, industries may adopt the contract-growing alternative, which, apart from eliminating heavy capital cost, also preserves the structure of existing land ownership patterns.

Need for Productivity Improvement Programmes—In the present setting, the participation of industries and extension is minimal. The current drive to increase grain supply to close the existing supply-demand gap requires industries to play definite roles in these areas. Industries should partake actively in the research process that yields new technologies such as the breeding of sorghum to impart specific desirable traits. As regards the role of industries in agricultural extension, the case of the Nigerian Tobacco Company Limited shows that a single farm can secure its raw material base for many years through product-specific extension programmes.

Diversified Input Base—Industries need to provide internal mechanisms for avoiding the negative effects of socio-economic shocks in the supply of grains. One way to achieve this is to plan a multiple-input production system. The presence of substitutabilities among grains makes this possible. A mono-grain industrial system will lead to large business failure in the event of disease outbreak to climatic problems that significantly reduce output of the particular grain.

Free-Enterprise Grain Market—A free-enterprise grain market as it now exists implies that increased competition will result between domestic and industrial users. This is expected to create an upward pressure on the market price of sorghum and other grains. Consequently, industries are expected to improve their individual and aggregate resource-use efficiency. This is necessary to enable them produce at competitive prices in the product market, which in turn determines the level of profit margins that accrue to the industries.

CONCLUSION

The demand and supply sides of the grain market have important implications for increasing the industrial utilization of sorghum. In particular, sorghum, which is traditionally a non-traded commodity, requires successful programmes in backwards integration and productivity improvements. The existence of substitutability among grains also offers the advantage of diversified input base to the industries. Finally, the present policy of non-intervention in the grain market demands high production efficiency in the use of sorghum as industrial raw materials.

REFERENCES

Akinwumi, J. A. (1970), 'Economics of Maize Production in Oyo Division' M. Phil. Thesis, Department of Agricultural Economics, University of Ibadan, Nigeria.

Ayoola, G. B. (1989), 'Agricultural Produce Price Policy Under the Structural Adjustment Programme: Proposal for Alternative Mode of Market Intervention', presented at the 25th Annual Conference and Silver Jubilee Celebration of Agricultural Association of Nigeria, Federal University of Technology, Owerri (limited distribution).

Nigeria, Federal Ministry of Agriculture, Water Resources and Rural Development, 1988. *Agricultural Policy for Nigeria*.

Table 1: Net Production and Net Availability of Major Grains, Nigeria 1961–1984

YEAR	MAIZE Net production	MAIZE Net Avail.	RICE Net production	RICE Net Avail.	WHEAT Net production	WHEAT Net Avail.	SORGHUM Net production	SORGHUM Net Avail.
1965	870.0	870.2	173.3	174.7	11.3	47.3	3176.3	3216.3
1966	731.3	731.4	88.5	89.9	9.8	186.5	2370.0	2410.0
1967	562.5	562.6	227.3	228.5	7.5	129.3	2541.8	2581.8
1968	844.5	844.9	206.3	207.7	7.5	113.2	2427.0	2467.0
1969	941.3	941.4	211.5	211.8	7.5	182.9	3039.0	3111.8
1970	1082.3	1091.1	209.3	209.9	5.3	264.0	3039.0	3079.0
1971	952.5	956.4	201.0	202.8	5.3	364.2	2845.5	2885.5
1972	478.5	480.9	333.8	334.0	5.3	301.9	1723.5	1732.5
1973	920.3	922.3	364.5	370.4	3.0	311.0	2343.0	2345.0
1974	396.0	398.0	401.3	402.9	4.5	322.5	3549.0	3550.0
1975	1022.3	1024.3	482.3	486.3	13.5	330.5	2194.5	2195.0
1976	806.3	823.3	163.5	170.2	15.0	750.5	2212.5	2212.5*
1977	582.8	619.6	306.0	351.0	15.8	735.4	2494.5	2494.5*
1978	368.5	425.5	127.5	540.5	15.8	894.6	1797.0	1799.0
1979	368.3	408.8	117.0	681.7	15.8	820.6	2097.3	2478.3
1980	483.0	651.0	75.0	642.9	18.0	1201.0	2477.3	2478.3
1981	479.3	624.0	135.0	1151.0	18.8	443.8	2651.3	2651.3*
1982	469.5	814.5	152.3	445.1	18.8	1624.0	3060.8	3060.8*
1983	770.0	870.0	74.3	774.3	24.0	1522.3	3173.3	3173.3*
**CV	33.7	28.4	52.6	63.6	50.5	82.9	18.5	18.0

Source: Underlying data from Federal Office of Statistics, Lagos.

* Quantities of sorghum imports are not available for these years. However, they are not expected to make significant differences in the net production figure because this crop is generally a non-traded commodity.

** CV means coefficient of variation =

$$\frac{\text{Standard deviation}}{\text{Mean}} \times \frac{100}{1}$$

Table 2: Selected Strategies of Agricultural Development with Substantial Relevance to Sorghum and Other Grains, 1960–1988

	Strategy	Relevant details
1.	Rice/Maize Project	A programme of the old Western Region
2.	National Accelerated Food Production Project (NAFPP)	Ongoing, designed to enhance the technical efficiency of the farmers in respect of selected crops, including sorghum policy. Instruments include subsidy, credit, and adaptive research as well as demonstration plots.
3.	River Basin Development Authorities (RBDA)	Originally established (first 11, later 18, now 11 in number) to undertake development activities in their respective areas. They are not crop-specific.
4.	Operation Feed the Nation (OFN) programme	A mass mobilization campaign towards increased agricultural production (launched in 1977). Grain production received encouragement under OFN.
5.	National Fertilizer Board	To undertake bulk purchase of fertilizer and distribute them to states (established 1977). It benefits grain production immensely.
6.	Nigerian Grains Board	One of the seven commodity boards (established 1977) to undertake the marketing and development functions of maize, sorghum, and other grains.
7.	Agricultural Credit Guarantee (ACGS)	A scheme to facilitate credit availability to the farmers administered by Central Bank of Nigeria (commenced 1977), it is not crop-specific.
8.	Green Revolution Programme	Launched in 1977 to accelerate the attainment of self-sufficiency in food, particularly grain production.
9.	National Grains Production Company (NGPC)	Established 1979 to expand food grain production, provide efficient storage and handling facilities, develop food grain processing as well as an efficient marketing system. Entered into joint venture production operation.
10.	National Cereal Research Institute	To conduct research into grains.
11.	Nigerian Agricultural Insurance Agency	To offer protection to farmers from the effects of natural disasters, established 1988. It is not crop-specific.

Source: Miscellaneous documents

16

Accelerated Soyabean Production in Nigeria: Strategies and State of the Art

Paper presented at the Soyabean Production and Utilization Workshop, organized by National Productivity Centre and held at Benue Hotels, Makurdi, 25–28 September 1991.

The current strategic importance of soyabean as a crop is the result of several forces arising from the structural adjustment policies. As a primary input in the vegetable oil, dairy, and feed industries, production of soyabean receives the impetus that is generally provided for local sourcing of raw material. The deregulation programme meant the freeing of soyabean production and marketing from the operations of the defunct Commodity Boards System, which could boost the morale of producers as income rises. The restriction of agricultural imports, especially the outright ban of grain import, leads to the determined exploration and exploitation of the available substitution effects among local products, including soyabean. These factors constitute the major ways in which soyabean fits into the overall framework of the current reform, which has the concept of self-reliance as the underlying theme of economic recovery.

This paper highlights the strategies in use toward sustainable increases in soyabean production and also describes the current status of the soyabean economy of Nigeria. Policy implication of the issues is drawn to provide the basis for some recommendations.

STRATEGIES FOR ACCELERATING SOYABEAN PRODUCTION

Technology Inducement

The International Institute for Tropical Agriculture (IITA) is in the forefront of efforts to develop appropriate processing technology (see IITA 1990). A number of prototypes for threshing soyabean are already at mass production state.

Other important agencies in this regard include the Rural Agro-Industrial Development Scheme (RAIDS), the universities and research institute. The Federal Ministry of Science and Technology (FMST) provides suitable backstop and financial facilities for the research institutes to investigate and develop small-scale machines for reducing the labour intensity of soyabean production. A 'Nationally Coordinated Research Projects on Soyabean' was launched through these means as the main thrust behind genetic improvement of the crop and general technology development concerning its production, processing, and utilization. This project is executed under the auspices of Nigerian Soyabean Association in close collaboration with some research institutes, universities, and the Agricultural Development Projects (ADPs).

A number of varieties have been released which undergo continual evaluation for the retention of desirable traits and development of superior traits with time in varied agroecological environments. Table 1 shows samples of promising varieties based on their yield performances under evaluation experiments in 1991. The University of Agriculture Makurdi (UAM) was one of the institutions that displayed prototype machines for soyabean threshing at the 1991 conference of NSA held at Benin City. UAM also has a formidable soyabean research team, which conducts investigations into genetic improvements, utilization, and socio-economics of soyabean enterprise. The governments of the northern states have jointly approached UAM for its team to reactivate the Yandev research station for research in soyabean production, processing, marketing, and consumption. When this proposal is actualized, it will surely supplement other efforts in generating product and process technologies for enhanced efficiency of the soyabean industry.

Demand Pull

The awareness campaign that accompanied the discovery of strategic importance of soyabean was substantial. Government mounted **sensitization** programmes of that soon made soyabean a household name. The uses of the commodity, as a proteinous crop, were established for reducing malnutrition among growing children, pregnant women, and lactating mothers. The **various**strategies of effecting awareness include school programmes, radio jingles and discussions, posters as well as market information dissemination channels.

A remarkable effort was also made to diversify domestic and industrial use of the crop. The IITA and agriculture ministries distributed booklets that show various recipes for preparing soyabean-based household menus such as the soyabean-maize pap (soyogi), soyabean cake or paste (soy-akara, soy moin-moin), to mention a few. In particular, positive response was substantial among women in the use of soyabean for milk production as a substitute for conventional milk products, which were scarce following economic reforms. Government sponsors promotion of these products through displays and training sessions at field days, conferences, trade fairs, and other meetings in the urban and rural areas. The Better Life programme now facilitates the employment of women in small-scale soyabean -based enterprises.

Considerable industrial capacity has been generated through promotion of private enterprises and joint venture investment involving government to increase the value-added utility of soyabean. This has found uses for the crop in the dairy feed and vegetable oil production systems. The largest production capacity was create through the establishment of Taraku Limited in Benue State. This plant has capacity to process 72,000 metric tonnes of soyabean into about 10,000 metric tonnes of refined soyabean oil, 61,000 tonnes of soymeal, and 9,000 metric tonnes of soap per annum (Riedel 1991). At the present, Taraku Golden Soya is a famous high-quality premium vegetable cooking oil based on established international standards.

Ecological Specialization

The principle of ecological specialization seeks to tap special production potentials of particular areas and special environment niches as a matter of policy (FMAMNRD 1988). This involves the concentration of efforts on places where comparative advantage is greatest. This is the logical basis for the sharp focus on Benue State that has demonstrated immense capabilities and potential in soyabean production in the country. According to FOS (1983) for instance, 4,000 ha. (80%) was put to soyabean in Benue State (1980/81 session) out of the total of 5,030 ha. put to the crop nationwide. The outputs of soyabean in that year were 2,000 tonnes and 22.40 tons, respectively. Therefore, a strategy of accelerated production should explore the potential of the farming systems in the state to generate large incremental soyabean output.

An extra-budgetary allocation of five million naira was granted to the state recently, following its being officially declared as a special soyabean-producing area by the Federal Government. Although there is large room for further actions to give practical impact to the ecological specialization principle, this effort could enhance the competence of the state in the national and international soyabean market.

Production Support

The Federal Department of Agriculture (FDA) provides the lead to give facilities to soyabean farmers. In general, this forms part of the regular programmes of the Agricultural Development Projects (ADPs) but could also take the form of special production subsidies and schemes. FDA has an ongoing programme of soyabean seed production through the National Seed Service (NSS) and select universities and research institutes. The programme is aimed at sustaining the tempo of seed demand consistent with the expansion of land area under soyabean cultivation.

A number of states have been identified as soyabean-producing states where support efforts are primarily concentrated to generate new soyabean production initiatives. The ADPs in these states provide extension services, infrastructure and input services to farmers. The FDA is also undertaking a national survey to provide information needed for continuous assessment of soyabean production in the country. Other areas of FDA support functions include training programmes, which are conducted in collaboration with specialized agencies (see UAM 1991), as well as donation for holding soyabean conferences.

Group Action, Collaborative Effort

The National Soyabean Association (NSA) provides a forum for interaction of individuals, enterprises, and agencies involved in the soyabean industry. In this medium, for instance, the scientists working on soyabean are able to derive their research priorities directly from the problems of farmers expressed through the ADPs experience or by farmers themselves, and the felt needs of processors, marketers, and consumers through events taking place in the newsletters and proceedings of conferences/meetings. Lately, the *Tropical Oilseed Journal* has been established by the NSA to provide means of communication with the international scientific community on soyabean matters.

To corroborate national efforts and extend the benefits of experience beyond national boundaries, an initiative is under way to form the African Regional Soyabean Network (ARSN). A planning workshop on this subject was jointly sponsored by the Food and Agricultural Organization (FAO), International Institute for Tropical Agriculture (IITA), and National Cereals Research Institute (NCRI) in January 1991. At full establishment stage, the ARSN could pool mental resources together for more efficient soyabean production and utilization systems in the African region. Scientists could also forge effective collaborative relationships that would yield fast results based on cross-fertilization of ideas and sharing of international experiences.

Conflicts of Demand and Supply

Observations made on soyabean market in the past five years indicated the occurrence of temporal and regional conflicts between demand and supply. In Oyo State, for example, where production efforts were successful in 1986 season, the market demand was relatively low as it was mostly based on atomized household demand. The dairy industries in the surrounding urban centres discriminated against the available supplies, which failed, in most cases, to meet standards of seed colour, purity, and varietal specifications. The resulting market situation compensated households, who enjoyed cheap products and penalized farmers, who sold at low prices. Government of this state had to come to the rescue of farmers, as buyer of last resort when farmers did not find outlet to dispose of soyabean produce. The government also offered floor prices only; production initiatives of the subsequent years suffered immensely, leading to low level of output.

About the same time, a similar development was taking place in the soyabean market of Benue State. There, farmers who produced with the enthusiasm of having the Taraku plant close to them were disappointed on grounds of specifications. Consequently, price fell because of glut at the same time when Taraku blamed its low-capacity operation on lack of raw materials. This situation has improved somewhat, owing to a contractual growing scheme which matches farmers' production decisions with specific demand of the processing industry.

Profitability Assessment

Available evidence indicates that soyabean production is associated with modest financial gains. Ayoola and Ibrahim (1989) report some estimates of cost and returns of soyabean production in the Gboko area of Benue State. The profitability assessment indicator varied with farm size category. Gross margin per unit of inputs was higher for larger farm sizes (greater than 10 ha.) except for labour. Gross margins are also different from one input to another in each farm size category.

On average, planting of improved seed and correct seed rate is most promising among the class of inputs. The apparent low level of fertilizer application to soyabean could explain the lower level of return to fertilizer. The labour-intensive nature of production, especially in the aspect of harvesting, accounts for the lowest return per man-day. The latter situation implies the presence of large room for enhancing income from soyabean production through the provision of suitable machines for harvesting.

Miscellaneous Issues

In the present state of the art about soyabean production and utilization, several issues call for attention for resolutions. First is the short shelf life of the milk produced from soyabean. Women are not encouraged to engage in the time- and energy-consuming process of making soyabean milk for home use, which could only be stored for 2–3 days even in the fridge. In fact, soymilk becomes slimy and unfit for consumption within hours after production at ordinary temperature. The attitude seems to be that the small quantity of milk needed to table in one or two mornings is not worth the efforts of long cooking, grinding, and serving. The acceptance of soymilk is also complicated by the presence of bean odours in the product.

Another issue is that of composite nature of domestic soya productions. Preliminary observations on the attitude of households to soyabean indicate that the inability of soyabean to be consumed alone is important in its domestic utilization. It cannot be simply cooked and eaten on its own, unless mixed in varying proportions with maize or cowpea. The practical effect of this property is that soyabean use is only common in enlightened homes where the health and nutritional qualities are known. The rural people, it appears, are not too convinced about the need for using composite mixtures for pap, akara, moin-moin, and other items when whole maize and whole cowpea are readily available, sometimes at less cost and less trouble. Therefore, there is more work to be done towards the use of soyabean in homes with greater simplicity and independence comparable to other food items.

Furthermore, it seems that the presence of anti-nutritional factor in soyabean is little appreciated by users. Boiling of soyabean, which removes the nutritional inhibitor, reduces its competitiveness for use as livestock feed at the small-scale levels compared to maize and other crop concentrates. It is common to observe the feeding of soyabean without boiling, which automatically equates the feed value of the crop to that of roughage as the bioavailability of the protein is greatly inhibited in this mode of use.

POLICY IMPLICATIONS

The sustainability of increased soyabean production in Nigeria depends on three important dimensions of policy. The first concerns the role of demand management policies to keep consumption growing without falling over time. Demand, being a function of income, will be sustained if and only if the ruling price is affordable. In this connection, the crucial issue is whether soyabean will remain cheap relative to substitutes as an input for domestic and industrial products. Even at the current high rate of protection of the economy, the Golden Soya cooking oil, for example, is being sold at prices too unaffordable for demand to be a strong driving force of high-capacity utilization. Therefore, the raw material crisis faced at Taraku Mills may be due more to the absence of aggressive sourcing of the crop, usually necessitated by excess demand for final output, rather than the inability of farmers to respond to policy incentives to increase supply capacity of soyabean.

Unless appropriate policy instruments are employed to sustain an upwards trend in domestic and industrial consumption of soyabean, idle capacity will induce a vicious cycle whereby low demand for final product will cause low-capacity utilization at home and industries; an excess supply situation is inevitable in the long run which could dampen farmers' morale and, hence, forcibly reverse the trend of increasing national soyabean output.

Another issue of the current state of affairs concerning soyabean is the policy implication of competition resulting from increased soyabean production. Competitive situation arises from the limited available resources for producing all agricultural commodities. It stands to reason that the land area under other crops will diminish to accommodate increased soyabean production. This is most likely so because most small-scale farmers either displace other crops in the mixture or plant soyabean solely on land that was originally put to other crops. In both cases, the farmer undergoes partial budgeting only, which does not necessarily involve capital expenditure on new land clearing. On the other hand, observation in Benue State suggests that the collective output of large-scale farmers, who could fault this mode of production, is relatively small.

The major policy implication of the displacement of other arable crops by soyabeans is that the benefit stream of the soyabean programme is implicitly discounted by the opportunity costs in terms of reductions in the output of other crops. Therefore, the farmer and society may be better or worse off, not only to the extent of the emergence between market price and shadow price of soyabean, owing to direct and indirect subsidies, but also the magnitude of opportunity cost of pushing soyabean production in terms of these other crops. In the light of this issue, production policies that stimulate the production of soyabean in new areas should be accompanied by compensating input services, including land development fertilizer and agrochemicals.

The last issue of policy concerns the distribution consequences of increased soyabean output. In the short run, it appears that soyabean-based cropping systems could generate substantial income increments of farmers. In that case, and considering the concentration of potential in a region of the country, policy efforts might be required to ensure redistribution of the gains of soyabean production across crops and regions. Otherwise, there will be a strong tendency of farmers to hop from crop to crop in disregard of the relative comparative advantages associated with certain ecological zones of the country. Apart from the wasteful duplication of resources evident in regional and temporal shifts in the production functions of specific regions, the policy of ecological specialization could easily be frustrated in the process.

CONCLUSIONS

In the past five years, various strategies have been implemented towards accelerated production of soyabean in Nigeria. These include technology inducement, awareness campaigns, promotion of ecological specialization, and the traditional support services such as extension and input services. In effect, these strategies have modified the status of the soyabean economy considerably, especially in regard to the structure of demand and supply sides of the market, profitability of soyabean enterprises among other aspects.

However, the sustainability of further progress in the soyabean industry plays a major role to policy to manage demand for final products efficiently, resolve the trade-offs between soyabean and other essential crops effectively, and address the distributional consequences of increasing soyabean production equitably.

REFERENCES

1. Ayoola, G. B. and J. E. Ibrahim (1990), 'Soyabean Production in Benue State of Nigeria', paper presented at the Annual Conference of the National Soyabean Association, Badeggi, Nigeria.
2. National Soyabean Association (1990), 'Report on Evaluation of Soyabean Varieties', presented to the General Meeting at the 1991 Conference, Benin City, Nigeria.
3. IITA (International Institute of Tropical Agriculture) (1990), Soyabean Research at IITA, GLIP Research Monograph No. 2, Ibadan.
4. University of Agriculture, Makurdi (1991), 'Soyabean Production and Utilization Training Workshop', Cooperative Extension Centre (CEC) Training Workshop Series No. 2.
5. FOS (Federal Office of Statistics) (1983), Nigeria National Integrated Survey of Households, Lagos.

Table 1: Some promising varieties of soyabean and their yield performances under evaluation experiments, 1991

Variety	Yield Performance (kg/ha.)	Maturity Days
	EARLY	
TGX 536-02D	1606	97
TGX 1497-1D	1440	94
TGX 1485-1D	1387	92
TGX 1019-2EB	1348	101
TGX 1566-2E	1345	92
	MEDIUM	
TGX 1448-2E	1821	111
TGX 1440-1E	1734	114
TGX 1448-1E	1656	112
TGX 1458-2E	1616	112
TGX 1455-2E	1595	111
Sampsoy 2	1504	110
TGX 1447-2D	1486	113
M-351	1475	112
TGX 1437-1D	1451	111
TGX 1447-1D	1438	110
TGX 1113-3D	1269	110
TGX 849-313D	1240	109
	LATE	
TGX 1596-2E	1282	117
TGX 923-2E	1232	122
TGX 1445-3E	1195	120
TGX 1445-2E	1173	124
TGX 1470-1D	1124	124
Sampsoy 1	921	116
Malayan	508	129

Note: Less than 100 days = early maturity, 100–120 days = medium maturity, above 120 days = maturity of varieties

Source: Nationally Coordinated Soyabean Project, 1991

17

Nigeria's Rural Infrastructure: Past and Present

Invited paper presented at the Workshop on Social Mobilization, Poverty Alleviation and Collective Survival in Nigeria, organized by Society for International Development, Ibadan chapter, Nigeria; Conference Centre, University of Ibadan, 24–27 September 1989.

In spite of the preponderance of rural infrastructural development in the programmes of the present government, still pessimism is widespread. The pessimism prevails because significant improvements in the living standards of the rural people have not been observed in order to justify the huge budget and non-budget support to the rural sector. However, the general belief remains in the efficiency of the rural-oriented strategy of developments, particularly the ability to provide contemporaneous solutions to the numerous problems of the rural people through infrastructure support.

The general purpose of this paper is to investigate the basis for pessimism by critically examining the past and present rural infrastructure development efforts in the country. First, the rural infrastructure strategy is characterized (Section I). Second, the past and present programmes are briefly surveyed (Section II). Third, the persistent problems and issues are highlighted (Section III). Finally, some guidelines for effective rural infrastructure strategy are enunciated.

CHARACTERIZATION OF RURAL INFRASTRUCTURE STRATEGY

The discernible trend in development economics is the gradual movement away from growth and other non-poverty-oriented performance criteria towards the people, particularly rural people, as the focal point of any development strategy. In particular, the worldwide approach is to launch massive attack on rural poverty, which will also benefit the urban economy in the long run. In this connection, rural infrastructural build-up is construed as the primary requirement of the rural people to manifest their full economic potential. Therefore (and this is worth noting), the rural infrastructure strategy is only a subset of the overall economic development strategy, which in turn is an integral part of the general economic development strategy of the developing countries. In essence, the rural infrastructure strategy is not an alternative to other poverty-alleviation strategies of economic development, but an extension and natural revolution.

The general notion underlying the rural infrastructure strategy is that it is difficult for the rural sector to contribute significantly to economic progress in the absence of basic facilities that also enhance their living standards. These facilities and their roles are generally described under three categories:

i. **Rural physical infrastructure**

- Rural roads cause accelerated delivery of farm inputs, thereby reducing transportation costs and enhancing spatial agricultural production efficiency.
- Storage facilities help to preserve foods in the forms that consumers need them and to the time they need them; on-farm storage also helps to stabilize inter-seasonal supplies.
- Irrigation facilities assure farm water supply and stabilize food production by protecting the farm production system against uncontrollable and undesirable fluctuations in domestic food production.

ii. **Rural social infrastructure**

- Clean water, decent housing, environmental sanitation, personal hygiene, and adequate nutrition help to improve the quality of rural life.
- Formal and informal education promotes rural productivity by making the farmer able to decode agronomic and other information, and carry out other desirable modern production practices; basic education also promotes feeding quality, dignity, self-respect, and sense of belonging as well as political integration of the rural people.

iii. **Rural institutional infrastructure**

- Farmers' unions and cooperatives facilitate economics of scale and profitability of rural enterprise;
- Agricultural extension improves the technology status of the farm business.

The observed emphasis on agriculture is based on the universal fact that life in rural regions is based mostly on agriculture. This explains the initial concentration of efforts on those infrastructure which facilitate agricultural processes directly, albeit not exclusively. The presence of substantial externalities, chain effects, spillover effects, and linkage efforts among rural activities ensure that the general focus on agriculture will also benefit other aspects of rural life in due course. These infrastructure items are essentially a public obligation owing to five reasons: First, the lumpiness of expenditure required is generally beyond the reach of ordinary individuals; second, there is the presence of free-rider problem which discourages private individuals; third, infrastructural facilities are durable stocks of capital which require regular maintenance costs and hence life-long intergenerational consequences; finally, the provision of infrastructure cuts across different ministries and disciplines which are under the control of government.

OVERVIEW OF PAST AND PRESENT PROGRAMMES

In Nigeria, the traditional approach towards rural infrastructural development is consistent with the idea of Mosher (1976); that is, it is simultaneity that is most important, but not administrative integration. The arrangement based on this idea led to the initial multi-ministerial efforts. Under this arrangement, the Ministry of Education is to establish schools in rural areas. Works Ministry is to construct rural roads. Rural markets are the responsibility of the Trade Ministry. Agriculture Ministry is to carry out extension and associated works, etc. Parallel ministries exist at the federal and state levels to carry out the numerous activities of providing infrastructure simultaneously. The local governments further the activities through their functional cells.

In the context of underlying philosophy and policy, the traditional approach is rooted in the colonial administration. For what they are worth, historical data and impressionistic evidence suggest that in the colonial era of development, rural infrastructural build-up was not a deliberate policy, but merely represented the residual outcome of the surplus extraction theory.

That is, the road, the railway, the rural markets, and other infrastructure of the period were unavoidable necessities for sourcing export produce, and not the results of people-oriented development efforts.[30]

After independence, two limitations of the traditional approach were soon realized: firstly, for selected infrastructure, simultaneity was less important than integration; and secondly, passive rural infrastructure build-up could not generate fast progress in solving the poverty problem. Consequently, integrated rural development was conceptualized, and a deliberate programme in this regard was mounted in the form of a farm settlement scheme.[31] The rural infrastructural components of the scheme include physical (rural roads, rural housing, markets, etc.), social (primary school, healthcare, etc.), and institutional farmers' cooperatives, cooperatives, cooperative tailoring societies, etc.). Farm settlements exist to date in many states of the federation, but merely now consist of ill-maintained and inaccessible buildings and other structures with few or no settlers. In practical terms, they have petered out from human memories as a component of the integrated rural development policy.[32]

Another stage in rural infrastructure programming consists of the collaborative efforts of federal and state governments to establish, in series, a number of integrated Agricultural Development Projects (ADPs) under the funding assistance of the World Bank. The first generation of ADPs includes the enclave types which commenced officially as follows: Funtua (April 1975), Gusau (April 1975), Gombe (November 1975), Ayangba (1979), Ilorin (1979), Ado-Ekiti (1981), and Oyo North (1982). Subsequent projects were mixed in features, some with no initial World Bank support (ADPs), others called phased ADPs. Presently each state of the federation has one ADP at the statewide level. Table 1 illustrates the infrastructural development achievements of the ADPs as at 1986.

The River Basin Development Authorities (RBDAs) also had a brief spell of life as an agency for rural infrastructural development in the country. Originally, eleven RBDAs were established (Decree No. 25 of 1976) for the primary aim of developing the economic potential of the available water bodies. Later, a major departure in their functions took place with additional

---

[30] For concrete evidence over a long time, see annual reports of the colonial Agricultural and Forestry Departments between 1900 and 1954.

[31] The concept of farm settlement was Israeli type of rural development strategy, which was introduced to the old Western Region of Nigeria immediately after independence. See Western Nigeria (1963).

[32] For thorough explanations of the failure of the farm settlement scheme, see Adegboye et al. (1969) and Olatunbosun (1971).

responsibilities to perform direct agricultural and other rural development activities. The subsequent enabling decree also proliferated the authorities to eighteen in number at the same time when 'rural' became part of their name: River Basin Rural Development Authorities (RBRDAs). Before the bodies were reversed to the original number and functions (1985), they had a number of diversified water supply projects to their credit, some of which Table 2 describes.

The latest stage in infrastructure programming commenced in 1986 with the establishment of the Directorate of Food, Roads, and Rural Infrastructure (DFRRI). Its nomenclature epitomizes the areas of immediate attention: 'food' specifies the particular problem area; 'infrastructure' is to constitute the main policy instrument; the emphasis on roads is based on its relative importance among the class of infrastructure. This nomenclature on the whole was intended to narrow the scope of the rural development problem down so that efforts would be directed toward the most effective areas.

The other distinct feature of DFRRI is its location in the presidency, as against the usual ministry. The idea is to fasten progress in decision-making and ensure uninterruptible fund flow. Initially, a sum of ₦450 million out of the projected ₦900 million realizable from the newly introduced second-tier foreign exchange market operation was expected to be used by the directorate to provide about 60,000-km length of rural feeder roads. The subsequent budget allocations to DFRRI are ₦360 million (1986), ₦400 million (1988). The original motive of this relatively generous funding is not for DFRRI to set up its own bureaucracy for doing things directly, but to cause things to be done, using the existing machineries.

PERSISTENT PROBLEMS AND ISSUES

Three recurring problems in the rural infrastructural development process are (i) dwindling commitment of programme sponsors, (ii) incessant perturbations in the institutional framework, and (iii) low level of in-house operational research and planning of projects. An equal number of selected unresolved issues may also be discussed as done by Streeten Burji (1978) within the context of basic needs: (i) the resources required for meeting infrastructure requirements, (ii) participation of the local people, and (iii) the trade-off between rural infrastructure supply and other objectives.

COMMITMENT

The problem is not so much absence of projects, as relatively ineffective projects. One reason for ineffectiveness of rural infrastructure projects is the lack of sustained commitment of the sponsors. Commitment can best be illustrated in terms of funding. Table 3 shows that the actual fund allocations to ADPs are fairly lower than the budgeted amounts (consider the total and average commitment values). In practical terms, poor commitment increases in the order of federal government (86%), state governments (81.7%), other sources (71.4%), and World Bank (47.5%).

As pointed out somewhere else (Ayoola and Idachaba 1989), the effect of poor funding is midstream abandonment of key components, as well as poor execution of some others. This fact is related to the great number of unfinished roads and bridges, building structures, water projects, and other infrastructure distributed all over the country.

PERTURBATIONS

Two levels of perturbation exist in the infrastructure process. First, at the policy level, changing perception of the role of government has pronounced effects on the continuity of the overall rural development strategy. During the colonial period when infrastructure build-up occurred as a residual effect in the pursuit of other objectives, meaningful progress could not have been made compared to the long period of years. In the post-independence period too, when conscious policy efforts had been made, lopsided emphasis was earlier placed towards urban development projects in relation to rural ones, particularly, in the oil-boom era. However, at the present time, rural infrastructure is accorded a good level of prominence among development programmes. If the role perception of government continues to fluctuate in the future, the rural development pathway will become uneven, and at each point of change, some part of the recorded achievements of the preceding efforts will definitely be lost.

Second, at the organizational level, perturbations arise mostly from frequent personnel changes in leadership positions. Let us examine the state agriculture ministries, which play significant roles in rural infrastructure supplies, to substantiate the point (Table 4). The evidenced high rates of turnover in the political, administrative, and professional leadership positions create adverse effects. They end up in particular programmes being given more or less emphasis, redesigned, reintroduced, or the implementation pace speeded up or slowed down so as to reflect the new political, philosophical, ideological, and occupational biases of the new people involved. Again, the rural development pathway consequently becomes uneven.

OPERATIONAL RESEARCH AND PLANNING

There is need to examine the inner mechanism of the agencies responsible for rural infrastructural development. The principal elements of these can be succinctly described as follows:

(i) **Infrastructure Budgeting and Costing, Screening, and Evaluation:** Using the feeder road plans under the ADPs as an example, the key elements of these processes can be highlighted. There is the involvement of local government authorities. The usual selection criteria are the road condition and the agricultural productivity of the area served, as well as geographical spread and access to other socio-economic factors. The engineering and cost data are to be combined with these and matched against economic benefit estimates to produce a ranked order of candidate roads.

(ii) **Tendering and Force Account Implementation:** The World Bank as a co-financier of ADP-based projects favours executing rural infrastructure works through contracting by tender. The tendering procedures are a part of loan agreements and include mutually exclusive methods such as International Competitive Bidding (ICB), Limited International Bidding (LIB), and Local Competitive Bidding (LCB). Sometimes, Direct Contracting (DC) and Fixed-Price Contracting (FPC) methods may be used. Force account implies abstinence from contracting, or simply execution by direct labour.[33]

(iii) **Cost Control:** This involves the provision of checks on expenditure against budgeted amounts in order to discover possible divergence between budgeted and actual expenditures so as to prevent cost overruns. Quality control generally goes hand in hand with cost control, with the aim of not sacrificing quality for costs. In a strict sense, cost control of infrastructural projects applies only to force account works, because contractors

---

[33] For detailed descriptions of the various methods of implementing rural infrastructure projects, see Ayoola (1988); also see World Bank (1985) for supporting information.

have the responsibility to do their own internal control that will ensure the fulfilment of the contract obligations with agreed terms. But this is not always the case in practice as contractors fail to play this role effectively, which leads them to request for substantial contract additions.[34]

The technical processes involved in rural infrastructure supply as outlined above contain inherent problems. First there exists a great paucity of data necessary for good infrastructure planning. Some key areas of data inadequacy as regards rural road planning are existing road network and density, traffic flows, produce movements by types and quantities, to mention but a few. For example, the 1986 data on cost was used for planning the project year two-feeder road improvement programme in Benue State ADP in 1987, owing to unavailability of updated figures (FACU 1988). This is grossly inadequate, judging from the unstable price movements in the economy.

The second problem area is shortage of operational research activities to back up rural infrastructure planning. This results in the use of inappropriate generalizations in concepts and measurements. Several examples abound in the cost-benefit analysis of roads (see Ojukwu 1987). One, it is usually assumed that there would be no measurable increase in crop production without rural road improvements; this has been done so as to allocate the entire incremental crop production to road construction efforts in the absence of suitable criteria for benefit allocation. Two, it is usually assumed that farm gate prices would not change between pre- and post-project period, owing to lack of information about the effect of increased accessibility to farm produce markets on competition. Three, it is usually assumed that hired labour rates will not change between the pre- and post-project periods; this is done in spite of the fact that better accessibility to rural areas causes upward pressure on hired labour rates, owing to increased possibility of off-farm employment. Four, it is usually assumed that incremental production will totally be exported from the areas of production on the ground that income elasticities of demand for food in rural areas are low. Lastly, sometimes the Kenyan model and other such models have been borrowed for use in computing operating cost

---

[34] One interesting example suffices concerning the Sokoto ADP. The contractor handling the Farm Service Centre at Lolo made request for 'additional transport claim' in October 1985 because 'the ferry broke down and he had to smuggle his materials over the closed Niger border, cross the river bridge between Niger and Benin-Nigeria Republic and smuggle them back over the closed border between Benin and Nigeria at Lolo' before reaching the project site (Liang 1988).

of rural roads in Nigerian ADPs. In general, unrealistic assumptions cause great divergence between models and reality.

The third problem area consists of the lax supervision and cost control outfits of rural infrastructure agencies, which has led to huge wastes in money and time. The lack of supervision usually leads to non-lasting structures, such as the farm service centres of Sokoto ADP which had their roofs easily blown off by wind (Liang 1988), and the great number of non-functioning handpumps in the rural borehole programme of Bauchi State ADP (Kwara 1988). DFRRI, on the other hand, demonstrates the clearest evidence of inefficiency in cost control mechanisms. As it turned out in Benue State DFRRI for example, a low percentage of the total job as percentage for the total job was done with all budget cost (Idachaba 1988). The problem with contractors as regards the commonly poor quality of infrastructure work is also related to ineffective supervision efforts.

Furthermore, there is the problem of wasteful delays in the implementation process. It is quite common to observe project sites in the country being temporarily deserted by workers while construction is still going on. The usual reason, apart from failing funds, is the delay in procurement of construction machinery and other materials, especially the offshore components. Bauchi State ADP reported a delay of one year in the arrival of road construction machinery after the international competitive bidding process had been completed (Kwara op. cit.).

The last but not the least of problems to be touched upon in the context of inner processes of rural infrastructure agencies is the frequent occurrence of role confusion. The cases that stand out in this regard concern RBDs and the original location-specific, river-based objective to include direct agricultural production activities that caused a thin spread of limited resources and the consequent inefficiency of programme performance. This informed the recent revision of the underlying policy concerning the authorities, which in itself is not without substantial social costs. DFRRI is on the verge of repeating the same experience by virtue of its reported direct agricultural production efforts in some states of the federation.

Resources

As earlier explained, the projects of the rural infrastructure strategy draw largely on public resources. The question of appropriate sharing formula to meet the resources requirement among the federal, state, and local governments then becomes pertinent. This notion is based on the fact of differential distribution of money, men, and materials among the three tiers of government. For instance, while the federal government may have huge financial resources at its disposal that can be disbursed to facilitate infrastructure works in rural areas, the state government is better disposed to mobilize equipment and personnel, while the local government is closest to the rural communities in terms of sourcing for local labour and requirements.

The following issues, therefore, deserve attention:

- an estimation of the required resources to meet specified levels of rural infrastructure by a specified date;
- an indication of the means of mobilizing, allocating, and using these resources efficiently;
- the determination of the changes that must be made in policies at the three government levels towards meeting the rural infrastructure needs.

Local Participation

The satisfaction of rural infrastructure requirement does not rest on supply management only; it also depends on effective demand management. The latter presumes that generation and articulation of demand for public services through local participation is quite important. In the case of rural infrastructure, the particular aspect of local involvement that is of immense importance is maintenance. It is in the area of maintenance management that strong cooperation of the local organizations has an important role to play.

For instance, if water flowing from rural standpipes is permitted to drip, the puddles can become the breeding ground for mosquitoes and other vectors. Under this condition, the welfare gain from the water supply project is neglected by welfare loss from debilitating diseases. Consequently, the siting of these pipes and their efficient use depend on local cooperation and on the social changes that ensure it. Particular experiences under the ADPs show that representative local bodies can minimize wastes and handle maintenance work. They can also minimize the amount of benefits of rural infrastructure services accruing to the privileged groups. Some of the questions are:

- how can the commitment of the local people be mobilized and local participation be secured and strengthened?
- in what ways can local community-based programmes be incorporated into the rural infrastructure strategy?

Balance of Objectives

The provision of basic infrastructure is an important objective, but it is not the only objective. Some others, which may sometimes compete with it are high income to farmers and affordable prices to consumers, diversified rural economy, protection of the environment, to mention a few.

An illustration will suffice: The completion of a new feeder road network opens up the rural community concerned to outside markets. But it is also capable of increasing food and other commodity prices because of the increased traffic and the sudden demand pressure. Consequently, in the short run, welfare of people may fall. However, of course, the productivity of agriculture may also be enhanced owing to improved income positions. In this example, the two-pronged price effect suggests that the rural infrastructure strategy must attain the balance of objectives, in such a way that there is a positive incremental benefit to the society at large. The following issues, therefore, arise:

- What are the trade-offs, if any, between rural infrastructure projects and the alternative objectives and policies?
- Specifically what, in the light of the experience of about thirty years, is the relationship between economic growth and the policy of satisfying rural infrastructure needs, in empirical terms?

CONCLUDING REMARKS AND RECOMMENDATIONS

The shift of policy emphasis from mere growth to basic rural infrastructure is a movement from abstract to concrete objectives, from a concentration on means to focus on ends, and from certain double negative (e.g. reducing unemployment) to positive (e.g. providing infrastructure facilities). However, policy and operational problems as well as a number of unresolved issues inherent in the existing delivery systems limit the degree of success.

The following specific recommendation may be made to provide guidelines for effective rural infrastructure strategy:

(a) The sponsors of rural infrastructure projects must maintain a close link between planned and actual fund allocations and other aspects that determine their enduring commitment over the entire project life.

(b) There should be built-in stabilizers in the rural infrastructure strategy to limit the adverse effects of perturbation arising from frequent organizational and personnel changes on the implementation performance of projects.

(c) The rural infrastructure strategy requires provision for contemporaneous operational research and planning studies to perfect the various technical processes involved in the implementation of projects; some of the processes that are presently inefficiently carried out are infrastructure budgeting, costing, screening, and evaluation, as well as tendering, direct labour utilization, cost control, monitoring and supervision.

(d) Appropriate responsibility sharing formula must be found to allocate resources among the different tiers of government in the infrastructure development process.

(e) Efforts should be intensified to involve the local people in the infrastructure development process, including provision of assistance to the target groups, towards instilling confidence in them to utilize the autonomous forces within the individual and the group for ensuring their cooperation in the execution and maintenance of the facilities;

(f) The rural infrastructure development objective must constantly be balanced against other desirable economic objectives with a view to applying complementary policy instruments for shifting the numerous trade-off curves in the rural sector at each point in time.

REFERENCES

Adegboye, R. O., A. C. Basu, and D. Olatunbosun (1969), 'Impact of Western Nigeria Farm Settlement on Surrounding Farmers', *J. Econ. and Social Studies* 1 (2).

Akinyosoye, V. O., *River Basin Development Authorities and the Nigerian Food Economy: An Assessment*, NISER Agricultural Policy.

Ayoola, G. B. (1988), 'Rural Infrastructural Development in Nigeria', Invited paper, Bi-annual Conference of Nigerian Institute of Quantity Surveyors, Durbar Hotel, Kaduna 13–15 Oct. 1988.

Ayoola, G. B. and F. S. Idachaba (1989), 'Technology and Nigerian Agricultural Development', Invited lead paper, International Conference on Third World Strategies for Technological Development, Federal University of Technology, Yola, 20–26 Aug. 1989.

FACU (Federal Agricultural Coordinating Unit) (1988), *Project Year Two Feeder Road Improvement Programme (Benue ADP)*, Ibadan, 1988.

Idachaba, F. S. (1989), 'Strategies for Achieving Food Self-sufficiency in Nigeria', Keynote Address, 1st National Congress of Science and Technology, University of Ibadan, 16 Aug. 1989.

Kwara, A. D. (1988) 'Success/Failures of the Rural Infrastructure Section of the ADP – Bauchi ADP Experience', Workshop paper, FACU, Kaduna, 6–10 June 1988.

Liang, A. D. (1988), 'Success/Failure of the Rural Infrastructure Section of ADP – SARDA Experience', Workshop paper, FACU, Kaduna, 6–10 June 1988.

Mosher, A. T. (1976), *Thinking About Rural Development* (New York: Agricultural Development Council, Inc), 35–56.

Ojukwu, C. (1987), 'Estimating Crop Benefits for Rural Road Programme', Workshop paper, FACU, Ibadan, 12–15 May 1987.

Western Nigeria (1963), White paper on Integrated Rural Development Official Document No. 8, Ibadan.

World Bank (1985), *Guidelines, Procurement under IBRD Loans and IDA Credits* (Washington DC).

Table 1: Physical achievement of the World Bank assisted Agricultural Development Projects (ADP) in rural Infrastructural Development as at December 1986*

Name of ADP	Period existence	Rural Roads (km) Constructed	Maintained	Rehabilitated	Bridges and culverts	Dams	Boreholes	Wells tube or open	Farm service centres	Training hall/ workshop
Recent Project										
1. Anambra State ADP	1986	14.5	0	0	-	-	-	-	52.3	33.3
2. Bendel State ADP	1986	-	0	0	-	-	-	-	-	-
3. Benue State ADP	1986	-	9.1	0	0	-	-	-	-	-
4. Cross River State ADP	1986	-	0	-	-	-	-	-	-	-
5. Imo State ADP	1986	14.7	47.8	73.0	50.0	-	-	-	33.3	58.3
6. Plateau State ADP	1986	30.6	5.3	-	0.1	-	66.7	14.0	-	-
7. Ogun State ADP	1986	-	-	-	-	-	0	-	100.0	-
Older Project										
8. Bida ADP	1981–86	69.6	114.6	329.7	-	-	-	100.0	-	89.3
9. Borno State ADP	1983–86	23.7	-	-	51.5	-	100.0	163.8	126.5	140.0
10. Sokoto State ADP	1983–86	70.8	-	-	-	-	-	-	-	40.0
11. Goingola State ADP	1983–86	22.6	-	-	31.3	-	-	-	41.7	40.0
12. Ilorin ADP	1981–86	26.7	91.3	-	-	-	400.0	-	33.3	75.0
13. Oyho North ADP	1983–86	80.6	-	48.4	21.9	25.7	-	83.5	-	-
14. Kaduna State ADP	1985–86	34.3	25.0	-	-	-	-	-	33.3	2.4
15. Kano State ADP	1983–86	55.7	58.3	40.0	-	-	100.0	81.6	-	-
16. Ekiti Akoko ADP	1983–86	73.9	41.6	-	-	-	-	-	-	-
Average	-	43.2	35.7	67.3	25.8	12.9	133.3	88.6	60.1	59.8

Means not available; Source: Underlying data from APMEPU Project Status Report, Kaduna 1987.

Table 2: Selected water supply infrastructural works of the River Basin Development Authorities, circa 1984

RBDA		Project Name	Proposed Purpose
1. Anambra-Imo	(i)	Lower Anambra Irrigation Project	- Pumping station - Irrigation facilities for 5,000 ha. rice production - 35,000 tonnes of paddy rice expected in one season
	(ii)	Agbala Market garden project	- Horticultural project to produce okra, tomatoes, eggplant, cucumber, lettuce, pepper, green African spinach, and onions
	(iii)	Egbema/Oguta Project	- Fisheries venture - Crops production including rice and maize
	(iv)	Agbala/Lokpanta/Mgba-kwu Int. Poultry Projects	- Poultry products including eggs, chicks, broilers, etc.
	(v)	Odekpe Project	- Farmer participation irrigation scheme, crop production including rice and maize
	(vi)	Otucha/Umaeze Anah/Aguleri Product	- Maize cultivation
	(vii)	Ntigha/UmuNwa Project	- Commercial fish production - Rice production
2. Benin-Owena RBDA	(i)	Poultry project	- Eggs and other poultry products - Production of several crops
3. Chad Basin	(i)	Underground water development project	- Underground water development for food and livestock

		(ii)	Alau Dam and Jere Bowl Scheme	-	To two water of the seasonal river near Maiduguri for storage to be released to Maiduguri after treatment
		(iii)	South Chad Project	- - -	Irrigation of 9,000 ha. of land Irrigation of 67,000 ha. of land within Lake Chad water Settlement for 1,600 farm families
		(iv)	Baga-Kirenowa project	- -	To irrigate 20,000 ha. of land Settlement of 5,000 farm families
		(v)	Yedseroam Reservoir and Irrigation Scheme	-	To irrigate 35,000 ha. of land
		(vi)	Agro-Allied Industries	-	To set up agro-allied industries for proper utilization of the irrigation schemes
4.	Cross River Basin Development Authority			- -	Integrated irrigation Agric. and livestock projects
5.	Hadejia-Jama're	(i)	Kano River project	-	To irrigate 20,000 ha. of land from Tig Dam
		(ii)	Onallawa Gorge Dam	-	To supply irrigation water, generate electricity, and supplement water supply
		(iii)	Hadejia Valley Irrigation Project	-	To irrigate over 28,000 ha. of land
		(iv)	Jama're Valley Project	-	Valley and town water supply
		(v)	Agro-allied project	-	Establishment of processing plant

		(vi)	Minor Irrigation Schemes	-	Establishment of processing plant
6.	Lower Benue	(i)	Doma Dam and Irrigation Project	-	To supply Doma with drinking water
				-	Irrigation facilities
		(ii)	Crop Project	-	Crops production in several products
		(iii)	Livestock Project	-	Production of livestock products
7.	Niger Delta	(i)	Reramabiri rice project	-	8,000 tonnes of mill rice production
		(ii)	Community fish and commercial fish farm project	-	To establish fish farms
		(iii)	Sagbama River Project	-	4,000 ha. of oil palm, 7,000 ha. of rice, etc.
		(iv)	Isampou Rice Project	-	6,000 ha. of rice to be produced
		(v)	Flood Control and Reclamation Project	-	To control floods
		(vi)	Rural Water Supply Project	-	To sink boreholes for rural water supply
		(vii)	Yenegoa Oil Palm Establishment	-	To produce oil palm on 4,000 ha. of land
		(viii)	Adona River Project	-	Large-scale vegetable scheme
		(ix)	Nun River Project	-	Irrigation project to establish village rice scheme, raphia palm project, jetty construction, and land reclamation
		(x)	Orashi River Project	-	Integrated project involving village agriculture, animal husbandry, etc.

8.	Niger River	(i)	Kainji Lake	-	Irrigation schemes
		(ii)	Upper Kaduna Project		
		(iii)	Lower Kaduna Project	-	To supply irrigation water from Kaduna area
		(iv)	Kampo River Village Project		
		(v)	Oshin-Ogun River Valley	-	Irrigation scheme
		(vi)	Gbako-Gbakogi Valley Project	-	Irrigation schemes
		(vii)	Gurara River Valley Project	-	Irrigation schemes
		(viii)	Niger River Valley Project	-	Irrigation schemes
		(ix)	Livestock/Poultry and Fishery Project	-	Livestock production
9.	Ogun-Oshun	(i)	Upper Ogun Farm Project	-	Cropping and fish farming
		(ii)	Sepeteri Farm Project	-	Irrigation and crop production
		(iii)	Oyan Dam River	-	To supply 701 litres of water to Lagos and Abeokuta
				-	To provide irrigation water for 12,500 ha.
				-	To generate electricity
		(iv)	Ikere Gorge Dam	-	Multipurpose
		(v)	Upper Oshun Basin	-	Seed multiplication
		(vi)	The Lower Ona Basin	-	Seed multiplication
		(vii)	Itoikin Farm Project	-	To train farmers in irrigated rice growing and processing
		(viii)	Headquarters farm complex	-	Crop production

		(ix)	Vegetable Garden		
		(x)	Mokolokin Farm Project	-	Irrigation of 12,000 ha. through Oyan River Dam
		(xi)	Poultry and Fisheries Project	-	Animal products
10.	Sokoto-Rima	(i)	Bakolori Dam Project	-	To supply canal irrigation for 23,000 ha.
				-	Generation of HEP
		(ii)	Gogoronyo Dam Project	-	Irrigation
		(iii)	Middle Rima Valley	-	Irrigation schemes
		(iv)	Niger Valley	-	Irrigation for 25,000 ha.
		(v)	Karadawa River Basin Development Project	-	Irrigation and drinking water supply
		(vi)	Zauro-Birnin Kebbi Polder Project	-	Irrigation expected to cost ₦140,372,000
		(vii)	Kurfi Ruma Fakuma-Cheranohi water supply scheme	-	Construction of small dams for drinking water
		(viii)	Gada-Bonye Water Supply Irrigation Project	-	Drinking and irrigation water supply for 133,000 people and 2,500 ha. of land
		(ix)	Jabija Irrigation/Water Supply Project	-	Irrigation and drinking water supply for 37,000 ha. of land
11.	Upper Benue	(i)	Dadin Kowa Dam Project	-	Irrigation schemes
		(ii)	Tallum Irrigation Project	-	To develop 280 ha. of farmland from Kiri Dam
		(iii)	Mambilla Irrigation Project	-	Livestock production
		(iv)	Crop Production	-	Cereal production

	(v)	Gassal Irrigation and Pilot Farm	- Irrigation schemes
	(vi)	Donga Irrigation Pilot Farm	- Irrigation of 150 ha.

18

The Institutions and Processes of Rural Infrastructural Development in Nigeria

The various types of physical, social and institutional infrastructure are desirable because of their important roles in the economy.[35] In particular, it is recognized that among the available policy instruments of rural development, the provision of basic infrastructure constitutes the most useful method of providing the solutions contemporaneously to the problems of the rural people. This recognition is amply demonstrated by the current policy emphasis on provision of rural infrastructure as a means of achieving sustainable economic growth and development in Nigeria.

The general objective of this paper is to examine the roles of the various institutions responsible for the provision of rural infrastructure and to discuss the processes involved in the performance of these roles. The paper is arranged as follows: Section II briefly surveys the past and present efforts aimed at solving the rural development problem. Section III identifies and discusses the technical processes of rural infrastructural planning and implementation, laying particular emphasis on the general procedures in the Agricultural Development Projects (ADPs). Section IV enunciates the practical problems in the processes and those of the institutional aspects, as well as presents recommendations for effective rural infrastructural development programmes. Section V concludes the paper.

---

[35] See Idachaba, F. S. *Rural Infrastructures Development in Nigeria*, Ibadan, University Press, 1985.

INSTITUTIONAL FRAMEWORK FOR RURAL INFRASTRUCTURAL DEVELOPMENT

Even though both private and public efforts can generate new infrastructure, four reasons are available to explain why the task of infrastructure provision in rural areas is generally a public obligation. First, the lumpiness of expenditures on rural infrastructure is beyond the reach of ordinary individuals. Second, there is marked divergence between infrastructural benefits that accrue to an individual sponsor and those benefits that accrue to the rest of the society, which leads to a free-rider problem. Third, infrastructural facilities are durable stocks of capital that yield future income streams and which therefore require regular stock maintenance costs. Finally, the provision of infrastructure cuts across ministries and disciplines.

Traditionally, provision of rural infrastructure was decentralized among the numerous government ministries. Under this arrangement, the Ministry of Education is to establish schools in rural areas. Works Ministry is to construct roads. Rural markets are in the hands of the Trade Ministry, while the Agriculture Ministry is to carry out extension works and conduct outstation research. In the end, the bureau that is saddled with the responsibility to further these activities is the Local Government Authority. But because this is the lowest and poorest tier of government, rural infrastructure provision was consequently relegated to the lowest background.

However, a slight deviation from the sectoral approach under the traditional arrangement is worth noting. This refers to the deliberate effort of the old Western regional government to address the rural development problem specifically through the establishment of a farm settlement scheme in 1963.[36] The scheme involved the settlement of young school leavers in specific rural locations. It also includes the organization of cooperative tailoring societies to execute a school uniform project in rural areas, and also a rural broadloom weaving programme. However, because the primary attention of the government was to solve an impending employment problem, the resultant rural infrastructural facilities put in place were only a residual outcome. By any standard judgement, the farm settlement scheme failed woefully, having petered out into near nothingness and leaving as its main legacy today a system of ill-maintained and inaccessible building structures with few or no settlers, which are scattered over a number of states in the country.[37]

---

[36] See Anonymous, White paper on Integrated Rural Development, Western Nigeria, Official Document No. 8 of 1963.

[37] For detailed discussion of issues in the Farm Settlement Scheme, see R. O. Adegboye, A. C. Bagu, and D. Olatunbosun, 'Impact of Western Nigeria Farm

The permissive atmosphere around rural infrastructural development in the early days was actually as a result of constitutional neglect. There was no ministry at the federal level initially to coordinate agricultural development activities, which were conceptually subsumed in the overall rural development problem. Agriculture appeared on the residual legislative list, which made it a regional responsibility. However, following a recommendation of the Food and Agricultural Organization of the United Nations at the instance of the Nigerian government, a Ministry of Natural Resources and Research was created in April 1965. Initially this ministry had limited functional scope and capabilities, essentially because it was not immediately backed up by necessary statutory powers and instrument. This explains the deliberate omission of the word 'agriculture' in the name of the ministry, which was done so as not to offend the political sensibilities of the regions. Finally, this ministry has variously transformed in name, scope, and size to become what is now known as the Federal Ministry of Agriculture, Water Resources, and Rural Development.

The most determined thrust behind rural infrastructural development was made in March 1976 when the Federal Department of Rural Development (FDRD) was established. This department was immediately charged with the responsibility for the design, formulation, and implementation of the integrated Agricultural Development Projects (ADPs). The first generation of ADPs were the enclave types which commenced as follows: Funtua (April 1975), Gusau (April 1975), and Gombe (November 1975). Newer enclave-type projects later developed in Lafia (1977), Ayangba (1979), Ilorin (1979), Bida (1980), Ekiti Akoko (1981), and Oyo North (1982). Presently each state of the federation has one ADP either in the enclave stage or at the statewide level.

The ADP concept is a package of rural infrastructural development works distinguished by the following elements.

- an input delivery and credit supply system through a network of farm service centres which ensures that no farmer travels more than 5–15 km to purchase needed inputs;
- a massive rural feeder road network that has opened up new areas for cultivation and marketing of farm produce and timely delivery of farm inputs;

Settlements on Surrounding Farmers', *J. Econ and Soc. Studies* XI (2) 1969. Also see D. Olatunbosun, 'Western Nigeria Farm Settlements: An Appraisal', *J. Developing Area* 5 (3) (April 1971).

- a revitalized, intensive, and systematic extension and training system backed up by timely input supply and adaptive research services; and
- autonomous project management together with inbuilt monitoring and evaluation.

See Table 1 for the summary of selected infrastructural achievements of the Nigerian ADPs as at December 1986.

The slow pace of rural infrastructural development is of utmost concern to the present administration. This led to the establishment in December 1986 of the Directorate of Food, Roads, and Rural Infrastructure (DFRRI). Its nomenclature epitomizes the areas of immediate attention. 'Food' in this name specifies the particular problem area, and 'rural infrastructure' is to constitute the main policy instrument. The particular reference to roads connotes that this category of rural infrastructure, among the others, possesses the greatest explanatory power in the variability of rural welfare standards in the short run. Therefore, concerted efforts are to be directed at them immediately by the directorate. This nomenclature, on the whole, helps to narrow the scope of the rural development problem down so that the directorate will direct its limited resources, within a limited period of time, to the most effective areas.

The establishment of DFRRI is generally informed by the need to hasten progress in the rural areas of the economy by locating the decision-making authority at the highest level of political authority. Truly, the idea seemed to have worked well to the extent that huge sums of money were immediately released to the directorate, using the uninterruptible executive orders resident within the Presidency. The stream of budgetary allocations to DFRRI confirms the government's commitment to address the rural development problem squarely as follows: Initially a sum of ₦450 million out of a possible ₦900 million realized from the newly introduced Second-tier Foreign Exchange Market operation was expected to be used by the directorate to provide about 60,0000-km length of rural feeder roads. Consequently, budgetary allocations were ₦360 million (1986), ₦400 million (1987), and ₦500 million (1988) for DFRRI operations. In addition, the directorate was organized in such a way as to be free from the mainstream of sluggish civil service bureaucracy. Its affairs are to be directed from the apex and down the line by top military officers who, by the nature of their profession, are expected to mobilize money, materials, and manpower to achieve concrete results within a short period of time.

Over the period of its existence, DFRRI has been a focus of intense media and academic criticisms. Some of its obvious shortcomings have been ably described by Idachaba[38] (1988) as the 'lack of programme focus in DFRRI and the fatal failure of DFRRI to monitor its programmes, resulting in programme atrophy and colossal waste of resources'. Specifically the directorate has been indicted for lack of programme accountability and being shrouded in 'role confusion and programme fuzziness'. Since a culture of monitoring and evaluation has not evolved within the DFRRI outfit, then its activities could only be coagulated from the bits and pieces of uncoordinated media reports. (See Table 2 for identifiable DFRRI activities as at August 1987.)

TECHNICAL PROCESSES IN PLANNING AND IMPLEMENTATION OF RURAL INFRASTRUCTURE PROGRAMME

The ideal practice is to harmonize the technical processes of rural infrastructural development works under the different institutional arrangements, which is to be followed by a comparative analysis. This will not be possible presently because DFFRI does not generate necessary data within and outside itself to do such analysis. In addition, the traditional arrangements under the various ministries are not well articulated to separate the efforts into clear processes. However, the existence of the Federal Agricultural Coordinating Unit (FACU) and the Agricultural Projects Monitoring, Evaluation, and Planning Unit (APMEPU) together with the inbuilt monitoring and evaluation capabilities of the individual ADPs makes it possible for us to know in good detail what are being done there and how they are being done. We shall therefore draw heavily from the activities of the ADPs in order to highlight the technical processes in infrastructure provision.

It is worthwhile to note that the ADP itself is probably not exhibiting the desirable virtue of programme accountability of its own volition. But there exists a force majeure in the corporate person of the World Bank as a principal financier. The loan agreement requires that a number of tested processes be followed in the implementation of such programmes in which the bank provides financial assistance. The processes are discussed as follows:

---

[38] See F. S. Idachaba, 'Strategies for Achieving Food Self-sufficiency in Nigeria', Keynote address delivered at the 1st National Congress of Science and Technology, held at the University of Ibadan, 16 August 1988.

Infrastructure's Budgeting and Costing, Screening and Evaluation

The Staff Appraisal Report (SAR), a document of the World Bank which culminates from all initial identification mission report, followed by a preparation study report and then followed by a series of negotiation meetings and wrap-up meetings, during which officials of government, the bank, and consultants featured severally, is the working frame for rural infrastructural development in each ADP. The SAR sets out clearly the project objectives and the long-term targets of specific infrastructural items to be provided for meeting the objectives. These targets are further broken down into annual infrastructural budgets in which the quantities of the various infrastructures are sequentially allotted for execution from project year 1 to the termination period. Then the monitoring and evaluation machineries are set in motion against these annual targets.

Generally, the procedure for feeder road planning is more elaborately developed than the other infrastructure. In the case of buildings, prototype designs of the farm service centres, workshops, and staff houses are simply to be executed. The situation is similar for the designs and costings in respect of boreholes and dams. As for roads, the process involves pre-screening and evaluation works with implications for costing before decisions can be made about implementation.

The first step is to identify agriculturally productive areas of the project and take an inventory of available rural infrastructure there. The roads for improvement are primarily selected in collaboration with the LGAs. The Chairman of Local Government is often asked to provide a list of roads that require reinstatement, to be arranged in some order or priority. This list is then subjected to the proper screening by FACU in conjunction with the relevant ADP. The usual criteria for this exercise have been provided in the staff working papers of the World Bank, and these essentially include the road condition and agricultural productivity of the area served. Geographical spread and access to other socio-economic factors are also considered before reducing the list to a manageable number, referred to as candidate roads. The evaluation of the candidate roads begin with the Engineering Division of the project, where the engineering data involving the terrain characteristics, earth-moving qualities, etc. are produced together with line diagrams and the determination of the nature of required work. This will provide the basis for costing, and three categories of costs are necessary at this stage: (i) road improvement base cost, (ii) annual routine maintenance cost, and (iii) periodic maintenance cost. The standard improvement works required in the new ADPs on which costing is based are as follows:

- widening of existing width
- removal of topsoil on widened section
- filling of low-lying areas and culvert embarkment
- culverting across channels
- provision of side drains
- backfilling of culverts
- extension of existing culverts
- provision of town-out culverts.

Other aspects of the costing are the cross-section and permanent structure standards of the roads. These in respect of recent ADPs include clearance width of 12 mm; width of ditch centre to centre of 9 mm; width of granular surfacing of 5–6 mm and surfacing thickness of 150 mm or 200–250 mm over soft subgrade soils. Drainage should be literate ditches, corrugated metal pipes of concrete pipe culverts or gibbons.

Ranking of candidate roads is achieved on the basis of economic cost-benefit analysis. At this stage, the technical support facilities of FACU are very useful, including the perfected computer programmes for deriving the usual discounted measures of project worth. The popular ranking criteria are the net present value (NPV) and the benefit-cost ratio (BCR) of the candidate roads. The groups of benefits often estimated in the process include the following:

(a) Benefits to farmers as a result of marketable incremental crop production at the farm gate after adoption.
(b) Benefits to transporters, markets, and processors of the incremental production owing to the improved business activity
(c) Benefits to road users from savings in operating costs of vehicles as a result of improved road conditions.

Two additional cost items, which would make the cost-benefit analysis of roads complete, are the extension services cost and the commercial services cost within the zone of influence. The aim of the cost-benefit analysis is to ensure that the infrastructure promises substantial economic and financial benefits and to ensure that priorities are given to roads within greater potential benefits than others.[39]

---

[39] The detailed procedures to follow are contained in World Bank working paper and several FACU documents, the latter including Infrastructure Methodology and Organization and Feeder Road Planning, Screening and Economic Ea, Procedures.

Tendering and Force Account Implementation

A rural infrastructural development programme is normally executed in one or a combination of two ways: contracting or force account. The position of the bank is that programmes be executed generally through the competitive bidding process. For works that are eligible for financing or that will significantly affect the satisfactory execution of the bank-assisted projects in terms of costs, quality, and completion time, the bank's favourite is execution through the International Competitive Bidding (ICB). This practice is justified on three grounds. One ground is the need for economy and efficiency in the execution of the project, including procurement of the goods and works involved; second is the bank's interest as a cooperative institution in giving all eligible bidders from developed and developing countries, an opportunity to compete in providing goods and works financed by the bank; third is the bank's interest, as a development institution in encouraging the development of local contractors and manufacturers in the borrowing country. When ICB is discovered not to be the most economic and efficient, alternative methods provided for are the limited international bidding (LIB) and the local competitive bidding (LCB). Two other methods of infrequent occurrence in Nigeria are the direct contracting and the fixed price contracting.

The first step in international competitive bidding is to prepare the bidding documents, which state the type and size of contracts together with the proposed contract provisions. The most common types of contracts provide for payments on the basis of a lump sum, unit prices, cost plus fees, or a combination of these.[40]

Cost reimbursable contracts are acceptable to the bank only in exceptional circumstances such as conditions of high risk or where costs cannot be determined with sufficient accuracy. If the project requires a variety of works and equipment such as earth dams, boreholes, and buildings, separate contracts may be awarded for the works and for the supply and/or installation of different major items. For a project such as rural road reinstatement which

---

[40] For the purpose of this discussion, tendering and bidding shall have the same meaning. See World Bank, Guidelines, Procurement under IBRD Loans and IDA Credits, Washington DC, 1985. The bank does not provide for the method of fixed price contracting. A striking example of this method was practised in Sokoto ADP and was justified by the director of engineering for 'contracts of short duration, less than 26 weeks, and buildings of about ₦100,000/₦50,000 in value'. In this method, tenders are simply invited from those contractors who find a predetermined value acceptable to them.

requires similar but separate civil works or items of equipment, bids may be invited under alternative contract options that would attract the interest of both small and large firms, who should be allowed to bid for individual contracts (slices) or for a group of similar contracts (package) at their option. In a feeder road project, slices may include culverts and bridges or sections of the road length with varying degrees of roughness, construction, rehabilitation and maintenance requirements. It is allowed in certain cases for the design and engineering and construction information to be provided under one contract. Alternatively, the project may be responsible for the design and engineering, and invite bids for a single responsibility contract for the works required.

The bidding documents should furnish all information necessary for a prospective bidder to prepare a bid for the works to be provided. While the detail and complexity of these documents will vary with the size and nature of the proposed bid package and contract, they should generally contain information on the following aspects: form of contract; conditions of contract, including general and special; technical specifications, bill of quantities and drawings, and necessary appendices such as formats for various securities. The mode of tender invitations has been standardized for the ADPs by the bank. For instance, the advertisements should appear in both local and international press, and reference must be made to the bank as financier. There are also adequate guidelines on validity of bids, conditions of contract, preparation of bidding documents, disbursement procedures, currency issues, terms of payment, etc. Similarly, the bid opening and evaluation, and pre-qualification of bidders are all guided.[41]

The limited international bidding (LIB) method differs from ICB only in the fact that bidding is done by direct invitation without open advertisement. This may be appropriate when the amounts of contracts are small or when there are only a limited number of contractors or other exceptional reasons, which may justify departure from full ICB procedures. However, it is also maintained that bids would be sought from a list of contractors broad enough to assure competitive prices. In LIB, domestic or regional preferences are not applicable.

The local competitive bidding is also an acceptable method of tendering. It is permitted when the nature or scope of works are unlikely to attract foreign competition, especially because of small contract values, or when works are scattered geographically or spread over time like in rural road maintenance, or when works are labour-intensive, or when workers are available locally at prices below the international market, or finally, when

---

[41] For detailed guidelines, see World Bank, op. cit.

advantages of international competitive bidding are clearly overweighed by the administrative or financial burden involved. These qualifications provide large enough leverage for any work to be eligible for LCB. This explains why, despite the bank's preferences for ICB, LCB remains the most popular means of tendering for rural infrastructure development works. In LCB, advertising may be in a local language, and local currency will generally be used for the purpose of bidding and payment. Nevertheless, the bank still requires that LCB should accommodate interested foreign firms that wish to participate under local circumstances, and the procedure should provide for adequate competition in order to ensure reasonable prices and that the methods used in evaluation of bids and award of contracts should not be applied arbitrarily, but should be made known to all bidders.

Direct contracting which means elimination of competition in the contract process may sometimes be used. This can be justified on the ground that an existing contract for works has been properly awarded and needs to be extended for construction or provision of additional works of a similar nature. In this case, the bank wishes to be satisfied that no advantage could be obtained by further competition and that the prices on the extended contract are reasonable.

Force account implies abstinence from contracting or execution by direct labour. It is the use of in-house task force for construction, which is generally justified on grounds of cost reductions. The bank recognizes the method, but demands that the project procures the earth-moving plants through ICB. In feeder road construction, for instance, a team is mounted, comprised of a fleet of equipment and people. An example of force account utilization for rural road works is recently reported in Kaduna State ADP. The road construction team is made up of equipment (four scrapers, one each of 300 HP bulldozer, 200 HP bulldozers, 180 HP motor grader, tractor with trailer, fuel tanker, low loaders, two each of water tankers and tippers). Two culvert teams were also mounted, each comprised of one each of culvert foreman, labourer, concrete mixer, concrete vibrator; two masons and twenty-five carpenters. The combination of these teams was capable of doing 0.6 km road per day together with a number of culverts all valued at ₦17,000 per km. This contrasts drastically from the usual contract sums of between ₦50,000 and ₦60,000 per km of road excluding pre-award overheads on road alignment, survey, and design of drainage structures.

The use of force account is gaining more ground with two recent developments in the rural economy. One development is the gradual emphasis that is being placed on maintenance of existing roads rather than construction of new ones. In particular, the quantities of work involved in annual routine and the five-year periodic maintenance of roads are not easily defined in advance for definite contracting. In this regard, the director of engineering in the Sokoto ADP recently declared that there was little point in constructing laterite feeder roads unless we adopted a planned maintenance scheme. The maintenance instincts worry him so much as to say that the new MSADPs[42] should change the M from Multi to Maintenance in the present circumstances.

The other development is the mounting of general unemployment. In this regard, the ADPs seem to be receiving some official pressures requiring them to mop up some part of the surplus labour market. Thus, the new concept of more labour-based construction is evolving rapidly, in which the usage of plants is to be reduced to accommodate more people.[43] The condition of adjusting capital-labour ratio cannot be practically imposed on contractors.

Besides these developments, the bank accepts the use of force account units, which are defined to include any government-owned construction unit that is managerially and financially autonomous. This acceptability is based on four grounds:

(a) when quantities of work involved cannot be defined in advance;
(b) when works are small and scattered or in remote locations where mobilization of costs for contractors would be unreasonably high;
(c) when work must be carried out without disrupting ongoing operations;
(d) when the risk of unavoidable work interruption is better borne by the borrower than by a contractor, and
(e) when no contractor is interested in carrying out the works.

Cost Control Methods

The essential element of a cost control system is to provide checks on expenditure against targets, which have been set out in a budget. The purpose is to discover what extent outturn is actually deviating from targets so that action can be taken in reasonable time to bring the flows into line with the desired objective. A control system normally involves the regular preparations of statements of these flows. An effective system requires up-to-date

---

[42] MSADP means Multi-State ADP which is a generic name given to the recently negotiated ADPs comprising several states at once.

[43] For details see S. S. Rajah, 'Viability of Labour-Based Construction System for New ADPs', Paper presented at Workshop of Engineers (Northern Projects) 22–24 August, 1988.

information to be rapidly available so as to take any corrective action in sufficient time to have the desired effect.

Quality control generally goes hand in hand with cost control. The statements of cost monitoring must reflect the quality standard attained so that quality is not simply traded for low costs. The act of cost control generally applies to force account works as contractors have the responsibility to do their own internal control that will ensure the fulfilment of the contract obligations. The Kaduna ADP example can be provided to illustrate the control methods of direct-labour construction. The ADP provides for culvert construction and casting yard fuel or lubricants for plants at zonal headquarters, which are sent to road construction sites through transfer vouchers. At site, there are two storekeepers, one for construction and the other for fuel or lubricant. Site store ledgers and daily uses are made through authorized side issues vouchers requisition. In addition, the following records are left for management surveillance: (i) site attendance book, (ii) daily plant hours of operation, (iii) site store ledgers for culvert materials and fuel or lubricant, (iv) daily progress logbook, (v) plant or vehicle logbook, and (vi) site maintenance or service record book.

RECOMMENDATIONS FOR SOLVING PRACTICAL PROBLEMS IN RURAL INFRASTRUCTURAL DEVELOPMENT PROGRAMMES

Recommendation 1: There exists a great faculty of data in the process of planning and implementation of infrastructural programmes in the country. A close examination of the technical processes reveals inadequacy of data in several areas, including road network and density, existing traffic flows, produce movements by types and quantities, etc. For example, the 1986 data on cost was used for planning the project year two-feeder road improvement programme for Benue State ADP in 1987 as the project could not submit updated figures at the time of analysis. This is grossly inadequate, judging from the unstable price movements in the economy. It is therefore recommended that the agencies concerned, particularly FACU and APMEPU, should accelerate their system of reliable data inflows to improve the planning and implementation efforts.

Recommendation 2: The technical processes discussed incorporate a number of swapping generalizations and unrealistic assumptions owing to either the lack of data or the inability to make measurements. A number of such assumptions call for mention. One, it is usually assumed that there would be no measurable increase in crop production unless rural roads were improved. This has been made so as to appropriate all incremental crop production to road construction efforts in admission of the fact that there are no acceptable criteria to disaggregate the benefits among other possible causes. Second, sometimes it has been assumed that farm gate prices would not change between pre- and post-project period. This is frequently done in spite of the fact that better accessibility has an effect of increasing hired labour rates because of the increased possibility of off-farm employment. Fourth, it is often assumed in the analysis that incremental production will all be exported from the area of production on the grounds that income elasticities of demand for food items are low. This assumption of no change in home consumption is particularly spurious. Lastly, sometimes the Kenyan model and other such models have been used in computing operating cost for Nigerian ADPs. In general, unrealistic assumptions cause great divergence between models and reality. It is therefore recommended that the relevant agencies should mount intensive micro studies to focus on identifying and definitizing the unknown variables in the rural development problem. In-depth planning studies are required at this stage to assign meaningful values to the undetermined coefficients.

Recommendation 3: Our supervision and cost control outfits are lax. As an illustration, when the Farm Service Centres of Sokoto ADP were taken off by wind, the director of engineering was satisfied because the same thing happened to the building roofs of the Sokoto airport, which flew in a similar 'intercontinental jet' fashion (his words quoted). The engineer should have known that such problems could have been solved at the design and construction stages if effective supervision existed. Lack of quality control also led Bauchi ADP to incur an additional sum of ₦324, 000 when a significant number of handpumps in a rural borehole programme failed to eject water. Therefore, it is recommended that the various programmes need to be complemented with effective supervisory capabilities so that the various borehole taps would not dry up, and the Farm Service Centres and other buildings would not be short-lived. For DFFRI in particular, the case for an overall consultant to oversee progress is worth considering. Furthermore, lack of adequate supervision and quality control embedded in the contract agreements will soon push the heavy equipment, for which no indigenous maintenance capacity exists, into the junkyards.

Recommendation 4: Significant delays in project start-ups and execution exist. For example, the zonal road units in Bauchi ADP experienced major problems when the procurement of construction machinery took about one year to arrive after ICB. Even though competition is desirable, project authorities should ensure that it does not cause costly time wastages. Unless the projects have been treated as mutually exclusive types at the preparation stage on a time basis, project authorities should avoid delays in implementation that will lead to unnecessary and avoidable cost overruns.

Recommendation 5: The performance of the contractors leaves much to be desired. A number of rural infrastructural works have had to be abandoned and contracts revoked in the past because of the inability of contractors to perform in line with agreed obligations. The tenderers sometimes stake all their energies to win contracts for providing infrastructure, with little attention to their capacity to cope. Consequently, such contractors turn around midway in execution to request for cost adjustments on flimsy grounds. The reason for 'additional transport' claim of a contractor working on the Farm Service Centre at Lollo in Sokoto ADP in October 1985 is particularly interesting. According to the project engineer, the contractor made extra contract claim because 'the ferry broke down and he had to *smuggle* his materials over the closed Nigeria/ Niger border, cross the river bridge between Niger and Benin Republic, and then *smuggle* them back over the closed border between Benin and Nigeria at Lollo' before reaching the project site. So, he wanted extra transport' payments above the contract sum. In some other cases, failure of contractors to meet the specifications of project designs have led to huge adjustment expenditures incurred by the project authorities. One is therefore constrained to recommend that contract agreement should allot substantial remedial costs to the suppliers and contractors.

Recommendation 6: We have learnt that Sokoto DFFRI and the ADP collaborated somehow to pass some new dam construction equipment belonging to the ADP over to DFFRI for use. We have also learnt sometime ago from the press that there was confusion over an ADP road in a state, which the state DFFRI simply planted their signpost on to claim credit. These events suggest that role conflicts are imminent among rural infrastructural development institutions. What is recommended is a synchronized functional collaboration of a close type between DFFRIs and ADPs.

Recommendation 7: There is urgent need to amend the DFFRI decree to separate the functions of programme design and formulation from implementation, monitoring, and evaluation. For instance, it took DFFRI Lagos several months after its creation to discover that Benue DFFRI, among others, had completed only 7% of roads, for which not less than 100% of the ₦8 million had been received for doing the work. There is the need for DFFRI to build up its own management information system on resource flows and programme performance. As at now there is no consistent record of DFFRI activities and achievements all over the country. This is why DFFRI claims to have constructed several roads, through the illusion that it could do everything, everywhere in rural development, at the same time. The directorate seems to be spreading limited resources too thinly in diversified directions such as direct agricultural production, which is not in accordance with the new agricultural policy provisions. There is absolutely no need for large-scale farms involving DFFRI, at the same time when such other bureaus as River Basin Development Authorities are being disengaged from the same practice. What is recommended for DFFRI is visionary priority setting in which rural feeder road construction should rank most superior.

CONCLUSION

The nature of the rural development problem needs to be properly understood before an appropriate policy instrument can be correctly formulated, designed, and executed. Among the class of available solutions, persistent emphasis on the provision of rural infrastructure is the quickest means of enhancing the welfare standards of Nigeria's rural dwellers.

The inner mechanism of the various institutional means of providing rural infrastructure will only work well if the authorities demonstrate clear programme vision, are target oriented, and have the ability to prioritize the policy instruments well. In addition, the database needs to be tidied up to be able to play a meaningful role in the various technical processes involved in rural infrastructural development. There is also the urgent need to develop uniform technical support capabilities in all the affected institutions to ensure programme accountability and cost effectiveness in the performance of their roles.

REFERENCES

Adegboye, R. O., A. C. Bagu, and D. Olatunbosun (1969), Impact of Western Nigeria Farm Settlements on Surrounding Farmers', *J. Econ. and Soc. Studies* XI (2).

Anonymous (1963), White paper on Integrated Rural Development, Western Nigeria, Official Document No. 8 of 1963.

Idachaba, F. S. (1985) *Rural Infrastructures Development in Nigeria* (Ibadan: University Press).

Idachaba, F. S. (1988), 'Strategies for Achieving Food Self-sufficiency in Nigeria', Keynote address delivered at the 1st National Congress of Science and Technology, held at the University of Ibadan, 16 August 1988.

Olatunbosun, D. (1971), 'Western Nigeria Farm Settlements: An Appraisal', *J. Developing Area* 5 (3) (April 1971).

Rajah, S. S. (1988), 'Viability of Labour-Based Construction System for New ADPs', Paper presented at Workshop of Engineers (Northern Projects), 22–24 August 1988.

World Bank, Guidelines, Procurement under IBRD Loans and IDA Credits, Washington DC, 1985.

Table 1: Stock of Selected Rural Infrastructures provided by Nigerian ADPs/ADAPs* as at December 1986**

ADP and Period	Rural Roads (km) Constructed (1)	Maintained (2)	Rehabilitated (3)	Bridges & Culverts (No.) (4)	Dams (No.) (5)	Boreholes (No.) (6)	Wells (tube or open) (No.) (7)	Farm Service Centres (No.) (8)	Training Hall/ Workshops /Dev. Centres (No.) (9)	Remarks (10)
Recent Projects										
Anambra State ADP (1986)	150 / 22 / (14.7)	450 / 0 / (0)	1000 / 0 / (0)	NA / NA / (NA)	NA / NA / (NA)	NA / NA / (NA)	NA / NA / (NA)	35 / 19 / (52.3)	3 / 1 / (33.3)	Farm Service Centres include stores
Bendel State ADP (1986)	NA / 41.5 / (NA)	900 / 0 / (0)	1000 / 0 / (0)	NA / NA / (NA)	NA / NA / (NA)	NA / NA / (NA)	NA / NA / (NA)	NA / 20 / (NA)	NA / NA / (NA)	Existing Farm Service Centres were only renovated
Benue State ADP (1986)	NBA / NA / (NA)	1100 / 100 / (9.1)	20 / 0 / (0)	30 / 0 / (0)	NA / NA / (NA)	NA / 1 / (NA)	NA / NA / (NA)	NA / 35 / (NA)	2 / 0 / (NA)	Renovation of old Farm Service Centre
Cross River State ADP (1986)	NA / NA / (NA)	1400 / 0 / (0)	NA / NA / (NA)	NA / NA / (NA)	NA / NA / (NA)	NA / NA / (NA)	NA / NA / (NA)	NA / 1 / (NA)	NA / 3 / (NA)	
Imo State ADP (1986)	150 / 22 / (14.7)	400 / 191 / (47.8)	100 / 73 / (73)	50 / 25 / (50)	NA / NA / (NA)	NA / NA / (NA)	NA / NA / (NA)	30 / 10 / (33.3)	12 / 7 / (58.3)	Renovation of existing Farm Service Centres and workshops

Plateau State ADP (1986)	250 76.6 (30.6)	1020 55 (5.3)		735 1 (0.1)	6 0 (NA)	3 2 (66.7)	100 14 (14)	NA NA (NA)	NA NA (NA)	
Ogun State ADP (1986)	NA NA (NA)	NA NA (NA)	NA NA (NA)	NA NA (NA)	NA NA (NA)	100 0 (NA)	NA NA (NA)	10 10 (100)	NA NA (NA)	Renovation of existing Farm Service Centres only.
Older Projects										
Bida ADP (1981–86)	620 431.8 (69.6)	500 573 (114.6)	30 98.9 (329.7)	5 NA (NA)	NA NA (NA)	NA NA (NA)	NA NA (NA)	53 53 (100)	NA 3 (NA)	
Borno State ADAP (1986–86)	190 45 (23.7)	NA NA (NA)	NA NA (NA)	66 34 (51.5)	NA NA (NA)	2 2 (100)	NA NA (NA)	49 62 (126.5)	28 25 (89.3)	
Sokoto State ADP (1983–86)	1900 1345 (70.8)	NA 1090 (NA)	NA 953 (NA)	NA 925 (NA)	NA 10 (NA)	NA 1665 (NA)	500 819 (163.8)	NA 203 (NA)	5 7 (140)	
Gongola State ADAP (1983–86)	235 53.1 (22.6)	NA 1156 (NA)	NA NA (NA)	NA 195 61 (31.3)	NA NA (NA)	NA NA (NA)	NA NA (NA)	60 25 (41.7)	5 2 (40)	
Ilorin ADP (1981–86)	270 72.2 (26.7)	195 178 (91.3)	NA 238 (NA)	NA NA (NA)	NA NA (NA)	7 7 (100)	NA NA (NA)	NA 45 (NA)	5 2 (40)	
Oyo North ADP (1983–86)	396 319 (80.6)	0 22 (00)	31 15 (48.4)	594 130 (21.9)	35 9 (25.7)	NA NA (NA)	255 213 (83.5)	27 9 (33.3)	4 3 (75)	
Bauchi State ADP (1983–86)	NA 1455 (NA)	NA 794 (NA)	NA NA (NA)	NA 41 (NA)	NA 94 (NA)	NA 1205 (NA)	NA 28 (NA)	NA 187 (NA)	NA 22 (NA)	

Kaduna State ADP (1985–86)	134 46 (34.3)	120 30 (25)	NA NA (NA)	NA 47 (NA)	NA NA (NA)	NA NA (NA)	NA NA (NA)	NA NA (NA)	NA NA (NA)
Kano State ADP (1983–86)	1000 577.1 (55.7)	600 350 (58.3)	30 12 (40)	NA NA (NA)	NA NA (NA)	100 100 (100)	750 612 (81.6)	15 5 (33.3)	41 1 (2.4)
Ekiti Akoko ADP (1983–86)	180 133 (73.9)	380 158 (41.6)	320 152.1 (47.5)	NA NA (NA)	NA NA (NA)	NA NA (NA)	NA NA (NA)	NA NA (NA)	NA NA (NA)
Completed Projects									
Ayangba ADP (1978–84)	NA 1561 (NA)	NA NA (NA)	NA NA (NA)	NA NA (NA)	NA 1 (NA)	NA NA (NA)	NA 22 (NA)	NA 34 (NA)	NA NA (NA)
Imo State ADAP (1983–85)	NA NA (NA)	NA NA (NA)	338 725 (214.5)	30 52 (173.3)	NA NA (NA)	NA NA (NA)	NA NA (NA)	41 41 (100)	1 7 (700)
Lafia ADP (1980–85)	260 469 (180.4)	300 327 (109)	645 757 (117)	NA NA (NA)	5 2 (40)	3 3 (79.3)	440 349 (79.3)	NA NA (NA)	NA NA (NA)

* ADP = World-Bank Assisted Agricultural Development Project; ADAP = Purely internally funded imitation of ADP

** The first of trio of numbers is the appraisal target, the underlined figure is the actual, the figure in brackets is the physical achievement in percentage.

NA mostly implies not applicable, meaning that the appraisal report does not provide for the specified infrastructure in the ADP concerned. It seldom implies that data are not available.

Source: APMEPU, Kaduna

Table 2: Activities of DFRRI as at 31 August 1987

State	Km of Roads Constructed	Other Infrastructural Provision
Abuja	-	180 hand-dug wells, 30 boreholes, 20 handpump wells
Anambra	3,900	25 rural water schemes completed
Bauchi	1,123	Water provided for 256 communities out of 156 water scheme
Bendel	-	-
Benue	-	-
Borno	-	-
Cross River	2,111	-
Gongola	1,271	216 wells were dug and 36 ponds located
Imo	1,808.6	Provision of 250 UNICEF type of boreholes
Kwara	1,854 new roads, 534 rehabilitated	-
Niger	-	-
Lagos	653	5 out of 16 electrification schemes covering 25 communities completed
Ogun	Over 1,000	-
Ondo	1,207	Contract awarded for 250 wells fitted with handpumps
Oyo	2,400	-
Plateau	1,361	Boreholes and tube wells in all 14 local government areas
Rivers	222 (tarred)	Installation of 9 pumps at Segbara Local Government Area
Sokoto	96	-

Source: Collated from newspapers and magazines

19

Proposals for Alternative Mode of Market Intervention

Paper presented at the Silver Jubilee Conference of the Agricultural Society of Nigeria, Owerri, 3–6 September 1989.

An apparent confusion ensued in the Nigerian agricultural produce market immediately after the abolition of commodity boards as a component of the new policy of deregulation under the Structural Adjustment Programme (SAP). The economic managers have been generally unresponsive to make new reforms, probably because the beneficiaries are comprised of the farmers who enjoyed massive windfall of incomes from sudden price increases. But how much of the higher incomes is due to increased production (which is sustainable) and how much is due to market phenomena (which is ephemeral) remain unanswered empirical questions. However, the fact that the new income status of the cocoa farmers is already receiving stress signals from the unstable world prices, so soon after the much-celebrated windfall proceeds, suggests that the event is routed in the market and may not necessarily endure.

Therefore, it appears unrealistic to posit, as alternative to intervention, totally a free-market system that would leave agricultural prices to be determined solely by market forces. This paper therefore, proposes an alternative market intervention policy to the commodity boards system. The background is first surveyed before the main elements of the alternative policy are discussed. The data used were collected through personal interview of the relevant managers of price policy, as well as the reading of their official documents.

BACKGROUND

Evolution of Intervention Boards

The traditional agricultural produce price policy in Nigeria has consisted of three generations of marketing or commodity boards. These agencies acted as intervention buyers to forestall pronounced price variations and maintain farm incomes. The first generation of boards existed between 1947 and 1954 and operated under national mandates along commodity lines: Cocoa Marketing Board (1947), Palm Produce Marketing Board (1949), Groundnut Marketing Board (1949), and Cotton Marketing Board (1949). The second generation of boards existed from 1954 and operated under regional mandates on multiple-commodity basis: Western Nigeria Marketing Board (1954), Eastern Nigeria Marketing Board (1954), Northern Nigeria Marketing Board (1954), and Mid-Western Nigeria Marketing Board (1963). The third generation of boards were established in 1976 (Decree No. 29 of 1977) to operate under national mandates along expanded commodity lines: Cocoa Board (Ibadan), Groundnut Board (Kano), Grain Board (Minna), Cotton Board (Funtua), Palm Produce Board (Calabar), Rubber Board (Benin), Roots and Tuber Board (Makurdi). All of these were scrapped in 1986 and their assets sold under receivership.

Intervention Arrangements

An important ancillary body in the commodity boards system was the Technical Committee on Produce Prices (TCPP). This was the real price policymaking body on behalf of the boards, which drew its broad-based membership from the boards, agriculture ministries (federal and states), the finance ministry, and a host of other agencies. The task of TCPP was to recommend produce prices of export crops, guaranteed minimum prices of food crops, and the operational expenses under which the boards operated. The price recommendations were to be made to the Price Fixing Authority (PFA) in the presidency, which approved and announced.

Two types of prices were fixed: one for export crops and the other for food commodities. A producer price was fixed for an export crop, at which the relevant commodity board was obliged to buy. Commodities so bought were to be sold later in the overseas market at higher prices. But no arrangements were made for restitutions. Itinerant purchases of export produce were carried out by appointed licensed buying agents whose regimes of operational expenses were also recommended by TCPP and fixed by PFA. These arrangements were aimed at protecting the farmer from the erratic price fluctuations in the world markets.

As regards food crops, the fixed prices were called guaranteed minimum prices (GMPs). A GMP was a floor price at which the relevant boards were obliged to buy directly from farmers. This arrangement was aimed at mopping up the excess output immediately after harvest to disallow prices from falling below desired limits. The purchases would be held as buffer or strategic stocks, which would be resold at hitherto unstated prices at excess demand periods, in order to prevent prices from rising too high. The arrangement was expected to be capable of maintaining farmers' incomes at desired levels, and to prevent price increases from exceeding affordable limits.

Problem Areas

Three problem areas have been identified in the intervention process, which finally led to the abolition of the boards. These are functional, administrative, and political problems, all of which can be empirically substantiated. The last one (i.e. political intransigency) is well established (Idachaba 1973; Heleiner 1964; FAO 1966). The effect of political interference mainly included the use of huge marketing board reserves meant for stabilization purposes of extra-budget revenues. In essence, the intervention policy turned out to be a way of transferring farm incomes to the execution of government projects, most of them non-agricultural. This practice was also open to great abuses as revealed in the need to probe the role of top political office holders and the consequent confiscation of certain personal properties for public uses (Western Nigeria 1963).

The functional and administrative problems that primarily led to the failure of the last set of boards in the country will be substantiated by examining the structure and activities of the TCPP. These sets of problems manifested in three deficiencies: the fixing of prices that did not reflect market characteristics, poor relevance of fixed prices to farm production decisions, poor relevance of fixed prices to the buying decisions of the commodity boards. Beginning with the quality of fixed prices, the twin factors that led to this are the composition and considerations. These are to be assessed against the determinants of prices, which the TCPP originally desired to follow. These are listed as follows:

(i) world market developments with particular reference to trends in world prices;

(ii) world current market situation as determined by the interplay of production, consumption, stock values, supply and demand for the commodity;
(iii) local (domestic) market prices as determined also by the forces of demand and supply;
(iv) activities and influence of international commodity arrangements on the commodity;
(v) local production costs of a defined unit of the commodity;
(vi) prospect of the commodity in the coming season with particular reference to expectations of changes in the variables listed in i–v;
(vii) previous years' producer prices;
(viii) effect of past and current producer prices on production level;
(ix) effect of producer prices on the rest of the economy, including

(a) cost of living,
(b) cost in relation to industrial raw materials,
(c) intercrop variations, e.g. changes in grain prices which may result in shift in cotton production,
(d) need to check smuggling;

(x) inflation rate;
(xi) financial implications of recommended prices for government, e.g. the budgetary burden of subsidy that may be involved;
(xii) prevailing economic situation in the country.

Needless to say, the above listed points are objective determinants of price in the agricultural market, but the available evidence suggests that they were generally disregarded in the price fixing process. This fact is first reflected in the composition of the membership of the TCPP. Instead of having professional economists from the university and other such bodies, the committee was comprised of mainly seasoned administrators from government ministries and the boards. This probably led to the need for the committee to attempt at certain times to commission studies that would inform the price-fixing process. However, because the consultants were not integral parts of the committee, such studies were set aside for their limited relevance to the problems under focus.[44]

---

[44] An example of this was the study on the role of the Naira exchange on the low level of agricultural production, which was commissioned to the University of Calabar (team led by Professor E. Toyo). See TCPP minutes file for the meeting of 26 July 1983.

As regard considerations, a significant deviation from the objective determinants was observed. Reading of the minutes of the meetings, which led to the recommendation of prices, shows that rather than following the preset determinants, the issues of importance were the geopolitical types. Discussions at the meetings revealed that despite the national mandates given to the boards, each of them was loosely affiliated to states or sections of the country, for example, Cocoa Board (West), Groundnut and Cotton Boards (North), Oil Palm (East), Rubber (Mid-West), etc. In essence, representatives of boards and state ministries merely attended the TCPP meetings to argue cases for upward revisions of prices of commodities that were predominantly produced in their geographical areas[45] (see growth rates of prices in Table 1). Therefore, the joint effect of inadequate composition and considerations of the TCPP led to prices that distorted the agricultural market, rather than those that showed proper relativities.

Another question was whether the fixed prices were made relevant to production decisions. The question is answered by examining the dates of announcements of produce prices to farmers against the production seasons of the respective commodities. Table 2 shows significant delays in this regard, thereby suggesting that farmers in their production decisions did not employ the fixed prices. Consequently, any increase in output could not be directly associated with price signals as originally intended. The reasons for the lack of relevance of prices to farmer production decisions can be attributed to the significant delays observed in the sequence of administrative steps that culminated in price announcements. According to Table 3, the convening of meetings, holding of meetings, seeking of and obtaining approval of recommended prices all lagged behind normative schedules. In this circumstance, the lateness of farmers to receive price information was quite inevitable.

The last aspect of depicting the failure of the price policy is that of the determination of degree to which the fixed prices were also relevant to the buying operations of the boards. This is done by matching the dates of official information with the commencement of the commodity buying seasons. Table 4 shows that in general the official information of the boards lagged behind normative schedules. The resultant effect would be lack of adequate time for preparation and hence the rushing of procedures and lateness in

---

[45] This fact could be substantiated using the concerns of the official agencies, where the relevant commodities are produced in the price fixing process. For example, the governor of Ondo State wrote a letter to the Chief of General Staff to cause revisions of cocoa prices. Similarly, the Kaduna State Ministry of Agriculture protested on behalf of the prices of groundnut and allied commodities.

the commencement of actual buying. Farmers showed discontentment for produce wastage in storage and the incurring of substantial holding costs.

The ultimate consequence of poor and irrelevant prices in quality and usage was the failure to induce farm output and to maintain farmers' incomes as originally desired. In addition, prices could not also be stabilized in the food market because the boards lacked adequate competitive competence in relation to the open markets. All these led to the widespread bankruptcy of the commodity boards (see Table 5). This feature made them easy to abolish under the new deregulation policy of SAP.

Next is a proposal for an alternative market intervention policy replacing the commodity boards system. The proposal draws from the Common Agricultural Policy (Pearce 1981), which is supplemented with the lessons of experience from the commodity boards system to conceptualize the new intervention arrangement.

MECHANISM OF ALTERNATIVE MARKET INTERVENTION POLICY

As earlier expressed, it is unrealistic to posit a free-market system that would leave agricultural prices to be determined solely by market forces, not in the least under a developing agricultural economy in which most producers are weak competitors owing to their small sizes. The current events under SAP have indicated that in the absence of a central policy, farmers would not be protected against wide income fluctuations, the consumers would not be protected against wide price fluctuations, and the market would not be protected against inefficiency factors, in particular, sharp practices such as produce adulteration and poor quality. These have led to low performance of the overall economy in terms of foreign exchange earnings, living standards, and international credibility after deregulation.

Objectives

The goal of the proposed intervention policy is not necessarily different from the one hitherto pursued. This in general is to use price as an instrument for achieving increased agricultural production and fair living standards among producers and consumers. The main specific objectives are as follows:

(i) to set remunerative prices for enhancing incomes of farmers and security of agricultural supply;

(ii) to pursue agricultural market stability through price stabilization; this means that wide price fluctuations are to be prevented;
(iii) to ensure that consumers obtain their supplies at reasonable prices; reasonable implies affordability, which means that the income status of average consumer is considered.

Mode

The mode of the proposed intervention policy rests on two provisions: internal price support and external protection. The internal price support will work on the setting of a price for each scheduled product on an annual basis, which helps the farmers to plan production, and gives economic guidance to all market users. Depending on the nature of the particular product, this price may be called target price, guide price, or norm price. It is the price at which the purchases should take place in the market of the specified commodity in a given year. At a certain percentage below this price, there is to be a floor price or intervention price at which intervention-buying takes place to clear the market of any production that cannot be sold at a higher price.[46]

The two-price mode of operation sets the regimes of prices where the market operates. The purpose, as different from the one-price mode as in the past, is to protect the farmers and the consumers simultaneously. The use of a common price all over the country ensures ecological specialization in agricultural production, whereby low-cost producing areas will outcompete high-cost ones. This will make for optimum utilization within the country as is recently desired (FMAWRRD 1988). In addition, support buying enables the farmers to roughly determine their expected incomes in advance of production, so that good production decisions are arrived at.

Structure and Finance

The exposition on the commodity boards system reveals that the key problem areas in the implementation of the price policy are the constituent organs and the efficiency of their operations, rather than the principles underlying the system. Therefore, the proposed market reform introduces the most radical departures in the structural organization of the system. The

---

[46] In the future, choice may have to be made between imposing similar levies on food imports and imposing an outright ban on them, on the basis of cost and benefits.

important problem areas to be eliminated, as earlier identified, are political, functional, and administrative, which rendered the structure ineffective. The assumed principal causes of these problems are the dominance of government, composition of the personnel involved, and the deviation from objective norms of price determination. The proposed intervention arrangements are directed toward eliminating these problems.

First is to remove the predominance of government in the structure, which had caused the infusion of political considerations in the policy, and explained the presence of wasteful delays and bureaucracy. In the previous arrangement, the PFA was located in the presidency, TCPP was purely a government body located in the Central Bank, and the commodity boards themselves were government buying agencies. These are to be replaced by other agencies under the proposal. The PFA is eliminated to shorten the chain and remove the political connotation of the Cabinet Office. The new body, which will combine the work of recommending and approving the produce prices as earlier done by TCPP and PFA respectively, may be called Agricultural Produce Price Commission (APPC). As in the past, the Central Bank shall be the secretariat of the commission, but the chairman will not be government-appointed; rather it shall be appointed democratically within the commission itself. The membership of APPC shall be broad-based and must include assorted cadres of professional economists, serving on a permanent basis.

In the case of external protection, there are definite departures to be proposed. First, there is no need to distinguish between traded and non-traded commodities. That is, all scheduled commodities are set against their corresponding world markets. However, given the self-sufficiency posture that has led to the outright ban of food items, exportation can be temporarily prevented by imposing a levy on food exports to ensure that they are not sold below the desired price.[47] The second means of external protection is the encouragement of exportation in specified commodities by paying a refund or restitution to bridge the gaps between national prices and lower world prices when they exist. This is to be determined in the light of world prices, the size of the domestic surplus, and expected future trends. Should world prices rise above national prices, an export tax may be imposed, which deters exports and restrains national prices from following world prices up. Therefore, export protection, using variable levy or restitution serves to encourage or discourage imports and exports as desired.

---

[47] In the future, choice may have to be made between imposing similar levies on food imports and imposing an outright ban on them, on the basis of cost and benefits.

The APPC will establish the Commodity Economics and Policy Research Network, which will constantly inform its decisions in the price fixing process. This body should be university-based with a wide geographical spread. Its work shall be the provision of service to the commission as an independent source of critical analysis and evaluation of commodity programmes and their price effects. This will ensure the consideration of all necessary factors in price determination, including world market situation.

The cutting edge of the proposed price policy shall consist of several bodies of Cooperative Intervention Buyers (CIBs). A CIB essentially replaces the commodity board but not as a government participation. As a cooperative venture, it will have shares to be subscribed mostly by the farmers who are the sources of produce to be bought. It shall have representation in the commission and abide with the later authority as regard market operations. Apart from others, the most important advantage of the CIB system is that the gains of intervention will be internalized and be refunded through dividends on shares. This arrangement, which enhances the incomes of farmers, also removes the possibility of using retained funds meant for stabilization purpose as revenue sources for government. However, its operations shall be subjected to taxation.

Financing of the proposed intervention policy will be based on both equity and support funds. It is expected that each CIB should be capable of meeting its own operational expenses right from the outset. The principal component of this is the refund or restitutions out of profit from stabilization operations and share capital. The support financing to be provided by government is similar to a subsidy commitment. This is to be operative over a five-year development phase, and it shall cover the initial capital cost outlay, the expenses of the APPC, as well as the cost of establishing the research network. This government role will be reviewed after five years when the system will have become securely established.

SUMMARY AND CONCLUDING REMARKS

The swift abolition of the commodity board systems at the initial stage of the Structural Adjustment Programme was based on the observation that the inherited price policy was operating like a dog in the manger; it was not only failing to resolve the issues of market instability, insecurity of supply, unstable farm incomes, but also was preventing their solution by other means. Analysis now shows that the inherent problems are political, functional, and administrative in nature.

However, it is unrealistic to posit a free-market basis for determining agricultural prices. The present atmosphere, which is devoid of a concise central price policy, is also inadequate for the country. Recent evidence confirms this through the collapse of cash crop prices in the world market, which wipes away the expectation of continued increases in farm income. It is probable that the recent income windfalls which have accrued to farmers initially are due to the joint effects of the abolition of the commodity boards and deregulation of the foreign exchange market, and not to the former alone. Therefore, a new produce market intervention policy is sought which removes the problem areas of the old commodity boards system and is capable of meeting the desired objectives.

REFERENCES

Akintomide, M. A. (1977), 'Comparative Analysis of the Marketing Boards System and Other Arrangement for Commodity Marketing', International Conference on the Marketing Board System, Nigeria Institute for Social and Economic Research, Ibadan, 29 March – 3 April.

Antonio, Q. B. O. (1984), 'Marketing Development in Nigeria: A Review of Relevant Government Policies', in J. P. Feldman and F. S. Idachaba (eds.), *Crop Marketing and Input Distribution* in Nigeria, Ibadan, Federal Agricultural Coordinating Unit, December.

FAO (Food and Agriculture Organization) (1966), 'Government Marketing Policies in Latin America', *Report of the FAO Seminar*, Bogotá, November–December, Rome FAO.

Federal Military Government (1977), 'Commodity Board Decree 1977' Supplement to Official Gazette 8 (64) (21 April), Part A.

Food Strategies Mission (Team Leader: F. S. Idachaba) (1980), *The Green Revolution: A Food Production Plan for Nigeria*, 2 vols. (Lagos: Federal Ministry of Agriculture), May.

Helleiner, G. K. (1984), 'The Fiscal Role of the Marketing Boards in Nigeria's Economic Development 1947–1961', *The Economic Journal* LXXIV (September): 560–610.

Table 1: Growth rate in fixed prices, 1977/78–1986/87

Commodity	Prices, selected years			Average Annual Growth Rate %
	1987/89	1982/83	1986/87	1977/78 1986/87
Cocoa	1030	1300	3500	+17.3
Coffee/arabica/robusta/liberica	977	1025	3567	+22.9
Groundnuts	275	450	1000	+18.4
Benniseed	NA	315	396	+4.9
Soyabean	130	NA	550	+21.0
Shea nuts	80	NA	287.5	+15.0
Ginger (split/peeled)	350	NA	875	+16.1
Cotton (Grade 1)	160	NA	1000	+16.9
Kenaf (Grade 1)	NA	NA	431	+7.5
Palm Oil (special/technical)	340	468	900	+14.3
Palm Kernel	150	230	600	+22.6
Copra	200	245	400	+10.6
Rubber (processed/unprocessed)	NA	NA	1400	+10.2
Tea	N.A	700	700	+0.0
Sorghum	110	220	575	+18.7
Millet	110	231	575	+19.5
Maize	130	210	520	+18.8
Rice (paddy/milled)	320	450	1000	+15.4

Table 2: Relevance of price information to production decisions, Nigerian Grains Board, 1981–1985

Year	Date of meeting to recommend grains prices	Modal date for planting grains	Expected	Actual	Relevance of price information in production* (days behind/before schedule)
1981	Mar 17	Apr 15	Feb 1	Jun 16	-136
1982	Feb 18	Apr 15	Feb 1	Apr 23	-82
1983	Feb 25	Apr 15	Feb 1	Apr 28	-87
1984	Mar 29	Apr 15	Feb 1	Jun 5	-125
1985	Mar 29	Apr 15	Feb 1	Jun 17	-137
Mean	-	-	-	-	-142

* Negative figures mean number of days behind normative schedule; Positive figures mean number of days behind normative schedule.

Source: Field survey, 1987

Table 3: Timeliness of administrative process leading to the approval of agricultural prices

Year and the month of meeting		Convening meeting	Seeking approval of prices from PFA	Giving approval of prices from PFA
1918	1st	-33	-	-
	2nd	-10	-27	-14
	3rd	+22	-24	-
	Mean	-7	-26	-14
1982	1st	-15	-24	-37
	2nd	-13	-57	-28
	3rd	-2	-11	-41
	Mean	-10	-31	-35
1983	1st	+18	-16	-42
	2nd	-22	-26	-46
	3rd	-10	-20	-32
	Mean	-17	-21	-40
1984	1st	+47	-12	-51
	2nd	+17	-13	-60
	3rd	+4	-60	-40
	Mean	+11	-28	-50
1985	1st	-11	-47	-29
	2nd	-8	+2	-84
	3rd	-8	-26	-84
	Mean	-2	-24	-65
Grand mean		-5	-26	-45

Note: Negative figures mean number of days behind normative schedule, while positive figures mean number of days ahead of normative schedules. Normative schedules are based on (a) time allowance of four weeks between convening of meeting and holding it, (b) time allowance of two days to write to seek approval after the meeting, (c) time allowance of two days to issue approval.

Source: Underlying data of dates from TCPP secretariat

Table 4: Relevance of price information to intervention buying operations boards, 1981–1985

Board(s)	Year	Earliest commencement of buying season	Date of price information expected writing letter	Actual date of writing letter	Relevance of price information (day behind/before schedule)**
Grains	1981	Jul. 1	Jun. 1	Jul. 21	-51
	1982	Jul. 1	Jun. 1	May 25	-88
	1983	Jul. 1	Jun. 1	May 3	-64
	1984	Jul. 1	Jun. 1	Jun. 8	-8
	1985	Jul. 1	Jun. 1	Jun. 25	-25
	Mean	-	-	-	-29
Cotton Groundnut	1981	Nov. 1	Oct. 1	Jul. 21	+71
	1982	Nov. 1	Oct. 1	May 25	+128
	1983	Nov. 1	Oct. 1	May 3	+150
	1984	Nov. 1	Oct. 1	Jun. 8	+114
	1985	Nov. 1	Oct. 1	Jun. 25	+94
	Mean	-	-	-	+111
Cocoa	1981	Oct. 1	Sep. 1	Sep. 9	-8
	1982	Oct. 1	Sep. 1	-	-
	1983	Oct. 1	Sep. 1	Oct. 18	-47
	1984	Oct. 1	Sep. 1	-	-
	1985	Oct. 1	Sep. 1	Nov. 5	-65
	Mean	-	-	-	-40
Rubber	1981	Nov. 1	Oct. 1	-	-
	1982	Nov. 1	Oct. 1	Jan. 19*	-110
	1983	Nov. 1	Oct. 1	Dec. 29	-89
	1984	Nov. 1	Oct. 1	Dec. 12	-72
	1985	Nov. 1	Oct. 1	Mar. 28*	-157
	Mean	-	-	-	-112
Palm Produce	1981	Jan. 1	Dec. 1	-	-
	1982	Jan. 1	Dec. 1	-	-49
	1983	Jan. 1	Dec. 1	Jan 19*	-49
	1984	Jan. 1	Dec. 1	Dec. 29	-11
	1985	Jan. 1	Dec. 1	Dec. 12	-114
	Mean	Jan. 1	-	Mar. 28*	-56
Grand Mean		-	-	-	25

* Month in the following year
** Negative number of days behind normative schedule, positive figures mean number of days before normative schedule.

Source: Field survey, 1987

Table 5: Indebtedness of commodity boards to Central Bank of Nigeria as at 30 November 1986

Board	Total loan granted 1976–1986	Total loan plus accrued interest	Payments, i.e. sales proceeds plus refund	Outstanding balance
Cocoa Board	2,233.9	2,275.9	2,275.9	0
Palm Produce Board	662.2	853.9	325.1	528.8
Cotton Board	359.7	449.7	289.3	160.4
Rubber Board	257.5	309.4	154.5	154.9
Grains Board	184.0	212.5	58.1	154.4
Groundnut Board	42.2	53.4	38.2	15.2
Total	3,739.5	4,154.9	3,141.1	1,013.7

Source: TCPP records

Management Issues in Financing the Graduate Farmers Scheme in Nigeria

Paper presented at Annual Conference of the Farm Management Association of Nigeria, Abeokuta, August 1988.

The null hypothesis that is being tested under the new Graduate Farmers Scheme in Nigeria is that the calibre of the person that is farming does not actually matter. The relevant dimensions of the test statistics must include profitability criteria, credit recovery rate, and the endurance level of the participating graduates, all of which are too early to assess. However, available information suggests that discussion of the salient issues in the administration of this credit programme is not necessary.

The new policy attention on graduate farming seems to have been informed at three levels. First, there has emerged a problem of general unemployment in the country. This is a direct consequence of below-capacity production and reduced growth in the major sectors of major economy, which have been accentuated by the balance of payment and associated problems since early 1980s. In particular, school leavers and university graduates increasingly found it difficult to secure jobs in the preferred sectors such as banks, oil companies, manufacturing companies, consultancy firms, and public service. Probably, the source of immediate worry of government is the fact that the bulk of the unemployed individuals are urban-based, being a category of the population that are vocal and could destabilize governance. Consequently, agriculture was quickly discovered as a source of engaging the youth, thereby leading to extra budgetary provisions to provide necessary incentives. The end result of policy planning is the Graduate Farmers Scheme, which is borne out of a wider programme of self-employment for school leavers.

Secondly, the policy authorities expect that agriculture graduates will be more successful farm managers than the peasants. This notion is based on the wealth of academic knowledge, albeit short of much practical experience, of the former, which is expected to make pronounced differences. Besides, the graduate can adopt innovations fast, safely, and he possesses the ability to decode complex information, all of which should facilitate production and productivity. In addition, he appears to be generally more distinguished, which enhances acceptability and free access to the sources of improved inputs, including the banks and government agencies.

Thirdly, the new scheme is thought to be suitable for gradual replenishment and replacement of the ageing peasant farming population with the more vigorous youthful population. The former is a slow learning and farms smallholding, with limited capacities for expansion. This, together with the first two reasons, provides the rationale for attracting graduates to practical farming. But whether or not these expectations can be fulfilled in the context of the present scheme remains a question that must be discussed. This paper provides the framework for such discussions. The main features of the Graduate Farmers Scheme are highlighted in Section Two. The key issues involved for a successful credit programme are surveyed in Section Three. Section Four contains the conclusions.

The Initial Ideas, New Dimensions, and Selected Events

The original of a special credit package for agriculture graduates is not clearly established yet. What is known is that the scheme was fully operated in Oyo and Lagos states in the 1986 fiscal year. There had been advertisements on radio in both states shortly after the budgets have been read. Interviews were later conducted to select the initial participants among several applicants who were unemployed graduates of agriculture from the universities, polytechnics, and the colleges of education. The main elements of the Oyo State model are described succinctly as follows:[48]

1. Each graduate receives cash of ₦5,000 in the first year, to be followed by another sum of ₦5,000 (later adjusted upwards to ₦8,100), i.e. ₦1,310/ha., in the second year, both to be disbursed directly by the state's Agricultural Credit Corporation; cheque for the first instalment was personally distributed by the governor at a

---

[48] There was no formal memorandum of conditions or credit guidelines released; the main elements were based on personal contacts.

commissioning ceremony. The conditions are quite liberal, including the deposit of degree certificate and provision of two guarantors by each borrower.
2. Each graduate is allocated 5 ha. cleared land in the first year and another 5 ha. cleared land (upon good performance) in the second year at one of five farm settlements (including Ipapo, Akufo, Ogbomoso, Eruwa, and Esa Oke farm settlements); all activities are placed under the supervision of the Ministry of Agriculture and Natural Resources, whose commissioner launched the programme at the Ipapo Farm Settlement. No other location is permitted.
3. Participants are to enjoy the benefits of existing state facilities' functions, including input delivery system and marketing services.
4. The original repayment schedule was as follows: the first ₦5,000 was to be fully repaid with interest (12½%) within the very year of disbursement as a condition for obtaining the second instalment; strong agitation by graduates caused a rescheduling as follows: one year moratorium, repayment of the entire capital and interest to be spread over five years ending 1990.
5. Participants are tenants on the cleared land for two years only, after which they are to become regular settlers who pay rent on a 10-ha. area, or they are allowed to seek alternative sites on their own; this condition also met with resistance from the participants on the ground that it shortens their business horizon somewhat.

This package was immediately attractive to the newly established National Directorate of Employment (NDE), a body that has been set up to tackle the unemployment problem all over the federation. Presently, the Graduate Farmers Scheme had spread to many states as a major component of the self-employment programme of this directorate. The NDE has opened an office in each state, working together with the respective ministries of agriculture in putting the scheme in place. However, state roles are now reduced, being limited to provision of land and the recruitment of one hundred graduates per state. Funding is purely federal, and the directorate does the supervision and disbursements in most states. A few variations exist in a number of states. The Oyo State Agriculture Credit Corporation assists in the disbursement of the cash and kind forms of the loan. The steering committee draws members from the state ministries, who usually include the Permanent Secretary and another professional. In Kwara State, the state involvement is performed by a Deputy Permanent Secretary (Special Duties). Sokoto state has not commenced the scheme yet because of poor response of graduates to the advertisement.

In essence, the NDE package differs from that of Oyo State in two main respects, which are:

1. Both crop and livestock activities were covered (₦11,500 for 5 ha. crops, i.e. ₦2,300/ha., and ₦13,500 for livestock); the credit is to be disbursed in partly cash and partly kind forms.
2. The participants are not compelled to particular sites. They are free to choose sites in their Local Government Areas or elsewhere. Usually, government assists in the acquisition of whatever area. Four types of land were identified in Gongola State. These are family land, government allocation, personal land, and rented land.

To judge this scheme in the present position, let us examine the several submissions made at the annual review meetings in Oyo State, which turned out to be mere fault-finding sessions between the participating graduates on the one hand and the government side on the other hand. First, officials alleged that certain participants were not serious, showing lack of concentration on and commitment to the programme. One basis of this allegation was the information that a number of the graduates had taken up other jobs and other engagements, most suspiciously post-graduate programmes. Another basis was the fact that most graduates did not live permanently on the farm but were resident in the urban centres.

In return, the graduates blamed several aspects of government for the failure in the first year of operation and mostly the poor weather for the failure of the second year. Specifically, the input delivery system was generally indicted, with particular reference to the vices of Agricultural Inputs and Services Unit (AISU). The absence of a coherent pricing policy when the grains market was glutted in 1986 was also mentioned. The excess supply which created a downwards pressure on the general price level (e.g. maize price falling as low as ₦200/ton when production cost was high, up to ₦700/ha.) was the result of good harvest coupled with sudden release of stock following prior abolition of the commodity boards. In another claim, the complex lack of facilitating marketing assistance was a major source of business failure. There were no structures on many farm settlements where produce would be temporarily held (except at Akufo farm settlement, where living houses were allocated). On-farm processing and general transportation were very difficult, leading to huge crop wastages. Lastly, participants also alleged that their programmes were given excessive publicity, and they cited the effect of this on local house rents and family demands. This allegation and the accusation of red tape or bureaucratic delays in major decisions (particularly loan disbursements) were of no less importance in the success of this programme than the ones earlier mentioned.

To sum up, the implementation of the initial ideas and subsequent modifications has faced a number of problems. Practical efforts to nip these in the bud must consider a number of fundamental issues of credit administration to which we now turn.

Issues in the Administration of Graduate Farmers' Credit

The several issues to be discussed in the context of the graduate farmers' credit package have been separated into three groups. These are

(a) characteristics of credit,
(b) the borrowers' characteristics, and
(c) miscellaneous issues.

Characteristics of Credit

Two characteristics of this credit programme will be discussed. The first is adequacy and the other is form. Adequacy can be examined at two levels, which include the adequacy of cash to meet the necessary input requirement of the land size and the adequacy of the total package to satisfy both the implicit and explicit objectives of the programme. The latter shall be deferred to the discussion of borrowers' characteristics because these affect the suitability of the entire package.

The notion of inadequacy of the cash provision is based mainly on the grounds of agitation of the graduates in Oyo State which caused the second year cash instalment to be increased from ₦5,000 to ₦8,100. Participants have argued strongly that their expenditure profile in the first year averaged ₦700/hectare of maize land, thereby leaving only ₦1,500 available for exigencies, processing, transportation, and general upkeep. This balance does not also leave room for acquisition of simple tools and machines such as spraying pump (about ₦1,250) etc. The rationale for acquiring small-scale machines is based on the poor performance of the general input delivery system.

This table shows that the prices of inputs and operations have increased substantially between 1986 and 1988. Private owner charges have generally increased higher than government prices. For instance, government charge for ploughing has increased by 20% between 1986 and the current season, while the increased recorded in the private owners market was over 33%. The resultant effect of such increases in price is to decrease the real value of credit.

Inadequacy of loan both in absolute and relative terms will lead to inefficient farm ventures. What appears to be a solution to the problem of inadequacy is for government to require would-be participants to submit a project proposal. This will eliminate locational differences in enterprise and price structures. Greater adequacy standards will also be achieved by making liberal allowances for acquisition of necessary capital items, general upkeeping of graduates, and possible cost overruns.

The second characteristic of credit to be discussed is form. In general, a mixture of cash and kind forms is desirable. Credit in kind eases input procurement, provides some economies of scale, and generally limits diversion. Nevertheless, cash must also be available for prompt payments of labour and other services. The Oyo State model is devoid of kind forms. This has been improved upon by NDE, which supplies some inputs in kind. The question being incremental cost of credit administration, which results. This question carries small weight in government credits, which should be seen as part of general services. What is most important is to avoid the kind of delay in input distribution which the NDE participants are now complaining about.

Borrowers' Characteristics

An important question is, how does one see a university graduate in his professional and social statuses? Professionally, his ambition seems far higher than the opportunities provided by present package. The low level of capital provided in this package appears quite incongruous with the exposure of the graduate to efficient machineries of farm operations in the curricula which have been recently undergone in the university. Therefore, the participants would continue to feel that circumstances have only (probably temporarily) pushed them to take part in the scheme as a second best option.

The social status of these participants is equally important for discussion. A number of characteristics abound in this regard. First, a typical Nigerian graduate is deeply used to urban life. In this setting, he is familiar with supply of pipe-borne water, electricity, and tarred roads. Even though he has studied agriculture, his exposure to a practical farm situation is limited to a few brief visits to big farms and experimental holdings in the neighbourhood. Suffice it to say that the environment into which the credit scheme is throwing the young graduate is drastically different from his average experience. On the existing farms, social life activities are at the lowest ebb, including the unavailability of hygiene, food, and water. Communication is also very poor, owing to bad roads and lack of transport facilities.

An attempt to reduce the effect of this condition on health and cost of production has led to a rather expensive pattern of living of the graduates, relative to the size and scope of their credit package. Typically, there is a four-point pattern of settlement among the current participating graduates (see Fig. 1). The graduates were originally scattered over the state, living either permanently or waiting to secure jobs. Most of them have their parents and families located in the towns, which serve this purpose. They all moved to Ibadan when this scheme started. In Ibadan, they waited with relatives until they were posted to the various farm settlements. There also arose the need to be semi-permanent in Ibadan for linkage with the Ministry of Agriculture and the Agricultural Credit Corporation. This linkage was later to become a stable feature of the programme as the experience of participants has shown. After allocation to farm settlement, each graduate settled at the nearest urban centre to his own farm. The various towns which have served as secondary urban centres are Iseyin (Ipapo Farm Settlement), Ilesha (Esa Oke Farm Settlement), Eruwa (Eruwa Farm Settlement), Ibadan (Akufo Farm Settlement), and Ogbomosho (Ogbomoso Farm Settlement). The participants visit the respective secondary urban centres daily for food and farm labour while they go to the primary urban centres (Ibadan) at least once a week to attend to some official matters with government agencies, including input procurement and loans disbursement matters. Visits to the original places of abode are usually made at weekends. The real action site is the farm settlement where participants rent single rooms in the nearby villages to sleep and keep the farm tools.

The cost implication of this pattern of settlement is immense. A lot of expenses are incurred in moving from one point of abode to another, as well as in maintaining several houses in several places. One graduate who wanted to reduce the length of this chain by bringing his family to the secondary urban centre rented a three-bedroom flat and paid rents in advance. This ate so deep into the available fund that he consequently became incapable of performing basic operations such as fertilizer application and weeding on the maize farm, thereby causing crop failure.

Ogun State Government seems to recognize this characteristic of the borrowers more than the others. Another fund was secured to build houses on site for the participants in this state. This will reduce the length of the chain substantially, thereby reducing the societal pressures on the borrowers. An alternative suggestion is to incorporate building expenses into loan package, which also gives the programme a more permanent picture. The fact is that graduates must have adequate compensating inducements to adapt them to life on farm. In the present setting, they are not likely to close their eyes to urban jobs. What may result in the present circumstances in the end is gradual withdrawal from the scheme as the employment situation improves in the preferred sectors.

Miscellaneous Issues

Equity considerations are important issues in the graduate farmers' credit package. The stance of equality maintained by government in the value of loan and other treatments of participants is the basis of the observed inequity at the implementation stage. First, the production environment is largely different with respect to costs of labour, transportation, living, among others. Equal treatments to participants presuppose that a unit of produce costs the farmer the same amount in all the locations of a state and in all states of the federation. This practice disregards the existence of internal comparative advantages and disadvantages. In this regard, it is suggested that each farmer be treated on his own merit as regards the peculiar environment in which he farms. There is need for careful study of the production process in each location, which will require a great deal of flexibility in the total value of credit. There are several examples of the environmental inequity in Oyo State scheme to justify differential treatments. One example is that some farm settlements are located by the side of main roads such as Ipapo, which enables the farmer to transport fertilizer, seed, and other bulky items easily. The situation is different at the Eruwa farm settlement, where about fifteen kilometres need to be covered on foot. Therefore, transportation cost is higher in the latter environment than the former. Another example of inequity is the differential presence of infrastructural facilities at the farm sites. For example, while participants in Akufo farm settlement enjoy the unique opportunities of being allocated farmhouse on lease, such facilities are not available in the other sites in Oyo State. In complete contrast to the Akufo situation, the Ipapo farm settlement is totally bare, with no place to rest in the sun, hide in the rain, keep farm implements, or hold produce briefly. Substantial costs are attached to the inevitable wastages that commonly result in these circumstances.

The differential incidence of weather disasters is another issue of equity. In the event of drought, for instance, the farms located in derived savannah areas are likely to be more affected than those in the rain forest areas. The reverse is expected in the case of a flood. Differential effects of disease outbreaks cannot be so easy to predict. The point being made is that natural disasters, which are common events in agriculture, will not affect farmers equally. Therefore, there is the need for differential assistance to participants at such times. Government needs close monitoring of the scheme, which will enable the correct assessment of situations when help is required. The basis of the suggestion for participants' support is that government lured them and it has a purpose to sufficiently attract them and keep them there. Be that as it may, protection at business infancy will reduce frustrations and desertions and lack of repayment ability. Such frustrations are now appearing on the faces of pioneer participants of Oyo State. The strange weather of 1987 when both drought and flood occurred in the same year was reported to have caused huge failures on the farms. This has led the participants to ask for government support in one way or a combination of three. First, government can design an income support scheme to cushion the effect of the bad weather. The second way is to grant new loans because the original awards have been sunk and lost to bad weather. The third suggestion is for government to write off substantial parts of the loan as a relief for the participants to remain in business. Whichever way is convenient for government, it is imperative that assistance be given during bad weather so as not to lower the morale of participants who are being encouraged to farm. In general, farmers cannot be over-pampered with direct assistance of this nature.

Another important issue is the noise effect of government roles on the scheme. The government, while trying to inform the public about its activities, often goes too far into unnecessary details which may create new problems. Good credit administration involves the protection of the debtor. Dissemination of information about credit matters needs careful selective exposure to the public. At a time like this, when average family incomes are very low, reading of the names of beneficiaries on radio and television to come and receive relatively big sum of money, as it happened in Oyo State, might not be in the business interests of the recipients. To worsen the matter, participants in this scheme were watched on the television while shaking hands with governor to receive their open cheques. Discussions with participants established that this practice created two problems. One was the unintended demand on the money from friends and relatives who came to borrow and take their own share of the cake. Some participants confessed in this regard that they required extra self-discipline to resist such pressures as the requests to use part of the money to renovate old houses, settle old family debts, among others, which were some proposals made to them at emergency family meetings. This kind of societal pressure is a feature of a largely illiterate setting like ours, where understanding of purpose is minimal and the ability to decode information heard on radio or television varies widely. This factor must be having a strong explanatory power in the explanation of low recovery rates of most government agricultural credits. The second problem caused by undue publicity of credit facilities was the measurable rises in house rents and general costs of living in the towns and villages close to the farm settlements. Participants claimed that this could be sourced to the speculative knowledge of the local populace that a new category of farmers would be arriving in their midst who carried a lot of government money.

Further on the role of government in the credit administration are a number of other important issues. One is the lack of a comprehensive memorandum of scheme. The memorandum needs to state government objectives, details of credit package, all conditions attached, and the specific roles of all parties involved in the scheme. The evidence of the possible problems if a memorandum is lacking occurred in Oyo State. The supervising ministry and credit corporation, in the absence of clear guidelines, have been opposed to the granting of moratorium and non-removal of interest at source. Consequently, before operations commenced, the interest had been deducted before disbursing the balance to participants. Suddenly, the governor made contradictory statements on this at a public outing: 'the credit was interest free in the first year and a moratorium is granted'. This led to stiff agitation from the graduates, which caused the governor express order for refunds. This type of confusion among government agencies can be wasteful and shows lack of proper coordination that is required for a scheme like this to succeed.

The usual problems of bureaucratic delays and administrative red tape have also been encountered in the graduate farmers' scheme. The negative effects of these problems on agricultural activities cannot be overemphasized. In Oyo State, the scheme dragged very long on the ground before actions were actually recorded on the farm. The idea was first expressed during budget announcements in January 1986. The briefing of the Commissioner of Agriculture took place on 17 April 1986 when farm activities were supposed to have long started. There were several shifting of dates before cheques were actually disbursed in May. The farm operations of that season must have been rushed up. Enough time allowance was not made for input purchases, organization of production and settlement, which might lead to unexpected cost overruns and inefficiency.

Finally, there were several reports of poor performance of the various government agencies which are to play important roles in the scheme. The most important support function is the input delivery system. The channel of input delivery exists to perform this role, such as the Agricultural Inputs and Services Unit (AISU) in Oyo State, which had been seriously incapacitated because of lack of adequate numbers of tractors and other machines. In addition, several machines are usually out of order. This has caused several disruptions in the work programme of participants, which creates new expenditure burdens. Therefore, there is need for total revamping of the various input delivery organizations so as to enhance the productivity of the graduate farmers.

Conclusion

The present form and function of the graduate farmers' scheme will likely produce a new generation of debtors because the characteristics of the loan have not been perfectly synchronized with those of the borrowers. From the standpoint of stemming an imminent unemployment problem, the scheme will succeed only to the extent that graduates will be temporarily engaged until new vacancies exist in the preferred sectors of the economy. However, it seems to be a panicky measure to solve the food shortage problem, as it is presently constituted. In this regard also, there is no evidence to suggest that the calibre of person that is farming will actually matter.

If government wants to attract graduates to the farm through a credit programme, the ceiling of such credit must be flexible and should reflect the peculiar environments where agricultural production takes place. The value of package must be adequate in size and scope, it must involve minimum publicity, the support services must be readily available, and participants must be assisted promptly at bad times.

REFERENCES

Adegboye, R. O., A. C. Basu, D. Olatubosin (1969), 'Impact of Western Nigeria Farm Settlement on Surrounding Farmers', *J. Econ. and Social Studies* 1 (2 & 9).

FAO (Food and Agriculture Organization) (1966), *Agricultural Development in Nigeria 1965–1980*.

Idachaba, F. S. (1985), 'Integrated Rural Development in Nigeria: Lessons from Experiences', Paper presented at the Workshop on Designing Rural Development Strategies, Federal Agricultural Coordinating Unit, Ibadan, December 1985.

The Marketing of Agricultural Pesticides in Nigeria: Organization and Efficiency

Invited paper presented at the National Workshop on the Pesticide Industry in Nigeria, conference Centre, University of Ibadan, Ibadan; 24-27 September, 1990.

Agricultural pesticides are a wide spectrum of chemical products including insecticides, acaricides, molluscicides, rodenticides, nematicides, fungicides, herbicides, and plant regulators. They have contributed as modern inputs to the achievement of Green Revolution objectives in many parts of the world. Well-organized and efficient marketing of pesticides is required in Nigeria to generate large incremental food and non-food output, commensurate with the growing demand of the population for these items. There is indication that a large excess demand capacity for pesticides exists in the country. For instance, relative to optimal land-based herbicide requirement estimates of 5,232.8 thousand litres for maize production alone in 1985 (Ayoola 1988), the main private suppliers sold 382.1 thousand litres only (APMEU 1987). It is believed that even if direct user-imports of herbicides were added in that year, the planned herbicide requirement of maize cultivation would not have been met, let alone the requirements of other crops. This paper views this situation as largely a marketing problem. Therefore, the organization of pesticide marketing system is examined towards analyzing the parameters of the degree of efficiency attained.

GENERAL ASPECTS OF PESTICIDE MARKETING

In the Nigerian situation, in which physical shortages of pesticides seriously affect agricultural productivity, the strengthening of the marketing system is particularly urgent. The effectiveness of marketing organization depends on several factors. Among these, product characteristics, externalities, and government policies should be illuminated.

Product Characteristics (Toxicity Factor)

Pesticide marketing involves special elements due to the toxic nature of the products. According to Apeji (1990), a pesticide is a poison *sua generis*, which is used to disrupt some aspect of the pest organism's vital function so as to inactivate the pest or even kill it. The biological activity of the pesticide is due to the constituent active ingredients, which may enter the body of plant or animal pests through skin contact (dermal toxicity), ingestion (oral toxicity), and breathing (inhalation toxicity). The implication of the toxicity factor is that there is additional need for pesticide safety in the marketing chain. The consumer needs protection from the exposure to treated materials; the farmers and workers need protection during pesticides' application; factory workers need protection during production and packaging.

Therefore, pesticide marketing should normally take place under certain rules and regulations which ensure safety of handlers and end users. The International Group of National Associations of Manufacturers of Agrochemical Products (GIFAP) provides some guidelines, many of which have implications for marketing functions in Nigeria (Ogunyadeka 1988). In particular, marketing information must be clear and simple to guide marketing agents and users about choice of product, handling of packs and repacks, prevention of contamination, storage techniques, dosage, mixing, empty containers, and equipment as well as first aid measures. The environmental consequences of pesticide residues should be part of marketing information.

Externalities

The application of pesticide is a good example of the presence of externalities in agricultural resource use. On the benefit side of the profit equation, a benefit accruing to the farmer as a result of pesticide application is a benefit to the society and there are no positive externalities in the form of additional benefits to society. But on the cost side, society often bears costs over and above those borne by the farmer. For instance, a single application of pesticide to the pod-sucking insect of cowpea may obstruct the fulfilment of objectives of other enterprises in the immediate environment: honey production farmers whose bees were killed, pesticide residues in water and food, the possible death of wildlife which hurts the recreation business.

Thus, there arises a new stream of costs to society as a result of the farmers' action to increase productivity through pesticide use. Some of these include the increased costs of treating drinking water to make sure residues are removed, the non-market cost of decreased wildlife, business failure in honey production, among others. The role of marketing in limiting the impact of negative externalities associated with pesticide use is to strengthen the quality and quantity of advisory and information services to farmers. The marketing activities must be carried out under government regulations and close monitoring and supervision. The active ingredients and externality effects of the new pesticide formulations need to be assessed and regulated.

Pesticide Subsidy

The critical instrument of government intervention in pesticide marketing is subsidy administration. There was a strong need to generate massive pesticide uptake by farmers through subsidized prices. The first comprehensive and consistent farm input subsidy programme was the cocoa pesticide subsidy scheme of the old Western Region (Idachaba 1981). In this scheme, annual subsidy outlay was ₦1,116,694 (1959/60–1976/77). The defunct Nigerian Cocoa Board subsequently administered another stream of cocoa pesticide subsidy in the producing states of the federation, amounting to ₦3,655,155.39 (1978/79–1980/81). A number of other pesticide subsidy schemes have been operated directly and indirectly at both state and federal levels.

The present situation is that government intends to gradually withdraw subsidy from farm inputs, including pesticides. There is also a concurrent effort to privatize the distribution system. Consequently, within the context of sustainability and continuity of existing agricultural development efforts, the objective of full-cost recovery has been set as an initial step toward privatizing the input distribution network of certain agricultural parastatals. Toward this end, the retail end may be reorganized to emphasize farmers' cooperatives and other accredited bodies in the supply of pesticides and other inputs directly to farmers. Some of the ways in which the subsidy factor affects pesticide marketing are highlighted as follows:

1. Consuming farmers have become mentally dependent on pesticide subsidy to the extent that the fiscal burden required to sustain it at current levels of use is overwhelming; the market for these products therefore fluctuates with the unstable budget allocations.

2. The presence of subsidy might be largely responsible for distortions in the pesticide market, whereby the price mechanism fails to convey the preferences of the consumers to the producers through the marketing systems; the inevitable but not physically observable consequence is misallocation of scarce productive resources.
3. The presence of subsidy in the marketing system might have greatly reduced the level of competition among sellers and consequently the accruing incentives, which in turn dampens private enterprise initiatives in the distribution of pesticides; substantial value-added losses might therefore exist.
4. Subsidy could introduce undesirable elements into the marketing system, especially if international border price advantages exist against the country, which administers heavy subsidy on farm inputs.

In the alternative, well-organized and functioning market will reduce distribution costs significantly enough to offset the level of subsidy support. In particular, the adequate transport function of marketing will enhance free flow of pesticides and deep penetration of the farmers. A massive feeder road network will therefore lead to substantial cost savings in terms of vehicle maintenance and traffic hours necessary to distribute a unit of pesticide.

ORGANIZATIONAL ASPECTS OF PESTICIDE MARKETING

The typology of pesticide marketing organization is presented in Figure 1. Domestic supply of pesticides is satisfied in two ways. One way is by the importation of active ingredients by agrochemical companies, which subsequently undertake investments in blending. In the blending process, the local formulation depends on 70% of imported raw materials (Ikemefuna 1988). The second way is through importation of finished pesticides by either the agrochemical companies, or possibly certain government agencies. This constitutes the largest channel of supply, accounting for about 90% of total supply (Ikemefuna op. cit.). The agrochemical companies operate a chain of wholesalers, while the government agencies sell to farmers directly. The critical elements of market organization are structure, conduct, and performance.

Structure

The structure of pesticide market organization concerns the degree of sellers' concentration, product differentiation, as well as freedom of entry and exit. However, sellers' concentration seems to be the most critical structural element of pesticide marketing. In the case of product differentiation, it is only necessary to ensure that the wide variability of the agrochemical products needs a market intelligence operation, which will aim at preventing the exploitation of the farmer's generally low educational level to the advantage of suppliers. The more differentiated the agrochemical product, the greater the possibility to sell the same type of pesticide at higher prices under different labels. As to freedom of entry to and exit from the pesticide market, the theoretical argument is that it is when such freedom exists that market clearing price can be arrived at. In this condition, excess profit is mopped from the market, and the sellers obtain satisfactory incentives to stay in the market. On the demand side, the farmers who can pay for the product obtain supplies.

In regard to sellers' concentration, the number and size of sellers come under focus. A survey conducted by APMEU (1987) indicates that five agrochemical companies are the major suppliers of pesticide in Nigeria. They are (i) National Oil and Chemical Marketing Company (NOLCHEM), (ii) Swiss Nigerian Chemical Company (formerly Ciba-Geigy), (iii) Chemical and Allied Products Limited (CAPL), (iv) Rhone-Total Nigeria Limited, and (v) Dizengolf West Africa (Nigeria) Limited. To these can also be added Unichem Nigeria Limited, which is the authorized distributor of Bayer AG's products in Nigeria. Table 1 shows the relative shares of the first five agrochemical companies in the pesticide market. A lopsided concentration is indicated. The combined effect of the presence of few sellers and high concentration of sales is the tendency toward monopoly, whereby sellers can collectively or individually manipulate the price or quantity at will to increase their profits at any given time.

Conduct

The behaviour of a market as it affects decision-making about price has a number of possibilities: price dictation which is akin to monopoly, dominant price leadership which reflects high concentration of sellers, barometric price leadership which suggests that a firm has an edge, and collusive price leadership which is a feature of a few sellers. As small and concentrated as the pesticide market presently is, the possibility exists for firms to engage in overt or collusive pricing. An informal cartel can be formed through which pesticide prices can be fixed or influenced.

The presence of subsidy on pesticides has probably strengthened the collusive behaviour. As subsidy discourages aggressive marketing stint, the pesticide suppliers need to present common pricing posture to obtain the maximum subsidy claim from government.

Performance

How good the job of the pesticide supply companies is can be highlighted as follows:

(a) **Products capabilities**—There is wide range of pesticides suitable under the different ecological environments. There is also indication that high degree of specificity and selectivity has been obtained. But there is large scope for market information and extension to properly inform the farmers in this regard.

(b) **Technological progress**—The innovativeness of the existing manufacturers and marketers is called into question by the presence of large volumes of local raw materials not utilized in pesticide formulation. After many years of pesticide use in the country, locally available raw materials should have been prospected and exploited towards the substitution of the active ingredients and solution media needed to formulate the products. Apart from the negative implications of foreign dependence (Ayoola 1980), the high import content of pesticides means that much of the gains of production and marketing fail to accrue to Nigerians. In addition, the situation makes it difficult or impossible for the existence of appropriate feedback mechanism to incorporate farmers' experiences in the formula of pesticides.

(c) **Level of output**—The high prices of pesticides are suggestive of large excess demand situation in the country. Table 2 shows the distribution of different pesticides by ADPs compared to their estimated quantities. The indication is that large increases in supply are necessary.

(d) **Unethical practices**—Product adulteration has been variously reported in the pesticide market. This suggests the presence of weak intelligence and surveillance activities in the market. If unchecked quickly, farmers' confidence will decrease, which will definitely affect their level of productivity.

PARAMETERS OF PESTICIDE MARKETING EFFICIENCY

The organizational aspects, including structure, conduct, and performance play important roles in determining the degree of efficiency attained in marketing. The critical parameters of pesticide marketing efficiency are demand estimates, delivery system, promotion and advisory services, pricing, margins and incentives, financing, and technical support service.

Estimation of pesticide demand

The accurate estimates of pesticide demand are necessary not only to ensure that farmers' demands are met by timely supplies but also to avoid excess stocks together with associated inventory-carrying costs. Given that the extension system effectively demonstrates to farmers the presence of incremental physical and monetary benefits from pesticide usage, another factor creating the need for forward demand estimates is the need to make necessary arrangements for transportation and credit which will make it easy for farmers to buy the inputs.

The accuracy and timeliness of agrochemicals demand estimates requires improvements to improve the effectiveness of their marketing. Usually, the right quantities of pesticides are not delivered at the distribution points at the right time. This results from underestimation of farmers' demands and causes enormous direct and indirect social costs. Correct and timely estimates will enhance government's allocative efficiency in terms of foreign exchange priorities and fiscal responsibilities.

Delivery Systems

The marketing system for pesticides involves two delivery systems: competitive marketing and government distribution system. In the former, private agrochemical companies are supposed to be under intense pressure of new entries and should therefore be inclined to offer good services in the forms of genuine products and support functions.

The presence of parallel government delivery system greatly influences the operations of the private bodies. At present, pesticide distribution by government agencies is part of a generalized input distribution system integrated with other services such as extension and irrigation services. In the case of the defunct marketing or commodity boards system, pesticide sales were integrated with produce marketing, which had the advantage of non-fragmentation of delivery systems and, consequently, reduced operating costs. There is basis to undertake the harmonization of the government and private delivery systems, as the possibility of abuse exists, whereby the cheaper government delivery channel is exploited to obtain supplies of pesticides only to be resold through the private system. The development of arbitrage is a direct consequence of price discrimination in the two parallel markets that are not spatially separated.

Pesticide promotion and advisory services

The wholesaler and retailer have great potential to promote pesticides use. The heavy reliance on government extension services often gets frustrated for lack of funds, transport, personnel, and other facilities. Therefore, the official extension efforts require the support of the private marketers for greater effectiveness. In particular, the pesticide retailers can become effective change agents upon training. Facing private establishments, according to Fagbamiye (1988), the problem is that agricultural extension is basically expensive in terms of personnel, transportation and infrastructure. It is 'not a profit-oriented service and therefore cannot be commercialized'. Nevertheless, the development of consumer-oriented promotion and advisory service through the marketing system can strengthen the existing strategies of agricultural development.

Pesticide marketing margins and incentives

The option of fixed retail margins is now a popular idea in the gradual privatization of the farm input distribution system. The rationale for fixed marketing margin is to control prices during scarcity. Gradual privatization will probably begin at the retail end of the chain with the involvement of cooperative societies and other accredited farmers' bodies who earn fixed margins for their retail operations. The explanation is to prevent traders from making excessive profits or to protect agricultural production system from operating at high cost. However, there is doubt about the possibility of enforcing fixed prices, especially if they fall out of line with the realities of demand and supply. In such a situation, according to Mittendorf (1986), 'the market pressure to evade them become overwhelming; a black market develops, particularly under conditions of inflation, and price policies have to be reviewed incessantly and adjusted accordingly'.

Fixed marketing margin is likely to be associated with a kind of equalization policy, whereby the same pesticides are supplied at the same prices throughout the country. This disregards differential transport and other costs involved. The implication of such a policy is that the nearby farmers finance part of transport costs of farmers located far away from a given point of origin of the pesticide. This may lead to an implicit discouragement of the use of pesticides where costs of transport are too low, apart from being unfair and the possibility of poor input allocative efficiency that results.

In addition, the effect of fixed margins on sellers' morale in the long run is negative. In the long run, retailers and wholesalers have to finance adequate stocks and promotional and technical advisory services, for which fixed margins would not be adequate. Therefore, fixed margins cannot provide sufficient incentive to improve marketing and promotional services in the long run. Consequently, conditions of limited competition in pesticide marketing will require additional incentives. Such incentives are implicitly available in competitive systems through the market mechanism. But when competition is impaired as necessitated by the need for marketing margins, exogenous rewards may be necessary to encourage innovative proposals, cost reduction, and improved customer services.

In this regard, examples of practices to be encouraged among pesticide sellers include seasonal price differentiation and quantity rebates. For instance, lower prices at off season of farm production when demand pressure is low would encourage the early purchase of pesticides and therefore even out seasonal demand for transport. The reduced pressure on transport at peak of production would lead to savings in annual average cost of transport. In addition, volume discounts would encourage purchases by farmers' groups. Consequently, seasonal price variation and quantity rebates enhance efficiency of marketing and production, leading to lower produce prices. However, if competition is not impaired, these practices need not be exogenously induced as the market provides necessary incentives to ensure that they are performed.

Pesticide marketing finance

There is a large scope to improve access to institutional credit for the private and cooperative enterprises to finance stocks and investment in storage and transport. Both the private and government delivery systems underestimate the need for short-, medium-, and long-term credit for farm input dealers. The main problems of institutional credit are non-payment of loans and high administrative costs, including delivery and recovery costs. In addition, the branch network of banks is inadequate to service small farmers' groups at the same time when the presence of bureaucratic procedures remains a major constraint for them to get formal credit. Therefore, extension of the branch network where viable and greater emphasis on group loans are plausible suggestions for reducing the risk of credit non-repayment and administrative costs. Furthermore, the agrochemical marketing companies, which obviously enjoy greater access to institutional credit, should offer credit facilities to farmers as a means of generating additional sales.

RECOMMENDED ACTION

The product characteristics, particularly the toxicity factor, the presence of substantial externalities, and other factors suggest a major role for government in planning, monitoring, and evaluation services in the pesticide market. The objective is to continuously review the marketing organization and efficiency toward the distribution of greatest volume of pesticides at the minimum cost.

1. There should be regular evaluation of the preceding year's marketing performance. In the evaluation exercise, the actual outcome will be compared with original plan to indicate critical factors, such as availability criteria, adequacy of stocks and support services such as credit and extension, as well as cost effectiveness. Critical areas for parastatals include the appropriateness of cost accounting procedure, and economic sustainability of delivery system.
2. Annual standardized surveys of costs and margins of pesticide marketing should be conducted towards creating the required cost awareness and calculation methods, and promoting the analysis of options for reducing marketing costs commensurate with required services.
3. There is need to provide advice on the preparation of marketing development plans needed to restructure the pesticide marketing system. The necessary areas of advice are outline of future strategies, policies, instruments, and training necessary to strengthen the marketing system. There should also be constant advice on the nature and degree of more integrated delivery systems that are adequately linked with agricultural practice of fragmented systems that are not cost and service effective.

SUMMARY AND CONCLUSIONS

The toxic nature of pesticides, presence of substantial negative externalities, and the rationale for subsidy are important issues in designing appropriate organization for pesticide marketing. The desirable feature of organization has structural, conduct, and performance dimensions. The degree of pesticide marketing efficiency depends on accurate and timely estimation of demand, appropriate delivery systems, promotion and advisory services, remunerative market incentives, and accessibility to credit. There is wide scope for government activities in support services such as planning, monitoring, and evaluation of the organized pesticide marketing in Nigeria.

In conclusion, the minimal contribution of pesticides to green revolution objectives in Nigeria arises largely from defects in structure, conduct, and performance of the marketing system. Therefore, there is urgent need to expand the quantity and enhance the effectiveness of pesticide use among farmers through a marketing reform. The objective of this is to promote the technical and organizational efficiency of marketing so as to deliver the maximum volume of the inputs to farmers at minimum costs. While government is required to take initiatives for the market reform, collaborative efforts of the private sector participants are crucial to appropriate solutions of the marketing problems. In particular, the role of private agrochemical needs review to emphasize marketing research and development objectives.

REFERENCES

Apeji, S. A. (1990), 'Pesticide Toxicology and Environmental Pollution', Workshop on Pesticide Usage and Environmental Pollution, Abeokuta, 25–27 June 1990.

Ayoola, G. B. (1988a), 'An Optimal Time Path for Self-Sufficiency in Maize: A Production Control Approach', PhD thesis, Department of Agricultural Economics, University of Ibadan, Ibadan.

Ayoola, G. B. (1988b), 'The Dependency Problem in Food Grains and Rationale for Self-Sufficiency Policy in Nigeria', *J. Rural Dev. Nig.* 3 (1).

Fagbamiye, A. I. (1988), 'Agricultural Extension by CAPL', *Daily Times*, 26 April 1988.

Idachaba, F. S. (1981), 'A Farm Input Subsidy Policy for the Green Revolution Programme in Nigeria', Paper for Ministerial Committee on Farm Input Subsidies, Jos, 8–9 December 1981.

Ikemefuna, R. M. (1988), 'Marketing of Pesticides in Nigeria', *Daily Times*, 26 April 1988.

Mittendorf, H. J. (1986), 'Input Marketing Systems' in Dieter Elz (ed.), *Agricultural Marketing Strategy and Pricing Policy* (Washington DC: World Bank).

Oguyadeka, A. (1988), 'Safe Handling of Pesticides', *Daily Times*, 26 April 26 1988.

Table 1: Market Shares of Major Agrochemical Companies, Combined 1985 and 1986

Name of Company	Herbicide	Insecticide	Fungicides	Seed Dressing Rodenticides
Nolchem	*13.4%	*37.3%	*100%	*95.8%
Swiss-Nigeria	51.6%	*2.13%	0	0
CAPL	*82%	*55.7%	0	*4.2%
Rhone	5%	*2.9%	0	0
Total		6.3%		
Dizengolf	43.4% *3.8%	*2% *93.7%	0	0

Source: Underlying figures from APMEU survey files

* Means the sales were recorded in metric tons
Unmarked figures are sales in klt
mt = metric tonne
klt = thousand litres

Figure 1: Typology of Pesticide Marketing Organization

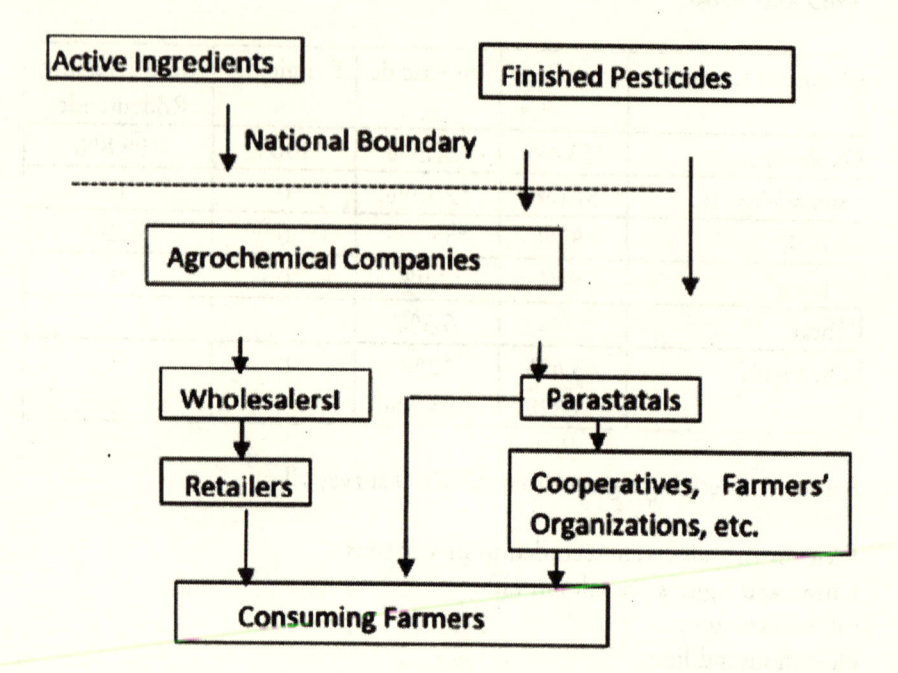

Table 2: Targeted and Actual Distribution of Agrochemicals of Selected ADPs, 1988

Name of ADP	Targeted Distribution	Actual Distribution	Actual as % of Targeted Distribution
Bendel State ADP	300000 Lt	4934.5 Lt	1.64%
CRS ADP	7000 Lt	925 Lt	13.21%
Kano ARDA	594 Lt	266 Lt	44.78%
Lagos State ADP	28,280 Lt	1060 Lt	3.75%
Niger ADP	819000 N	149521.50 N	18.26%

Source: Assembled from APMEPU, Digest on Agricultural Development Projects, 1988

The World Bank-Assisted Agricultural Development Project as a Model for Public Sector Financing in Nigeria

Paper presented at the National Workshop on Financial Management, Socio-economic, and Administrative Policy Implementation at Hill Station Hotel, Jos, 23–26 April 1991.

An era of agricultural strategy commenced in Nigeria in the early 1970s as a result of unprecedented post-war shortage of food. The essential point of departure of this strategy is the involvement of the World Bank in financing command-area projects, officially called Agricultural Development Projects (ADPs), in conjunction with the Federal Government and State Government of domicile. This strategy has involved the commitment of huge sums of domestic and international capital.

There is the widespread notion that ADPs have attained remarkable degrees of financial management efficiency compared to other productivity-enhancement efforts in Nigeria. This notion is based on the casual observations of no celebrated case of missing money among the ADPs, unlike similar earlier and contemporary projects. A high level of financial discipline is assumed to exist in the ADPs, which possibly prevents the manifestations of such syndromes as embezzlement, corrupt enrichment, misappropriation of money, and abuse of funds.

The aim of this paper is to highlight and substantiate the factors possibly responsible for the level of financial management status attained in the ADPs. These factors will be discussed to

(a) assess the degree to which the claim of high financial management performance is true, and

(b) evaluate the claim in terms of conflicts and trade-offs inherent in the procedure for financing the projects.

We shall briefly survey the concept and evolution of ADPs, and illuminate the financial modes as a prelude to examining the critical factors of financial management performance.

CONCEPT AND EVOLUTION

The concept of ADP revolves around how to enhance the technical and economic efficiency of the smallholder farmers on a sustained basis. Based on lessons of experience in public agricultural project implementation, the factors which had limited extension efforts in enhancing efficiency at the farm level include excessive bureaucracy, financial indiscipline, slow rate of technology build-up, inaccessibility of farmers to productive inputs, and poor state of rural infrastructure. Therefore, in the original setting, the ADP concept sought to achieve the following:

(a) revamping and revitalizing the agricultural extension work together with a system of on-farm and on-station adaptive research;
(b) putting in place a network of input supply points that ensures timely supply of the critical inputs at the farm gate;
(c) provision of basic rural infrastructure, particularly massive networks of rural feeder roads and potable water supply;
(d) establishing autonomous project administration and financial management, severed from the mainstream agricultural ministries;
(e) establishing a built-in project monitoring and evaluation as well as other technical support services in planning and start-up of projects.

The original conception was for an ADP in a given state to start up at the enclave level first, before becoming statewide. The enclave stage would take five years, during which the learning process lasted that would facilitate implementation at statewide stage. Aspects of the envisaged learning process include reorientating the traditional agricultural administration to become sensitive to innovations, improving the proficiency of technical and professional staff, among others. Lately, the concept of enclave ADP has been done away with, and statewide projects have emerged directly in a number of instances. A new learning process introduced takes the form of time phasing of statewide projects. The project is to be redesigned, reappraised, and renegotiated at the end of successive phases.

It soon became necessary to accelerate the rate of project establishment with the realization that it was difficult for certain states to wait for others in experimenting with a development concept. The concept of Accelerated Development Area Project (ADAP) evolved (see Food Strategies Mission, 1980), which was fashioned closely after ADP except for the non-involvement of World Bank in the financing arrangement. ADAP was merely an interim arrangement pending the establishment of ADPs in the affected states. Subsequently, groups of projects were to be prepared together as multi-state ADPs (MSADPs) with joint appraisal process and bulk financing negotiation with the World Bank. Three such loans have been approved to date, comprised of MSADP I, MSADP II, and MSADP III. Table 1 describes the evolution of projects in the country. The first generation projects were enclave types at Funtua, Gombe, and Gusau. At present, there are twenty-two projects, all at statewide levels including Abuja.

FINANCIAL MODES

A typical ADP is operated on a tripartite funding arrangement as earlier indicated. The World Bank fund is usually for meeting offshore obligations. The federal and state governments provide counterpart funding. Other sources of funding which are quite negligible include the participating farmers. The Staff Appraisal Report indicates specific roles of the joint financiers as well as function-oriented financial schedules over the life of each project. Table 2 illustrates the financial plan of an ADP, specifying the contributions of each funding agency.

The mode of financing an ADP contains important elements that are worth highlighting. Firstly, organization of financing is based on the finance department of the ADP. The financial controller reports directly to the project manager on fund allocation, disbursements, and expenditures. A broad-based project management unit (PMU) exists to make major decisions and provide general directions in regard to commitment of project funds. The World Bank showed preference for expatriate financial controller. A project account is opened to remove financing away from the regular procedure of the main ministry. The financial controller is responsible for disciplined financial procedures, which ensure that expenditures are made according to stipulated rules, and for specified project tasks.

Secondly, drawdown of the World Bank loan is effected in Washington DC through an assigned project officer. The intent to commit offshore expenditures on the project is communicated to the World Bank. The sending of a 'no objection' message back to the ADP depends on adherence of the project authority to guidelines, particularly as contained in the SAR. Subsequently, procurement of offshore components is carried out using laid-down processes (Ayoola 1993). In some cases, the World Bank insists on retroactive finance whereby the project has already committed the money to be reimbursed later.

FACTORS OF FINANCIAL MANAGEMENT PERFORMANCE

This section examines the role of certain inherent factors of the ADP concept in the financial management performance of the projects. As indicated above, the critical ones among these factors are

(a) autonomous management status,
(b) counterpart financing,
(c) expatriate staff, and
(d) technical support services.

Autonomous Financial Management Status as a Means of Reducing Political Interference

The presence of bureaucratic delays in fund allocation to implement time-sensitive stages of projects has adversely affected the agricultural development process in the past. In addition, the degree of political interference was high not only in the aspect of fund allocation but also appointment of staff, procurement of project materials, and award of contracts. Therefore, the setting up of ADPs under autonomous management units was informed by the need for timely release of funds and reduction of political interference. There are rigid rules for appointing key project staff, particularly the project manager, acceptable to the World Bank. In addition, the involvement of the ministry is to be limited to its participation in the PMU and release of its own share of fund as it falls due.

However, it is not strictly correct to say that there is no outside interference in the financial management of the ADPs. A number of instances abound, including interferences in location of projects, appointment matters, and use of project facilities. In some extreme cases, the main ministry issues an order to influence the normal course of project activities.[49]

---

[49] The abuse of project facilities is particularly disturbing; it is commonplace to

Interference has substantial implications for project financing, especially in terms of using project funds for unbudgeted activities and delay in project start-up and implementation. Inevitably, these practices constitute leakage elements to limited project funds. It also creates additional burden on project financial control system, thereby preventing the achievement of preset project objectives.

Counterpart Financing Affects Rate of Loan Drawdown

The underlying motive of providing for counterpart funding arrangement is for the federal and state governments to have stakes in the development of agriculture in their country and states respectively. The principle of retroactive financing further ensures the commitment of government at both levels. The drawdown of the bank's loan for a given project is tied to the release and commitment of local funds. Consequently, loan drawdown is not full as a result of poor funding commitment of federal and state governments. Table 3 reveals the low drawdown positions of projects.

A recent proposal to enhance loan drawdown was put forward during midterm review of the MSADP I projects. An incentive scheme will be put in place by the World Bank, whereby the undrawn sums in respect of poorly committed projects will be released for use by other more committed projects. The adverse effects of poor counterpart funding commitment are non-fulfilment of project objectives in time and quantity, as well as wasteful loan servicing arising from payment of commitment fees on an approved loan portfolio. In this connection, the ADPs cannot be associated with unqualified prudence in financial management.

Expatriate Personnel Towards Ensuring Financial Accountability

The foreign content of the top management staff of the earlier ADPs was quite high. The interpretation of expatriate involvement featured in the several debates in the early days of the ADP concept in the country. On the bank's side, probably the financial control of the capital intensive ADPs could not be left entirely in the hands of local staff, not necessarily because

---

observe assignment of project cars to political authorities and other functions that are unrelated to a project. A work study is necessary to assess the degree of project facilities for project works and non-projects works.

they were not competent or experienced enough but as a means of ensuring financial accountability. To make this argument, it was thought that the expatriate would not be easily involved in the traditional relationships that could cause the project to be defrauded. Furthermore, the expatriate would be less sympathetic to the poverty conditions of the local environment that explains substantial parts of most public sector funds. Also the expatriate staff has his international credibility to protect in order to secure further assignments from the bank after the expiration of his current contract.

On the other hand, the strongest argument against the dominant presence of top-management expatriate staff concerns its macroeconomic implications, especially in the contexts of self-reliance and self-sufficiency policies. The expatriate personnel earn exorbitant remuneration dictated at levels higher than the country. Their salaries have high hard currency components, which are not generally spent inside Nigeria. The first source of worry about this arrangement is the conflict between the desire to increase food self-sufficiency on one hand, while we become more dependent on other countries for expatriate management on the other hand (Ayoola 1986). According to Oyaide (undated), this situation only makes the greater part of the benefits of our domestic development efforts accrue to industrialized countries. In addition, such situations lead to sagging morale on the part of available local personnel. Therefore, an empirical investigation is necessary to determine the relative extent to which the ADPs, under existing financial arrangements, have created development benefits at home and abroad.

Sustainability Aspects of Financing

The important issue now centres on the need to guarantee the sustainability of project activities in the post–World Bank financing period. One way to tackle the issue is to favour proposal for the Agricultural Development Fund (ADF) to be used as a means of maintaining the tempo of specific activities after the formal termination of individual project lives. Our immediate reservation is that this is a means of continual trapping of the country in the cycle of loan negotiation, loan servicing, and probably perpetual use of debt instrument in the search for agricultural progress. If truly the ADPs have made the envisaged impact over some twenty years, there should be no need for new debt financing of agricultural development efforts. From this viewpoint, it would be possible to sustain the project achievements with local funds.

Another way to tackle the sustainability problem of the ADP financing is the device to involve financial participation at the local level. To this end, the bank has evolved a sustainability matrix, which principally includes beneficiary contribution to improve cost recovery by forming users' associations in each community to run their own facilities. For instance, they would charge fees to cover the investment cost and routine maintenance of the pumps in the rural water supply scheme.

Financial sustainability is a critical issue concerning the ADPs. The projects will only be a good model for public sector financing to the extent they can retain their virtues and maintain their tempo in the period after the investment life.

The Role of Technical Support Service

The presence of built-in monitoring and evaluation capabilities is unique as a feature of the ADP system, which can explain a substantial part of financial performance among the projects. In addition to the presence of in-house monitoring and evaluation (MEU) in the administrative structure of each project, there is a virile system of technical support services. The Federal Agricultural Coordinating Unit (FACU) with headquarters in Ibadan was established to take over the project preparation function from the ad hoc consultants of the World Bank. The unit has since performed this role on a permanent basis for subsequent ADPs in succession. It also mounts a Project Facilitating Team (PFT) to effect successful take-off putting the basic structure in place, including the financing system and staffing. The Agricultural Projects Monitoring and Evaluation (APMEU) with headquarters in Kaduna was established to provide technical services in project monitoring and evaluation.

Between these two agencies, which are also financed under World Bank loans, and in association with the World Bank, the financial plans of each project are drawn up, appraised, monitored, and evaluated. The two units have developed immense technical and professional capabilities in analysis and reporting of ADP activities in the country. There are radio and other telecommunication linkages with all ADPs to obtain financial and other information quickly. The presence of micro and mainframe computers enhances information processing, analysis, and retrieval. Each ADP produces monthly reports, quarterly reports, and annual reports. APMEU publishes midterm review reports, project completion reports, project status reports, and occasional reports. Each report has a full complement of financial information about the project at given points in time. In addition, there are standard publications for reporting price movements and special studies.

In our own opinion, most of the financial respectability earned by the ADP system appears to be the effect of its open reporting of its transactions concerning the project, but it may conceal a great deal too. For example, the numerous reports fail to incorporate time with quantities of money released to individual projects to enable us to evaluate the role of each financier in relation to targets set for particular activities. For another example, there is also no basis in the existing reporting formats to estimate the financial implications of non-project services, such as vehicles used by directors general and commissioners in the mainstream ministries. To the extent that these instances have serious implications for overall project objectives, it is somewhat difficult to accept without reservations the proposition that the financial system of ADPs qualifies them as the best model for public sector financing.

SUMMARY AND CONCLUSION

The assessment of the financing mode of the World Bank–assisted ADPs reveals important features as follows:

(a) the counterpart funding arrangement which is to ensure commitment of federal and state governments is not completely successful;
(b) financial management autonomy is largely abused;
(c) the involvement of expatriate financial managers has probably enhanced financial performance, but it imposes new forms of economic dependency on the country;
(d) sustainability issues have been late to be raised; and
(e) the built-in system of technical support in planning monitoring and evaluation is an essential means of ensuring financial accountability, but some important information is concealed.

The ADPs can serve as model for public sector financing subject to the ability of sponsors to smoothen the rough edges identified in the implementation process.

REFERENCES

APMEU (Agricultural Projects Monitoring and Evaluation Unit) (1988), *Project Status Report*, 4 volumes, July 1988, Department of Agriculture and Rural Development.

APMEU (Agricultural Projects Monitoring and Evaluation Unit) (1989a), *Project Status Report*, 4 volumes, July 1989, Department of Agricultural and Rural Development.

APMEU (Agricultural Projects Monitoring and Evaluation Unit) (1989b), *Digest on Agricultural Development Projects*, Department of Agricultural and Rural Development, July 1989.

Oyaide, O. F. J. (undated), 'External Financing for Agricultural Development in Nigeria: The Role of the World Bank Series', FC-C003.

Federal Ministry of Agriculture (1980), 'Food Strategies Mission, the Green Revolution: A Good Production Plan for Nigeria (Final Report)', May 1980.

Ayoola, G. B. (1988), 'Rural Infrastructural Development in Nigeria', invited paper, Bi-annual Conference of Nigeria Institute of Quantity Surveyors, Durbar Hotel, Kaduna, 13–15 Oct. 1988.

Ayoola, G. B. (1988), 'Rural Development in Nigeria: The Dependency Problem in Food Grains and Rationale for Self-sufficiency Policy in Nigeria', *J. Rural Development in Nigeria* 3 (1).

Table 1: Agricultural Development Projects in Nigeria (ADPs)

State	Description	Commencement	Termination	Budgeted cost (₦ Million)
Akwa Ibom	Carved out of Cross River State ADP	1988	1991	47.25
Anambra	MSADP I	1987	1991	38.90
Bauchi	Follow-up of the Gombe Project	1981	1988	192.80
Bendel	MSADP I	1987	1991	29.10
Benue	MSADP I follow-up Enclave Ayangba ADP	1987	1991	37.70
Borno	Merger of Southern Borno ADP and Borno State ADP	1988	1992	133.00
Cross River	MSADP I	1987	1991	27.60
Ondo	MSADP III	1981	1987	48.20
Gongola	MSADP II	1987	1992	220.09
Imo	MSADP I follow-up Imo State Accelerated Area Project	1987	1991	34.20
Kaduna	Follow-up of Enclave Funtua ADP	1985	1993	672.80
Kano	Statewide from inception	1982	1989	272.80
Katsina	Carved out of Kaduna State ADP	1989	1993	696.75
Kwara		1980	1988	41.80
Lagos	MSADP III	1987	1991	56.50
Niger	Offshoot of Enclave Bida ADP MSADP II	1987	1992	146.00
Ogun	MSADP I	1987	1991	8.16
Oyo	Follow-up of enclave Oyo North ADP	-	-	-
Plateau	MSADP I follow-up of enclave Lafia ADP	1987	1991	64.00
River	MSADP III	1988	1991	74.04
Sokoto	Follow-up of Enclave Gusau Project under ADP I for extended World Bank funding	1983	1989	274.30
Abuja	-	1990	-	-

Source: APMEU Digest on Agricultural Development Various Uses

Table 2: BIDA Agricultural Development Project Financial Plan 1980–1987 (in ₦ Million)

Year	World Bank	Federal Government	State Government	Total
1980	₦2,100 23%	2,700 30	4,200 47	9,000 100
1981	₦2,477 29%	2,180 25	3,900 46	8,557 100
1982	₦3,327 32.8%	2,700 26.6	4,150 40.6	10,147 100
1983	₦2,530 23%	2,080 19	6,270 58	10,880 100
1984	₦1,800 20%	2,100 23	5,200 57	9,100 100
1985	₦1,700 36%	1,000 21	2,000 43	4,700 100
1986	₦1,700 24.3%	1,000 14.3	4,300 61.4	7,000 100
1987	₦ -	1,750 36	3,114 64	4,864 100
1980–87	₦15,644 24%	15,510 24	33,104/5 52	64,258 100

ID# Deregulation and Decontrol of Nigeria's Fertilizer in the Context of Structural Adjustment Programme[50]

The agricultural economy of Africa as a whole depends so much on the fertilizer sector that the input has assumed considerable political visibility in several countries on the continent. Nigeria's fertilizer market, consisting mainly in the available institutional framework and mechanisms for procurement and distribution, was an obvious target of the drastic economic reforms introduced under the Structural Adjustment Programme (SAP) since 1986, owing to the evidenced failure to properly integrate supply distribution and consumption of the critical farm input.

This paper conducts a diagnosis of the pre-reform situation and evaluates the reform strategies, as preliminary steps toward future analysis of the implementation process and impact assessment. The goal is to provide a proper understanding of the forms and functions of the perennial fertilizer crisis in Nigeria as well as identify necessary actions to stem it. The reform of fertilizer market elsewhere in Africa such as Cameroon (Truong et al. 1992), Kenya, Uganda and Ethiopia (Sodhi 1992) may be faced with problems similar to the Nigerian situation from which this paper will provide useful lessons of country or regional policy experience.

---

[50] **Key Words**: Deregulation, Decontrol, Fertilizer Market, Structural Adjustment Programme, Diagnosis, Strategies, Preliminary Assessment.

ANALYTICAL FRAMEWORK AND DATA

The diagnosis of the pre-reform situation is carried out with respect to five aspects—organizational structure, ownership structure, market structure, institutional/policy framework and performance. The strategy of the reform is evaluated with respect to eight aspects—background, analytical framework used by decision makers, external support, objectives of reform, main characteristics of reform strategy, reform modalities, and complementary policies.

The data used involved secondary types in the forms of documentary materials such as study reports and service records obtained from relevant agencies. Considerable information also derived directly from the knowledge of the author as a member of the National Fertilizer Technical Committee (NETC), a think tank that advises the Honourable Minister of Agriculture on the fertilizer economy. The minutes of the meetings of National Council on Agriculture (NCA) and the monitoring reports at the Fertilizer Procurement and Distribution Division (FPDD) were particularly found useful as source of empirical information.

DIAGNOSIS OF PRE-REFORM FRAMEWORK

Institutional and Policy Framework

Two structural formations existed in the fertilizer market before reform. In the earliest formation (pre-1976), state agencies (e.g. ministries, corporations, etc.) ordered fertilizer using private importers; there was no local production capacity at that period. Commissioned sales agents and extension agents served as the retail outlets. Each state of the federation handled its own fertilizer operations independently, leading to a decentralized, and more or less uncoordinated system.

In the subsequent formation (1976–1983), the FPDD had been established in the federal agriculture ministry to undertake importation and distribution of fertilizer on behalf of the states. At this time of centralized procurement and distribution, the Federal Superphosphate Fertilizer Company (FSFC) had also been established as the first domestic production plant. Depots were established near the ports to serve as the take-off points for transportation to the hinterland and state agencies. Subsequently in 1979, distribution was decentralized, wherein the states lifted fertilizer directly from the ports for onward transportation to their respective channels.

The traditional fertilizer price policy in Nigeria involved the administratively fixed fertilizer price, which was subsidized heavily. The subsidy was applied to (a) reduce CIF of imported fertilizer at the port; (b) reduce production cost and markup of locally produced fertilizer in reference to world market price level; and (c) wholly absorb the cost of distribution including haulage, warehousing, and handling.

Ownership and Market Structure

The enterprise activities in the fertilizer market before reform were largely public owned. The limited aspects of private sector participation were

(a) the use of private importers and commissioned sales agents before 1976,
(b) the use of private transporters for inland distribution in the post-1976 period,
(c) the involvement of farmers' cooperatives at the retail end.

The participants in the fertilizer market might be grouped into three: suppliers, transporters, and retailers. In the period immediately preceding reform, the suppliers comprised FFDD as the only importer and procurer, and FSFC as the only domestic manufacturer. The transporters were private companies and individual enterprises that moved fertilizer at fixed charges. The officially recognized retailers were the World Bank–assisted Agricultural Development Projects (ADPs) through the farm service centre stores.

Performance

The dominant presence of government in the fertilizer market before reform had obviously stimulated consumption significantly. According to Falusi (1986), the greater part of incremental food production recorded in the ADPs was attributable to increased fertilizer use. The evidence is probably stronger in the case of the middle belt and northern (savannah) zones (FPDD 1987) compared to the southern forest zones. Nevertheless, crop response to fertilizer application was a far cry from that of developed agricultural economies (FAO 1986). Also, based on international standards, the per capita fertilizer consumption and average value-cost ratios were inadequate (Ogunfowora 1993).

As the level of consumption increased, distribution and marketing activities became too complex, creating immense pressure on the capacity of the controlled institutional and policy frameworks to cope with the demand of an efficient fertilizer market. The following elements constituted the observed inefficiency of the market pre-reform:

- Leakages, port wastage, short landing, and transit losses;
- Inadequate and untimely supply;
- Artificial scarcity, black marketing, and smuggling;
- Erratic importation pattern arising from untimely release of funds;
- Excess demand coupled with wasteful and cumbersome procurement procedure;
- Erratic and uncoordinated ship arrivals;
- Storage and stock control problems;
- Transportation bottlenecks including wrong deliveries, non-deliveries, and under-deliveries.

Given the dismal performance of the highly regulated market, the evidenced high transaction cost of fertilizer to farmers was a natural consequence. Besides, the fiscal burden of subsidy overwhelmed government at the expense of budget allocation to social services. Thus, the reform process that commenced in 1986 was socially and economically expedient.

STRATEGIES EMPLOYED

Reform Objectives and Analytical Approach

The reform sought to deregulate and decontrol (i.e. liberalize) the fertilizer market with the objectives to

(a) facilitate procurement operations,
(b) improve distribution efficiency, and
(c) enhance the value-cost ratio of fertilizer to farmers.

The goal of the reform, consistent with the philosophy behind SAP, was the promotion of financial performance of enterprises in the fertilizer market, thereby upgrading the social and economic performance of the overall agricultural economy.

It was generally assumed by both the analysts and policymakers that the poor performance of the fertilizer sector owed fundamentally to the absence of a competitive environment for distribution and marketing, which necessitated the gradual removal of public sector roles.

The main analytical approach used was to specify alternative options for achieving deregulation and decontrol. These were

i. that the new National Fertilizer Company (NAFCON) should market all domestic production of fertilizer and import the shortfall,
ii. that local manufacturers should market their products while an independent organization imported and marketed the shortfall,
iii. that each local producer should be allowed to market its products in addition to any number of registered and public organizations who could import and distribute fertilizer,
iv. that FPDD should continue under a reorganized structure and arrangement to import and distribute fertilizer,
v. that a single independent national marketing organization should market all domestically produced and imported fertilizers,
vi. that local producers should market their products and NAFCON or any government agency should import the shortfall,
vii. that NAFCON should market all domestic output of fertilizer and FPDD import the shortfall.

An initial external support of the fertilizer market reform was provided by the World Bank, which granted a special loan facility amounting to US$250 million between 1984 and 1986 for the purpose of enhancing fertilizer operations, subject to (a) the reduction of fertilizer subsidized over five years, (b) commercialization of fertilizer operations and retail business, among others.

The second loan was considered but was not implemented, based on disagreement between the World Bank and the Federal Government over privatization and other issues.

Reform Modalities

The elements of the deregulation and decontrol process undertaken were as follows:

(a) Government commenced the gradual withdrawal of fertilizer subsidy in 1986 through successive price adjustment; following, the official prices of fertilizer have changed from ₦15/50-kg bag (high analysis) and ₦10/50-kg bag (other types) in 1985 to the present (1994) levels of ₦150/bag and ₦110/bag, respectively.

(b) The National Fertilizer Company (NAFCON) was commissioned in 1987 as a private enterprise and the only supplier of nitrogenous raw materials (urea, DAP) to other plants as well as largest producer of compound products (installed capacity = 750,750 MT).

(c) The growth of privately owned command-area fertilizer blending plants was promoted to increase the number of producers in the market; up to seven of such plants have commenced operation in different parts of the country, e.g. Kaduna, Kano, Maiduguri, and Minna.

(d) The number of depots or distribution points was increased from six to forty to facilitate the delivery network.

(e) The arrangement for sharing the residual subsidy burden was changed from wholly federal government to even (1/3) sharing formula by federal, state, and local government.

Complementary Policies

The Structural Adjustment Programme was a multiple-instrument package containing a number of policies complementary to the current liberalization of the fertilizer market. The key ones among such policies are enunciated as follows:

i. Technical Committee on Privatization and Commercialization, TCPC (now Bureau of Public Enterprises): The main aim is to reduce the participation of government in businesses and stimulate the role of private sector in the structures of ownership and market; this was achieved through divesting of its government holdings in several companies. The attempt to sell government shares in NAFCON was frustrated by its strong link with the subsidy and other volatile issues in the agricultural sector.

ii. Foreign Exchange Market (FOREX): A liberalized foreign exchange market was established at the onset of SAP, which has meant continuous devaluation of the naira. Consequently, imports prices have significantly increased with the consequence that the envisaged effect of gradual subsidy removal through the successive administrative price increases was negated by declining value of domestic currency relative to foreign currencies at which fertilizer prices were quoted.

iii. Abolition of Commodity Boards: The six intervention boards (namely for grains, rubber, groundnut, palm produce, cocoa, and cotton) were swiftly abolished in 1986 as part of a general deregulation of the agricultural economy. Thus, given the degree to which allocative decisions of farmers reflect the current produce price regimes, fertilizer use efficiency could result from the absence of price interventions in the food market.

PRELIMINARY ASSESSMENTS

The mechanisms of implementation and more robust impact assessment will be the subjects of a follow-up study to be conducted when the reform process is fully matured. Nevertheless, certain off-the-cuff observations could be made at this stage to warrant some preliminary assessment of reform effectiveness and obstacles to success.

Reports after five years of the deregulation and decontrol process indicate that measures of market inefficiencies still take on high values, including the persistence of late supplies, high transaction costs, non-agricultural use of fertilizer, inadequate supplies, and artificial scarcities through hoarding and smuggling activities (Table 1). Moreover, the successive increases in officially fixed prices has been associated with increased subsidy burden as the resultant effect of progressively lower value of the naira, increased production costs incurred by local producers, and possibly, increases in the world market price level.

The deregulation and decontrol process has not had any perceptible influence on the phenomenon of unintended beneficiaries (Reutlinger 1988; Idachaba 1992) being the failure of policy benefits to accrue mainly to the target population of farmers. The various categories of unintended beneficiaries of government intervention in the fertilizer market include

(a) importers who enjoy abnormal profit arising from the separation of the world and domestic markets of fertilizers;
(b) transporters who earn excessive rent especially through sharp practices such as deliberate wrong deliveries (undercutting paid distances), enroute disappearance of fertilizer loads which later reappear through black market channels, and leakages through smuggling to the neighbouring countries;

(c) unrecognized middlemen (e.g. government officials or their friends, members of legislature, top military officers) who obtain huge allocations of fertilizer to be sold at prices higher than the official levels;

(d) surreptitious non-agricultural users of fertilizer for detergent, explosives, and such other industrial products in which, by that practice, they enjoy fertilizer subsidy to become low-cost producers;

(e) local producers who make substantial savings from the lack of need to undertake promotion activities because the market for their product is given.

Given this situation of the fertilizer market, major sources of the high transaction cost to farmers and low value-cost ration (Agner and Ogunfowora 1994) are established. By and large, the envisaged higher allocative efficiency applicable to farm inputs as a result of the complementary policies, especially the abolition of commodity boards, could not be traced to fertilizer use by farmers. This probably informs the recent thinking in the policy arena that half measures of liberalization might not be appropriate for the fertilizer market. And it now appears that the appreciation of the need for full-scale liberalization privatization is on the increase, but it is constrained by the widespread fear that fertilizer prices would become unaffordable to farmers subsequently.

CONCLUSIONS

The five years of limited deregulation and decontrol have meant little to nothing to the resolution of fertilizer crisis in Nigeria as most of the symptoms of aberrant market situation persist, including late deliveries, high transaction cost, unintended use and beneficiaries, supply inadequacy, and artificial scarcity. Derivable from this is the need to inject competition into the fertilizer market, giving the basis to recommend a full-scale liberalization programme whose critical elements are (i) elimination of government roles in the enterprises of the fertilizer market, particularly importation; (ii) disengagement of government from allocation of fertilizer to state agencies and other bodies; (iii) complete withdrawal of subsidy on fertilizer prices; (iv) removal of the role of government in fixing fertilizer prices and transportation charges; (v) infusion of private sector initiatives to undertake enterprise activities as producers, distributors, wholesalers, and retailers of fertilizer; (vi) divestiture of government holdings in the production enterprises.

The argument that fertilizer prices could become too high is reasonable but sounds unconvincing as it fails to consider the advantage of eliminating the huge markups currently imposed by the several irregular market participants. For such an argument to lead to a decision not to fully liberalize or privatize, the magnitude of these markups and other cost savings arising from budget and non-budget sources should be properly weighed against the anticipated high prices to be charged by private entrepreneurs. The simple fact that other farm inputs such as agrochemicals operate successfully in the free market in the country further reduces such fears to the barest minimum. In addition, there is practically no basis to expect that farmers, being generally poor though, who are intrinsically efficient in decision-making about resources use (Schultz 1964), will not undertake constant evaluation of their investment portfolio when any input prices rise relative to others, given their adequate knowledge of fertilizer and the fact that it is well tested in all agrochemical zones of the country. In the event, however, that the fertilizer market is made competitive based on this reasoning, government's role simply becomes that of facilitating, directing, and quality control.

REFERENCES

Gerner, Hemy and O. Ogunfowora (1994), 'The need for effective fertilizer market development in Nigeria', *Proceedings of 1994 International Fertilizer Association Conference for Africa*, Dakar, 1–3 February 1994, International Fertilizer Development.

FAO (Food and Agriculture Organization) (1986), *Fertilizer Yearbook*, Lome.

FPDD (Fertilizer Procurement and Distribution Department) (1987), *Towards Efficiency of Fertilizer Use and Development in Nigeria*, Proceedings of the National Fertilizer Seminar, Port Harcourt, 28–30 October 1987.

Falusi, A. O. (1986), 'Use of Fertilizers for Increasing Crop Productivity: ADP Experience' (Lagos: National Productivity Centre).

Idachaba, F. S. (1992), 'Policy Options for African Agriculture', in Jean Dreze and Amartya Sen (eds.), *Wider Studies in Development Economics: The Political Economy of Hunger*, Volume III: Endemic Hunger (Oxford: Clarendon Press).

Ogunrowora, Bisi (1993), 'Analysis of Fertilizer Supply and Demand in Nigeria', Paper presented at the Symposium on Alternative Pricing and Distribution Systems for Fertilizer in Nigeria (Ibadan: Federal Agricultural Coordinating Unit, Nigeria).

Reutlinger, Shlomo (1988), 'Food Security and Poverty in Developing Countries', in J. Price, G. Hinger, Joane Leslie, and Caroline Hoisington (eds.), *Food Policy: Interpreting Supply, Distribution and Consumption*, EDI Series in Economic Development, Longman (London: John Hopkins University Press).

Schultz, Theodore W. (1964), *Transforming Traditional Agriculture* (New Haven, Connecticut: Yale University Press).

Sodhi, Amar Jit Singh (1992), 'Privatising the Fertilizer Business: Some Experiences from East Africa', Paper presented at the Fifth Annual Meeting of African Fertilizer Trade and Marketing Information Network, Lome, International Fertilizer Development Centre-Africa.

Truong, Thom V., S. Tjip Walker, D. C. Moore, and Felix Mkonabang (1992), 'Cameroon Fertilizer Sub-sector Reform Programme', Paper presented at the fifth Annual Conference of the African Fertilizer Trade on Information Network, Lome, IFDC-Africa.

Table 1: Selected Feature of Nigeria's Fertilizer Market, Comparative Analysis of the Pre- and Post-reform Periods

	Pre-reform	Post-reform
Fertilizer requirement (MT) [1]	-	1,790,590
Fertilizer allocation (MT) [2]		
Planned	-	1,000,000
Actual	-	415,219
Fertilizer supply (MT) [3]		
Imported	781,237	500,100
Local	44,527	574,700
Total	825,764	1,074,800
Subsidy rates [4] (%)	67.4	81.6
Budget implications [5]		
Capital allocation to fertilizer (₦M)	215	909
Share in agric. allocation (%)	32.1	71.3
Share in crop allocation (%)	59.3	-
Production capacity [6]		
Installed (MT)	100,000	1,480,750
Utilized (%)	20	41
Fertilizer Pricing [7]		
Nominal (N/ton)	190	800
Real (N/ton)	190	160
- US$/ton	212.53	43.35

Notes

1. Requirement as contained on indents forwarded by states in 1989
2. As reported for the 1989 fertilizer year
3. 1981–1985 average (pre-reform) and 1988–1992 average (post-reform)
4. 1981–1985 average (pre-reform) and 1987–1993 average (post reform)
5. In 1985 pre-reform year and 1987 post-reform year respectively
6. In 1985 pre-reform year and 1993 post-reform year respectively
7. Nominal prices deflated by the Nigerian consumer price index (1985 = 100); conversion based on current ₦/US$ exchange rates.
  - = Not available.

A Synthesis of the Nigerian Livestock Policy

Paper presented at the Joint NISER-FDLPCS Conference on the Nigerian Livestock Industry and Prospects for the 1990s, Kaduna, 29 October – 1 November 1990.

This paper attempts to coalesce the bits and pieces of the livestock policy process from historic data into a synthesized whole. The aim is to highlight the inherent features of the process, towards informing future policy actions concerning the livestock sub-sector.

SYNTHESIS

Administration of Livestock Policy

The colonial agricultural administration broached livestock development through the first Forest Department, established at Olokemeji, near Ibadan in 1900. A flock of sheep and goats was kept at this location, as well as a herd of cattle. Animal husbandry was integrated into a mixed farm structure at the headquarters and the subsequently established forest estate, with reserves at Mamu, Ilaro, Ibadan, etc.

The first Agricultural Departments were established in 1910 (Southern Provinces) and 1912 (Northern Provinces), which were merged in 1921. Husbandry activities were intensified but still not clearly differentiated into functional parts. Subsequently, the Shika farm was established mainly for animal husbandry works, supplemented initially with cotton seed production.

By and large, primary attention of colonial livestock administration was focused on animal health and disease control. A veterinary research laboratory was established at Vom in 1922, with benches and sink fully in place by 1924.

Even though a Livestock Investigation Centre (LIC) was later put side by side with the laboratory, Vom soon became a centre for research in the production of livestock vaccines. Initial efforts concerned production of anti-rinderpest serum. Later, experimental works concentrated on other diseases such as pleuropneumonia of cattle, trypanosomiasis, tuberculosis, and contagious abortion. The investigation centre maintained a herd and handled routine husbandry works such as milk production and breeding. Sheep, goats, and pigs were also handled. A veterinary school was soon established, which, together with the laboratory and centre, formed the nucleus of the first Veterinary Department at Vom.

In the 1954–1967 period, only the regions had constitutional mandate to pursue agricultural development, including livestock. Therefore, livestock development, with veterinary component subsumed, was part of the goal of Regional Agriculture Ministries. However, following the creation of the first Agriculture Ministry at the federal level in 1967, the Federal Livestock Department (FLD) was established. As the overall administrator of the national livestock policy, FLD has expanded its sphere and functions in the country. Table 1 is a brief description of the existing organs in the administration of livestock policy in Nigeria.

Livestock Policy Objectives

The articulation of livestock policy objectives follows the evolution of the overall agricultural policy in the country. In the earliest phase of this evolution, policy objectives were not declared in advance, such that the implicit objectives could only be imagined in retrospect. In the next phase, disparate sectoral or regional objectives were indicated without an attempt to synthesize them together to form a single harmonious agricultural policy. Such policy objectives were present only in working files, annual reports, and other such documents as political manifestoes, brief write-ups, and memoranda.

The phase of conscious formulation of policy objectives began in 1937, when the first Forest Policy was prepared. The document was revised in 1945 and subsequently decentralized among the regions in 1952. The earliest articulated objectives of livestock policy was stated in the Agricultural Policy 1946. Although wordy and merely descriptive, there was the intent to improve 'the general standard of living of the people by inducing proper use of the resources available to them' (Nigeria 1946). Therefore, emphasis was laid on development of the Northern Provinces pastoral or livestock production area including northern Sokoto, northern Katsina, Daura Emirate of Kano Province, Borno Province, Bauchi, Plateau and Adamawa provinces.

Subsequent policy documents prepared during the regional control period (Western Region 1952; Eastern Region 1953) merely made tangential references to livestock development without stating quantitative objectives. For instance in the old Eastern Region, the objective was 'to develop the livestock industry which would eventually be essential for a balanced system of agriculture in most areas'.

The latest policy document for the entire agricultural sector was launched in 1988 (FMAWRRD 1988). This contains the finest disaggregating of objectives among various sub-sectors. The livestock policy objectives have been articulated within the context of the overall agricultural objectives, which in turn derive directly from the entire macroeconomic policy objectives and societal goals. Specifically, the livestock policy objectives have been stated as follows:

(i) to make Nigeria self-sufficient in the production of livestock products;
(ii) to improve the nutritional status of Nigerians through the domestic provision of high-quality, protein-rich livestock products;
(iii) to provide locally all necessary raw material inputs for the livestock industry;
(iv) to allow for a meaningful and efficient use of livestock by-products;
(v) to improve and stabilize rural income emanating from livestock production and processing;
(vi) to effectively protect the rural livestock farmer from the unpredictable vagaries and risks incidental to livestock production;
(vii) to provide rural employment opportunities through expanded livestock production and processing; and
(viii) to effect proper land use and maintenance of the ecosystem for expanded livestock production.

These objectives have been translated into concrete quantitative targets (FMAWRRD 1986). The self-sufficiency elements were phased out according to available resource endowment and nature of products—poultry meat (five years), poultry eggs (four years), sheep and goats (five years), dairy milk (four years), and others (unspecified).

Instruments of Livestock Policy

The generalized types of instruments used in the early times for livestock development include gardens/farms such as the Shika farm earlier mentioned, agricultural extension, agricultural education/training, agricultural research, pest and disease control services, agricultural cooperation, agricultural industries, rural infrastructural development, agricultural legislation, private sector initiatives/incentives, agricultural credit, input support services, public sector enterprise international business and assistance, settlement schemes, agricultural campaigns. Suffice it to say that there are many instruments to use in the pursuit of the stated livestock objectives. The detailed description of such of these in regard to the livestock sub-sector is suppressed in this paper in preference for a discussion on specific livestock policy instruments.

The recent policy document emphasizes the ecological specialization, sedentarization, and a host of other specific instruments of livestock policy, which are worth highlighting. They consist of instruments for making genetic and marked improvements in livestock and livestock products, as well as ensuring regular source of livestock feed and the provision of wholesome meat and other livestock proteinous products for human consumption.

1. **Animal Improvement Services:** Livestock Investigation and Breeding Centres (LIBC) are established at various places for the purpose of investigating the production potentials of local animals and to maximize their productivity by genetic selection and cross-breeding with improved genetic stock of exotic origin. For instance, Kaduna State has a swine improvement centre at Zonkwa and goat improvement centres at Kuka Aljana and Rima. In the same state, breeding of rabbits is performed at the centres located in Kagoro and Dambo. Other states also keep LIBCs in different locations—Borno (Nguru, Muna, Dahori, Bila; Kwara (Shao and Osara for Ndama cattle, Kabba for piggery and Fololo, New Lokoja for sheep). The role of FLD in livestock breeding includes the multiplication of improved local cattle to produce elite animal units. To this end, various centres were maintained in the states to supply technical services, breeding stock, and other inputs to farmers. These include Ohaozara (Imo State) for trypano-tolerant cattle, mainly Ndama and Keteku; Funafuna (Niger State) for Sokoto Gudali breeds; Jibiro (Gongola State) for white Fulani and Adamawa Gudali cattle. Sites for sheep and goat breeding and multiplication were established at Tuma and Ladinawa at the Katsina sheep farm (Katsina State) and Zugu (Sokoto State). There are pig multiplication centres established

by FLD at Okija (Anambra State), Iguobazuwa (Bendel State), Fumudu Manya (Rivers State), Kabba (Kwara State), Iperu Remo (Ogun State), Ondo (Ondo State), Kurmin Jibrin (Kaduna State), and Numan (Gongola State). In poultry, hatcheries are located in Ajura (Ogun State), Usefon/Nzukara (Cross River State), and Dawaki Tofa (Kano State), all totalling 5.4 million day-old chicks capacity annually.

2. **Livestock Production Services:** An important aspect of this service is demarcation of grazing reserves, towards eliminating the incessant clashes between farmers and livestock owners. An early effort concerns the Kuka-Jungaral project along the Sokoto and Katsina borders of the old Northern Region. Support services at grazing reserves include water reservoirs to prolong the grazing intervals into the dry season. Other services are access roads, spray races, and veterinary clinics. Reserves legislation is enacted for controlling activities of users such as breaking of grazing entities. Supplementary feedstuffs are normally provided, like groundnut cake, cotton seed, and salt licks. Prominent reserves are in Kwara State/Nweri (200 ha.), Kinikini (100 ha.), Utula (100 ha.), Gidan Magaji Shagari (50 ha.), Ayangan (50 ha.), Alapa, Lata, Olodan, Omi-eran (50 ha.). The Federal Government also encourages the development of ranches such as the purpose-specific ranches at Mokwa and Manchok for finishing operations of feeder steers on six-month cycle based on improved pasture, modern husbandry methods, and range management. Other ranches are established for beef fattening, dairy, and rabbitry.

3. **Relocation of the National Herd:** This instrument was initiated with the aim to secure proper distribution of livestock by destocking the overgrazed areas in Sudan zone to Guinea and derived savannah zones.

4. **Pasture Improvement Programme:** The regional pasture improvement programme commenced to assess pasture, seed adaptability, field establishment and utilization, and seed production centres. Pasture trials are conducted to test different varieties of tropical grasses and legumes at various rates of fertilizer application from semi-arid to humid areas. This has helped in distributing seeds to states, based on the empirical results.

5. **The Livestock Development Programme:** A programme in general livestock development was co-sponsored by the Federal Government and the World Bank. A distinct autonomous Federal Livestock Development Project was established to undertake smallholder schemes, investment schemes, animal feed programme,

and secondary school livestock farms as well as provide services in processing, marketing, and training.

6. **Pest Control, Animal Health, and Veterinary Public Health Services:** Initially, Tsetse and Trypanosomiasis Control Division was established under the FLD and later (April 1979) merged with the excised Crop Pest Unit of the Federal Department of Agriculture (FDA) to become the Federal Department of Pest Control Services (FDPCS). The former was hitherto responsible for the control of major livestock disease pests, particularly the tsetse eradication scheme (which had started since 1956) and trypanosomiasis control. FDPCS carries out control activities such as ground spraying, aerial spraying, biological control, among other methods. As regards animal health, the Veterinary Division of FLD undertakes programmes in treatment and vaccination of animals against endemic diseases such as rinderpest, contagious bovine pleuropneumonia, anthrax, black quarter, and rabies. In addition, FLD provides support services to the states through the cattle vaccination campaign, clinical services, as well as dipping and spraying facilities. The veterinary public health programme is to ensure that livestock products are wholesome for human consumption. Since the Meat Inspection Edict of old Northern Nigeria (1968), this function has been transferred from the Health Ministry to the Agriculture Ministry. Inspection is carried out at the main abattoir, slaughter slabs, and cold rooms. FLD complements state efforts in this regard by deploying supervision and training staff in meat inspection. The department also directly handles meat inspection at those centres involved in interstate trade, including Mokwa, Bauchi, and Nguru.

Implementation of Livestock Policy

Nigeria has implemented four official plans in the post-independence period, but it still fails to meet the state livestock objectives. Each plan has consisted of programmes and projects, which are articulated within the existing policy framework and financed systematically on annual basis through the budget. In many instances, the implementation is carried out directly by the livestock department of the agriculture ministry (state/federal). In some other instances, separate self-accounting parastatals may serve as implementation agency, having an autonomous management status. Certain government projects may also be implemented within a private enterprise framework.

By and large, the critical issues in implementation concern the organization for execution, personnel, and the financing character of programmes and projects. It turns out, in general, that the performance of the organizational and personnel structures is largely dependent on the financing function. The first stage of financing a given livestock project, like other projects, is the allocation of capital sums of money to agricultural programmes or projects in the current plan. The second stage is the allocation of funds to these programmes/projects in the annual budgets. In principle, the budget allocations are to be derived from the plan allocations, the former to be released within a given year on quarterly basis. Table 2 shows some examples of the financing character of the livestock sub-sector. Casual statistical examination of figures on this table indicates three marked divergences between plan allocation and budget allocation, revealing a poor funding or implementation commitment of government in the medium term; divergence between budget allocations and the actual releases, revealing the poor funding or implementation commitment of government in the short term; divergence between the actual release and the actual expenditures which suggests weak programme accountability among the executors. The combined manifestation of this divergence is the indication of the presence of a high degree of implementation indiscipline in the overall livestock policy process. There is an additional indication that the predictive powers of the available planning model are inadequate.

IMPLICATIONS FOR FUTURE MANAGEMENT OF LIVESTOCK POLICY

The foregoing synthesis contains important implications for future management of the current livestock policy in the 1990s. One of these is that administration of policy cannot be decentralized too far in a federal setting. An important reason for this inference is the need for harmony in the administrative process as well as consistency in the policy objectives. In effect, therefore, the policy environment needs to be closely monitored at both federal and state level to ensure that different priorities are not set from those contained in the current document. In particular, the choice of instruments requires a periodic assessment to identify location-specific limitations facing their effective use.

Another important implication of the synthesis concerns the problems of policy implementation. There is evidence that well-articulated objectives and presence of efficient instruments are only necessary, but not sufficient conditions for effective livestock policy. The overriding factor in this regard is finance of the constituent programmes and projects. In specific terms, there is need for high correlations among the plan allocation, budget allocation, actual release, and actual expenditure.

SUMMARY AND CONCLUSIONS

The historic synthesis of Nigerian livestock policy reveals the important features of the past livestock policy environment: the initial absence of properly articulated objectives led to the absence of suitable basis for setting meaningful criteria for evaluating the impact of policy; decentralization of entire policy administration among regions has created a policy atmosphere devoid of the vertical and horizontal relationships in policy objectives and instrument, thereby leading to wasteful duplication of policy efforts among the regions; the effective use of available livestock policy instruments is limited by poor implementation, whereby significant differences exist among plan allocation, budget allocation, actual release, and actual expenditure.

The successful management of future livestock policy, therefore, will depend largely on stability of the policy environment, particularly the degree of implementation discipline which includes, in the main, intense programme accountability on the part of executors as well as funding commitment on the part of sponsors.

REFERENCES

1. Annual Report of the Agricultural Department (1900–1960), Colonial period, Nigeria.
2. Annual Reports of Agricultural Ministry, 1960, Post-Colonial, Nigeria.

Table 1: Main Organs of Nigerian Agricultural Research Systems

S/NO	INSTITUTIONS	OWNERSHIP/HEADQUARTERS
	A. Research Institutes	
1.	Agricultural Extension and Research Liaison Service	Federal / Kaduna (Kaduna State)
2.	Agricultural Extension Research Liaison & Training	Federal / Umudike (Imo State)
3.	Cocoa Research Institute of Nigeria (CRIN)	Federal / Ibadan (Oyo State)
4.	Federal Institute of Industrial Research (FIIRO)[a]	Federal / Oshodi (Lagos State)
5.	Forestry Research Institute of Nigeria	Federal / Ibadan (Oyo State)
6.	Institute for Agricultural Research	Federal / Zaria (Kaduna State)
7.	Institute of Agricultural Research and Training (IAR&T)	Federal / Ibadan (Oyo State)
8.	International Institute of Tropical Agriculture (IITA)	International Body / Ibadan (Oyo State)
9.	Kainji Lake Research Institute (KLRI)	Federal / New Bussa (Kwara State)
10.	Lake Chad Research Institute	Federal / Maiduguri (Borno State)
11.	National Animal Production Institute (NAPRI)	Federal / Shika, Zaria (Kaduna State)
12.	National Cereals Research Institute (NCRI)	Federal / Badeggi (Niger State)
13.	National Horticultural Research Institute (NIHORT)	Federal / Ibadan (Oyo State)
14.	National Root Crops Research Institute	Federal / Umudike, Umuahia (Imo State)
15.	National Veterinary Research Institute	Federal / Vom (Plateau State)
16.	Nigerian Institute for Oil Palm Research	Federal / Benin-City (Bendel State)
17.	Nigerian Institute for Trypanosomiasis Research (NITR)	Federal / Kaduna (Kaduna State)
18.	Nigerian Stored Products Research Institute (NSPRI)	Federal / Benin-City (Bendel State)
19.	Rubber Research Institute of Nigeria (RRIN)	Federal / Yabal (Lagos State)
	B. Specialized Agricultural Universities	
20.	University of Agriculture, Abeokuta	Federal / Abeokuta (Ogun State)
21.	University of Agriculture, Makurdi	Federal / Makurdi (Benue State)
	C. Specialized Technology Universities[b]	
22.	Federal University of Technology	Federal / Akure (Ondo State)
23.	Federal University of Technology	Federal / Minna (Niger State)
24.	Federal University of Technology	Federal / Owerri (Imo State)
25.	Anambra State University of Technology	State / Anambra State
26.	Rivers State University of Science and Technology	State / Port Harcourt (Rivers State)
27.	Abubakar Tafawa Balewa University of Technology	Federal / Bauchi (Bauchi State)
	D. General University[c]	
28.	Ahmadu Bello University	Federal / Zaria (Kaduna State)
29.	University of Benin	Federal / Benin (Bendel State)
30.	University of Calabar	Federal / Calabar (Cross River State)
31.	University of Ibadan	Federal / Ibadan (Oyo State)
32.	Obafemi Awolowo University	Federal / Ile-Ife (Oyo State)
33.	University of Ilorin	Federal / Ilorin (Kwara State)
34.	University of Maiduguri	Federal / Maiduguri (Borno State)
35.	University of Nigeria	Federal / Nsukka (Anambra state)

Note:

(a) FIIRO conducts research into the industrial use of agricultural commodities.
(b) Technology universities conduct research into agricultural technology.
(c) General universities have faculties of agriculture, which research into all aspects of agriculture.

SOURCE: (1) *Agricultural Research Centres* Volume 2, pp. 624–630.
(2) Joint Admission and Matriculation Board Brochure 1988–89 Session, pp. 7–237.

Table 2: Financing Character of Selected Fisheries Projects, 1981–87

Projects	Plan Allocation N(m)	Total Budget Allocation N(m)	Actual Release N(m)	Actual Expenditure N(m)
1. National Accelerated Fish Project	12.00	7.52	6.00	5.44
2. Fishing Terminal Project	35.00	32.46	35.68	26.62
3. Inshore Fishing Project	10.00	4.94	4.81	3.33
4. Fish Storage, Processing, and Marketing Schemes	5.00	4.05	3.72	3.42
5. Fish Seed Multiplication Project	5.00	4.63	3.41	3.028
6. UNDP/FAO Freshwater and Fish-farm Demonstration Project	1.00	0.46	1.33	1.137
7. UNDP/FAO Artisanal and Inshore Fisheries Development Project	1.20	1.425	1.265	0.96
8. Fishing Terminal Management Company Project	2.00	0.58	0.71	0.167
9. Pilot Fish Farms	6.00	5.23	3.87	3.407
10. Fisheries Inspectorate and Equipment Supply Project	2.50	0.476	1.526	1.453

*Staistically significant

Source: Federal Department of Fisheries, Lagos (field survey)

The New Land Development Strategy for Accelerating Food Production in Nigeria

Revised paper previously submitted as memorandum to the Central Task Force on National Agricultural Land Development Authority (NALDA), Federal Ministry of Agriculture and Natural Resources, Abuja, Nigeria, May 1991.

Over the past thirty years, Nigeria has implemented several intervention programmes in the quest for self-sufficiency in food production and supply. Yet there is increasing concern over the vexing food problems and the apparent failure of past efforts to properly resolve them. This concern gives rise to the recent thinking that the problems possibly persist as issues in sustainable development, defined in terms of the inherent interactions of agriculture with population and environment.

The concept of land development was reintroduced in the 1991 budget to undertake development of 30,000–50,000 ha. of farmland in each sate of the federation over the 1991–1993 rolling plan period, and address the issue of underutilization of land and labour resources in the rural areas. A task force was first set up to prepare a blueprint in this connection, before the National Land Development Authority (NALDA) was constituted as the implementing agency under the presidency.

The objective of this memorandum is to briefly examine the new strategy of land development within the context of the Nigerian food problem. The two aspects of immediate concern are

(a) conceptualization of NALDA as a source of critical linkages in the pursuit of sustainable agricultural development, and
(b) the optimal role of NALDA, given the existing strategic intervention modes.

NALDA IN SUSTAINABLE AGRICULTURAL DEVELOPMENT

A brief exposition of the synergies of agriculture, population, and environment is necessary at this stage of conceptualizing the forms and functions of NALDA as a strategy of sustainable agricultural development. Essentially, we desire to identify pressure points in the nexus where interactions are most pronounced and new developmental interventions would presumably be most effective.

First, population dynamics underpin most of the issues in declining agricultural performance. The need to survive, particularly the need to meet human physiological requirements for good food and nutrition, affects human fertility decisions; the growth rate of population will continue to put pressure on agriculture. Therefore, Nigeria's population growth rate of about 3.1% per year, doubling the population in about 23 years, has staggering implications for the country's agricultural resources in particular, and the human and natural resources in general. The relationship between population growth and poor agricultural performance can also be viewed in reverse order: because traditional agriculture fails to evolve rapidly enough, fertility preferences of rural communities are affected towards increasing rates so as to ensure services of children on farms. In particular, women's fertility decisions are against reducing family sizes because of the need for more children for alleviating the burden of their multiple roles as farmers, household managers, and child barers/rearers. Thus, agricultural technologies and strategies which enhance the productivities of rural dwellers (women and men alike) will reduce fertility rates considerably. This vicious cycle has relevance in the conceptualization of a land and development strategy, which should seek not only to address a specific agricultural problem but also prevent the possibility of new stress conditions that may escalate the fertility problems. As NALDA may invariably contain a settlement component, particular attention is required on the strong synergies between agriculture and population as a factor of sustainable development.

Second, agricultural activities are a critical link between people and their environment. As people seek to husband their natural resources for their livelihood, environmental degradation mounts increasingly. This threatens the sustainability of yield from the available stock of resources in perpetuity and, hence, human survival over successive generations. The severity of this trend of negative interaction of the nexus is directly related to the pressure of population on arable land and the use of inappropriate agricultural technology. Perhaps the most critical environmental problems in Nigeria are soil degradation, desertification, water contamination, and deforestation. These problems, of course, have harmful impact on economic growth, distributional equity, and resource integrity and are probably the eventual outcomes of the effect of population pressure on shortening fallows, soil mining, and agricultural intensity. The direct relevance of this mode of interaction of agriculture, population, and environment to the new land development concept is grave. As much of the country, particularly the south is covered by heavily weathered acidic soils with low inherent fertility, uncontrolled cultivation will remove many of the nutrients directly and expose the soil to rain, heat, and sunlight to give way to rapid loss of soil productivity through leaching, acidification, erosion, and structural deterioration. While fertilizer application may enhance the nutrient status of the soil initially, excessive leaching makes it unavailable to crops and the structural deterioration continues. NALDA, therefore, needs to develop substantial capabilities for addressing these issues, without which sustainable agricultural development will remain largely illusory in the country.

NALDA AND CURRENT POLICY OUTLOOK

The current agricultural policy document has critical implications for the new efforts of NALDA, which must fit perfectly in place with respect to stated objectives and specified instruments. First, the new strategy must, as a matter of necessity and consistency, uphold the relative roles of government and private sector as contained in the policy. While government should provide support functions in technology creation and transfer, as well as infrastructure to the farming communities, private initiative remains the sustainable means of agricultural enterprises or industry. Suffice it to stress here that the persistent mistakes of role confusion in new development programmes, as experienced with RBDAs and DFRRI, should be avoided.[51]

The second critical element of the current environment is the need to pursue the principle of ecological specialization practically. This entails the use, to the best advantage, of the relative resource endowments and potentials of different regions of the country, as well as the peculiar environmental niches in agricultural production. The practice promises large incremental food output based largely on the comparative advantage theory. NALDA has large room to make an impact in a short time if the principle is properly operated. The specific tasks will include

---

51 RBDA—River Basin Development Authority. DFRRI—Directorate of Food Roads, and Rural Infrastructure.

(a) establishing a basis for improving the knowledge of situation about differential agroecological potentials not only as conditioned by edaphic and climatic features of specific areas, but also as revealed by consumer preferences and other market forces;

(b) emphasizing farm enterprise models that are most suitable for specific areas of the country, comprised of aspects of production technology modes (e.g. tractorization, animal traction, etc.), integrated crop livestock modes with possible benefits accruing from resource conservation (e.g. use of crop residues), and integrated crop-livestock-fisheries modes which could utilize marginal land for fish farming. Thus, NALDA as a new and priority development institution can provide leadership in designing efficient commodity programmes for different agricultural locations in Nigeria.

Lastly, the role of NALDA should be properly integrated within the framework of the activities of other development programmes and projects. In this regard, Table 1 describes the relative foci of existing agencies towards identifying the gaps to be filled and eliminating wasteful parallel functions. The critical functions of government in the new dispensation can be divided into six types: extension services, provision of rural infrastructure, credit, irrigation services, and input distribution. In these functions, the existing agencies have developed considerable expertise. Even though the participants in NALDA's programmes will require the same services, it is not necessary for the new body to set up new facilities for providing them. What is required is for NALDA to be properly integrated with the relevant agencies to only fill gaps in function that might exist in the process of providing the critical services. In most cases, it might only need slight modifications of the programmes of the specialized agencies to offer services under NALDA's programme compared to major efforts necessary for NALDA to provide the same services independently.

The issue of land clearing services leaves much to be desired. As it presently operates directly under the ministry, the activity of land clearing is at low ebb. It appears that the ineffectiveness of the programme actually underscores the rationale for setting up NALDA. Therefore the task of opening up new land can be a special function of the new agency, while the ministry should be divested of the execution of similar programmes.

CRITICAL COMPONENTS OF INTEGRATED LAND DEVELOPMENT PROGRAMME

Land Clearing

It appears that the opening up of new land is an inevitable component of the proposed action programme under NALDA. This will involve the use of heavy machines to remove trees and stumps for cultivation. Essentially, this will constitute a new element of agricultural subsidy as it reduces capital investment requirements of private farmers. However, it can be devoid of the usual problems of other forms of farm subsidy as it will likely accrue largely to intended beneficiaries and because market intermediaries are not necessarily to be involved in the delivery process as in the case of fertilzer; therefore, its distortionary consequences are limited. However, it constitutes new fiscal burden on government, but it can be easily rationalized as a means of changing the mode of government subsidy support for agriculture fom direct to indirect types.

A seemingly more important issue is whether or not beneficiaries should pay for NALDA's services in land clearing. This issue assumes some importace in the light of the market tendencies of current policy reforms. One opinon holds that, like extension, beneficiaries should not pay for land clearing. To make this argument, the role of government in developing the econom is thought to include meeting part of capital layout of production costs. Another opinion is for the farmer to pay, in full or in part, for the services rendered in land clearing. The basis of this idea is to make beneficiaries have a stake in the development of their land, without which indiscriminate clearing will tae place, associated with huge financial losses and environmental risks. The latr idea is in perfect consonance with the economic notion that free goods hae no value to users. Therefore, the support for payment for clearing services not necessarily to recover public costs, let alone generate returns, but only give value for cleared lands and introduce rationality to the clearing proce.

Another important issue relating to land clearing is how to attain hig enough efficiency of land clearing at the minimum damage to the soil. Th calls for in-house capability of NALDA to supervise the operations closel Supervisory effort should include the use of objective criteria to determine th necessity of clearing a particular piece of land. Clearing operations, becaus of environmental consequences, should only be performed where there critical need. The present notion is that much of cleared land area exists what may be lacking critically is access to it. In this connection, it appear necessary, as a first step in the implementation of the new programme, t take stock and make description of the status of agricultural land in the country, towards determining the cumulative effect of historical programmes on land development. The relevant agencies will include ministries, past and present corporations, among others. For instance, the RBDAs had opened up well over 140,000 ha. by 1986, much of which was held up in direct agricultural production ventures, before they were disengaged from such activities under the new policy. There is going to be little wisdom in clearing new, probably environmentally fragile land in some places where such already exists unutilized.

Settlement Component

The critical source of lessons to be learnt about settlement programme is the farm settlement scheme of the old Western Region (Adegboye 1969). The useful performance of this scheme was established upon

(a) the underage nature of beneficiaries, who, according to an FAO account (1966) were too 'immature' to take farming seriously; it was thought in this account that the philosophical motivation of young school leavers to run profitable farms under government assistance in training, credit, and infrastructure was an impracticable proposition to 'turn physically and mentally immature youths into serious-minded hard-working farmers . . . young men can only be expected to settle down and devote themselves seriously to productive work after they have passed the age of 20 and have married and thereby assumed responsibility for maintenance of other persons than themselves'.

(b) the too-low income-generating capabilities of the settlers to meet the repayment schedules compared to the expectations of the prototype plans, which consequently prevented the fulfilment of the anticipated demonstration value of the scheme on surrounding farmers.

(c) the enormous capital outlay per settler which rendered the scheme incapable of making any significant contribution to the unemployment problem in the country, given the rapid rate of population growth.

One issue that may also assume great importance in the settlement component of NALDA's programme is the state of land tenure in the country. This issue has relevance under two alternative modes of settlement. In the first mode, government may acquire large areas of land in certain locations, clear, and allocate to interested beneficiaries. This mode creates the greatest perturbation in the existing social system. It is therefore associated with heavy capital outlay comprised of expenditures in housing, potable water supply, and possibly health and education facilities, among others. We want to discourage this mode in the light of the following experiences:

(i) the heavy expenditure implications of social infrastructure will constitute a major diversion and be time-consuming, thereby delaying the impact of the new programme;
(ii) the functioning of social infrastructural facilities depends largely on other agencies than the agriculture ministry, which may be a new source of agitations from society and possibly an avenue for distraction;
(iii) high rate of desertion results from the unfulfilment of expectations of farmers to be rapidly transformed far up in the poverty line.

By and large, this mode of settlement tramples upon the existing land ownership rights, which can easily throw NALDA to the familiar cycle of land compensations, inertia to vacate acquired land and take over difficulties. This can also pose a severe obstacle to land development impact within a short time.

In the second mode, the status quo of society may be maintained by pooling contiguous pieces of land together to define a development unit. Land clearing services can be provided to each unit in a cooperative setting, which can also facilitate the accessibility of farmers to other services. This mode is attractive because of the possible avoidance of land acquisition problems, without affecting the chances of large-scale operators to also benefit from land development services.

In effect, the settlement component of NALDA's programme depends on four main decisions:

(a) the choice of beneficiaries on age and responsibility grounds towards limiting rate of desertion,
(b) the choice of allied services that will guarantee high rate of return on beneficiaries' investments,
(c) the optimal capital requirement to maximize the number of beneficiaries in order to solve the current unemployment problem in part,

(d) the choice between the acquisition-development-allocation mode of land development and the minimum perturbation mode so as to reduce difficulties associated with land rights.

CONCLUSIONS AND RECOMMENDATIONS

The proposed activities of NALDA are conceived within the general context of sustainable development, which contains the agriculture-population-environment nexus as the central issue. The critical components emanating from this issue for NALDA are land clearing services and the settlement of the targeted clientele. Optimal decisions are required to make choices among alternatives available, in the light of past experiences. Even though the information required for making the necessary decisions have been provided in the foregoing discussion, the need for a standing technical committee arises, given the amount of technicalities involved in the proposed activities of NALDA. The specific tasks of the technical committee will include the following:

(i) to monitor the appropriateness and observance of technical contents of projects, particularly land clearing specifications;

(ii) to monitor consistency of land development activities with provisions of national agricultural policy, such as relative roles of government and private sector, practical effect of agroecological specialization policy, etc.;

(iii) to monitor activities on the ground so as to avoid conflicts with and duplication of other development efforts;

(iv) to initiate the conduct of studies for providing relevant knowledge of specific problem situations, e.g. the need for deriving objective criteria for selecting commodity programmes in specific agroecologies, and collection of relevant information on the socio-economic and cultural aspects for designing suitable settlement schemes for specific places;

(v) to produce monitoring reports on the activities of NALDA on a regular basis, consistent with the desire for programme accountability.

REFERENCES

Adegboye, R. O., A. C. Basu, D. Olatubosin (1969), 'Impact of Western Nigeria Farm Settlement on Surrounding Farmers', *J. Econ. and Social Studies* 1 (2 & 9).

FAO (Food and Agriculture Organization) (1966), *Agricultural Development in Nigeria 1965–1980*.

Idachaba, F. S. (1985), 'Integrated Rural Development in Nigeria: Lessons from Experiences', Paper presented at the Workshop on Designing Rural Development Strategies, Federal Agricultural Coordinating Unit, Ibadan, December 1985.

Table 1: Normative Interaction Matrix Between NALDA and Selected Existing Agricultural Programmes

AGENCY/ PROGRAMME/ PROJECT	PRINCIPAL COMPONENTS	POSSIBLE AREAS OF COLLABORATION WITH NALDA
Nigerian Agricultural and Cooperative Bank (NACB)	Special focus on credit delivery to agriculture	Supply of credit needs of NALDA's clientele under Agricultural Credit Guarantees Scheme
Agricultural Development Projects	Agricultural extension, input distribution, rural infrastructure (emphasis on rural feeder roads and water supply schemes)	Support services to NALDA participants towards enhancing technical and allocative efficiency of farm production
River Basin Development Authorities	Irrigation infrastructure	Support services to make multiple cropping possible among participating farmers in defined land units
Land Use Act / Allocation Committees	Allocation of specified areas of land to farmers	Facilitation of NALDA in land allocation
Directorate of Food, Roads, and Rural Infrastructure	Infrastructure development—rural roads, water supply, electrification, etc.	Services to NALDA settlers and cooperators

The Structural Adjustment Programme (SAP) and the Nigeria Food Crop Sub-Sector

Invited paper presented at the National Conference on the Impact of Structural Adjustment Programme (SAP) on Nigerian Agriculture and Rural Life, organized by NISER and the Friedrich Ebert Foundation of Germany, Ibadan, 26–27, November 1989.

This paper examines some of the unresolved issues on the SAP agenda. Far from providing answers, we first briefly survey the antecedents, followed by a description of the underlying philosophy and the constituent instruments of the programme. Afterwards, concrete issues are discussed with the aim of drawing useful lessons of experience from the implementation of the SAP in regard to the food crop sub-sector. Lastly, specific guidelines are stated for enhancing food crop production in the future.

ANTECEDENTS

Among the several explanations offered for the worsening performance of the food system in the pre-1985 period, the role of the exuberant oil-boom is most illuminating (World Bank 1983). Firstly, petrodollar and the associated marginal propensity to import created immense urban boom. As a result, rural youths were attracted to towns, leading to increased proportion of aged farmers at the same time when higher demand pressures were made on the food production system by increasing number of city dwellers. The supply shortfalls were closed with continuous food imports, which were made possible by the oil-based foreign earnings. The great diminutions of the food production capacity were therefore inevitable.

However, following the acute balance-of-payments problem that greeted the latter part of the fourth development plan period, there was great scarcity of essential commodities, which consisted mostly of food items. Pending a suitable programme for solution, previous governments resorted first to an economic stabilization policy of palliative measures (Shagari's) and later to rationing of basic items including rice (Buhari's). In addition, a proposal for recovery was drawn up, based primarily on obtaining an adjustment finance of 2.5 billion dollars from the International Monetary Fund (IMF). This plan finally aborted on somewhat political grounds.[1]

As an alternative, the line of a self-reliant Structural Adjustment Programme (SAP) was chosen towards economic recovery. The implications of this approach and the means of using the approach to resuscitate the food system form the discussions of the next section.

PHILOSOPHY AND INSTRUMENTALITIES

Self-reliance is the underlying theme of the Structural Adjustment Programme. It is borne out of five main ideas as far as the food system is concerned. First, much of the hardships we suffered while in recession arose from outside the economy. In particular, they arose from the sudden shortage of foreign exchange. This is probably why the need for adjustment finance was immediately felt. The thinking is that such a loan would restore quickly the balance-of-payment equilibrium, thereby spreading the shock of adjustment over a long future so as to minimize the pains. Much as the approach is legitimate (Martin and Solewsky 1988), there exists the fear, as public opinion ascertained during the debate on IMF loan, that much money would be put in the care of the Nigerian public service system, which has a bad record of financial management and accountability.[52]

The second idea to support the self-reliance philosophy is the hazard of dependency relations. The most important element of this is the possibility of unexpected natural socio-economic discontinuities in the various import sources, thereby causing what has been called transitory insecurity (World Bank 1986). Such discontinuities include prolonged drought, large-scale

---

[52] The IMF loan issue spanned three governments. The Shagari administration applied for the facility in April 1983. It assumed greater political visibility during the Buhari era. Babangida quickly appointed a Presidential Committee on IMF loan to steer a national debate on the issue. Government finally followed the preponderant public opinion to break off negotiations with the IMF. (for details see Ashwe and others 1988).

pest infestation, as well as civil disturbances and war. The carry-over effect of sudden supply shortfalls in the importing country may range from simple public discontentment to massive starvation. Table 1 shows Nigeria's vulnerability to external shocks in selected sources of imports for the 1979 trade year. The high values of vulnerability index for USA mean that if, for example, 100 Nigerians had to die in this year from world grains shortage problem, about 27 and 28 of such deaths could be sourced to shortage in these countries respectively. The remaining deaths could be mainly sourced to shortage in other thirteen countries (i.e. only 3 deaths on average). We could then conclude that Nigeria's grains dependency was excessive on the USA and UK in 1979. If their economy sneezed in this year, ours would catch intense cold.

The third idea to support self-reliance is food provision of social utility. In this case, domestic availability of a commodity is conceived as a public good. This translates to saying that the higher the local content of a good or service in the food system the greater the social utility available to citizens from its consumption. In this regard, the utility is in the form of implicit gratification, which the citizens of a country feel because they are able to feed themselves.

The fourth idea is based on the infant industry argument. This considers that certain industries should receive extra protection until they become securely established. This protection already existed in respect of agriculture in the forms of monetary and fiscal concessions (e.g. preferential interest rates, tax concessions), tariff and non-tariff trade protection, and subsidies, among others. Under SAP, the agricultural protection has been reinforced with the outright bans of food imports. Such measures are aimed at shading the infant industry away from the outside influences of the more efficient producers until it develops adequate competitive competence. There is substantial merit in the notion that Nigerian agriculture is an infant industry because it is still characterized by technical and economic disequilibria, which are the result of its movement from old to new production surfaces.

Lastly, the theory of second best (Lipsey and Lancaster 1956) provides a rationale for the pursuit of self-reliance in food production. Simply put, it states that if one or more of the pareto-optimality conditions cannot be satisfied, it is neither necessary nor desirable to satisfy the remaining ones. In the context of the food economy, the evidenced defects in structure, conduct, and performance of the world markets are such institutional restrictions which make the best welfare positions unattainable; and it is quite irrelevant to pursue the satisfaction of the remaining equilibrium conditions of ideal world trade. Therefore, one cannot discredit self-reliance policies simply on account of violation of the equilibrium conditions for pareto-optimality.

Food Crop Programmes and Instruments under SAP

The programmes and instruments under SAP have been designed towards achieving the following broad objectives:

- to strengthen the demand management policies,
- to stimulate domestic production and broaden the supply base of the economy,
- to establish a realistic exchange rate system,
- to diversify industrial production,
- to liberalize trade and payments,
- to reduce administrative control and rely more on the market forces,
- to promote export.

With regard to the food crop sub-sector, a number of specific instruments and sub-programmes are worth highlighting. These include (i) rural infrastructure for stimulating food crop production, (ii) market reform for enhancing farmers' incomes, (iii) reduction of direct government production efforts to eliminate wastage, and (iv) withdrawal of subsidy to reduce distortions.

1. Rural infrastructure for stimulating food crop production

The Directorate of Food, Roads, and Rural Infrastructure (DFRRI) was established in 1986 to accelerate the process of enhancing the technical efficiency of the rural workforce, particularly farmers, through the provision of rural infrastructure. The distinguishing features of DFRRI are as follows:

- It is a supraministerial body with headquarters located in the presidency; this has been done for the purpose of accelerating decision-making and eliminating the usual wasteful delays inherent in the traditional ministries.
- It enjoys relatively generous funding; the sum of ₦450 million was initially set aside for its use out of the expected ₦900 million realization from the newly introduced Second-tier Foreign Exchange Market; the subsequent budget allocations are ₦360 million (1986), ₦460 million (1987), ₦500 million (1988), and ₦300 million (1989).

- DFRRI is not modelled to implant its own bureaucracy for executing projects, but to cause things to be done faster within the framework of existing machineries.

2. Market reform for enhancing farmers' incomes

The inherited commodity boards system was swiftly abolished with the commencement of the SAP—an effort towards deregulation of the agricultural economy. This was based on the realization that the constituent organs of the system were not only failing to resolve the issues of market instability, insecurity of food supply, unstable farm incomes, but also were preventing their solution by other means. The specific problem areas have been identified (Ayoola 1989) to include political, administrative, and functional elements. These problems culminated in huge bankruptcy of the commodity boards at the time of the abolition, except the Cocoa Board (Table 2). At the present time, relatively free enterprise prevails in the food crop markets. However, conclusive statements about the impact of the policy of non-intervention on food crop producers and consumers will await further investigations. If the observations of high grain and tuber prices withstand empirical scrutiny, it is possible to hypothesize that the introduction of the free enterprise food markets has benefited the producers by way of increased revenues, to the detriment of consumers. And depending on the relative magnitudes of increases in farm incomes and because farmers are fewer than food consumers, one may expect *a priori*, that the food crop market reform generated net welfare losses in the past four years. This, again, is not conclusive, and may be a short-run phenomenon only.

3. Removal of direct government production

The traditional agriculture ministries' River and Basin Development Authorities were hitherto actively engaged in direct agricultural production activities. SAP includes an effort to disengage them totally from the practice. Government is now limited in role to the performance of facilitating functions only. This implies that the business enterprise initiatives in agriculture will reside primarily in private sector participation. For one thing, the superior sensitivity of private investors to price signals and incentive means that the nation's scarce resource will be better allocated by them. For another thing, the small private farmers constitute the bulk of the population and the largest part of the labour force, which means that any competition that reduces their income statuses will produce adverse welfare results. Besides, government is fraught with unfavourable features, including

  (i) general ineptitude of its officials to business opportunities,
  (ii) lack of sensitivity to the usual economic variables,
  (iii) wasteful bureaucratic delays caused by strict adherence to official protocol, and
  (iv) widespread corruption, political intransigency, absence of financial and programme accountability.

4. Rationalization of input subsidy

Originally, farmers were considered to deserve being pampered with subsidy support, owing to their limited purchasing power to demand improved technologies. However, lessons of experience from the operation of direct subsidy programmes suggest that the policy might be self-defeating in many ways. These include the following:

- the development of dependency mentality among farmers on government subsidy, thereby making it a permanent obligation of government and a huge fiscal burden;[2]
- the absence of proper price relativities following distortions created by subsidies, which is needed to direct production decisions of the farmers along the lines of efficient resource utilization;
- the low level of morale of the private sector participants to build up substantial enterprise initiatives in the distribution of inputs and outputs of agricultural production, owning to price rigidities with the consequent limiting effect on GDP;
- the presence of substantial externalities whereby farm subsidy benefits illicitly unintended channels within and without the economy;
- the frequent exploitation of subsidy programmes to serve the selfish motives of civil servants, politicians, and their elite friends, to the detriment of the ordinary farmer.

The case is therefore presently strong in favour of indirect subsidy as the means of government support in agriculture. This is a fundamental premise on which the emphasis on rural infrastructure is based. Through this way, the problem of unintended beneficiaries is minimal, because once the infrastructure is installed, the farmer possesses absolute, albeit not necessarily exclusive right of access to them.

SELECTED EXPERIENCES AND ISSUES

Is a cheaper naira enough?

The preponderance of exchange rate reform component of the SAP presupposes that naira devaluation can restore our trade balance almost alone. Even though the presence of an outright ban obscures the possible effects of currency depreciation on the food crops sub-sector, the entire food system is also greatly influenced by opportunities in the rest of the economy. Therefore, the question of whether a cheaper naira is enough to improve the trade position is relevant to the food crop sub-sector.

Agreed that, had the naira not fallen since 1985, our financial position now would undoubtedly be much worse. But as we enter the fifth year of devaluation, initially through SFEM and later FEM, it appears that this policy does not have the capacity to haul us out of the predicament, to the extent we hoped. As Faux (1988) has noted, that heavy reliance on currency depreciation would result in lower real income and a larger foreign debt burden, both of which we have experienced. The implication for policy is not necessarily that the naira should be artificially stopped [53] from falling further. But we require strengthening our strategies with aggressive common sense trade management and policy instruments that do not require further reduction in Nigerian living standards. Some empirical questions therefore are

- What is the most appropriate mix of currency depreciation, which hurts all trading partners without regard to the relative sizes of their trade surplus against us on the one hand, and other protectionist instruments such as tariffs, which can be source-specific on the other hand?
- How can we properly resolve the inability of exchange rate protectionism to distinguish between imports that are necessities, including agricultural inputs, and imports that are luxuries, so as to arrest the escalating cost of food crop production?
- What is the elasticity mix of our exportables to warrant the expectation of an export-led growth through naira depreciation, which could compensate for lower-income consequences of import discouragement in agriculture, as suggested by Galbraith (1988)?

---

[53] Take the fertilizer subsidy as example; selected budget allocations are as follows: ₦5 million (1976/77); ₦7.8 million (1977/78), ₦31 million (1979/80), ₦66 million (1980), ₦105 million (1981). For details see Idachaba (1981).

- How flexible have the country's directions of trade been, to be able to adjust towards increasing import trade with those partners whose currencies have become relatively cheaper owing to the foreign exchange reform?
- Have we been able to avoid a currency war so much to expect that retaliation by other countries would not have neutralized the effect of naira devaluation significantly?
- How vigorously have we pursued the bilateral option to be able to adjust our trade strategy to the wide variety of trade regimes that exist in the real world?

Surprises of Deregulation

We follow Alfred Kahn (1988) to characterize surprises as a product of mistaken expectations and unforeseen outcomes. Confining our prescience to the area of agricultural products, the income windfalls of cocoa farmers at the onset of the SAP was quite expected. This immediately produced pronounced effects on cocoa-based food products at home. The apparent correctness of the view that the middleman role of the commodity boards trapped substantial incomes of the farmer was demonstrated in the high positive income effect of deregulation by abolishing the board. What was not so much expected was the quick emergence of bad trade and other sharp practices that accompanied the events. It so happened that Nigerian produce soon lost substantial credibility in the international market, owing to poor quality, which soon restricted further growth of the incomes. Besides, it is also surprising how swindlers and other dubious people came to the scene as a new generation of problems limiting farmers' incomes. This, no doubt, would have limited the growth of farm production that was normally expected with increased incomes. Besides, the sustainability of these income levels is quite doubtful, as the movement of world prices shows (Table 3) the within-month deviations in cocoa price are high.

However, issues have been raised as to whether the observed windfall incomes that accrued to farmers were really due to the removal of commodity boards only (Ayoola 1989); it is unlikely. In particular, the joint effects of linear and non-linear combinations of the variables of the proscribed commodity boards and devaluation of the naira will also be significant.

The real surprise from crop sub-sector was the immediate collapse of grains prices soon after market deregulation began. When we expected an upsurge in grains prices by the removal of the floor prices (guaranteed minimum prices, GMPs) as it happened following the removal of producer prices, what actually happened was the fall in price of maize to less than half of GMP. That was probably the unanticipated result of sudden release of stock during the unregulated winding-up programme of the grains board. The dampening effect of this on production in the following year (1987) was further complicated by bad weather, which included both widespread drought and flood disasters in the same year. All acting together could significantly explain away the inability of the grains market to resist blowing the lid off inflation since 1986 (Table 4). The composite inflation rate for major food crops is 20%.

Even if surprises and complications have not occurred, the question remains unanswered: How far can we carry the free market policy in agriculture? According to Pearce (1981), it is quite unrealistic to posit a free market system that would leave agricultural prices to be determined solely by market forces. Consistently, alternative intervention mechanisms are already being suggested (Ayoola 1989; Abdullahi 1989).

Adjustment and Unemployment

It is probably inevitable to experience structural unemployment with structural adjustment. After all, if a sector of the economy is constricted while another is expanded, labour must be released by the former and absorbed by the latter. However, the transition period features the presence of idle labour owing to slow rate of absorptive capacity increase relative to the rate of release. In agriculture, the emergency of structural unemployment has complicated the other forms, particularly seasonal unemployment problems.

The effort of the National Directorate of Employment (NDE) towards alleviating the pains of adjustment as regards the resultant low-income statuses and the associated welfare losses is significant. The NDE has mounted formidable agricultural employment programmes with substantial implication for food crop production in particular. However, it is disturbing to observe that these programmes, as in the case of similar panic measures in the past, lack substantial long-run postures.

Agricultural history of Nigeria is replete with attempts by government to solve unemployment problems by mounting emergency agricultural programmes, only to have the participants desert when the initial problem peters out. In the case of the farm settlement schemes, for instance, huge resources were committed to settle the products of the Western Nigerian free primary school education (Western Nigeria 1963). In the end, the farmers deserted the settlements in search of more remunerative work in cities. Also, there was limited demonstration effect on surrounding farmers, and it remains doubtful if school graduates could perform any better than the traditional farmers (Adegboye et al. 1969).

In an earlier paper (Ayoola and Idachaba 1989a), we have identified six similarities in the operation failure for the ongoing employment programmes of NDE. These can be listed by asking pertinent questions as follows:

- how appropriate are the characteristics of the credit package, in form and adequacy, to justify the expectation that the farm business will pay the borrower in the long run?
- how well do the characteristics of the borrowers in their professional and social settings conform with the low-technology farm enterprise, compared to their recent experiences at university and other farms and their familiarity with city lives?

Our optimism about the long-run effects of the ongoing graduate farmers and similar schemes will depend on the answers to the above questions. As presently constituted, the employment programmes might not be able to prevent high rate of desertion to the preferred sections when the economy picks up again. As for macroeconomists, the challenges of high unemployment have not been adequately met. Blinder (1988) prescribes four main stages as follows: (a) define it, (b) explain it theoretically, (c) explain it empirically, and (d) devise policies to reduce it. These stages imply that solving the unemployment problem extends beyond the mounting of emergency programmes in food production, to painstaking analysis of the nature of the problem. NDE may therefore wish to collaborate with university and other researchers towards proffering lasting solutions.

GUIDELINES FOR FUTURE DEVELOPMENT

The food crop production has a remarkable swath of prosperity in the new agricultural policy (FMAWRRD 1988). Given the appropriate institutional framework, three provisions of the policy will be emphasized here. These are ecological specialization, technological progress, and rural infrastructure support.

Ecological Specialization

Ecological specialization means the permission and encouragement of the regions, which differ in climatic, edaphic and socio-economic factors, to use to the best advantage any particular benefits in skills and resource endowments. Through specialization of this kind, wasteful duplication of efforts is avoided, thereby creating high resource use efficiency and internal comparative advantage. Agroecological specialization also extends to the use of special environmental niches such as the temperate enclaves of Jos Plateau, and others like Mambilla grains production efforts may require being concentrated in the derived savannah middle belt and certain parts of the upper north so that the tree-crop economy of the southern part would receive greater attention for output maximization.

The practice of agroecological specialization, however, has definite implication for marketing policy. The relevant issue in this context concerns the practice whereby commodity trade is sometimes restricted by some state governments. There are reports that certain states of the federation sometimes impede the outflow of food items to other states, usually as a panic measure during shortages.[54] This practice in itself is capable of working counter to the policy of agroecological specialization. This happens as every state will attempt to grow everything everywhere for the fear that specially endowed states might later obstruct free flow of the relevant commodities. In the end, the whole country loses out in terms of resources use efficiency, efforts wastage, and consequently low volume of national output. Therefore, the practical implication of ecological specialization as a national strategy of food crop production is that one cannot speak of self-sufficiency in any product on a state-by-state basis. Nigeria should operate, as one large indivisible market in which the surplus output of specially endowed producing areas should move freely to deficient areas without hindrance. This requirement of the production and market structures is made highly necessary by the spatial distribution of industries in relation to raw material sources in Nigeria.[55] It is therefore necessary to zone the country according to commodity productivities, along which development efforts will be applied, using efficiency criteria.

---

[54] A copious example is Kaduna State where the government issued instructions 'to prevent out-of-state traders from buying off and evacuation of this season's harvest of grains'. Consequently, the state-owned Farmers' Supply Company (FASCOM) instituted a programme 'against all those who will buy grains cash down and carry away from the state' (Anonymous).

[55] For example, while the production of grains is concentrated in the north, the poultry industry and the breweries that mainly consume the products are located in the south.

Technological Progress

The second need for future food crop production development is aggressive technological progress in agricultural production and marketing. We shall characterize agricultural technology as consisting of the nature and types of available inputs (e.g. seed, fertilizer, chemicals, tools, machines, farm power, etc.) and the ways in which they are combined (e.g. land-fertilizer ratio, labour-machine ratio, etc.). Technological progress is desirable because it reduces average unit costs of output, with input prices held constant. The relevant areas where this is needed in Nigerian agriculture include the following: (i) technology of land preparation; (ii) technology of planting, seed and seedling; (iii) technology of farm management practices; (iv) technology of harvesting, (v) technology of on-farm haulage, processing, and storage. These categories clearly imply that agricultural development is *sine qua non* for industrial development through which the technological drive can be provided and the momentum of progress be maintained.

There are three phases through which technological progress in agriculture could be initiated. These are (i) agricultural research for technology creation, (ii) diffusion of agricultural technology, and (iii) development of technological intellect. For meaningful progress to take place, determined efforts are required in these areas, which specially include the following:

- the adoption of a technology policy that is all-embracing and complete;
- the creation of effective agricultural technology through a virile national agricultural research system;
- the transfer of agricultural technology through effective extension system;
- the preparation of existing technology level through the development of technological intellect;
- the adoption of appropriate agricultural technology on merit criteria, including ecological and sociocultural considerations, simplicity, relative availabilities of capital and labour, divisibility and riskiness for the particular application of small-scale farmers.

Rural Infrastructure Support

The view is widely held that the provision of rural infrastructure is the essential foundation for sustained increases in food production. The general notion underlying the rural infrastructure strategy is that it is difficult for the rural sector to contribute significantly to economic and industrial progress in the absence of basic facilities that also enhance their production activities as well as their living standards. These facilities and their roles are generally described under three categories.

(i) Rural physical infrastructure, including

- rural roads which cause accelerated delivery of farm inputs, reduce transportation costs and enhance spatial agricultural production efficiency;
- storage facilities which help to preserve foods in the forms that consumers need them and to the time they need them;
- irrigation facilities which assure farm water supply and stabilize food production by protecting the farm production system against uncontrollable and undesirable fluctuations in domestic food production.

(ii) Rural social infrastructure, including

- clean water, decent housing, environmental sanitation, personal hygiene, and adequate nutrition, which help to improve the quality of rural life;
- formal and informal education, which promote rural productivity by making the farmer able to decode agronomic and other information and carry out other desirable modern production practices; basic education also promotes feeding quality, dignity, self-respect, and sense of belonging as well as political integration of the rural people.

(iii) Rural institutional infrastructure, including

- farmers' union and cooperatives which facilitate economies of scale and profitability of rural enterprise;
- agricultural extension which improves the technology status of the farm business.

However, the rural infrastructural development process of Nigeria contains a number of persistent problems and unresolved issues. Ayoola and Idachaba (1989b) have highlighted some of these, including the following:

i. Poor commitments of the programme sponsors;
ii. Incessant perturbations in the institutional framework;
iii. Low level of in-house operational research and planning of projects;
iv. The resources required for meeting infrastructure requirements;
v. The trade-off between rural infrastructure supply and other objectives.

SUMMARY

The Nigerian SAP has been formulated within the context of self-reliance strategy of economic recovery. This programme predicted on social utility provision, infant industry argument, socio-economic discontinuities, and the theory of second best. In regard to the food crop sub-sector, the instrument of direct and indirect effects is comprised of market deregulation, exchange rate reform, reduced government direct production efforts, and rationalization of input subsidies.

Lessons of experience from the implementation of the SAP suggest that

(i) we need to strengthen the exchange rate reform with policies that enhance the country's trade management and competitiveness but do not further reduce the living standards of people;

(ii) we need to be more flexible so as to provide automatic stabilizers that will negate the unforeseen adverse consequences of policy management process;

(iii) the mounting unemployment problem requires a longer-run approach to solution than is presently the case.

The food crop production sub-sector has a remarkable swath of prosperity in the new agricultural policy. But this cannot materialize unless the appropriate institutional framework is guaranteed, and the basic provisions are followed to the letter. The key aspects of the agricultural policy for the enhancement of food crop production include the adoption of ecological specialization, ceaseless pursuit of technological progress, and massive rural infrastructural support.

REFERENCES

Adegboye, R. O., A. C. Basu, D. Olatunbosun (1969), 'Impact of Western Nigeria Farm Settlement on Surrounding Farmers', *J. Econ. and Social Studies* 1 (2).

Abdullahi, Ango (1989), Keynote address, 25th Annual Conference of Agricultural Society of Nigeria, Owerri, August 1989.

Anonymous (undated), 'Format for 1984 grain purchase by FASCOM', Kaduna.

Ashwe, Chi (ed.) *The Nigerian Economist*, 23 June 1988 edition.

Ayoola, G. B. (1988), 'The Dependency Problem and Rationale for Self-sufficiency in Food Grains', *J. Rural Dev. Nigeria* 3 (1).

Ayoola, G. B. (1989) 'Agricultural Produce Price Policy under the Structural Adjustment Programme: Proposal for Alternative Mode of Market Intervention', Pro. 25th Annual Conference, Agricultural Society of Nigeria, Owerri, August 1989.

Ayoola, G. B. and F. S. Idachaba (1989a), 'Self-sufficiency in Food Production as an Essential Foundation for Economic and Industrial Development', Proceedings, 9th Annual National Training Conference of the Industrial Training Fund, Bauchi, 7–10 November 1989.

Ayoola, G. B. and F. S. Idachaba (1989b), 'Nigeria's Rural Infrastructure: Past and Present', *Proceedings of Workshop on Social Mobilization, Poverty Alleviation and Collective Survival in Nigerian Society for International Development*, Ibadan Chapter, University of Ibadan, 24–29 September 1989.

Blinder, Alan S. (1988), 'The Challenge of High Unemployment', Richard T. Ely Lecture, AEO Papers and Proceedings, *Am. J. Agricultural Economics* 78 (2) (May 1988).

Faux, Jeff (1988), 'A Cheaper Dollar Is Not Enough', *Challenge*, May–June 1988.

FMARRD (Federal Ministry of Agriculture, Water Resources and Rural Development (1988), *Agricultural Policy for Nigeria 1988*.

Galbraith, James K. (1988), 'Let's Try Export-Led Growth', *Challenge* May–June 1988.

Idachaba, F. S. (1989a) 'Strategies for Achieving Food Self-sufficiency in Nigeria', Keynote address, 1ˢᵗ National Congress of Science and Technology, University of Ibadan, 16 August 1989.

Idachaba, F. S. (1989b), 'A Farm Input Subsidy Policy for the Green Revolution Programme in Nigeria', Paper presented at the meeting of the Ministerial Committee or Farm Input Subsidies, Jos, 8–9 December 1989.

Lipsey, R. G. and K. Lancaster (1956), 'The General Theory of Second Best', *Rev. Econ. Studies* 24.

Martin, Ricardo and Marcelo Selowsky (1988), 'External Shock and the Demand for Adjustment Finance', *The World Bank Economic Review* 2 (1) (Jan. 1988).

Western Nigeria (1963), White Paper on Integrated Rural Development Official Document No. 8, Ibadan.

World Bank (1983), *World Development Report* (Washington DC, USA) 1983 issue.

World Bank (1986), *Poverty and Hunger: Issues and Options for Food Security in Developing Countries* (Washington DC, USA)

The Food Sub-Sector and Structural Adjust Programme in Nigeria: Some Matters Arising

Invited paper presented at the National Symposium on the Nigeria Food Question, University of Agriculture, Makurdi, 11–13 December 1989.

This paper briefly examines the major economic events over about four years of adjustment and highlights the important matters that have arisen in regard to the food sub-sector contemporaneously. Some guidelines are provided towards resolving the issues involved at the same time.

THE ADJUSTMENT PROCESS

The ongoing Structural Adjustment Programme (SAP) culminates from a series of palliative policy measures to resuscitate the Nigerian economy, following the balance-of-payments disequilibria that set in the early 1980s. These measures initially included an economic stabilization policy in Shagari's administration. This was followed by the rationing of the imported essential commodities during Buhari's administration. A concrete proposal for economic recovery was drawn up to be based on an adjustment loan financing by the International Monetary Fund (IMF) amounting to 2.5 million US dollars. This plan finally aborted on somewhat political grounds.

Consequently, the line of self-reliant structural adjustment was chosen towards economic recovery. The rationale for self-reliance as a philosophy of economic development is commonly based on social utility provision, infant industry argument, socio-economic discontinuities, and the theory of second best (see Idachaba 1984; Ayoola 1988).

The instrumentalities of the SAP have been designed to achieve the following broad objectives:

- to strengthen the demand management policies
- to stimulate domestic production and broaden the supply base of the economy
- to establish a realistic exchange rate system
- to diversify industrial production
- to liberalize trade and payments
- to reduce administrative control and rely more on the market
- to privatize and commercialize the public enterprises substantially
- to promote export.

In regard to the food sub-sector, the specific instruments and programmes include the following:

(i) Rural infrastructure projects for stimulating food production and enhancing the general living standards of the rural populace; in particular, the Directorate of Food, Roads. and Rural Infrastructure (DFRRI) was established as a supra-ministerial body to facilitate the provision of important infrastructure.

(ii) Farm investment projects to stimulate food production; in particular, the National Directorate of Employment was established to provide (inter alia) soft loans for establishing new farmers, including agriculture graduates and other categories of people.

(iii) Market reform for enhancing farmers' incomes; in particular, the existing commodity boards were scrapped to improve market stability, security of food supply and farmers' incomes.

(iv) Reduction in government's direct production efforts, to eliminate wastage; in particular, the functions and sizes of the River Basin Development Authorities (RBDAs) have been streamlined, and their non-water assets offered for sale to the public.

(v) Rationalization of agricultural input subsidy to remove distortions, reduce fiscal burden to manageable size, reduce negative externalities, including smuggling and unintended beneficiaries; in particular, fertilizer subsidy was partially withdrawn.

In the following sections, we shall examine the key events that have taken place in the food sub-sector in the process of executing the above programmes.

The important matters arising for each event will also be highlighted pari passu.

EVENTS IN FOOD PRODUCTION

Food production is to be directly stimulated through the DFRRI and the NDE, both of which have been evaluated before. The effectiveness of DFRRI is initially limited by multiple objectives and programme accountability (Idachaba 1988; Ayoola 1989) while the NDE lacks a basis for sustainability in the long run (Ayoola and Idachaba 1989). In addition to the efforts through DFRRI and NDE, government also pays special policy attention on what production. This is based on the fact that among a class of food imports, particularly grain imports, wheat impacts, special efforts are required to stimulate wheat production at home under the self-reliance policy.

Matters Arising

First, the question is being asked as to whether it is economically wise for government to be pushing such a non-native crop as wheat so hard. This question is premised on the apparent absence of comparative advantage in wheat production in Nigeria, which will create a large inefficiency in resource utilization in the aggregate sense. The question, however, could be evaluated in the context of the infant industry argument. On this basis, the counterargument is that there is the need for a hard initial push so as to assist wheat production to attain adequate competitive competence in the grain market. The several past production policies have offered substantial assistance to the production of other grains as well. Therefore, wheat production would be structurally disadvantaged without a special policy attention as is being provided. This argument gives the basis to encourage the present emphasis on wheat production in the country.

Trade-off in Wheat Production

Second, investigations indicate that wheat production does not compete for resources with other grains (Adeyeye 1989), but it does with vegetables (Igben 1989). The reason is that the bulk of tomatoes, peppers, and other vegetables are produced in the north as dry-season crops using irrigation water. Also, wheat is a dry-season crop requiring the same irrigation water. Therefore, wheat production displaces these vegetables in terms of water and land. This situation promises two possible effects, and therefore requires attention. One, it is capable of reducing the supply of vegetables, thereby raising their prices. Two, wheat itself might never be produced at the minimum desirable if the market reference favours vegetables. In essence, the new emphasis on wheat has implication for available capacities of the irrigation and land development schemes.

Thirdly, the wheat production efforts negate the principle of ecological specialization. The observed trend is the desire for wheat to be produced in all the states of the federation, which is at variance with the need for ecological specialization as entrenched in the new agricultural policy (FMAWRRD 1988). Ecological specialization policy demands that the regions which differ in climatic edaphic and socio-economic properties be encouraged to use to the best advantage any particular benefit of its skill and resource endowment. Through specialization of this kind, wasteful duplication of efforts is avoided, thereby creating high resource use efficiency and internal comparative advantage. In this connection, therefore, it is suggested that the effort in wheat production should be continued along the lines of ecological specialization.

EVENTS IN FOOD PROCESSING INDUSTRY

The SAP has led to creation of substantial value added in the food processing industry. This results from two major events. The first is that the reduced importation of final food products creates a challenge that leads to the emergence of new form utilities. The second is that the unavailability of imported raw materials has necessitated local raw material sourcing among industries, including programmes in backward integration.

Value-added Creation

Some manifestations of this event are illustrated in the following activities in the food industry:

(i) Sorghum milling to produce whole flour for making biscuits and wafers (Priyolkar 1989).
(ii) Dry milling or grains and tubers to produce composite flour such as sorghum/wheat, maize/wheat, cassava/wheat mixtures for

making bread and other bakery products (Aluko and Olugbemi 1989; Orewa and Iloh 1989; Obiana 1989).
(iii) Sorghum malting to brew larger beer and also to produce malted and weaning foods (Malomo 1989; Ogundurin et al. 1989; Igbedioh and Dumade 1989).
(iv) Cassava processing to produce starch, glue, etc.
(v) Soyabean processing for assorted food items.

One important matter is arising in the process of creating new value added in food. This concerns the continued presence of banned imported grains, particularly wheat and barley. Since this cannot be due to stockpiles of the pre-SAP period, as is often explained, it must be due to smuggling activities. Stockpiles of malted barley, for instance, have only between nine to twelve months of shelf life. How, then, can the stockpiles exist after three years of ban? In any case, the presence of hypothetical stockpiles of banned raw materials produces adverse effects in the food processing industry. Products made from them have clear edges over those made from their domestic equivalents. For one thing, people are accustomed to their tastes and, hence, give preference to them in the market, thereby outcompeting the local products. For another thing, the local products are still undergoing perfection, which may also be discouraged by the presence of superior products made from imported raw materials. This logic holds importantly in the case of bread, biscuits, and drinks. For instance, the marketability of Mayor, a publicized maize-based lager beer probably suffers from the continued presence of barley-based brands. Similarly, composite flour-based bread could also not be marketed successfully in the presence of traditional wheat-based bread.

The resultant effect of simultaneous presence of imported and local raw materials easily explains the problem of capital flight in the food processing industry. Reports have it that several bakeries which closed down in Nigeria have reopened in the neighbouring countries such as Benin Republic, where they now produce wheat bread for export to Nigeria. Therefore, it is worth suggesting that wheat bread itself be placed under ban, and the avenue for smuggling of affected items be closed.

Backward Integration

Several food industries have addressed the local sourcing of raw materials through the establishment of grain farms. These efforts are to be encouraged because backward integration promises reductions in working capital requirements, elimination of prohibitory transaction cost, and greater price competitiveness as a result of avoiding successive profit markups. But this event needs to be properly evaluated in the light of a few matters that have arisen.

Firstly, backward integration places demand on the industrial capital structure, particularly owing to the need to acquire farm machines and other fixed items. In this regard, how well is the event of backward integration supported by the events in the capital markets? A practical example will illustrate the case in point: The banking industry now offers up to 30% interest rate on fixed deposit. This means that 1 million naira yields ₦300,000 per annum or ₦25,000 per month if lodged on irrevocable order on withdrawal. It therefore pays investors to lodge money in the capital market rather than undertake the trouble of farm production, which yields very low and slowly, and involves the troubles of procurement, organization of farm, and the risk of natural hazards. Consequently, backward integration cannot thrive among food industries in the present circumstance.

Secondly, backward integration provides only a small proportion of individual installed capacities and is therefore not worth the efforts to the affected industries. For practical instance again, a 30-ha. maize farm (requiring about 2 million naira investment outlay), may yield only 60 tonnes of maize over the 3–4 months production season. But, this is only enough to satisfy the requirements of a flour mill of medium size for one to two days (Ogunlade 1989).

The low rate of return on agricultural investment in relation to the opportunity cost of capital and the small proportion of producible output in relation to installed capacities explain the reason for the exit of the early adopters from backward integration programmes, and the reluctance of others to adopt. Reports have it that Guinness Nigeria Limited has backed out and that Nigerian Breweries will follow suit in the New Year. However, one suggestion is possible, which is that the contract growers' method of backward integration be given greater emphasis among industries. Apart from eliminating heavy capital costs and the preservation of existing land ownership pattern, the contract growing method is a form of insurance to small farmers.

EVENTS IN THE FOOD MARKET

A major event in the food market is the increased market price of items. This is illustrated with a table. The possible causes of this event include the following elements:

- an upwards trending supply price is a result of increasing cost of farm production;
- an upwards trending demand price is a result of increased competition to purchase food items among a greater number of users.

Particularly in the agro-based industries, following the scarcity brought about by import restrictions:

- sympathetic price movements among close substitutes, particularly including the grains (maize, sorghum, rice, etc.) and tubers (yam and cassava) following scarcity of some of them brought about by import restriction;
- the income effect emanating from nominal statutory increases in public sector wages, first as a means of income adjustment and later as a means of adjustment relief;
- the absence of a central intervention buying and selling mechanism which could have stabilized food prices, particularly following the abolition of the commodity boards system.

An important matter has arisen from the event in the food market which concerns the appropriateness of the free market mechanism. This matter will be briefly discussed in the right perspective.

The Non-Intervention Policy for Food Market

The abolition of the commodity boards system under the SAP was based on the premise that it was operating like a dog in the manger—it was not only failing to resolve the issues of market instability, insecurity of supply, unstable farm incomes, but also was preventing their solution by other means. Analysis has shown that the inherent problems are political, functional, and administrative in nature.

However, the events in the food market since adjustment began suggest that it is unrealistic to posit a free-market basis for determining agricultural prices. These events include the quick emergence of bad trade practices such as adulteration of produce and swindling. Besides, it is also doubtful if one should attribute the massive windfall of incomes enjoyed by cocoa farmers to the absence of the intervention boards only. As it has been opined elsewhere (Ayoola and Idachaba 1989), the simultaneous substantial revaluation of the major currencies in which cocoa was being sold, particularly the dollar and pound, against the naira, would also be significant.

In addition, it is also possible that the domestic losses in terms of local cocoa industries that halted production for their inability to offer similar prices as was accruing to farmers from abroad, have negated the windfall of incomes that have accrued to the farmers. In particular, several cocoa processing plants were shut down, creating compensating income losses at home, to offset income gains from the sale of cocoa abroad. Therefore, a modified intervention is necessary only, not a very free market. And in this regard, a proposal has already been made (Ayoola 1989) which makes reforms in mode, structure, and finance of the intervention process. Essentially, the government dominance is to be enhanced, and the administrative chain is to be shortened.

CONCLUSION

Several economic events have taken place in the food sub-sector in the process of structural adjustment of the Nigerian economy. In turn, a number of matters have arisen from these events in the aspect of food production, processing, and marketing.

In general, the tempo of activities should be maintained but specific corrective measures are necessary. The particular issues to be addressed include:

(a) Trade-off between new farm products and traditional products,
(b) The mode of observing the principle of ecological specialization in agricultural production,
(c) The enhancement of the effectiveness of the ban orders,
(d) Devising the appropriate modes of backward integration, and lastly,
(e) Review of the policy of non-intervention in the food market.

REFERENCES

Adeyeye, V. A. (1989), 'The Impact of Structural Adjustment Programme (SAP) on Guinea Corn Production and Producer Returns: A Case Study in Katsina State of Nigeria' presented at the National Conference on the Impact of the Structural Adjustment Programme (SAP) on Agriculture and Rural Life, NISER, Ibadan, 26–29 Nov. 1989.

Aluko, R. E. and L. B. Olugbemi (1989), 'Sorghum as a Raw Material in the Baking Industry', presented at the symposium on the current status and potentials of industrial uses of sorghum in Nigeria, Kano, 4–6 December 1989.

Ayoola, G. B. (1988), 'The Dependency Problem and Rationale for Self-sufficiency in Food Grains', *J. Rural Development Nigeria* 3 (1).

Ayoola, G. B. and F. S. Idachaba (1989), 'Self-sufficiency in Food Production as an Essential Foundation for Economic and Industrial Development in Nigeria' presented at the 9th Annual Training Conference of the Industrial Training Fund, Bauchi, 9–10 November 1989.

FMAWRRD (Federal Ministry of Agriculture Water Resources and Rural Development) (1989), 'Agricultural Policy for Nigeria 1989'.

Idachaba, F. S. (1984), 'Self-reliance as a Strategy for Nigerian Agriculture: Cornucopia or Pandora's Box?' Proc. 25th Anniversary Conference, NES, 23–24 April 1984.

Igben, M. S. (1989), 'The Impact of the Structural Adjustment Programme (SAP) on Wheat Production in Kano State, Nigeria', presented at the National Conference on SAP and Nigerian Agriculture and Rural Life, NISER, Ibadan, 26–29 Nov. 1989.

Igbedioh, S. O. and V. B. Dumade (1989), 'Use of Sorghum in Infant Weaning Practices—The Experiences of NAERLS' presented at the ICRISAT and IAR, ABU Symposium on the current status and potentials of industrial uses of sorghum in Nigeria, 4–6 Dec. 1989.

Malomo, O. (1989), 'Sorghum in Lager Beer Production' presented at ICRISAT on Industrial Utilization of Sorghum, Kano, 4–6 Dec. 1989.

Obiana, W. A. (1989), 'Dry Milling of Sorghum' presented at the ICRISAT/IAR Symposium on Industrial Utilization of Sorghum in Nigeria, Kano, 4–6 Dec. 1989.

Ogundiwin, J. O., M. O. Ilori, and A. Okeleye, (1989), 'Brewing of Clear Lager Beer from Sorghum Grain SK59112 without Addition of Enteral Enzymes to Achieve Saccharification—A Case Study' presented at a symposium organized by the International Institute for the Semi-Arid Tropics, Kano, 4–6 Dec. 1989.

Ogunlade, Ayo (1989), personal communications.

Orewa, G. O. and A. A. Iloh, (1989), 'Current Status of Composite Flour Technology/Prospects with Particular Reference to the Production of Biscuits'.

Priyolkar, V. S. (1989), 'Use of Sorghum Flour in Biscuits and Wafers Production—NASCO Experience'.

Table 1: Nigeria estimates of dependency ration grains 1965–1983

YEAR	MAIZE	RICE	WHEAT	SORGHUM
1965	0.017	0.600	240	0.940
1966	0.01	1.00	1356	1.200
1967	0.013	0.390	1221	1.100
1968	0.035	0.510	1060	1.200
1969	0.008	0.100	1759	1.800
1970	0.610	0.210	3674	0.980
1971	0.300	0.670	5098	1.000
1972	0.370	0.040	4213	0.390
1973	0.160	1.000	7719	0.060
1974	0.370	0.290	5317	0.020
1975	0.140	0.620	1766	0.010
1976	1.500	3.000	3686	0.000
1977	4.700	10.000	3425	0.000
1978	11.00	240.00	4184	0.000
1979	8.200	362.00	3830	0.080
1980	26.00	569.000	4941	0.030
1981	22.00	565.000	17	0.000
1982	55.00	144.000	6420	0.000
1983	9.700	708.000	4693	0.000
***CV	-0.107	0.002	0.480	0.020

$$\text{Dependency ratio} = \frac{\text{import}}{\text{Total supply}} \times \frac{100}{1}$$

Source: Underlying data from Federal Office of Statistics

Table 2: Growth of means annual prices of major food crops, Benue State, 1986–1989 (naira per kilogram)

Food Crop	1986	1987	1988	1989	Mean annual growth rate %
Maize	1.19	0.89	1.48	2.13	28.3
Millet	1.68	1.08	1.28	2.14	16.7
Sorghum	1.39	0.95	1.67	2.06	22.5
Rice (threshed)	4.30	3.52	3.92	6.29	17.9
Cowpea	3.39	3.59	4.16	5.06	14.5
Yam	1.02	1.03	1.51	2.13	29.6
Cassava flour	1.94	1.40	1.80	2.40	11.4

Source: Underlying data from Benue ADP

Of Privatization and Public Agricultural Enterprises

The economic wisdom of disengaging the State from participation in direct production is a central issue in the modes of structural adjustment in Nigeria. In the agricultural sector, the adjustment operation commenced with the swift abolition of six commodity boards. Subsequently, government is consistent in the pursuit of market-oriented agricultural economy through a process of gradual privatization and commercialization. The goal, which is essentially to promote development of private enterprise initiative, has now been entrenched in the newly articulated national agricultural policy (FMAWRRD 1988).

However, there is apparent dearth of empirical information to justify the current reforms. Therefore, this paper circumscribes and illuminates public agricultural enterprise in the country. The objective is to provide necessary information for enhancing knowledge about the main features of government participation in these enterprises. This knowledge is needed not only to facilitate answers to the questions that are being raised on the subject, but also to articulate relevant lessons of experience that are crucial to the successful execution of current economic reforms.

The paper proceeds with brief conceptual clarifications and the statement of methodology (II). The main findings are then presented together with discussion of inferences (III). Section IV concerns conclusions drawn from the empirical results of the study.

II. CONCEPTUAL CLARIFICATION AND METHODOLOGY

Field experience suggests that the distinction between public enterprise and a bureau exists only in shades rather than in quanta. This means that there is in existence a wide continuum of public participation in enterprise.

Therefore, a public enterprise may be defined as an activity in which government pursues monetary benefits other than taxation and other forms of transfer payments. To fix ideas, such an activity basically includes (a) government's interests in bodies corporate with perpetual succession which are revenue-yielding or profit-earning; to these the names 'boards', 'corporations' often apply, (b) limited liability companies in which government participate through acquisition of majority (controlling) or minority shares and other interests, and (c) aspects of research and development (R&D) institutions concerned with pursuit of secondary objectives in revenue generation and possibly profit making; examples include the Agricultural Development Projects (ADPs) which also operate commercial units on cost-recovery basis, and universities running separate consultancy outfits and commercial farms. Defined in this general way, it is possible to extend the analysis deep enough to properly characterize the continuum of public enterprise.

A survey of these enterprises was conducted in the first half of 1988. Varied data were collected about them from the federal and state agriculture ministries, which acted as both initiators of public enterprise activities and supervising agency for the enterprises in the interest of government. Direct interviews were held with relevant staff to collect general data about the origin, functions, and performance of the various enterprises. This was supplemented with library and archival search involving administrative files and other documents such as official letters, memoranda, regular and irregular reports, statute books, newspapers, and others.

III. FINDINGS AND DISCUSSION

Table 1 presents a number of public agricultural enterprises together with brief descriptions of them.

In general, many federal and state enterprises made lacklustre performances only, compared to the huge investments in them. It appears that the enterprises fail to make desired impacts because of certain features inherent in government processes, including funding and planning problems, political interference and discontinuities, as well as corrupt practices, frivolous conflicts, and others.

Problems of Funds and Planning

The complaint of lack of adequate funding as the major problem of public enterprises is popular. In one concrete instance, the Kaduna State Agricultural and Marketing Company complained that the monies relied upon (totalling 35 million) were not received from government sources. Poor funding was also mentioned in Benue State, where the share capital subscribed by the local government and technical partners were not paid up to operate the cattle ranch.

The case of commercial services of ADPs illustrates the problem of poor funding properly (Table 2). Among the class of selected ADPs, the achievement index in regard to funding of commercial operations is only 47.3% on average. As these enterprises rely on usually uncertain annual budget allocations (Adelabu and Ayoola 1991), low physical performance in their various objectives becomes inevitable. Whereas flexibility in sourcing funds exists in private enterprises through the different financing instruments of the capital market, this practice is subject to approval of political authority in the case of public corporations, statutory boards, and similar enterprises.

Therefore, it appears that privatization is based on the thinking that withdrawal of public funds from these enterprises will enhance their productivity through increased funding commitment. At the same time, government will likely be disposed to a greater extent to allocate funds to basic services such as education and health. Private capital is more productive in use possibly because it is often tied to collateral, which entrepreneurs wish to preserve. On the other hand, operators of public enterprises are likely to be less sympathetic to forfeiting public properties used as collateral. They might also show the tendency to rely on the ability to use government influence against the exercise of right of the financiers (banks, etc.) to auction public properties in case of default. Such expectations may often be realized if the financial houses also have substantial government participation and interests.

Bad planning is another important problem of public enterprises. Evidence abounds for illustration in Ogun and Ondo states, where attempts were made in direct food production without definite enterprise organizations in place.

The extensive Ogun State grains farms ran out of hands. Severe labour shortage led to the use of office personnel, including senior staff, to perform farm operations. Subsequently, heavy overhead costs were incurred, which, together with other problems, made the venture unsuccessful. The Ondo State experience lasted for about five years probably because of the monitoring efforts of the bank which financed the project. Eventually, management and other problems led to failure of the enterprise, as field information established.

The Ogun and Ondo experiences border on inadequate planning of operations because carefully prepared resource-use schedules could have prevented the use of office personnel as emergency labour. These invariably caused the management problems, whereby the public enterprises could not compete for labour and other inputs in the market with private enterprises. Successful execution of farm plans also depends on speed of decision-making, which is very slow in the case of public enterprises, as the process often involves the highest level of political authority for release of funds and approval of proposed actions. On the other hand, private enterprises make faster decisions based on resource-use plans and act promptly to compete favourably in the input markets.

The issue of conflicting objectives is also an aspect of planning problem. Quite often, government enterprises do not pursue consistent objectives. While profit motive is desirable, there is always the political basis for restricting the extent to which this can be pursued. In the case of the Nigerian Grains Production Company, for instance, the profit motive is pursued, but subject to 'reasonable prices'* of output to the public. Similarly, the commodity boards were to enhance farmers' incomes through payment of remunerative prices at the same time when the items were to be resold to public at 'affordable prices' only. Consequently, high losses were sustained, owing to lack of flexibility in the pricing policy of the enterprises. Table 3 shows the magnitude of bankruptcy manifested by the commodity boards over their period of existence. Apart from the Cocoa Board, all boards accumulated debt to the tune of ₦1 billion. This outcome supports the thinking that welfare functions must be separately pursued from income-generating functions, and privatization is consistent with this thinking as a necessary condition.

Political Interference and Discontinuities

There are three variants of political interference in the operations of public enterprises which are relevant to this discussion. First is the subsidy question. Government justifies subsidizing agricultural production and consumption on grounds of stimulating input consumption and technology uptake by farmers, as well as promoting standard of living of general consumers along certain commodity lines. This may be necessary either because the existing income level cannot sustain production and consumption or because a weather-induced high production cost exists to make selling price of output exorbitant relative to current income level. Therefore, agricultural public enterprises are especially prone to subsidy intervention towards effecting government's welfare obligations.

However, the presence of subsidy explains a large part of the manifest market inefficiency of these enterprises. In a manner akin to setting conflicting objectives as discussed above, the presence of subsidy limits market flexibility through dependence of pricing policy of enterprises on government manipulations, thereby making profitability of the enterprises a lesser objective. This leads to permanent trapping of these enterprises in the cycle of uncertain government subventions, low revenue-yielding capacity, and non-sustainability.

The critical manifestation of the subsidy issue concerns the operations of Federal Superphosphate Fertilizer Company (FSFC) and the commercial units of ADPs which are involved in fertilizer production and marketing respectively. The attainable profitability of FSFC depends on the level of subsidy because it centralizes its sales to government hands which is not only carried out at predetermined prices agreeable to the latter, but also ties the receipt of its sales revenue to the usual government processes that are slow. The commercial units of ADPs cannot earn profit from their role in fertilizer marketing because they have no influence on wholesale and retail prices, coupled with the fact that government often fails to reimburse in respect of operating costs.

The second mode of government intervention is tantamount to intransigence, whereby the enterprises are subjected to government instructions outside their scheduled operations. Consider the case of Kaduna State FASCOM, which in 1984 was directed to embark on emergency grain purchase scheme in competition with 'outside' buyers. Such instructions are products of crisis management of the agricultural market during sudden scarcity. But they create new pressures on the enterprises during which their assets may be unprofitably overstretched, such as storage capacity and financial status, thereby affecting the normal operations adversely. In the case of FSFC, government is not only a big monopolist, but it also sets production quota in recent times, thereby further restricting the profit regime of the company.

Another form of interference concerns the appointment of chief executives and board members. As such appointments are usually not done on merit and expertise as in the case of the commodity boards system (Idachaba and Ayoola), management efficiency is generally low for public enterprises compared to private ones. Privatization involves gradual withdrawal of subsidy in preference for government to play facilitating roles only towards operational efficiency.

On a rather different plane of interference, government acts as primary source of discontinuities of the operations of public enterprises, through incessant reorganizations and political disturbances in the forms of changes in government, state creation, and civil war. Copious instances of these perturbations are available in many states. In the Cross River State ADC example, substantial discontinuity of operations resulted from the excision of Akwa Ibom as a separate state, owing to the emerging territorial protection of the plantations of the enterprise located in the new state. Similarly, in earlier times when Ogun, Oyo, and Ondo States were created from the old Western State, the operations of the then Western State Industrial Investment Corporation were temporarily discontinued pending reorganizations and assets sharing. The amount of difficulties encountered in the process is reflected in the example of the then Araromi/Lomiro/Ayesan Oil Palm Plantation, which had the Lomiro segment in the new Ogun State but the Araromi and Ayesan segments together with the oil mill in Ondo State. The effect of operational discontinuities includes the commitment of new streams of capital expenditure to operate redundant units in new states at the same time when excess capacities are idle in other states, culminating in low overall enterprise efficiency.

Corrupt Practices, Frivolous Conflicts, Etc.

Corruption is the bane of most public enterprises in the country. It is commonplace to read reports of missing monies in the form of embezzlement, misappropriation, lack of financial accountability, among other media captions of abuse of office. This together with frivolous disturbances in the management class constitutes an important problem of government enterprise initiatives. Concrete instances of financial probity are widespread, including the Benue State Cattle Ranch Limited, old Western Nigeria Cocoa Board, to mention a few. The Cross River State ADC is a case in point. According to a panel of enquiry (Anonymous 1980), conflicts at the top management level of the corporation were rooted in appointment and promotion matters in which a junior person was made the chief executive. Consequently, the financial controller, chief commercial officer, chief engineer, and the secretary failed to cooperate with the general manager.

Corrupt practices and unfriendly working environment are minimal in the case of private enterprises. Corruption is responsible for high rate of project failure of public systems, because of secondary scarcity of funds at execution stage. Also, frivolous conflicts invariably slow down the desired rate of progress as production is disrupted during investigations and reorganizations. On the other hand, private enterprises are not subjected to stop-go operations of this sort, as their geographical expanse has little or no bearing with the ownership structure. Continuity is required for systematic progress and build-up of financial and other assets of enterprise.

The last source of discontinuity observed is the effect of the civil war on public enterprises. Examples of this include Anambra, Rivers, and Cross

River experiences. The old Eastern Nigeria Development Corporation was practically discontinued at the onset of the civil war in 1967, with large expanse of its permanent crop estate laid waste. These estates were later to be reorganized into the Anambra Agricultural Development Corporation, Rivers State Agricultural Production and Marketing Corporation (defunct), and Cross River Agricultural Development Corporation. The war, therefore, produced negative effects on the enterprise potential of this region, which could be less in the case of private enterprise. Even though the latter might also be affected by the general conditions of war, it might be devoid of complete stoppage of activities as in the case of public enterprises.

IV. CONCLUSIONS

The current privatization programme is informed by the perpetual inability of public enterprises. Evidence abounds in agriculture to indicate the various dimensions of the problem, including poor funding commitment on the part of government, inadequate planning of operations, political interference and discontinuities, as well as corrupt practices and frivolous conflicts. Therefore, limited participation of government in these enterprises will possibly enhance their performance through increased efficiency and reduction of waste.

REFERENCES

Adelabu, G. B. and G. B. Ayoola (1991), 'Variability in Financial Allocations to Nigerian Universities', paper presented at National Workshop on Financial Management, Socio-Economic, and Administrative Policy Implementation, Hill Station Hotel, Jos, 23–26 April 1991.

Anonymous (1980), 'Proposals for the Re-organizations of Agricultural Development Corporation (ADC)', Calabar.

FMAWRRD (Federal Ministry of Agriculture, Water Resources, and Rural Development) (1988), Agricultural Policy for Nigeria, Lagos.

Idachaba, F. S. and G. B. Ayoola (forthcoming), 'Market Intervention Policy in Nigerian Agriculture: An Ex Post Evaluation of Technical Committee on Produce Price'.

Table 1: Typology of Selected Public Agricultural Enterprises in Nigeria

State/Fed. Govt.	Name of Enterprise	Objective/Activities	Relevant Features
Federal	National Grains Production Co. (NGPC)	Commercial production of grains	Incorporated 1975, federal as sole owner; authorized share capital ₦10 million
	Nigerian Agricultural & Cooperative Bank Ltd	Credit delivery to agric. sector	Development banking; concessional interest rates
	Commodity Boards	Market stabilization and maintenance of farmers' income	Established by Decree 29 of 1977 to succeed two earlier generations of boards. The six of them abolished, assets being disposed of.
	Federal Superhosphate Fertilizer Company (FSFC)	Production of phosphate fertilizer	Established 1976, 100% owned by federal. Capacity 100,000 tonnes (1991 quota = 10,000), problem of government control on pricing
Anambra	*United Palm Produce Ltd	Production of palm produce	Companies developed by splitting the Anambra Agricultural Development Corporation (ADC), which took over the estate of Eastern Nigeria Development Corporation after the war.
	* Cashew Industries Ltd	Cashew and livestock	
	* Integrated Livestock Ltd	Livestock respectively	
Benue	Benue State Ranch Ltd	To raise cattle for beef production	1,300-ha. ranch established 1980; technical partners and LGAs did not pay up shares. All 180 exotic cattle died from skin disease. Restocking with local breeds was subsequently undertaken.
	Agricultural Development Corporation	Direct large-scale production	Problem: top officials were implicated in case of embezzlement, leading to appointment of administrator.

Cross River	Agricultural Development Corporation	To reactivate ENDC estates	Established by Edict No. 3, 1969. Panel of enquiry discovered conflicts at top management level.
Kaduna	Agricultural Promotion & Marketing Company	Production of foodstuff	Established 1975; complaint of non-receipt of money from state government (₦3.5 m), NACB (₦1 m), and others ₦175,000.
	Farmers' Supply Company (FASCOM)	Procurement and distribution of farm inputs produce (cotton)	Initially a commercial unit of old Funta ADP (1974–79). Share equity capital ₦3m solely subscribed by Kaduna State Government.
Ogun	Direct Food Production	To demonstrate profitability of agric. production	1984/85, extensive grain farms and 300 heads of imported cattle; senior Ministry staff supplied their labour at critical farm operations. Soon discontinued following enterprise failure.
	Agricultural Development Corporation	Development of permanent crop estate	Ogun State's segment of defunct Western State Agric. Investment Corporation (which also originated from old Western Region Development Corporation). Part of estate is Lomiro oil palm plantation, which was part of Lomiro/Araromi/Ayesan oil palm plantation before Western State broke up into three states (Oyo, Ogun, Ondo States). The Araromi and Ayesan segments together with the oil mill fell inside Ondo State.
Ondo	Commercial Agricultural Development Project	Commercial grain production	1978–1983, loan (₦4m) from First Bank Ltd. Failure resulted from management problems and changes in policy.

Rivers	* Rivers State Oil Palm Co. (RISONPALM)	Separate production enterprises in oil palm, rubber, fish, etc.	Originated from old Rivers State Agricultural and Marketing Corporation (itself formed from old Eastern Nigeria Development Corporation (ENDC). War disrupted ENDC for 2 years.
	* Delta Rubber Company		
	* Pabod Fish Co.		
	* Pabod Food Co.		

Table 2: Funding Commitment of Commercial Services of Selected Agricultural Development Projects, Recurrent and Capital, Nigeria

ADP/Period	Budget (₦'000)	Actual	Commitment Index (%)
Anambra (1986–87)	4221.85	734.17	17.40
Bendel (1986–87)	3955.80	1289.40	32.60
Benue (1986–87)	9902.00	3159.90	31.91
Gongola (1982–87)	9034.00	8198.00	90.75
Ekiti Akoko (1986)	1265.21	683.46	54.00
Lagos (1987)	2417.26	22.92	0.95
Oyo North (1983–87)	12853.60	7338.00	57.09
Rivers (1987)	2823.54	795.10	28.16
Bauchi (1981–87)	32149.60	28640.10	89.08
Southern Bornu (1986–87)	968.10	687.50	71.02
Mean	7959.10	5154.86	47.30

Source: APMEU, *Project Status Report* Issue No. 2, July 1988

Table 3: Indebtedness of Defunct Commodity Boards to Central Blank of Nigeria as at November 1986 (₦ Million)

Board	Total loan 1976–86	Total loan interest	Payments proceeds plus refund	Outstanding
Cocoa	2233.9	2275.9	2275.9	0
Palm Produce	662.2	853.9	325.1	528.8
Cotton	359.7	449.7	289.3	160.4
Rubber	257.5	309.4	154.5	154.9
Grains	184.0	212.5	58.1	154.2
Groundnut	42.2	53.4	38.2	15.2
Total	3739.5	4154.9	3141.1	1013.7

Source: Culled from Idachaba and Ayoola (op. cit.)

Dilemmas of Sustaining Increased Crop Production in a Deregulated Economy

Volunteered paper presented at the 18th meeting of the National Agricultural Development Committee (NADC), Lomay International Hotel, Jos, 20–21 July 199

Given the importance of food and fibre in national economy, the recent policy efforts to increase crop production as part of the Structural Adjustment Programme (SAP) appear well placed. Nevertheless, the degree of success attained in this direction depends on such other components as macroeconomic deregulation, which has an overbearing influence on economic performance in general. Some elements of success recorded so far (CBN 1991) involved a growth in crop production index (1984 base year) from 103.5 in 1985 to 166.5 in 1991, incorporating positive changes between 1989 and 1990 (8.9%) and between 1990 and 1991 (6.6%). The staples and non-staples contributed to the incremental crop output observed during the period.

At this stage, it is crucial to examine the preconditions and dynamics of success in crop production, as well as the contingencies of sustaining it. On a modest scale, this paper will attempt such an exercise with a view to facilitating policy decisions in the crop sub-sector of Nigerian agriculture.

INFLUENCE OF DEREGULATION

A first measure at deregulating the agricultural economy involved the swift abolition of the six commodity boards existing. The defunct boards were used to exercise control on the crop market through price fixing and intervention buying. Reports (Idachaba and Ayoola 1990) later pointed out that the last set of boards performed as poorly as their predecessors in failing to attain the twin objectives of market stability and farm income maintenance. The initial observations indicated that the release of massive stocks of grains glutted the market, thereby dampening the morale of farmers who received very low prices compared to their expectations earlier in the season. Thus, output could not be sustained in the following one or two seasons, owing to low farm incomes consequent upon the abolition of boards. This might not be the case with cash crops, which actually enjoyed an initial boom that coincided with the disappearance of regulatory boards from the market scene.

Next was the deregulation of the exchange rate market, which led to the progressive devaluation of local currency. As a result, imported farm inputs have become very expensive, producing substantial cost effect in the domestic food prices (Ayoola and Mkpojiogu 1991). Concurrent with this was the presence of tariff and non-tariff policies preventing the international demand pressure from coming to bear upon the domestic supply of food crops. A particular case in point was the restriction imposed on agricultural trade in grains, roots, and tubers in the course of implementing SAP.

Another measure in deregulation concerns the interest rate policy. The preferential rate earlier enjoyed by agriculture was done away with, opening crop production enterprises to competition for credit with industries. A floating interest rate policy was adopted in the last quarter of 1987, resulting in sharp rises in lending rates. Although agriculture was a preferred sector under the new dispensation to enjoy the benefit or minimum credit portfolios from the banks, the allocation guidelines gave them great flexibility to discriminate against the sector. The Nigerian Agricultural and Cooperative Bank (NACB) notwithstanding, the relatively low financial returns associated with crop production put a ceiling on the sustained growth of the sub-sector.

The privatization and commercialization programme also represents important steps in deregulating the economy. The privatization or commercialization of public agricultural companies was aimed at enhancing their productivity and financial performance. Of course, side by side with this objective is the implicit assumption that the enterprises concerned would liquidate out of existence, should they be unable to flourish on their own. This implies a great challenge to the companies to strive to compete favourably within the private sector. The emerging situation quickly gave rise to positive response of private sector participants to tap the benefits of the backward integration policy. However, there is little to show for the efforts made as the various farms established to provide raw materials for the parent industries—flour, beverages, alcohol, etc.—soon closed down. Apparently the huge investments committed to crop production did not yield commensurate returns compared to the industrial components of the integrated firms.

Of particular interest is the privatization of the fertilizer market, which has evolved from complete government supply to the present state whereby the National Fertilizer Company (NAFCON) handles the procurement and distribution of domestic fertilizer production within the framework of subsidized price policy and government-owned infrastructure for storage and final sale to farmers. Year after year, the fertilizer market was in crisis owing to sharp practices and supply shortfalls occasioned by inability of government to sustain the fiscal burden of subsidy at the current level of demand.

INCUMBENCY OF CROP PRODUCTION SUCCESS

Developments during 1986–1991 seemed in large measure to validate the observation of increased crop output. As one would expect in the long term from the efforts to stimulate domestic production through increased fiscal allocations to agriculture (to implement new policy efforts (e.g. Directorate of Food, Roads, and Rural Infrastructure, emergency wheat production, etc.), value added of crop sub-sector changed in positive terms by 4.2% and 5.1% from 1989 to 1990 and from 1990 to 1991 respectively (CBN 1991). As regards prices, index figures (1975 base year) for cash crops have been somewhat unsteady. Rural market prices of most staples fell considerably between 1989 and 1990 (cassava -20%, gari -7.7%, yam, -5.3%, shelled maize -24.6%, sorghum -15.6%, millet -18.6%, husked rice -0.3%, shelled groundnut -6.3%). Only a few food items featured increased prices that period (white cowpea 3.9%, palm oil 87.8%).

Curiously, however, their achievements do not present a stable outlook in later years, thereby raising questions about their sustainability. The first observation was to the effect that the initial farmers' enthusiasm to stimulate the production of cash crops, cocoa in particular, was soon eroded, following the sudden collapse of world prices; on average, 1989 and 1990 index of world prices of cash crops fell by 11.9% for all commodities, ranging from -3.1 (cocoa) and -1.3 (soyabeans) to record high levels of 30.9% (copra) and 30.4% (palm kernel). The income consequences of this situation to farmers and government were significant as also those of higher rural market prices of most staples on consumers in 1991—garri (15.4%), shelled maize (61%), sorghum (114.2%), millet (97.1%), husked rice (19.7%), white cowpea (40.5%), and shelled groundnut (8.2%); in that year, only the price of palm oil fell by 24%, which was still a far cry from the increase experienced the previous year (87.8%). Politically, these trends contributed to the reasons behind public outcry for upward adjustments of public sector wages; the wake of sporadic industrial actions coincided with period of pronounced shortage of essential food commodities.

How can we address the sustainability question, concerning the ways to keep crop production growing continuously without falling? This paper argues that, apart from the ground efforts in land development, input supply and infrastructure build-up directed at facilitating farm production, further scope for sustaining the success attained in the crop production sector can be found through the optimal resolution of the dilemmas inherent in the deregulation process, particularly the aspects of parastatal reform, foreign exchange market, liberalized interest rate, and privatization.

Parastatals Sector Reform: What Degree of Market Freedom Exists?

The abolition of commodity boards meant that farmers should depend on market signals to make production decisions; the commodities considered for production and land area proportions should be determined basically through the preference revealed by household and industrial consumers through the market rather than the capacity of intervention buying of the boards. Market and income stability would be residual outcomes of the emerging market mechanism. The efficiency of this mechanism, however, would be determined by the degree of freedom prevailing.

Domestic market freedom lacks in two major ways as far as food commodities are concerned in Nigeria. First, the periodic panic measures of government pamper certain crop enterprises and create distortions in the true market situations. The preferential treatment involving free seed distribution and other direct production supports results in the improper decoding of the actual market signals, so that production of certain crops becomes cheaper than it otherwise should be. In such a situation, even the ecologically inefficient crop producers get involved, as revealed in the case of 'wheat round-up' programme. Those states that are not ecologically suitable for the accelerated wheat production have only wasted the seed and fund provided. Less visible is the opportunity of wheat production in terms of alternative better-adapted crops in the areas.

Second, removal of control is not total, as certain states take measures sometimes to put a wedge in the free flow of crop outputs. Among the possible reasons for such actions, occasional low output as a result of weather problems is most common. In addition, people themselves or state ministries may result to state self-sufficiency in food production at periods of uncertain political atmosphere. Whatever reasons prevail, obstruction of the freedom of the food supply market in the country runs counter to the objectives of market mechanism.

Deregulated Exchange Rate: Elasticity Constraints/Domestic Consumption Pressure

The obvious problem of using the exchange rate instrument to increase agricultural income is the slow response of output to demand pressures. Or, what else could explain the quick disappearance of the cocoa boom that emanated from the initial price effect of naira depreciation in 1986–87, other than the exhaustion of current output capacity? New output increases would take some time to be generated from the extensive rehabilitation of existing plantations and establishment of new ones. By that time, however, inflation would have caught up with the new cocoa price regimes to reduce the utility of the exchange rate policy to farmers, owing to higher compensating increase in food parity price. The situation is further complicated by the presence of pent-up demand in the home industries for processing cocoa so that, when released, could make government to even ban exportation as considered in 1990.

It thus appears that the observed high income elasticity of income from cash crops with respect to international price could not be sustained in the long run. Nevertheless, it is good for the economy since new production initiative resulted. What could then be suggested is to use the **gestation** period of the permanent crops affected to expand the domestic capacity for processing of exports. This will not only assure greater value added of exports but also results in substantial foreign exchange savings to be made from reduced importation of final products.

Liberalized Interest Rate: The Competitiveness of Crop Enterprises

A dilemma facing sustainability of incremental crop production arises from the poor competitive competence of the enterprises involved compared to the industrial sector. This is the primary basis of the heavy government support for agricultural production. Unless stimulants are available, large-scale and small-scale agricultural enterprises could not be sustained for a reasonable length of time. Credit, which represents a traditional mode of stimulating agricultural production, is a major element of the debate on deregulation during SAP. Being the price of a critical production input, interest rate charged on credit is the real focus of discussion. It is argued, in the classical sense, that preferential interest rates lead to distortion of capital market, hence inefficient use of scarce resources. This being so, and given the dilemma of poor competitiveness, crop farmers would then need compensating doses of government support to keep production going. The presence of ADPs, NALDA, DFRRI, and others could be justified on this ground while their functionality becomes critical.

Privatization: The Residual Subsidy Question

How to handle fertilizer subsidy remains a problem facing the privatization of the important crop production input. While procurement and distribution aspects were substantially privatized, government still bears freight costs and price subsidy, which produce negative consequences including high transaction costs, smuggling, and derailment of benefits. Within the context of a deregulated economy, these elements cannot be sustained in the long run. Yet farmers and free riders of accruing benefits have become so mentally dependent on subsidy that its withdrawal becomes such a crucial policy problem. Nevertheless, there appears to be no other way out of the heavy budgetary burden, international leakages, and inefficiency associated with fertilizer use than a disciplined graduated withdrawal of subsidy. If this is achieved, government then needs to plough back the budget proceeds into agriculture through the strengthening of safer means of providing support for crop production.

CONCLUSION

In conclusion, four dilemmas facing the sustained crop production as they emanate from macroeconomic deregulation have been identified and their resolutions attempted: (i) the limited utility of scrapping the commodity boards because market freedom is considerably curtailed; in this connection, the facilitation of market mechanism across geographical and political boundaries is recommended; (ii) the inadequacy of floating exchange rate system as an instrument of stimulating crop outputs owing to low price elasticities of their supply; in addressing this situation, it is recommended that expansion of domestic capacity for processing crop output be undertaken; (iii) the association of interest rate liberalization with reduced credit flow to crop agriculture; to alleviate the effect of this, farm production requires further doses of indirect support through existing institutional framework; and (iv) the frustration of privatization effort by the high political visibility of fertilizer subsidy; against this background, government is advised to undertake the planned gradual withdrawal of fertilizer subsidy, to be compensated for by new lines of support for farmers through appropriate programmes.

REFERENCES

Ayoola, G. B. and D. Mkpojiogu (1991), 'Analysis of food price variation under the structural adjustment programme: An explorative case study', *Journal of Agricultural Science and Technology* 1 (2): 81–85.

Idachaba, F. S. and G. B. Ayoola (1991), 'Market Intervention Policy in Nigeria Agriculture: An Ex-post Performance Evaluation of the Technical Committee on Produce Prices', *Journal of Development Policy Review* 9 (3): 285–299.

CBN (Central Bank of Nigeria) (1991), *Annual Report and Statement of Accounts*, Lagos.

Table 1: Some promising varieties of soyabean and their yield performances under evaluation experiments, 1991

Variety*	Yield Performance (kg/ha.)	Maturity Days
EARLY		
TGX 536-02D	1606	97
TGX 1497-1D	1440	94
TGX 1485-1D	1387	92
TGX 1019-2EB	1348	101
TGX 1566-2E	1345	92
MEDIUM		
TGX 1448-2E	1821	111
TGX 1440-1E	1734	114
TGX 1448-1E	1656	112
TGX 1458-2E	1616	112
TGX 1455-2E	1595	111
Sampsoy 2	1504	110
TGX 1447-2D	1486	113
M-351	1475	112
TGX 1437-1D	1451	111
TGX 1447-1D	1438	110
TGX 1113-3D	1269	110
TGX 849-313D	1240	109
LATE		
TGX 1596-2E	1282	117
TGX 923-2E	1232	122
TGX 1445-3E	1195	120
TGX 1445-2E	1173	124
TGX 1470-1D	1124	124
Sampsoy 1	921	116
Malayan	508	129

Note: Less 100 days = early maturity; 100–120 days = medium maturity; above 120 days = late maturity of varieties

Source: Nationally Coordinated Soyabean Project, 1991

PART 2

SECOND PHILOSOPHICAL DECADE
(1998 – 2008)

30

Micro And Macro Farm Business Data Requirements For Agricultural Planning And Policymaking In Nigeria

INTRODUCTION

Agricultural planning and policymaking are joint activities in forward looking for deliberate economic development. Typically, both activities are carried out by experts at three levels—aggregate or macro level, sector or meso level, and enterprise or micro level. At all levels, the outcomes in terms of choices and decisions of the experts depend on availability of reliable data as critical requirements for them to do the job of planning and policymaking well.

Planning is the purposeful and sustained allocation of limited resources to attain stated policy objectives. This definition implies, as follows,

- o All public planning is within the context of stated public policy objectives.
- o The resource limitation compels the planner to *rationally* allocate the scarce resources; thus there comes the need in planning for a set of decision rules and criteria for rational allocation.
- o Planning often evolves particular production and consumption patterns which would have otherwise not been the case without planning.
- o Planning often creates new institutions, usually implying need for government support, especially in developing countries.

Policy is a statement of intended action or inaction by the public authority in response to the need to achieve a predetermined purpose, whether formally declared or not. This definition also implies, as follows

o   That policy is essentially an intervention, usually in the market for goods and services, from which, like planning, the expected output or outcome would not have been the result without the policy intervention.
o   That policy may or may not be formally declared; this leads to typology of policy into implicit and explicit categories.
o   That in policy what the government does or says on a particular official issue (policy of intervention) is as important as what it does not do or say (policy of non-intervention); thus, even in the event of no statement or action there will be definite (policy) consequences of a passive stance taken by the government on any development issues.
o   That in policy there exists a purpose or objective underlying any action or inaction of government, whether it is formally stated and publicly declared or not; thus, the typology of public policy objectives includes two categories, namely implicit policy objectives and explicit policy objectives, so the real policy objectives may be revealed only after implementation has begun or has been completed, and these may be slightly or radically different from the initial statement of objectives.

The goal of this work is to highlight the activities involved in agricultural planning and policymaking as a joint activity, with a view to identifying and describing the need for farm business management data in the process of agricultural development. Section Two is a quick survey of agricultural planning and policymaking in Nigeria. Section Three is a highlight of activities involved in planning and policymaking, which indicates the need for certain types of farm business management data. Section Four highlights the preferences of agricultural planners and policymakers for different kinds of farm business management data. This, finally, is followed by conclusions and implications for the future of the FAMAS project.

AGRICULTURAL PLANNING AND POLICYMAKING IN NIGERIA

The planning process consists of different frameworks, namely perspective plan framework or visioning, medium-term planning framework (including rolling plans), and strategic planning framework or programming. The perspective plan takes a long-term view of the agricultural sector, ten years or more, sometimes singly or alongside other sectors of the economy; while the medium-term plans take shorter time horizons of four to five years in most cases. On the other hand, strategic planning framework involves design and formulation of programmes and projects to be implemented in meeting the plan or policy objectives. Among the several criticisms levelled against the early planning efforts in Nigeria's agricultural history, lack of **reliable** data was probably the most serious.

In Nigeria, agricultural planning is generally subsumed in the macroeconomic planning of the country. The earliest effort in development planning dates back to the colonial period, namely the Ten-Year Plan of Development and Welfare (1946), which was launched as a result of the need to rehabilitate the British metropolitan industries after the Second World War. The aim of this plan was to strengthen the colonial raw material base, namely groundnuts, soybeans, palm kernel, cocoa, among others. Then we had the Economic Development Plan (1955–1962), which was produced following the adoption of a federal system of government and the creation of three economically autonomous regions—Eastern, Western, and Northern regions.

The sequence of medium-term plans which incorporate agricultural development plans produced for the country is as follows:

1. First National Development Plan 1962–1968
2. Second National Development Plan 1970–1974
3. Third National Development Plan 1975–1980
4. Fourth National Development Plan 1981–1984
5. National Rolling Plans: 1990–1993, 1991–1994, 1992–1995, 1993–1996, etc.

Similarly, the series of articulated policies on agriculture and the attendant implementation strategies include the following:

o   Policies: Agricultural Policy for Nigeria (1987), National Seed Policy (1992), New Agricultural Policy Thrust (2003), National Policy on Integrated Rural Development (2001), National Fertilizer Policy for Nigeria (2006), etc.

o Strategies: National Accelerated Food Production Programme, River Basin Development Authorities, Integrated Agricultural Development Programme, National Special Programme on Food Security (2002), Presidential Initiatives on several commodities (2003), National Fadama Development Project (I, II, and III), Agricultural Development and Marketing Companies, National Strategic Food Reserve Programme, Fertilizer Stabilization Scheme (2000), National Root and Tuber Expansion Programme, Maize Doubling Programme, Community-based Agriculture and Rural Development Programme, National Programme on Food Security (2003), National Food Security Programme (2008), etc.

Thus, the need for data cannot be downplayed in the agricultural planning and policymaking process. Both activities depend on the conduct of field studies to determine the facts of the situation and to project for the future. The nature and sources of such data as they hinge on farm business management is the subject of the next section.

DATA REQUIREMENTS FOR AGRICULTURAL PLANNING AND POLICYMAKING

The three levels of agricultural planning and policymaking are aggregate or macro level, sector or programme level, and the micro or enterprise level. Agricultural planning and policymaking at the macro level entails analyzing the sector in the general economic development plans. The question at this level is how efficiently the natural, human, and financial resources would be utilized during a future time horizon. Finding the answer to such a question often necessitates the conduct of projection studies which help in estimating future quantities. Specifically, demand projection for food and fibre helps in estimating aggregate quantities of farm inputs and requisites required to produce such consumer items, while supply projection for food and fibre helps in estimating the gaps in quantities of these items to be filled through imports or through local substitutes.

For example, project studies treat demand for food and fibre (X) as a function of income (Y) and population (P), i.e. $X = f(Y, P)$; such that change in X is derived in relation to changes in Y and P by total differentiation, and wherein the proportional change in X is the elasticity parameter of X in relation to income (Y) times the proportional change in income Y plus the elasticity parameter of X in relation to population (P) times proportional change in population (P).

The main business management data required for this include the following cost categories:

- *Capital costs* such as cost of pumps, fertilizers, seed, working capital, land, replacement cost of items in the project after its useful lifespan, etc.

- *Operating and maintenance costs* such as labour costs, transportation cost, contingency allowances, etc.

- *Sunk cost* such as cost of items committed in the past; these are generally not considered in the project accounts for both social and financial analyses.

- *Secondary costs* such as externalities can translate into costs, e.g. the resultant flood effect of irrigation project.

- *Intangible costs* can be significant, hence should be taken into account through subjective evaluation, modification of normal benefit cost analysis to a least-cost type of analysis.

Agricultural planning and policymaking at the enterprise level involves analyzing farm enterprises packaged under programmes and projects.
The private benefit question is how to ensure that farmers will earn remunerative returns by undertaking enterprises under agricultural development programmes and projects. This necessitates the conduct of feasibility studies for estimating private returns through various analyses such as farm investment analysis, farm income analysis, fund flow analys and risk analysis.

For example, farm investment analysis entails several aspects, namely *Assessment of Financial Impact; Judgement of Efficient Resource Use, Assessment of is,Incentives, Provision of a Sound Financing Plan, Coordination of Financial Contributions and Assessment of Financial Management Competence.*

The data requirement for this purpose includes certain categories like the following:

> *Farm resource use* (land use map/calendar): total farm/cultivated area; labour use (annual labour requirement by crop operation for one hectare, e.g. land preparation and planting, harvesting etc.); labour distribution by crops and month (reckon only with hired labour, family labour ignored because it is the recipient of the incremental benefit)
>
> *Farm production* (yield and carrying capacity): total production, valuation of farm production and incremental residual value, net production available for sale/home consumption; applying projected technical coefficients, e.g. mortality/calving rates, culling rate, ratio of bull to breeding females, feed ratios
>
> *Valuation of farm production* (incremental residual value) included in the inflow, in the farm budget. It is important because not all utilities of an investment may be exhausted in the life of a project. It is usually recorded in the last year of the project.

The need for policy studies present in different forms

- *Design studies:* The main purpose of design studies is conceptualization focusing on conceptual issues such as diagnosis of causes, characteristics and consequences of policy problems, as well as ordered prioritization of policy problems, time lag between emergence and recognition of policy problem. The types and sources of data for this include scoping, inventorization, characterization, etc.
- *Formulation studies:* The thrust of formulation studies is specifications focusing. The concerns here are full specification of various components namely objectives, instruments, strategies, time phasing,

and financial implications. The types and sources of data for this form are farm surveys, household surveys, market surveys, etc.
- *Monitoring and evaluation studies*: Monitoring and evaluation studies focus on assessing performance by way of follow-ups, supervision, looking back, redesign, and reformulation. The types and challenges of data in this form include choice of appropriate methodology of evaluation and impact assessment, selection of impact indicators, availability and consideration of input from monitoring, identification of positive and negative impacts, conclusions on the extent to which policy has achieved its goal, hasty designs that make no provision for monitoring and evaluation, failure to specify performance indicators that will help track achievements and constraints during policy implementation, etc.

REVEALED PREFERENCE FOR MICRO AND MACRO FBM DATA BY AGRICULTURAL PLANNERS AND POLICYMAKERS IN NIGERIA

The public sector demand for farm business management data emanates from government agencies, international agencies, and NGOs in the agricultural sector. The preference for such data as revealed by agricultural planners and policymakers from these sources is highlighted in tabular format for each category of experts below.

1. Government Agencies—Federal Government

Agency	Selected Cases	Planning/ Policy Decision Involved	Micro Data Requirements	Macro Data Requirements
NARP— National Agricultural Research Project	Central Zone study under the World Bank–assisted National Agricultural Research Project (NARP)	Preparation of a research strategy plan for Nigeria	Enterprise profitability indicators	Indicators of comparative advantage in commodity production

FDA—Federal Department of Agriculture	Investment Profile of Crop Production and Processing Enterprises (2003)	Production of a guide to investors in agricultural sector	Capital requirements for farm establishment; structure of costs and revenues	Interest rate regimes; exchange rate regimes
	Fertilizer socio-economic study	Towards the preparation of a strategy for Nigeria	Data on knowledge, attitudes and practice of fertilizer use among farmers in Nigeria	-

2. Government Agencies—State Government (Delta State)

Agency	Selected Cases	Planning/ Policy Decision Involved	Micro Data Requirements	Macro Data Requirements
Ministry of Agriculture and Natural Resources	Design and Formulation of Agricultural Development Roadmap	Farmer support; Youth Empowerment through Agriculture	Farm modules; farm budgets; current land use pattern	Price levels; market supplies

3. International Development Community—Multilateral Agencies

Agencies	Selected cases	Planning/ Policy Decision Involved	Micro Data Requirements	Macro Data Requirements
FAO—Food and Agriculture Organization	NMTPF	Framework of assistance to the federal government based on common jointly agreed priorities, programmes and projects of agricultural development	Farm enterprise data	Institutional data

World Bank	Voice of the Poor	Qualitative poverty assessment	Focus group discussions	Poverty indicators
Common Fund for Commodities	Cassava Value Chain Project	Intervention to increase value addition to cassava in Nigeria	Production, marketing, consumption, and industrial use	International prices

4. International Development Community—Bilateral Agencies

Agency	Selected Cases	Planning/ Policy Decision Involved	Micro Data Requirements	Macro Data Requirements
DFID— (British) Department for International Development	Feasibility of an inventory credit scheme for grains	Consideration of establishment of an inventory credit programme for grains in Nigeria	Grain outputs, storage infrastructure, price regimes	International comparison
CIDA— Canadian International Development Agency	Agriculture sector study	Design of country programme for Nigeria	Farm and market enterprise data; input-output data; demand and supply data; environment data; institutional data (credit, gender, etc.)	Import/export data
USAID— United State Agency for International Development	DAIMINA	Development of agri-input market of Nigeria	Input use – fertilizer; seed, CPP; sales of inputs; etc.	Input import data, price level, exchange rates, etc.

5. Non-Governmental Organizations (NGOs)

Agency	Selected Cases	Planning/ Policy Decision Involved	Micro Data Requirements	Macro Data Requirements
Bill and Melinda Gates Foundation	AGRA seed study	Formulation of seed programme in Nigeria for the Africa-wide intervention AGRA	Demand and supply for seed, seed production and prices	Mandate institutions for genetic improvements of seed, seed requirements
Oxfam	Economic Justice Campaign in West Africa	Towards promoting justice in terms of farmer support, etc.		Role of stakeholders in the agricultural policy process
FIF—Farm and Infrastructure Foundation	Right to Food Campaign in Nigeria	Further to the Economic Justice Campaign, to sponsor a constitutional recognition of human right to food and initiate the RTF Bill for Nigeria	Food and nutrition survey data	Spatial distribution of food poverty, international positions on RTF

CONCLUSIONS

The need for reliable data for planning and policymaking cannot be downplayed in the quest for agricultural development in Nigeria. The historical sequence of activities in this regard attests to the fact. The requirements range from micro to macro data collected through planning and policy studies about cost and benefit layouts of agricultural business, about generating and sustaining farm incomes, and about uplifting the quality of life of the rural populace in general.

The implication for FAMAS pertains three needs arising. They are

- The need to further dimension FAMAS, laterally and vertically

- The need to mainstream best farm business management practices in farm enterprise, and
- The need for institutionalization of FAMAS activities at all levels of development governance, i.e. federal, state, and local.

REFERENCES

Ayoola, G. B. (ed.) (2002), *Agricultural Policy Networking: The Way Forward*, Full Proceedings of a CTA Workshop, Entebbe, Uganda, 6–10 November 2000, 193 pp., ISBN 92 9081 2583, CTA No. 1082, downloadable from the CTA website, *www.agricta.org/pubs/polnet/index.htm*.

Ayoola, G. B. (2000), *Essays on the Agricultural Economy I: A Book of Readings on Agricultural Development Policy and Administration in Nigeria,* Ibadan: TMA Publishers.

31

Emerging Issues For The Formulation Of Policy For Agri-Input Delivery System In Nigeria

INTRODUCTION

The production system is expressed in one economic jargon called production function, mathematically written as $P = f(x)$. In ordinary language, this means quantity of output is determined by quantity of input. This is how economists appreciate the role of inputs in the production process, implying that without inputs, there is no output. Therefore, the efficiency of the agricultural production system depends on the efficiency of the input delivery system. And, unless the problem with the input delivery system is effectively addressed, there is no basis to expect that the volume of output will grow as desired.

The problem with the agri-input delivery system of Nigeria pertains to inadequate quantity of supply, delays in supply relative to the needs of farmers, widespread nature of market sharp practices, among others. These problems form the original basis to justify policy intervention in the system. However, the situation is compounded by second-generation problems of implementing the policies in terms of perennial abuse of these policies coupled with the lack of sufficient implementation commitments.

In this essay, it is argued that poor policy formulation begets poor policy implementation. That is, good formulation is a necessary albeit not sufficient condition for good implementation of agri-input policies. Thus, the goal of this effort is to determine the key issues involved in the formulation of agri-input policy for Nigeria and to interrogate these issues somewhat, with a view to proposing their optimal resolutions. First, after the introductory overtures (Section I), we survey the background of such policy issues, followed initially by a highlight of the policy formulation process as regards agri-inputs, and finally by the discussion of the issues as they emerge from previous expositions.

Primer of Agri-input Policies

The need for policy intervention in the agri-input market predicates on three important arguments, namely the market failure theory, which holds that market for agri-inputs fails in certain respects that the market on its own is unable to correct for them on its own; equilibrium adjustment, which pertains to the long time of adjustment of the market from one equilibrium position to the next equilibrium position; externalities, in which there is visible divergence of private course from social course and the theory of second best, which states that once the conditions of *pareto-optimality* in attaining the first best equilibrium of the market is violated, as is generally the case, there is no basis to pursue the remaining conditions in attaining the second best option (Lipsey and Lancaster 1956).

The agri-input delivery system is conceived in the development communication context, which has as its main elements the specific package of inputs as a *message*, originating from a *source* and passing through a *channel*, to be delivered to a *target*. Within this context, we identify the various policy bottlenecks associated with each element as a *noise*, obstructing the free flow of the agri-input message as it passes from the source through the channel to the target. Elsewhere, the forms and functions of the noise elements in the policy pathway have been described in detail (Ayoola 2001). The manifest consequences of these noise elements on the performance of agriculture is reflected in the status of the country in terms of food import dependence, notwithstanding its status in terms of input import dependence.

Now, the African green revolution is under way, having gathered substantial momentum in the recent past. We recall the Africa Fertilizer Summit held in Abuja in June 2006, which has generated a twelve-point agenda for action among the countries (IFDC 2006). We also observe the follow-up collaborative activities of the Bill and Melinda Gates Foundation and the Rockefeller Foundation to form an Alliance for Africa Green Revolution. Both developments will definitely focus on a judicious use of agri-inputs in turning the situation around on the continent, with a view to attaining food security of the people in the shortest possible time. Thus, now is the time to get the policy environment right for agri-inputs in Nigeria, with a view to maximizing the advantage of the new assistance framework.

BACKGROUND TO POLICY INTERVENTIONS IN AGRI-INPUT MARKET

Policy Intervention Modes

The principal focus here is on the so-called green revolution inputs—seed, fertilizer, and CPPs (crop protection products), in that order.

Seed

Public intervention in the seed industry dates back to the colonial era, when the travelling teachers of agriculture used to carry planting materials from place to place, introducing 'new improved' crop varieties similar to the New Improved Blue OMO in the present time. This translated into organized extension work for the delivery of seed and other planting materials to rural dwellers.

At the moment, the National Seed Service (NSS) is the superintending agency for policy interventions in the seed industry. The National Crop Varieties and Livestock Breeds Registration and Release Committee was established by law (Decree 33 of 1987), for the purpose of regulating the activities of stakeholders in the seed industry. Specifically, the committee was charged with receiving and processing applications for the registration, naming and release of old and new crop varieties, and officially releasing the list of varieties recommended by the technical subcommittee (TSC) established for that purpose. Subsequently, the National Agricultural Seed Committee was established (National Agricultural Seed Act No. 72 of 1992), leading to the publication of a comprehensive list of all crop varieties released and registered in Nigeria, and giving mandate to NSS as sole source of foundation seed production in Nigeria, in collaboration with the National Agricultural Research Institutes (NARI).

In practical terms, the key modes of policy intervention in the seed industry at the moment consist of the following:

- The NARIs and universities as public institutions act as the original source of *breeder seed* of new varieties produced by them; the breeder seed of such public bred varieties should be released to NSS.
- The NSS is responsible for the production of *foundation seed* from breeder seed in collaboration with NARIs, universities, and private seed companies; the NSS would release the first-stage foundation seed to private seed companies while the second-stage foundation seed would be released to ADPs.
- The private seed companies and ADPs are responsible for the production of *certified seed* from the foundation seed provided by NSS, through companies' farms or contracted out to outgrowers / contract farmers. Both the private seed companies and ADPs are also responsible for selling certified seed to farmers to produce commercial grain.

Fertilizer

Policy intervention for replenishing the soil dates back to the era of regional control through to the first half of the second post-independence decade, when the regions/states engaged in separate importation of mineral fertilizer for distribution to farmers. At the federal level, the Fertilizer Procurement and Distribution Division (FPDD) was established as the sole agency for supplying mineral fertilizers to the Nigerian market. A loan was obtained from the World Bank in two stages for the purpose of massive importation and distribution at subsidized prices, which lasted till the 1990s. Also domestic supply capacity was enhanced by the establishment of two granulation plants (NAFCON and FSFC) followed by a series of blending plants. The sector has since undergone substantial reforms leading to the redesignation of the FPDD as FFD (Federal Fertilizer Department) and the systematic withdrawal of the government from importation and distribution activities and the reduction of subsidy on fertilizers.

At the moment, the main elements of policy intervention in the fertilizer sector consist of a market stabilization programme, as follows:

- The federal government undertakes limited purchase of mineral fertilizer, typically 250,000 MT through tendering in the local market, on a seasonal basis and at a uniform price; the consignments would be distributed to the states as indented.

- The state governments sell the allocations of fertilizer to farmers at 25% subsidy and at a uniform price throughout the country; the proceeds are remitted to the federal government.
- The states and local governments are also at liberty to support the farmers with additional subsidies on federal fertilizers or to procure additional fertilizer for distribution to farmers at subsidized prices.

Crop Protection Products

Public intervention in the market for crop protection products also has its roots in the defunct regional control era. The old Western Region launched the popular Cocoa Pesticides Scheme in the 1960s. Subsequently, the supply of CPPs has been mainstreamed through several projects of agricultural development, to facilitate access and judicious use. At the present time, the market for CPPs is considerably liberalized.

Lessons of Implementation Experience

The collective lessons of experience from the implementation of the policies are better perceived within the development communication context, which helps in examining the agri-input delivery system by treating:

a) The package of agri-inputs as a policy *message* to the farm population;
b) The policy authorities as the *source* of the message for its packaging and mobilization;
c) The market cum public extension system as the joint *medium* or *channel* for the smooth passage of the message, and
d) The farmers as *target* of the message for decoding and eventual use.

This framework helps in determining the nature of issues associated with the delivery system as a means of communicating the policy message about agri-inputs, thereby facilitating the determination of issues emanating therefrom. Thus, the following lessons of implementation experience can be discerned:

Lessons about the Message: The well-packaged policy message has the property that the availability of agri-inputs in desired quantity and quality holds the key to the efficiency of the agri-input delivery system. The fertilizer policy intervention accounts for large-scale awareness and adoption of fertilizers among the farmers. Nevertheless, the degree of availability falls short of the demand generated, which led to frequent crisis in the past. Even at the present time, substantial pent-up demand for fertilizers that is never met exists. The situation is similar with seed, in which case limited availability of improved seed provides the basis for the vast majority of farmers to continue with the traditional practice of using their own seed saved from previous harvests, season after season.

Moreover, the nature of agri-input demand is also important. As inputs, they are not demanded for their own sake unlike final products. Thus, the demand for agri-inputs is a derived type, in the sense that they are demanded for the production of other items than themselves. This implies that for their use to expand, the demands for the products that they are used to produce must first expand. That is, much of the sluggish uptake of the agri-inputs would be explained by the low production of the food items in the domestic market possibly emanating from policy-induced competition with food imports.

Lessons about the Source: The role of government as the source of the policy message initially involves the proper packaging of the agri-inputs in terms of proper articulation of the policy statement and proper formulation of strategies for implementing them, followed by sustained commitment of resources during implementation. This is not the usual experience with past implementation of agri-input policies. Budget provisions can no longer keep pace with growing demand for fertilizers through public distribution while farmers have also become mentally dependent on subsidy. Similarly, the public system for distributing improved seed has failed to meet the need of farmers based on insufficient funding.

Lessons about the Channel: The channel or medium of message flow in delivering agri-inputs to farmers represents the most crucial aspect from where most of the policy issues of the communication model emerge. For agri-inputs, the channel is comprised of the market and the public extension system. Indeed, the economist's viewpoint is that the extension system is just another parallel market for agri-input. In any event, the most important issue about the channel is the magnitude of noise present, which is any factor impeding the flow of the policy message, distorting it, or at least impairing its reception on reaching its destination. Thus, in the market for agri-inputs, the several noise elements affecting the policy message include

i. The unorganized nature of the market that makes it difficult to reach the agri-input dealer for policy participation and for enjoying economy of scale in their operations
ii. The sharp practices in the market, such as short bag weights, adulteration, general lack of truth in labelling practices, etc.

On the other hand, the noise elements in the public agricultural extension system include

i. Poorly trained and immobile extension workers
ii. Poor state of basic infrastructure in rural areas, e.g. poor rural road networks, dearth of rural water supply, inadequate rural electricity, etc.

Lesson about the Target: The disposition of the farmers as targets in deciding the policy message depends on their socio-economic circumstances and possession of voice in the policy process, among other factors. All too easily, the farmers are so vulnerable to cheats and interest groups who corner the part of the policy benefits meant for them. The fertilizer policy is a case in point, whereby the inputs arrive too late and too little for the need of farmers.

POLICY FORMULATION FOR AGRI-INPUT DELIVERY

Getting the policy environment right for an agri-input delivery system starts with proper policy formulation, in the first instance. Rather than a casual one, it is a serious analytical exercise, to be followed by a disciplined adherence to its implementation. The latter is often blamed for policy failures, which is not necessarily so in all cases. In this section, we first make the case for the adoption of a process approach to policy analysis before proceeding to highlight the requirements for policy formulation with reference to the agri-input delivery system.

Process Approach

In carrying out any type of policy analysis such as policy formulation, the choice of analytical framework is crucial; that is, between treating policy as a set of discrete events or as a sequence of events in process (Idachaba 2006). The process approach to policy analysis is superior because it brings out the real explanatory factors of policy behaviour much more clearly, including the roles of different stakeholders, and helps in subjecting such roles to efficiency tests while proffering more feasible solutions. The key stages in the policy process analysis are as follows:

- Articulation (of the policy problem)
- Formulation (of the implementation strategy)
- Appraisal (of the strategy document)
- Implementation (of the strategy as appraised)
- Evaluation and feedback

We are presently concerned with the first two stages that comprise policy formulation, i.e. proper articulation of the policy problem and proper formulation of the implementation strategy.

Articulation of policy on agri-inputs delivery

Formal *articulation* of government policies represents the first stage of any aspect of the agricultural *policy process*. This involves the formal recognition and proper definition of the policy problem in all its ramifications and the deliberate specification of the policy directions to follow in addressing the problem. In the practical sense, a policy is formally articulated when it is written down and explicitly and publicly declared in advance, following the due process. Otherwise, we have disparate policy statements about different elements of the policy tucked away in inactive and closed files or inside some grey literature within the ministry or the agencies at federal and state levels, constituting the implicit policies. It is obligatory for government to articulate its policies in a formal way, and such formally articulated policies become laws, more or less, that are binding on its agencies operating in the sector while also guiding the behaviour of stakeholders in the sector. Suffice it to say that at the moment a formally articulated policy on agri-input integrating all the inputs together is not in existence in this country. And now is the time for one, bearing in mind the looming commencement of the African green revolution.

Policy articulation in this sense is an attempt to answer the question of *what* the government position is on agri-input delivery, the critical elements being background of the agri-input policy environment, the challenges and objectives of agri-input delivery policy, guiding principles, policy directions, and instruments to be employed. These elements should be fully specified and succinctly presented and then published as small, handy quick-reference materials for use by government officials and stakeholders in the public and private sector, as done lately for the recently articulated National Fertilizer Policy for Nigeria.

Thus, consistent with the development communication paradigm in articulating the policy on agri-input delivery system for Nigeria discussed earlier, the *what* question borders on what the position of government is on the issues relating to the different aspects of the agri-input delivery as a communication system—the adequacy and completeness of the package of agri-inputs as a message, the role of government as the source of packaging the agri-inputs message, the noise level in the agri-input message delivery channel, and the disposition of farmers as the target of agri-input message.

Formulation of implementation strategy for agri-inputs delivery

While policy articulation attempts to answer the *what* question, strategy formulation attempts to answer the *how* question; that is, going by the policy statement articulated, *how* the government will deploy the *policy instruments* at its disposal and follow the *policy directions* predetermined in addressing the *policy challenges* identified and in meeting the *policy objectives* stated. The main elements of strategy design and formulation include elaboration of each instrument and how they will be utilized in various combinations, suitable institutional arrangements for implementation, logical framework, action plans and time phasing, phasing of activities and work programmes, financial plans and fundraising, among others. Thus, strategy formulation results in an elaborate document that contains the full specification of parameters for implementing policy as previously articulated. Again, such a formulation is presently non-existent for agri-input delivery system in the country at the moment, and one is urgently required preparatory to implementing the twelve-point agenda of the Africa Fertilizer Summit.

EMERGING POLICY ISSUES AND OPTIMAL RESOLUTION OF ISSUES

The foregoing expositions serve as veritable sources of the several issues to consider in formulating the policy on agri-input delivery system in Nigeria. We shall highlight such issues or pose relevant questions for discussion at this stage only, without attempting to resolve them ahead of the consensus to be built around them by the policy stakeholders themselves, and not by the analyst.

Some selected issues have emerged to be grouped in two categories as follows:

Political Economy and Governance Issues

Role of Government in the Agri-Input Delivery System: There are several options in supporting agricultural production, for example, direct and indirect roles of the government. Should government continue to provide direct subsidy in the agri-input delivery system in the presence of widespread abuse by its officials and the perennial leakages of the subsidy policy benefits to non-targeted individuals in the society?

Federal-State Relationship in Agri-Input Delivery System: The division of labour in agricultural development is clearly established in the constitution of the Federal Republic. Is the federal government permitted, under the relevant section of the constitution, to embark on market stabilization programme in the agri input-delivery system, like the one in operation for fertilizer?

Policy Due Process and Policy Best Practices in Agri-Input Delivery System: The buzz words of good policymaking include Inclusiveness, Consistency, Stability, Transparency, Openness, Programme Accountability, Participation, Professionalism, and Documentation, to mention a few. What are the elements of these in formulating the policy on agri-input policy delivery system for Nigeria?

Structural and Systemic Issues

Regulatory and Legal Frameworks: The nature of agri-inputs is very scientific and technical, implying that unsuspecting farmers can be easily deceived and exploited unless they are effectively protected from the sharp practices of agri-input dealers in the market. Are we satisfied that the existing regulatory agencies such as NAFDAC and SON are doing a good job in this area? Should the Agricultural Ministry establish or should not establish its own regulatory agencies, particularly for fertilizer?

Public-Private Partnerships Frameworks: The private sector operates in the agri-input market purely for profit motives while the public sector operates the agricultural extension system as a complementary policy to fill the gaps in service provision to farmers. What frameworks exist to maximize this complementari within the framework of PPP for the smooth working of the agri-input delivery system?

Small- versus Large-Scale Farmers: Both scales are desirable for agricultural development of the country. Is there or is there not a need for discriminatory instruments in the policy on agri-input delivery in respect of each category of farmers?

Organized Private Agri-Input Sector: The effort towards an organised agri-input sector through the IFDC/DAIMINA project is quite recent. However, the observed trend is to the effect that the Agro-Input Dealers Associations (AIDAs) have toed the line of traditional commodity and farmers' associations—political interferences, top-down structure rather than bottom-up, etc. What are the necessary safeguards required in the policy on agri-input delivery to eliminate these negative developments?

Technical Backup Support Services in Formulating and Implementing Policy on Agri-Input Delivery System: The need for certain supportive services has been recognized as the responsibility of the private sector and NGOs. These include professional services such as policy advocacy and brokering services that are critical to policy best practices and conducting policy (varietal) trials prior to large-scale adoption. How can we promote the private and other non-government sectors to render such services to the benefit of the agri-input delivery system?

Role of Agricultural Universities in Agri-Input Delivery System: The best way for agricultural universities to contribute to the functionality of agri-input delivery system, as is the case with the American land grant colleges that are our role models for the three agricultural universities in Nigeria, is through the Cooperative Extension System, which involves resource collaboration between the federal, state, and local governments under a single administrative umbrella of the universities. Are these universities involved in the input delivery system to that extent? If not, why not?

Regional Dimension: There is serious effort to harmonize the agri-input delivery policies of countries through the regional bodies such as ECOWAS and NEPAD, particularly within the contexts of Africa Fertilizer Summit and African green revolution. The main issue is the extent of participation of Nigeria and the degree of commitment of the government to international treaties in regard to policy on agri-input delivery system for the country.

CONCLUSION

The formulation of policy on the agri-input delivery system in Nigeria depends on the optimal resolution of the several issues raised. Specifically, the proper articulation of the policy statement on agri-input delivery system and proper formulation of an implementation strategy for the system require that the lessons of experience from past policies are considered in resolving the issues, subject to strict adherence to the tenets of policy due process and consistent with policy best practices in other parts of the world.

In conclusion, the most important lesson of experience from public intervention in the agri-input delivery system of Nigeria is that there have been certain instances in the past and at present. A ready example is the case of fertilizer wherein the subsidy policy is good but the implementation strategy has failed to work. Suffice it to say that from this point on, we desire a rule-based, evidence-led, and internally consistent policy articulation and strategy formulation for the agri-input delivery system of this country.

REFERENCES

Ayoola, G. B. (2001), 'Effective Communication with Women for Agricultural Development', in Ayoola, G. B., *Essays on Agricultural Economy I: A Book of Readings on Agricultural Development Policy and Administration in Nigeria* (Ibadan: TMA Publishers).

Idachaba, F. S. (2006), 'Agricultural Policy Process in Africa: Role of Policy Analyst', in Idachaba, F. S., *Good Intentions are not Enough: Collected Essays on Government and Nigerian Agriculture* (Ibadan: University Press Ibadan).

IFDC (International Center for Soil Fertility and Agricultural Development) (2005/6), 'Africa Fertilizer Summit: Abuja Declaration on Fertilizer for African Green Revolution' in *IFDC Corporate Report, USA*, Circular IFDC S29.

Lipsey, R. G. and **Lancaster K.** (1956), 'The General Theory of Second Best', *Review of Economic Studies* 24.

32

Analysis Of Budget Policy On Agriculture Under Different Governance Regimes

INTRODUCTION

The resurgence of analytical attention on the policy environments for agricultural development owes largely to the evidenced failure of the perennial focus on technology environments to yield desired results in developing countries. A copious instance is the failure of green revolution technologies to significantly contribute to food security in Africa as it did in Asia and other parts of the world (CTA 2000; Ayoola 2004). Particularly within this context, the role of public expenditure budget as an instrument of agricultural policy becomes visible, as an aspect of the ongoing debate on the nexus between governance and economic development, hence the concerns of the Food and Agriculture Organization (FAO) and African Union (AU) to stipulate some minimum floors (20% and 25% respectively) as mandatory allocations to agriculture in the national budget of developing countries. However, these and other recommendations bordering on the best budget practices and other non-budget governance issues have been generally ignored, thus accounting for the persistence of sluggish agriculture in these countries.

Specifically in the ensuing debate, the question is being asked whether or not the widespread cases of military governance in Africa in recent past had a role to play in the poor budget performance of the countries involved. Certainly this question has emanated in the case of Nigeria, based on progress recorded in terms of higher growth rates of agriculture following the regime change from a continuous fifteen-year military regime to the present civilian regime. Hitherto, the successive groups of military officers in the country had often predicated their perceived need for radical regime change on the need to improve governance, with particular reference to budget performance.

At the present stage, the main problem pertains to the analytics of these issues, especially the methodological aspects of formulating appropriate model structures for tracing regime change effects on the public agricultural budget. Initially, a stylized human development index of the UNDP type was constructed to track the reform-induced changes in the budget structure under the military regime (Ayoola 1992). However, the analysis was largely focused on determining the scope of the active budget restructuring in favour of social services rather than on tracking the accruing budget savings or waste reductions as sources of incremental allocations to agriculture in the public expenditure budget. Also, a more robust analytical model of agricultural performance of the public expenditure budget was subsequently formulated and applied to Nigerian data (Ayoola and Oboh 1999) based on the Stone-Geary utility functions (Henderson 1980), but the outcome of the analysis failed to attribute the observed effect of budget changes to regime changes explicitly as desired.

Therefore, this study proposes an alternative model of the analysis in terms of the factual-counterfactual trends, with a view to estimating the timeline of the relative attention accorded agriculture in the budget reform process. Specifically the model has the added advantage to answer the question of whether agriculture budget policy is regime-neutral or not, thereby yielding an empirical basis to address the implications of regime changes for agricultural performance of developing countries in financial budgets. It is envisaged that the results of the study would help in specifying the point of convergence of policy and politics in the development process and so, representing a definite contribution of agricultural economics to critical policy issues.

THE FORMAL MODEL

The factual-counterfactual model of budget analysis is akin to the popular before-and-after methodology which is popular in the economic literature for policy impact assessment (Kahnehman 1982; Ayoola 1994; Spellman 2001). Figure 1 describes the model structure, conduct, and performance in terms of the following elements.

1. Given

- A budget cycle in a particular year (t) for allocating and implementing public funds to agriculture and non-agriculture in the economy—during the cycle, the various agencies of the government submit budget proposals at some stipulated time for consideration by the apex government authorities.
- A reference regime or time frame in the past ($t < t_0$) under military governance, whereby proposed funds were allocated by *discretionary* approval without the involvement of a legislature—in the circumstance, the military authorities apply the rule of thumb in exercising the choice of projects and financial allocations for different purposes.
- A successive regime or time frame in the present ($t_0 < t$) the civil governance, whereby proposed funds were allocated after a debate and vote in the legislature—the civil authorities follow the democratic principles entrenched in the constitution in exercising the choice of projects and financial allocations for different purposes.

2. Assume

- That the budget process is the major source of fund to all economic sectors; other sources possibly include foreign aid, anticipated and unanticipated, as well as windfall revenues emanating from price increases of critical export commodities such as petroleum.
- That the different budget categories are additive and mutually reinforcing, i.e. the capital and recurrent heads make up the total allocation for the agricultural sector, without considering the possibility of applying a portion of these to other sectors.
- That the budget allocations truly reflect the preferences of a government authority for agriculture in such a manner that greater funds imply higher commitments to food security and other concerns of the people.
- That the revealed preference of the government for agriculture in the budget is a true reflection of the preference of the people for agricultural development, consistent with the theoretic social welfare function (Killick 1981).

3. Required:

- To determine if the preference of the public authorities for agriculture in the public expenditure budget during one regime or time frame is smaller or greater than the preference of public authorities for agriculture during another regime or time frame; that is, whether or not the budget allocations follow the same trend as from some time past ($t: t_{-1} < t < t_0$) through to the present time ($t: t_{-0} < t < t_1$);
- To determine if the preference of public authorities during a given regime or time frame is the same or different for different budget categories during the same regime or time frame, and during another regime or time frame; that is, whether or not the budget allocations follow the same trend within the same regime or time frame (either $t_{-1} < t < t_0$ or $t_{-0} < t < t_1$) or between two regimes or time frames.

4. Construct

- A different trend line for each of the two regimes or time frames (line AB, line BC); line AB* is an imaginary mirror image of the line B* representing another path that line AB could possibly follow instead of line AB.
- A projection of the trend line for the past regime or time frame (line AB) beyond the end point in time t_{-0} to the present point in time t_1, hence straight line BD as an extension of AB.

5. Proof:

- Compared—the successive points in time on line BC (which is known as the *factual*) with corresponding points in time on line BD (which is known as the *counterfactual*).
- Determined—the preference for agriculture in the public budget during a particular regime or time frame is higher or lower than the present regime or time frame, depending on the relative positions of the factual trend and the counterfactual trend.

The model performance depends on availability of secondary data on the budget as may be disaggregated in particular countries; that is, both the ex-ante and whether the data include initial allocations or actual budget expenditures. Such data are readily available and published in most countries by the public agency for government statistics.

In evaluating the performance of the factual-counterfactual model, we consider its simplicity as a measure of changes in policy variables resulting from economic reforms consequent upon regime changes, which is devoid of cumbersome econometric preconditions.

Empirical Application to Nigeria

In Nigeria since independence from British rule in 1960, the cumulative period of military regime is longer than that of civil regime: 1966–1979 (13 years), 1983–1999 (16 years). Under the present democratic rule (1999 till date), efforts have been made to undertake economic and political reforms based on democratic principles and good governance, which involves budget reforms and with an implicit motive to forestall further intervention by the military. In the reform process, agriculture takes pride of place through a budget policy that puts greater emphasis on farm production and export in terms of the special (presidential) initiatives for particular commodities. This is also consistent with policy changes in the donor community towards more 'ownership' in aid provision. In particular, this finds expression in the paradigm shift of the World Bank from structural adjustment lending to development policy lending as well as the EU in terms of the performance-based conditionality through budget support rather than the previous result-based conditionality oriented towards intermediate targets (Zattler 2005).

In this situation, an analysis of budget policy of the country is predicated upon the need to reveal the preference of the government for agriculture in the annual budget. This would help address the twin concerns of the policy authorities to generate democratic dividends and the affinity of donor community for fiscal responsibility, with special reference to the agricultural sector of the country.

DATA AND ANALYSIS

In applying the model, data were collected through secondary sources from the north-central geopolitical zone, particularly from Benue State, popularly known as the Food Basket of Nigeria, based on its vast agricultural potential. The data include published and unpublished material on recurrent and capital estimates. Agriculture in the budget comprises the provisions for different heads, namely crops and rural development, livestock, fisheries, forestry, the state Tractor Hiring Agency (BENTHA), the state Agricultural and Rural Development Authority (BENARDA), the state Agricultural Development Corporation (ADC), and the state college of agriculture. The summary of the data is presented on Table 1 with some descriptive features of the consolidated allocations to agriculture sector in terms of the disparities between initial and actual allocations and between recurrent and capital allocations, across two governance regimes, military (1994–1999) and civilian (1999–2003).

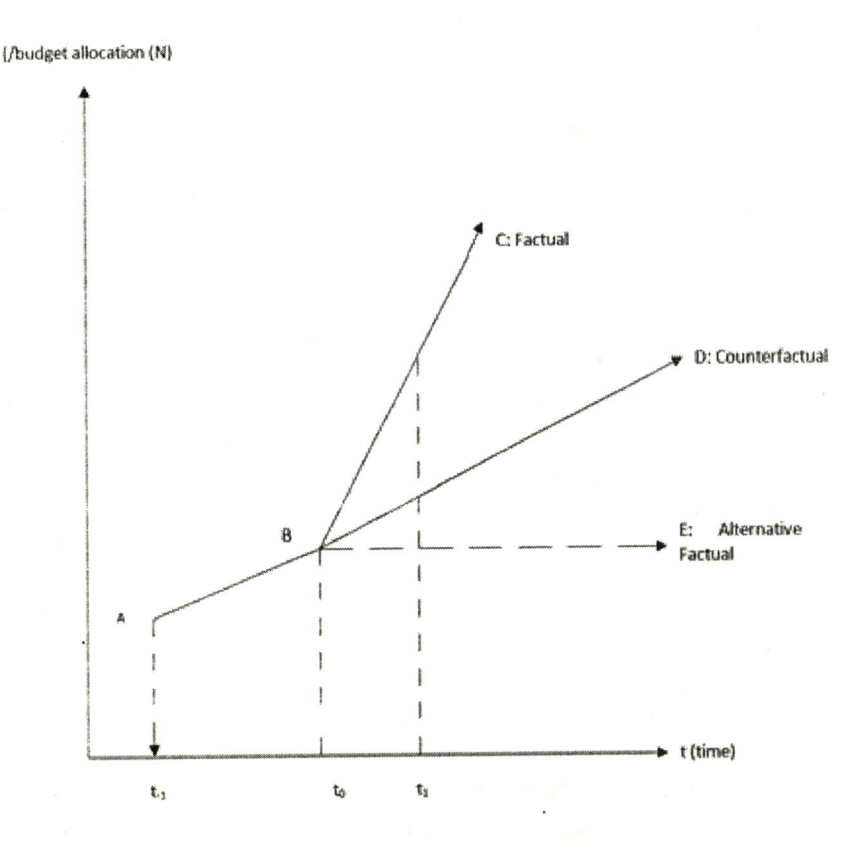

Figure 1: Factual-Counterfactual Model of Budget Performance under Two Governance Regimes

To test the null hypothesis of no difference in the allocations to agriculture between the military and civilian regimes, the formal model was specified and applied, first by estimating a set of trend equations based on simple linear regression model of the form $Y = a + bT$, where Y represents the budget allocation to agriculture of a particular class and T is the trend variable in years; *a* and *b* are the shift and slope parameters respectively. One of the trend equations represents the *factual*, which covers the current period of civilian governance (2000–2004), while the other trend equation covers the past period of military governance (1994–1999), which, when projected into the current period, represents the *counterfactual*. Thus, it is possible to draw comparisons in budget allocations between the periods for given episodes or between the episodes for given periods. Nevertheless, as generally recognized, mere extrapolation based on trends fails to incorporate the structural differences or changes implicit in the allocation behaviour of the policy authorities.

The choice of five years in each case is essentially to relate the results to recent events in the policy environment, such as the series of externally induced policy changes, particularly the World Bank / IMF–sponsored structural adjustment of the mid-1980s to mid-1990s. Besides, it is also of considerable analytical advantage that the analysis does not extend too much into the past, as trend projections perform poorly as we move further away from actual experience, owing to increasing margins of error associated with regression estimates (Kmenta 1971).

RESULTS AND DISCUSSION

The parameter estimates and predicted values are presented in Table 2. Upon projection, we arrive at the estimates of the shift and slope parameters for separate budget periods as well as the predicted values and divergences of the relevant variables. The regression estimates were statistically significant at 5% probability level. Judging from the signs of predicted divergences, it appears that the civilian regime has made greater fund provision to agriculture than the military regime in respect of the initial recurrent allocation, actual recurrent allocation, initial total allocation, and actual total allocation. On the other hand, it appears that the military regime has made greater provision for agriculture than the civilian government in terms of initial capital allocation and actual capital allocation to agriculture.

Table 1: Initial and Actual Allocations to Agriculture, Benue State, 1994–2003[56]

Year	Initial Allocation (N millions)		Actual Allocation (N millions)	
	Capital	Recurrent	Capital	Recurrent
1994	47.1	82.8	7.3	160.1
1995	73.4	110.0	51.5	158.5
1996	192.6	102.9	70.3	132.3
1997	110.4	108.6	75.1	89.7
1998	180.4	123.7	57.3	112.0
1999	313.5	280.7	20.9	293.2
2000	552.9	285.0	58.9	527.5
2001	661.9	592.4	146.8	675.8
2002	1166.4	746.3	106.3	703.9
2003	1189.0	874.0	303.3	638.1

Meanwhile, the greater recurrent allocations under the civil regime imply that budget policy at that time revealed greater preference for personal emoluments and overhead, *ceteris paribus*. This agrees with the observations during the current civil regime that is mostly preoccupied by reform measures to divest government from direct production ventures and to promote private sector initiatives. A notable example is the eventual privatization of the National Fertilizer Company (NAFCON) that represented a huge failure among similar public-owned enterprises in the sector financed from capital budget votes. Thus, the consideration of capital allocations to agriculture in the public budget has manifestly reduced from what used to be the case under the military. Rather, the burden of recurrent expenditure looms larger as efforts to downsize the public workforce became difficult in the face of civil opposition and agitations against retrenchment.

---

[56] **Source:** Benue State Government *Gazettes* (various issues, 1994–2003) Benue State Government *Budget Estimates* (various issues, 1994–2003)

Table 2: Regression Estimates and Predicted Parameter Values of the Factual-Counterfactual Model

Variable	Initial Capital Allocation	Actual Capital Expenditure	Initial Recurrent Allocation	Actual Recurrent Expenditure	Initial Total Allocation	Actual Total Expenditure
Shift Estimates:						
• Factual	0.176	2.760	0.126	0.133v	0.302	0.161
• Counterfactual	4.880	2.020	4.800	6.620	9.700	8.660
• Divergence	-4.704	0.740	-4.674	-6.487	-9.398	-8.499
Slope Estimates:						
• Factual	3.720	1.140	2.52	7.700	6.240	1.890
• Counterfactual	3.700	2.500	-7.8	-2.500	-3.900	-2.200
• Divergence	0.020	-1.360	10.32	10.200	10.140	4.090
Predicted Values:						
• Factual	33.656	13.020	22.806	69.433	56.462	17.171
• Counterfactual	56.680	37.020	-104.4	-28.382	-44.900	-22.140
• Divergence	-23.024	-24.000	127.206	97.813	101.362	39.311
t-values**	34.852	12.037	73.275	121.545	165.9	39.311

** = Significant at 1%

In any case, the greater volume of total allocation to agriculture under the civil regime probably suggests that the sector enjoyed high preference in terms of funding at that time far better than during the military era. Thus, the scope exists for budget restructuring towards better performance of the sector at the present level of available budget funds. For instance, a budget policy that directs more funds toward improving the extension information and rural infrastructure would stimulate agricultural growth faster than fund flows to non-performing public capital projects.

Finally, the results indicate that both regimes are consistent in budget implementation in the sense that for each budget category (capital/recurrent) the same regime has revealed preference for the initial allocation and actual allocation together. This suggests that there is no basis to infer that one regime exhibits budget discipline more or less than the other. Budget discipline means the ability of public authorities to achieve high correlations between the sets of initial budget allocations and the actual budget allocations. The issue emanates from the general observation that in several cases, the initial allocations and actual allocations have little or no bearing on each other, so the agricultural population suffers from what can be termed *budget illusion*.

Budget illusion is characterized by euphoria among the people who are satisfied with the government for making initial allocations but become subsequently disappointed with the same government for not following through with its budget commitment. The presence of budget illusion in both regimes is indicated by the observed divergences together between the factual and counterfactual estimates at the same points in time. The difference is due to the fact that the military regime, having no opposition, tends to suppress the manifestation of budget illusion, unlike the civil regime. Indeed in the current times the president has faced intense criticism from the National Assembly, sometimes bordering on threats of impeachment based on accusations of unfaithful implementation of the budget. This allows for budget accountability which thrives under the civil regime that seeks approval and reports back to the legislature on issues of budget.

SUMMARY AND CONCLUSIONS

the role of Agriculture in a developing country depends largely on budget policy of the government It is therefore important to determine the governance regimes that favour the sector more in the budget process. Towards this end, an analytical framework based on the factual-counterfactual helps to trace the trend of budget allocation behaviour of the government from past to the present with a view to drawing comparison on the basis of budget performance. In Nigeria results of the analysis indicate that the military showed greater preference for capital allocations to agriculture in the budget while the civil regime showed greater preference for recurrent expenditure. This reflects the presence of strong opposition during the civilian regime that tends to overblow the size of civil service workforce with the attendant recurrent commitments such as personal emoluments and general overhead.

On the whole, the civil regime reveals greater preference in terms of the total budget than the military regime. This suggests that the incremental recurrent expenditure during the civilian regime more than offsets the incremental capital expenditure during the military regime. Thus, the scope for budget restructuring in favour of agricultural growth through higher capital allocations in the public expenditure budget exists in the present democratic dispensation to a larger extent than during the previous dictatorship under the military regime.

REFERENCES

Ayoola, G. B. (1994), 'The Future of Agricultural Commodity Trade in Ijere', in M. O. and G. B. Ayoola (eds.), *Commodity Boards in a Liberalised Economy*, Proceedings of National Workshop, University of Agriculture, Makurdi (Makurdi: Fulladu Publishing Co.).

Ayoola. G. B. and V. U. Oboh (1999), 'A Model of Public Expenditure to Reveal the Preference for Agriculture in the Budget', *Journal of Rural Economics and Development* **14** (1): 56–72.

Ayoola, G. B. (2001), *Essays on the Agricultural Economy I: A Book of Readings on Agricultural Development Policy and Administration in Nigeria* (Ibadan: Farm and Infrastructure Foundation).

CTA (Technical Centre for Agricultural and Rural Development) (2000), *Agricultural Policy Networking: The Way Forward*, Summary of Report and Recommendations of CTA Workshop, Entebbe, Uganda, 6–10 November.

Henderson, J. M. and R. E. Quandt (1980), *Macroeconomic Theory* (London: McGraw-Hill International Book Co.).

Kahnehman, T. (1982), 'Background to Counterfactual Thinking,' Paper presented at EAESP Small Group Meeting on Counterfactual Thinking, uvic.ca/psyc/demanded/aix.html accessed 15 September 2003.

Killick, T. (1981) *Policy Economics: A Text Book of Applied Economics on Developing Countries* (London: Heinemann Publishers).

Kmenta, J. (1971), *Elements of Econometrics* (New York: Macmillan Publishing Co. Inc.).

Spellman, B. A. (2001), 'Wine, Women and Wells: Why Thinking about More (Consequent-Changing) Counterfactual Leads to Greater Attributions of Causality,' Paper presented at EAESP Small Group Meeting on Counterfactual Thinking, web.uvic.ca/psyc/dmandel/aix.html

Zattler, J. (2005), 'Reviewing Conditionality', *Development and Co-operation* **32** (July 2005).

33

Agriculture And National Development

INTRODUCTION

In this exercise, a background in terms of the imperatives of agriculture in national development, namely manpower, technology, and infrastructure, followed by a decade-wide critical analysis of the agricultural policy process governing the success of the country in fostering these imperatives is provided. Next is an examination of the role of agriculture, indicating the contribution of the sector to national development as well as discussing its performance of that role over the years, then a highlight of the main constraints and challenges facing Nigerian agriculture in maximizing its potential to play its role follows. Last is a consideration the efforts of the ministry to chart a course for national development through agriculture.

Background of Agriculture and National Development

Agricultural development is a primary condition for poverty reduction in the quest for national development. This is particularly so as it is the source of food security for the people, for their employment and income generation. The stated condition consists in three imperatives, namely manpower development, technology development, and infrastructural development, which upon convergence translates into national development through agriculture.

Manpower development is an aspect of human development that deals with building up the minds of individuals and sharpening their knowledge for development works in agriculture. In this regard, the country operates a network of middle-level institutions for agricultural training, such as schools and colleges of agriculture which build on general knowledge of agricultural science provided in primary and secondary schools for specialized extension and other services in agriculture. Further to this, several faculties and three universities of agriculture exist to provide specialized training for the sector to contribute to national development. Manpower development represents the reservoir of physical and mental energy required for making enterprise initiatives for farm production and other aspects of the agricultural value chain.

Technology development deals with constant generation and regeneration of the technology base for agricultural development. These include product and process innovations that bring about new methods associated with cost reductions such as technologies of land preparation (e.g. tractors, ploughs, harrows); planting, seed and seedlings; management practices (e. g. fertilizers, herbicides, etc.); as well as transportation, processing and storage. The institutional framework for generating technologies for agricultural development involves the several commodity-focused research institutes and the existing three agricultural universities, acting together with the relevant organs of the ministry to link research and extension systems together. The importance of technology for national development is demonstrated by the historic green revolution in Asia, where the emergence of new varieties of wheat quickly changed the food security situation of India and other countries from frustration to hope.

Infrastructural development is also germane to national development through agriculture, in terms of the stock of physical, social, and institutional facilities. In this connection the Seven-Point Agenda of the present administration ranks energy infrastructure as a priority area of policy attention of national development. Similarly the road network, railways, and waterways are important items of infrastructure for agriculture and national development. Therefore, agricultural development depends on the stock of rural infrastructures comprising as follows:

(i) **Rural Physical Infrastructure,** including

- Rural roads which cause accelerated delivery of farm inputs, reduce transportation costs and enhance spatial agricultural production efficiency,
- Storage facilities which help to preserve foods in the farms up till the time when consumers need them,

- Irrigation facilities which assure farm water supply and stabilize food production by protecting the farm production system against uncontrollable and undesirable fluctuations in domestic food production.

(ii) **Rural Social Infrastructure,** including

- Clean water, decent housing, environmental sanitation, personal hygiene, and adequate nutrition, which help to improve the quality of rural life;
- Formal and informal education which promote rural productivity by making the farmer able to decide agronomic and other information and carry out other desirable modern production practices; basic education also promotes feeding quality, dignity, self-respect and sense of belonging as well as political integration of the rural people.

(iii) **Rural institutional infrastructure,** including

- Farmers' unions and cooperatives which facilitate economics of scale and profitability of rural enterprise,
- Agricultural extension which improves the technology status of the farm business.

These imperatives combined, culminate in slow or speedy national development depending on the performance of agriculture. Therefore, we need a quick survey of policy attention to agriculture in the national development process in post-independence Nigeria. Although the colonial heritage since 1900 together with the activities of European merchants that built up to it represented a vantage point for take-off of the country, the goal of development before independence was totally different from the goal of national development afterwards. It is a known fact that the implicit motive of colonial development that mattered at that time was the surplus extraction from the country in terms of massive export of forest and agricultural products as well as other items required as raw materials for the Industrial Revolution in Europe and later to rebuild the metropolitan industries after the world wars. It was for that purpose, not necessarily in the interest of national development, that infrastructure such as the road and railway networks required to move the products from the hinterlands to the sea for transportation to Liverpool and other European ports were built all over the country.

Nevertheless, the visibility of agriculture during the colonial development period cannot be easily overlooked. This includes the following highpoints of policy action:

- The establishment of a botanical garden in 1893 at Ebute Metta in the colony of Lagos;
- The establishment of a Forest Department at Olokemeji near Ibadan in 1900;
- The establishment of the first Agricultural Department for the South at Moor Plantation, Ibadan in 1910. followed by another Agricultural Department for the North in 1912, both of which were amalgamated as one Agricultural Department for Nigeria in 1921;
- The establishment of the first generation of marketing boards with national mandates, namely Cocoa Board (1947), Palm Produce Marketing Board (1949), and Cotton Marketing Board (1949) and
- The creation of Ministry of Agriculture and Natural Resources (MANR) for each region, following the introduction of the federal constitution, accompanied with the establishment of a second generation of marketing boards with regional mandates, namely Western Nigeria Marketing Board (1954), Eastern Nigeria Marketing Board (1954), and Northern Nigeria Marketing Board (1954).

Regarding the cumulative policy attention to agriculture in the post-independence period, a period analysis will provide an appropriate framework, wherein there are five development decades: 1960–1970, 1970–1980, 1980–1990, 1990–2000, and 2000–2010. Meanwhile, let us examine an important political economy issue that emerged as part of the legacy of colonial agricultural administration of the country.

The inherited institution had generated an issue about the jurisdictions of the old regional government and the federal government for agricultural development, based on the federal constitution that placed agriculture on the **residential** legislative list, which implied that it was a regional responsibility. As soon as the new nation settled down, the federal government moved to establish a federal ministry for agriculture, but this was vehemently opposed by the regions on the strength of their constitutional right. The regions acted in this way against the backdrop of immense commodity boom at that time accompanied by foreign exchange revenues accruing from the export of farm produce through the operation of the individual marketing boards, the federal government invited FAO to study the situation, which recommended that it was necessary to establish a ministry of agriculture at the centre. Apparently to circumvent the constitutional provision, the federal government went ahead to establish a pseudo agriculture ministry, by name Federal Ministry of Natural Resources and Research (FMNRR), which carefully avoided the word *agriculture* so as not to offend the political sensibilities of the regions. Finally, in 1966, the new military government, having suspended the constitution swiftly created the Federal Ministry of Agriculture and Natural Resources (FMANR) by administrative fiat (Nigeria, *Official Gazette* vol. 55 No. 10 of 7 February 1966). That singular effort laid the foundation for the subsequent placement of agriculture on the concurrent legislative list in the 1979 constitution and later in the current 1999 constitution, implying that it is presently a joint responsibility of the federal and state governments.

This issue, as innocuous as it appeared, was later to play important roles in the national development process as the usual bone of contention about the optimal allocation of state responsibilities during constitutional reviews and intermittently at other times. The point to be made here is that the issue underscores the political economy importance of agriculture in the national development process. That is, the national development process involves areas of friction that need to be resolved before progress can be made. Otherwise the development process itself can become self-limiting if such frictions are allowed to degenerate into unhealthy rivalry and destructive conflicts of interest among the organs of state involved in national development actions. For instance, the absence of a suitable mechanism for resolving such issues of political economy created the need for a civil war in the early period of national development. That war necessitated the commitment of substantial resources of the nation to prosecute.

At this stage, we are in a position to outline the policy attention to agriculture in the national development process on a decade-by-decade basis, as follows:

1960–1970

o The regional MANR played a dominant role as the key agricultural development institution in the individual regions, coupled with the regional marketing boards as the major intervention agencies; also the Farm Settlement Scheme had been launched in the Western Region as the main strategy of intervention in the sector. The immense commodity boom of this period, featuring the Cocoa West, Oil Palm East, and the Groundnut North, represented a significant contribution of agriculture to national development, as each region translated the foreign exchange revenues accruing through exports into massive economic infrastructure.

o The establishment of Federal Ministry of Agriculture and Natural Resources led to the establishment of Federal Department of Agriculture (FDA) as the operational organ for implanting agricultural development programmes. The period of civil war between 1967 and 1970 represented a lull in the activities of the ministry, until the launch of NAFPP (National Accelerated Food Production Project) afterwards.

1970–1980

o In consolidating the federal ministry soon after the war, the federal government negotiated a loan package with the World Bank, which led to the sequential implementation of the series of integrated Agricultural Development Projects (ADPs). Command-area projects were established in sequence, beginning with Funtua, Gusau, and Gombe projects in 1975. Newer enclave-type projects were Lafia (1978), Ayangba (1978), Ilorin (1979), Bida (1980), Ekiti Akoko (1981), and Oyo North (1982). Finally, each state of the federation had one ADP, all at statewide levels. In the process, the issue arose about the slow rate of establishing these projects in other parts of the country under the World Bank assistance. This led to home-grown national Accelerated Area Development Projects (AADP) particularly in the eastern part of the country. The issue also arose about the desirability of enclave projects that implied that some parts of a state would wait while development was taking place in other parts. Initially, the implementation of ADPs was backed up service by two specialized units which were Federal Agricultural Coordinating Unit (FACU headquarters, Ibadan), which was responsible for preparation, negotiation, take-off, and planning of projects, and Agricultural Projects Monitoring and Evaluation Unit (APMEU headquarters, Kaduna) which was responsible for monitoring and evaluation. Later, both of them were merged to form the Projects Coordinating Unit (PCU).

o The potential of available water bodies was explored through the establishment of River Basin Development Authorities (RBDAs) Decree No. 25 of 1976. Eleven authorities were first established; the number grew to eighteen upon the redesignation as RBRDA (River Basin and Rural Development Authorities) before a reversal to the original number, 11, upon disposal of their non-water assets during the liberalization accompanying SAP. Along the line, six (UPN-controlled) state governments had taken the (NPN-controlled) federal government to court over the activities of RBDAs, particularly for carrying out rural development works in their jurisdictions such as sinking boreholes and providing other rural infrastructures. Curiously, this took place on the eve of a general election, which was capable of influencing the pattern of voting in rural areas in disfavour of the plaintiffs. The course of action for this case was premised on the previous constitutional rights of the states as entrenched in the initial federal constitution, thereby representing an instance of conflict between policy and politics in the process of national development.

o The Operation Feed the Nation (OFN) campaign was launched in 1979 by the military government. It was first drew attention to a looming food crisis in the country. The programme strengthened the NAFPP in terms of provision of inputs and increased tempo of extension services. A fertilizer loan was secured from the World Bank and the federal government established the Fertilizer Procurement and Distribution Division (FPDD) as the sole importer of fertilizer into the country for distribution to states at subsidized prices.

o The federal government acquired the regional marketing boards and reorganized them into a third generation of seven commodity boards in 1976 (Decree No. 29 of 1977) to operate under national mandates along expanded commodity lines: Cocoa Board (Ibadan), Groundnut Board (Kano), Grain Board (Minna), Cotton Board (Funtua), Palm Produce Board (Calabar), Rubber Board (Benin), Roots and Tuber Board (Makurdi). Also, the agricultural research system was reorganised with the establishment of commodity-based research institutes in different agroecologies in the country.

1980–1990

o Liberalization of the agricultural market took place in 1986 when the Structural Adjustment Programme (SAP) was launched, leading to the abolition of the commodity boards system. It was observed that the commodity boards had become dependent on government subventions for their operations, unlike the earlier period of regional ownership; and at the time of abolishing them, all the boards except

- one, Nigerian Cocoa Board, were indebted to the Central Bank of Nigeria.
- o A number of establishments accompanied the introduction of SAP in the agricultural sector. First, the Directorate of Food, Roads, and Rural Infrastructures (DFRRI) was established in the presidency, initially financed by the revenues accruing from the liberalization of the foreign exchange market. Second, two universities of agriculture were established at Makurdi and Abeokuta; third, the National Agricultural Land Development Authority (NALDA) was also established as a parastatal of the ministry.

1990–2000

- o Both DFRRI and NALDA were subsequently scrapped. A democratic economy was ushered in with the inauguration of a civilian government in 1999, after about 15 years of military rule.
- o Further reform of the agricultural sector involved the liberalization of the fertilizer market. The federal government withdrew first, from distribution operations and later from importation, in order to increase the role of the private sector in the fertilizer market. The Fertilizer Market Stabilizer Scheme was put in place by the federal government, to purchase and distribute fertilizers to states at 25% price subsidy level.

2000–2008

- o The FAO-supported National Food Security Programme (NFSP) was launched, followed by a series of commodity-based Presidential Initiatives in agriculture. The expanded phase of NFSP covering three sites per senatorial district was launched.
- o The reorganization of FMAWR started off; this involved creation of the National Food Reserve Agency (NFRA), and a food security strategy document—National Food Security Programme—was produced.

Against this background, it is obvious that the policy process for agriculture relates significantly to the overall national development process. Specifically, within the context of that process, agriculture is called upon to facilitate the speed of national development by performing certain roles.

ROLE OF AGRICULTURE IN NATIONAL ECONOMIC DEVELOPMENT

Food security, upon which the livelihood of the people depends, is the most important role of agriculture for sustained national development. The main dimension of this is the attainment of self-sufficiency in the production of staple foods, including maintenance of food reserve for possible emergency as well as positive net export of food items. Further, food security implies adequate supply of food entitlements of the people in terms of energy, protein, and other nutrients, including water. According to international standards, the agriculture of a nation is expected to ensure availability, accessibility, and affordability of these items, with a view to ensuring a decent living of the people. The other roles for agriculture in national development are secondary in nature, namely provision of employment, generation of incomes and foreign exchange required to pay for imports. Suffice it to say that the performance of these roles since independence leaves much to be desired for a steady national development.

At independence in 1960, agriculture including farming and herding accounted for the largest component of Nigeria's GDP, employing over 70 per cent of the economically active population. Before 1970, the sector had contributed more than 75 per cent of export earnings, but since then, however, it has stagnated, partly due to neglect and poor investment, and partly due to ecological factors such as drought, disease, and reduction in soil fertility. By the mid-1990s, the share of agriculture in exports had declined to less than 5 per cent. This led to the changing fortune of the country from an exporter of food to nearby countries to a net importer of food and fibre from other countries of the world. The food import bill is ever-increasing, reaching $24 billion in 2001. Presently, agriculture contributes about 41 per cent of the GDP and still employs about 70 per cent. The share of agriculture in the GDP is made up of crops 85 per cent, livestock 10 per cent, fisheries 4 per cent, and forestry 1 per cent.

The traditional cash crops produced in the country include oil palm, cocoa, rubber, and cotton, all of which were once exported but are now sold mostly locally. The food security crops grown in large quantities are sorghum, millet, maize (corn), rice, yams, and cassava, which are now widely sold for cash as well. These constitute the main staple food items of the populace. The major livestock includes cattle, sheep, goats, poultry, and pigs. The output of agriculture depends on smallholders who are generally poor and use cutlasses and hoes as the basic tools in the technology of production. The effort of government towards agricultural transformation of the country include provision of agricultural extension services together with farm input support, implementation of irrigation projects, investments in rural infrastructure, and introduction of modern seed varieties and chemicals. Although large-scale, machine-based farming is being promoted, it accounts for only a fraction of total production at the moment.

The farm population is dominated by small-scale operators cultivating between 0.5 ha. and 4 ha. as modal range of farm size. The yield of crops is low at about 4 tonnes of agricultural output per hectare compared to about 14 tonnes per hectare in other countries. Crop output in 2006 reveals the comparative advantage of the country in the production of several crops, such as cassava (4,572.89 MT) and yam (28,890.42 MT), which puts Nigeria in the lead among other countries of the world producing these items. Livestock production is inadequate, with about 30 per cent of slaughters being imported from neighbouring countries. Fish production was estimated at about 600,000 MT out of a domestic demand of 2.6 million MT.

Thus, in terms of performance, the county's status has changed from that of a net exporter of food to a net importer. The country has also its previous position as a leading supplier of agricultural produce into the world market with cocoa production stagnating at around 180,000 tonnes compared to 300,000 tonnes 25 years ago. The trend is the same for other commodities of historic importance, such as groundnut and palm oil, which have also witnessed dramatic decline in production. Furthermore, the country which used to be the biggest poultry producer in Africa has receded in status from 40 million birds annually to about 18 million.

The recent profile of crop and livestock commodities in terms of production and trade statistics is presented in *Table 1* for different years.

Table 1: Nigeria—Profile of Selected Agricultural Commodities and Forest Products in Recent Years

	Production (2007)		Trade (2005)	
Selected Crop	Yield	Output (tons)	Imports (tons)	Export (tons)
Food Crops (kg/ha.)				
Maize	1,659.5	7,800,000	17,668	2,226
Rice (Milled)	1,559.1	4,677,400	1,040322	4,367
Cassava	11,883.1	45,750,000	-	-
Wheat	875.000	70,000	3,714,683	168,355
Beans	-	-	1,701	-
Millet	1,316.2	7,700,000	0	504
Cash Crops (kg/ha.)				
Cocoa	450.400	500,000	0	267,900
Rubber	-	-	84	25,000
Groundnut	1,720.0	3,835,600	7,100	87
Cashew	2,000.0	660,000	0	17,277
Sesame seed	510.2	100,000	59,600	117
Cotton seed	-	-	5,100	15,700
Livestock (carcass weight, tons)				
Chicken	1000.0	233,100	39	-
Pig meat	-	6,730,000	33	-
Cattle	-	16,258,560	47	3
Sheep meat	110.0	105,570	9	-
Goat meat	127.0	148,830	-	-

Source: FAO STAT

CONSTRAINTS AND CHALLENGES FACING AGRICULTURAL DEVELOPMENT

The inadequate performance of agriculture for national development is at variance with the presence of a huge natural resource potential of the country. The country has a broad natural resource base for agriculture, the ecological variability ranging from tropical forest in the south to dry savannah in the far north, which implies a diverse mix of plant and animal life. The economy is largely rural and based on the productivity of the land, 30 per cent of which is cultivated. Soil fertility varies considerably, with a high concentration on areas around the river valleys with alluvial depositions. The arable land is 79 million hectares, out of which only 32 million hectares or 46 per cent is under cultivation. Nonetheless, the realization of this potential is constrained by certain factors, namely weather, labour, technology, infrastructure, and policy, which jointly pose general and specific challenges to agricultural development in the country.

Weather constraints consist of the increasing threats of desertification coupled with erratic pattern of rainfall. Desertification is reinforced by massive water impoundment and irrigation schemes particularly in the northern part of the country. Uncontrolled grazing and livestock migration put tremendous pressure on the environment in some areas, while other environmental threats include poaching and settlement within protected areas, brush fires, increasing demand for fuel wood and timber, road expansion, and oil extraction activities. The petroleum and natural gas industries have contributed in large measure to damaging the land, vegetation, and waterways in the Niger Delta, through their frequent oil spills, burn-off of natural gas, and clearance of vegetation at a fast rate.

The challenge of labour shortage on farms is also real. This has escalated in more recent times, owing to higher remunerations accruing to unskilled labour from public- and private-sector wage employment as well as the lingering youth restiveness in the Niger Delta region, among others. The origin of farm labour shortage is traceable to the windfall oil incomes that accrued to Nigeria in the 1970s, which created an immense urban boom that led to influx of youth into cities and the consequent sharp increase in the proportion of aged people on the farms. At the moment, the typical Nigerian youth will offer to do any job—such as menial jobs in the oil-producing companies (as day/night guards, clerks, or militia) or in the communication sector as vendors of recharge cards and in the transportation sector as commercial motorcyclists etc.—but farming. The situation in the country is further complicated by the scourge of HIV/AIDS, which depletes the stock of available farm labour supply, with an estimated 10 per cent of the adult population of about 75 million people already infected.

The *technology constraints* emanate from a number of sources, namely

o   The persistent failure of the national agricultural research system to generate new agricultural technology that is sufficiently adapted to different local environmental niches or conditions,
o   Failure to transmit proven technologies available on the shelf to farmers' fields where they are needed because of weak or non-existent linkages between the extension delivery system and the national agricultural research system,
o   Failures of existing extension services to effectively relay the farmers' field problems to researchers with minimum delays, etc.

The principal components of the technology environment include production technology such as farm machinery, technical inputs such as fertilizer, seed, and crop protection products (CPP). About technology of production, the use of farm inputs is sub-optimal, namely fertilizer (7 kg/ha.); irrigation (10 per cent of irrigable land under irrigation); tractors (30,000 tractors available all over the country), coupled with a low extension worker-farmer ratio of 1 to 25,000 farm families, as well as inadequate supply of farm credit. Furthermore, the use of improved agricultural seed is not popular, owing to low supply, which makes farmers stick to the traditional practice of sowing the seed saved from previous years.

On the other hand, *policy constraints* emanate largely in terms of persistent instability and discontinuities which characterise the policy environment. The consequence of this is uneven pathway of agricultural development as a result of frequent changes in organizations and personnel, thereby creating frequent perturbations in the process of national development.

The solution to policy instability is legislation of the strategy document and the organizational structure, which will make it difficult for such changes to take place arbitrarily in future.

Finally, the *infrastructural constraints* emanate from an environment characterized by poor status of basic needs of the rural people, including physical infrastructure (rural road networks, rural water supply schemes, rural storage and processing facilities, irrigation facilities, etc.); social infrastructure (for knowledge creation, healthcare, information, etc.) and institutional infrastructure (thrift and credit societies, development associations, cooperative societies, etc.). The poor state of infrastructure puts a low ceiling on farm output, marketable surplus, and volume of consumption of agricultural products possible. This affects the critical elements of the agricultural value chain, such as storage and processing, which results in low value addition in the sector.

THE WAY FORWARD

The way forward to enhancing the role of agriculture in the national development process is marked already in terms of two events that recently took place in the Federal Ministry of Agriculture and Water Resources. The first is the organizational reform of the ministry through the creation of National Food Reserve Agency (NFRA), which is currently in the process of being legislated. The second is policy reform through the production of a food security strategy document, which has been adopted by the National Council on Agriculture (NAC) and subsequently approved by the Federal Executive Council. It is believed that these recent events of government will greatly enhance the ability of agriculture to express its potential more fully in contributing to national development.

In this regard, NFRA will have the technical capacity to implement programmes and projects of agricultural development in line with the current policy thrusts as contained in the National Food Security Programme. They are as follows:

Import Substitution: That import substitution should be achieved for the five commodities which cost Nigeria an annual average of $2.68 billion in foreign exchange;

Substantial Food Security: That government should intervene to significantly enhance food security of its people consistent with gainful employment and wealth creation;

Promotion of Modern Agricultural Practices: That agriculture should shift from subsistence production to modern technologies of production;

Natural Resources Conservation: That the issues of declining soil fertility should be tackled;

Commodity Focus: That policy emphasis will be on select commodities at appropriate agroecological locations in the county;

Private Sector Participation: That the partnership between government and private sector will be strengthened in several areas for the benefit of agricultural enterprise growth and development;

Successor Farmer Generation: That the ageing population of farmers will be replaced on a constant basis by promoting youth development and empowering young people to embark on farming and other agricultural enterprises;

Participative Policy Process: That design and implementation of agricultural projects will be all-inclusive and fully participatory;

Policy Advocacy and Brokerage: That government will promote credible organizations to render professional services in advocating policy best practices for agriculture and rural development in the country, and in brokering policies for piloting and experimentation where necessary.

Safety Net Considerations: That government will pay special attention to vulnerable groups in society to make them food-secure as well.

In addition, some other notable features of the Strategy Document include the following:

Time Phasing: Intervention instruments were demarcated into short-term, medium–term, and long-term measures;

Role Definition: The roles of stakeholders were delineated among the federal government, state governments, local governments, private sector, and development partners.

Target Setting: Targets have been set for production of critical commodities for the country.

Towards this end, there have been reassuring indications demonstrating the high level of government commitment to agriculture. Given this level of commitment, it is envisaged that in no distant future, the targeted agricultural growth rate of 10 per cent to be accompanied by improved role of agriculture in national development, with particular reference to food security of the nation and employment for the people, will be achieved.

CONCLUSIONS

The foregoing account reveals the crucial role which agriculture has played in the national development of Nigeria from time immemorial. The performance of this role obviously leaves much to be desired, owing to certain political economy issues that emerged and the several constraints and challenges encountered at critical stages of the process. Nevertheless, the process has exhibited considerable resilience, judging from its stability over time and the absence of a biting food crisis that goes beyond control.

Therefore, the new efforts of the ministry, particularly the creation of NAFRA accompanied with production of a food security strategy document, has raised the hope of steady growth of the agricultural economy for maximum contribution to national development.

REFERENCES

Ayoola, G. B. (2001), *Essays on the Agricultural Economy I: A Book of Readings on Agricultural Development Policy and Administration in Nigeria* (Ibadan: TMA Publishers).

Ayoola, G. B. (2007), *National Council on Agriculture: Analytical Profile of the Highest Policy Advisory Authority on Nigerian Agriculture and Rural Development* (Abuja: Farm and Infrastructure Foundation).

FMA (Federal Ministry of Agriculture and Water Resources) (2008), *National Food Security Programme*.

34

Policy Advocacy: The Missing Link In Nigeria's Quest For Agricultural Transformation

Invited lead paper presented at the 17th Annual Conference of the Nigerian Rural Sociological Association held at the National Root Crops Research Institute, Umudike, 19 August 2008.

INTRODUCTION

The notion of agricultural transformation connotes that upgrading the traditional sector is crucial to attaining sustained progress of developing economies of the world. The central issue in agricultural transformation involves promoting productivity and competitiveness of the agricultural economy consistent with livelihood improvement in terms of food security, employment creation, and wealth generation. Thus, the need arises to examine the efforts in this direction in Nigeria with a view to identifying the missing link in the quest for agricultural transformation of the country and to proffering some feasible solutions. The main theme is that while technology matters in the quest for agricultural transformation of the country, policy probably matters the more, and specifically that the policy process is itself incomplete, with the perennial absence of credible actors to undertake the conscious advocacy for policy best practices in the agricultural sector.

Thus far, the policy process has been conducted under the tenacious assumptions that government will always act or fail to act in the best interest of the people, and that the communication gap between policy authorities and agricultural population is practically non-existent. In this exercise, it is argued that the promotion of policy best practices by non-government actors holds the key to the scientific transformation of the agricultural sector. This argument is supported by the following discernible trends in the provision of technical backup support for agricultural policy process:

- o In the traditional approach, policy advice flowed freely from the findings from agricultural research in published journal articles and papers that drew policy implications and recommendations. As this approach treated policy advice as a residual matter, the incessant failure of agricultural policies resulted from improper formulation and implementation, hence poor transformation of the agricultural sector.
- o In the subsequent approach, policy problems themselves formed the focus of scientific research, leading to valuable results and findings fed directly as recommendations and advice usually through technical briefs. As the outputs from this approach were not properly assimilated by policy authorities, policy failure manifested in terms of instability and discontinuities that generally marred the transformation process.
- o In another approach, policy backup support took the form of technical assistance for articulation on paper in order to pin government down to its policy commitments and to subject public authorities to the need for policy responsibility, policy transparency, policy accountability, and policy due process. Yet still, government failed to honour such commitments as evident in the attitude to discountenance the fertilizer policy and other policies recently articulated.

Against this background, this work is structured into four sections. First, we define policy within the context of agricultural transformation. Second, we highlight the stages of the policy process for agricultural transformation in order to show the weak points responsible for policy failure in the process. Third, we describe the main elements of policy advocacy as the missing role in the process. Last, we illustrate with a policy advocacy project recently undertaken by Farm and Infrastructure Foundation, an NGO for promoting policy best practices in agriculture and rural development.

POLICY FOR AGRICULTURAL TRANSFORMATION

To begin with, the technical meaning of policy itself is at the centre of people's understanding about agricultural transformation. The agreed working definition is as follows: that policy is a statement of intent, action, or inaction by the public authority in response to the need to achieve a predetermined purpose, which may or may not be formally declared. Thus, policy is a *statement* of the mindset of government, i.e. a matter of *what?* This contrasts with strategy as a *formulation* of how to implement the policy, i.e. a matter of *how?* Thus, a policy provides the overall framework which determines a government's aim and activities, wherefrom the different modes of strategy formulation emerge as applicable, in terms of plans, programmes, and projects as the means of implementing policy. In this context, the policy objectives follow a hierarchy from the highest level of aggregate (societal goals) to lower levels of aggregate (macroeconomic objectives) and to much lower and lower levels of aggregate, such as sectoral (e.g. agriculture sector) and sub-sectoral (e.g. fertilizer sub-sector) objectives. According to Kentish (2002b), 'when contributions are not expressed in a policy format, there is the very real likelihood that certain critical areas that could make or break the policy process are ignored.' In practice, however, the elements of the strategy formulation or planning are taken as not only consistent with the current policy but also synonymous with the policy itself.

For agricultural transformation to take place, certain elements of the definition of policy should be taken seriously. The first is that policy is essentially an intervention, usually in the market for goods and services, and for which the expected output or outcome would not have been the result without the policy intervention. The second feature of the definition adopted is that policy may or may not be formally declared, which leads to typology of policy into implicit and explicit categories. The third element to note is that the word 'inaction' in the definition of policy is not redundant, and it recognizes the fact that on a particular policy issue and at a particular time or stage in the policy process, what the government does or says (policy of intervention) is as important as what it does not do or say (policy of non-intervention). Thus, even in the event of no statement or action, there will be definite (policy) consequences of a passive stance taken by the government on any development issues. The fourth element to note in the working definition is the existence of a purpose or objective underlying any policy action or inaction, whether it is formally stated and publicly declared, or not, hence the categorization of the typology of public policy objectives into two, namely implicit policy objectives and explicit policy objectives. For instance, for good or for bad reasons, the government might choose not to reveal the real motive behind its policy stance at a particular time if it perceives a potential negative effect on public opinion.

Thus, the real policy objectives may be revealed only after implementation has begun or has been completed, and these may be slightly or radically different from the initial statement of objectives. For example, substantial policy benefits meant for the farming population may, in practice, accrue instead to politicians, civil servants, and the military, and other urban elites, to the knowledge of the government, but policy authorities may not act against it if the next general election is close by. This poses a problem to analysts about how to identify the unwritten, implicit things in the mind of government in order to unravel the rationale and build a useful database on such policies.

On the other hand, strategy means strategy for implementing the policy. As in a military strategy, the steps towards the formulation involve charting the course of action and staying on the course of action, with concentrated focus on the target of action, with the motive of meeting the stated policy objectives. Critical to this is the careful design of the course of action in great detail, including a series of *plans* to be designed, such as entry plan, work plan, action plan and exit plan, as highlighted below.

- **Entry Plan**: How to properly enter the implementation stage (phasing of implementation, stakeholder involvement, security of fund for implementation, guaranteeing full participation, etc.)—a good entry plan will ensure a safe landing upon entry in order to avoid the usual delays in project start-up through unexpected problem with beneficiary communities.
- **Work Plan**: How to properly mobilize resources (money/men/materials) in the long/medium term (perspective characteristics, activity schedules, timeline, log frame/s, etc.). A good work plan helps in properly sequencing the activities and facilitates monitoring and evaluation work.
- **Action Plan**: How to engage money, men, and materials in the immediate/short term (budget, safeguards, role assignment, etc.). A good action plan helps to guide the role of actors and to nip unexpected implementation problems in the bud.
- **Exit Plan**: How to rescue the plan when trouble looms ahead, i.e. in the event of a severe stress condition and the danger looms large that a plan will not survive till the end, what arrangements are in place for it to possibly recover or for damage control? The options include retirement option, bailout option, termination option, liquidation options, etc. A good exit plan will ensure that project sites are not abandoned prematurely without recovering the salvage value of structures or substantial portion of the sunk costs one way or another.

The issue here is that our agricultural transformation strategies are often formulated without a good entry plan, which sometimes leads to a project being dead on arrival. Also, they are often formulated without an exit plan, leading to severe damage to beneficiaries when the unexpected problems of implementation emerge as usual. Poor articulation or complete absence of strategies with clear action steps that ought to have been well simulated under different/several built scenarios (with assumptions) and properly analyzed with prospective outcomes before actual implementation has been the norm.

Further to conceptual matters, the need arose to make a technical distinction between the discrete approach and the process approach to policy analysis. The former reflects the traditional view of policy as a sequence of discrete events in decision-making, while the latter represents the contemporary view of policy as a sequence of continuous events in the process. Making this distinction represents a paradigm shift from the previous focus on the static properties of policy decisions to incorporate the dynamic character of policy into public decisions, which creates an opportunity to explore agricultural policymaking for causes and consequences of persistent policy failure in terms of the interactive relationships at the various stages of the policy process. As noted by Idachaba (2002b), 'these failures have persisted from year to year, regime to regime and country to country, as if policymakers are not capable of learning from their own mistakes or from mistakes of others.'

POLICY PROCESS FOR AGRICULTURAL TRANSFORMATION

The process approach reveals the need to focus more on both demand and supply sides of policy. Traditional policy analysis is based on physical, economic, and social research as sources of policy recommendations as a residual matter whereby the emphasis is put on sources of supply of policies namely political leaders, bureaucrats, and consultants rather than on the demand side of policies such as the small and large-scale farmers, agro-processors, and consumers. On the other hand, investigation anchored on the policy process involves considerations about the demand side, with focus on stakeholders such as supply farmers, food processors, food marketers, etc. and explores issues about them in greater depth.

Thus theoretically, policy process is brought about by a sequence of some (earlier) events and is followed up by other (later) events. Each event represents a stage in the policy process, and the various stages interact with one another in forward and backward linkages, leading to thrusts and feedbacks that produce a cobweb of activities that are non-linear and inherently cyclical. Hence, policy cycle or a policy web is a technical term describing the interrelated actions of the public sector, linked with actions of the private sector, to bring about certain production and consumption patterns in the economy, which otherwise would not have been the case without such actions.

The highlight of the five stages of the policy process for agricultural transformation is as follows:

- **Identification:** Policy identification involves recognition of the existence of an agricultural problem requiring an appropriate policy response or action; the issues include improper diagnosis of causes, characteristics and consequences of policy problems, lack of ordered prioritization of policy problems, lack of recognition of a problem, and a long time lag between a problem emerging and recognition of the existence of the problem. This depends on several factors, such as availability of information relevant to the policy, availability, and quality of mechanisms for sensitizing the policymakers to the existence of the problem; capacity of stakeholders to influence the policy agenda; and capacity and independence of the media as sources of information about emerging policy problem areas.
- **Formulation:** Policy formulation is a highly technical stage consisting of the articulation (design and formulation) of the various components of the policy, including objectives, instruments, strategies, time phasing, and financial implications. The issues are clarity of objectives; tenability of policy options (including their underlying assumptions); choice of strategies and instruments; coherence of plans, programmes, and implementation stages; and construction of monitoring and evaluation indicators. The overriding issues include the time lag between recognition of a policy problem and formulation of an appropriate policy response, which depends on several factors, such as technical capacity for policy formulation in ministries of agriculture and other government agencies as well as independent sector agencies, consultative arrangements, and policy information exchange between ministries of agriculture and other government agencies, on the one hand, and farmers' organizations, research institutions, and independent consultants on the other, as well as mechanisms for sensitizing policymakers to identify problem areas requiring a policy response.
- **Appraisal:** Policy appraisal involves verification or review of objectives, assumptions and instruments, review of both intended and unintended effects, review of resource availability and risk analysis. The issues peculiar to this stage are: the failure to appreciate the place and role of policy appraisal and verification before implementation; the lack of clear information on the benefits of policy appraisal before implementation, and the tendency for governments to be in a hurry

to show that they are responding to an identified problem before the expiration of their political tenure.
- **Implementation:** Policy implementation is the action stage wherein resources are mobilized to realise the policy objectives. The most important issues at this stage are effective mobilization of human and financial resources, appropriate institutional arrangements, specification of performance indicators, development of a work programme and a budget, and the role of monitoring and ongoing evaluation. The overriding issues are the non-linear nature of the path from policy design into policy decision, the lack of strong and smoothly functioning partnerships and bridges between policy analysts, designers, and formulators during implementation on the one hand, and government decision makers on the other and the difficulty in getting decisions made through policy advice, lobbying and advocacy as implementation progresses.
- **Evaluation:** Policy evaluation (including policy impact assessment) entails looking back at different points of implementation. The challenges are suitable methodology, impact indicators, input from monitoring, identification of positive and negative impacts, and conclusions on the extent to which policy has achieved its goal. The common issues are hasty policy designs that make no provision for monitoring and evaluation, failure to specify performance indicators that will help track achievements and constraints during the policy implementation stage, inadequate identification of sources and types of information required for monitoring and evaluation, and the general lack of a monitoring and evaluation culture in the agricultural policy process. Other issues include lack of appreciation by key stakeholders; the need for policy evaluation and impact assessment; failure to distinguish policy, programme, and project outputs from policy impact and analytical difficulties in conducting impact studies such as attribution and long time frame; and failure to generate and transmit the needed information for policy impact assessment.

ROLE OF POLICY ADVOCACY

The following can be said about policy advocacy in theory and practice:

1. **Concept of policy advocacy:** The key words include support, encouragement, backing, sponsorship, promotion, activism. The issues include given its position as an interested party in governance,

the government cannot be expected to advocate policies in the best interest of the people at all times. Therefore, a neutral body is required as partner to government and as watchdog of government. A platform must be created for farmers to express their policy interests and to give them a voice and a vote in the policy process.
2. **Steps in policy advocacy:** The calculated steps include analysis work in terms of background research; regular consultations in terms of close interactions and good relationship with policy authorities and policy stakeholders; negotiations with policymakers; brokering services in terms of policy articulation, formulation, verification, implementation, monitoring and evaluation; lobbying (which involves using the power of persuasion to push for decisions and actions for policy adoption); and sometimes pressure in terms of mass action by peaceful means to tie the hands of policy authorities behind their back.
3. **Lessons learnt:** That there are several enemies out there; that government and the people harbour many conflicts of interest; that international agencies act too cautiously for fear of running on a collision course with host government; that farmers and other associations have a weak voice and suffer leadership problems; that professional bodies are non-committal, grandstanding in posture, and self-promoting in their activities; that the private sector engages mostly in patronage and pursuit of business interest; and NGOs, which have high potential to address the policy problem, are unconsolidated in their efforts and weak in their capacities.
4. **The risks involved:** That the terrain for policy advocacy work is like a banana peel, very slippery. The critical question is, If and when the policy advocate runs into trouble with powerful policy authorities, who bails him out?

AN ADVOCACY PROJECT

The Farm and Infrastructure Foundation (FIF) organized an advocacy event in 2006. It was a mass action for the establishment of a fertilizer regulatory system in Nigeria (Figure 2). For the purpose, the farmers and agri-input dealers' associations were mobilized to stage a peaceful demonstration at the 2006 meeting of the National Council on Agriculture in Abeokuta, Ogun State. For the purpose, two *masquerades* (otherwise known as mascots) were hired to carry placards before the Minister of Agriculture at three locations. The peaceful demonstration was a success at Oba's palace, where the minister paid a courtesy call on the paramount ruler, the Alake of Abeokuta, who also lent his word in support. The demonstration was not so successful at the Governor's Office, where the minister was also expected to pay a courtesy call on the state governor, as the mascots were prevented by policemen from entering the protected premises. The demonstration turned sour at the venue of the NCA meeting proper, where the advance party of security detail in the convoy of the governor and minister were directed to dislodge the demonstrators.

Figure 1: A policy advocacy event for the establishment of a fertilizer regulatory system in Nigeria, Abeokuta, Ogun State 2006

CONCLUDING QUESTION

Obviously, this case helps in determining the potential role of policy advocacy in agricultural transformation in Nigeria. However, the all-important question is, Suppose the Abeokuta experience had degenerated into the arrest and detention of the policy advocate with his peaceful demonstrators who were there to bail him out?

35

Trends And Observations In The Funding Of Agricultural Research In The Naris And Fcas

INTRODUCTION

In Nigeria, like many other countries of the world, the inherited institutions for agricultural research were established and operated with the public budget as the norm. This is largely premised on the nature of the typical output of agricultural research as a public good. The theory holds true that such an output exhibits a non-rival consumption which means that one person's consumption of the output does not reduce its availability to anyone else at an extra cost, hence the presence of free riders in the market for research output, which constitutes a disincentive to the private sector to invest in agricultural research. Also, such an output has the characteristic of non-excludability, which means that the producer is unable to prevent anyone else from consuming it, hence the inability of the seller to make every consumer pay for the product as required, which further inhibits the development of private market for agricultural research. The combined effect of non-rival consumption and non-excludability makes the funding of agricultural research what it is, a traditional line item in the public budget.

However financing of agricultural research through the public budget in Nigeria is fraught with problems of inadequacy and irregularity. As a result, the institutions established to conduct agricultural research in the country have entered a process of progressive decay in terms of infrastructure and staffing over the years. Towards addressing this undesirable trend, the Federal Government implemented the National Agricultural Research Project (NARP) with World Bank assistance between 1994 and 1999, with a view to revamping the facilities in the National Agricultural Research System (NARS). Nonetheless, the initial impact of this project on the National Agricultural Research Institutes (NARIs) has started to wane.

In discerning the trends and articulating the observations about the funding of agricultural research in NARIs and FCAs, this paper draws from three past research works to provide the empirical basis. The first research work was conducted at the University of Agriculture Makurdi under the ASTI survey (Agricultural Science and Technology Indicators) published by IFPRI/ISNAR (Beintema and Ayoola 2004). The second is the earlier research conducted also at the University of Agriculture Makurdi under the MLAE (Middle Level Agricultural Education) project. Though this project was inconclusive, some aspects of the information generated have been published (see Ayoola 2004). The third is the TES (Techno-Economic Surveys) on the food industry conducted under the auspices of the Raw Materials Research and Development Council (RMRDC) (Ayoola et al 2000).

On these bases, this paper highlights the main features of the funding system for financing agricultural research in Nigeria and explores alternative mechanisms for consideration by the authorities of Agricultural Research Council of Nigeria (ARCN). In the paper, it is argued that unless innovative ways of financing agricultural research are quickly found, the public policies for achieving the research mandates of NARIs and FCAs will be mere platitudes. The main conclusion reached is that a forum is urgently required to bring together all the stakeholders in the financial market for agricultural research, with a view to identifying available opportunities and building cooperation and consensus among them about the way forward.

The first section of the paper deals with empirical trends and patterns of funding of the NARIs. The second section deals with some empirical observations about the funding of FCAs. The third section highlights the alternative mechanisms for funding agricultural research, followed by the section on the way forward for NARIs and FCAs. In conclusion, we make a clarion call for the convocation of a *Fund Sourcing and Revenue Generation Conference* together with some inherent policy implications.

Discernible trends and patterns of agricultural research funding in the NARIs

The findings from the ASTI study (Beintema and Ayoola, op. cit.) indicate that

1. The real burden of agricultural research in the country weighs more heavily on the NARIs than the agricultural universities and faculties, accounting for two-thirds of the country's total researchers. The 22 research institutes studied employed a total of 840 full-time equivalent (FTE) researchers representing 38 FTE researchers on average; in terms of distribution, five institutes employed more than 50 FTE, 16 employed between 10 FTE and 25 FTE, while one employed fewer than five FTE.[57] On the other hand, the burden of agricultural research weighs less on the higher education institutions comprising the universities of agriculture (UAs) and faculties of agriculture (FAs) in conventional universities, which collectively accounted for one-third of the total researchers and spending. The 59 institutions of this category studied employed 513 FTE researchers in agriculture, representing 10.3 FTE researches on average; the universities employed 10.53 FTE researchers on average, with only three out of them (older ones—UI, ABU, UNN) having employed between 25 and 50 FTE researchers in agriculture; more than half of them employed fewer than five FTE researchers, and nearly a quarter employed fewer than just two FTE researchers.

2. Funding for agricultural research primarily comes from the (federal) government; total spending on agricultural research declined from about $130 million in the mid-1970s to less than $50 million in 1990. Spending per scientist declined by half from $171,000 in 1971 to $78,000 in 2000; research investment declined from $0.81 per $100 of agricultural GDP in 1981 to $0.38 in 2000; in 1995, the ratio was $0.16, which was considerably lower than the average ratios for Africa and developing world ($0.85 and $0.62 respectively). During 1991–2000, staff salaries in 14 NARIs studied accounted for 55% while operational costs accounted for 22% and capital costs accounted for 23%, on average. In 2002 for instance, NIHORT's operational funding amounted to about N700,000 but its yearly electricity bill was N550,000, indicating gross inadequacy of recurrent funding. This led to many of the institutes resorting to using their capital budget meant for conducting experiments, training, and other research expenses, to pay utility bills and other fixed costs.

---

[57] One full-time equivalent researcher is one scientist whose full-time employment as a researcher has been adjusted for non-research engagements such as teaching and administrative duties.

3. The contribution of internally generated revenue to spending in NARIs is very limited. In the year 2000, six out of 14 institutes studied for this purpose contributed to their own income, but only in small amounts averaging 6 per cent.
4. The involvement of private sector in financing agricultural research is negligible. Results of the techno-economic survey indicate that R & D is practically inactive in respect of the FBT (Food, Beverage and Tobacco) industries (RMRDC 2000 op. cit.).

The problem with the funding of agricultural research in Nigeria is that when public budget allocations decreased, there were no viable substitutes for keeping the NARIs functioning properly as before. Thus, the major challenge posed is how to diversify sources of funds and increase the volume of funds for NARIs for conducting meaningful research.

Funding of FCAs

The survey of middle-level agricultural education institutions which include the so-called federal colleges of agricultural (FCAs) under focus was conducted in 1991 (Ayoola 2001 op. cit.). The middle-level institutions studied comprise schools and colleges of agriculture awarding National Diploma and Higher Diploma in different programmes, like General Agriculture to Agricultural Extension/Farm Management, Crop Production Technology, Animal Health Production, Home Economics, to mention a few. Prominent among the colleges are located at Ibadan, Samaru, Kabba, Lafia, Bakura, Umudike, Akure, Ikorodu, and Yandev.

The findings from the MLAE study indicated that the knowledge infrastructure in the agricultural colleges was in a bad state. This reflects poor funding by budgetary instruments and lack of alternative modes of funding through the private sector or internally generated revenues (Table 1). Thus, financial provisions for these colleges in the public budget are grossly inadequate. This prevents them from acquiring the critical mass of infrastructure for effective training of middle level technical staff required for research in the NARIs.

Table 1: Knowledge Infrastructure in Selected FCAs, 1991

	Schools/Colleges of Agriculture*					
	Ikorodu (400)	Ibadan** (320)	Akure (100)	Kabba** (480)	Lafia	Samaru (407)
Farm facilities: Meteorological station	-	n.a.	+	+	n.a.	+
Survey equipment	-	n.a.	+	+	n.a.	+
Pest control equips.	-	n.a.	+	+	n.a.	+
Farm machinery/tool	-	n.a.	+	+	n.a.	+
Irrigation equips.	-	n.a.	+	-	n.a.	+
Nursery facility	-	n.a.	+	-	n.a.	+
Greenhouse	-	n.a.	-	-	n.a.	+
Crop farm	-	n.a.	+	+	n.a.	+
Animal farm	-	n.a.	+	+	n.a.	+
Fish pond	-	n.a.	-	+	n.a.	=
Library facilities: No. of seats	70	92	120	256	27	85
Space: Classroom No.	4	2	7	6		
Lecture	0	0	0	0		
TheatreNo.	1	0	20	16		
Office Rooms	5	4	1	11		
Others						
Staffing: Principal Lecturer***	0	3	0	0	1	0
Senior Lecturer	1	2	3	1	1	2
Lecturer	0	4	4	4	4	4
Asst. Lecturer	0	1	0	0	5	6
Senior Technical	0	5	8	8	2	8
Farm Manager	0	0	0	0	0	0

*: Figures in parentheses are the student populations

Note: + = Available; - = Not Available; ** = Awards HND in addition to ND certificates, while others award ND only; *** = Additional staff required for HND.

ALTERNATIVE FUNDING MECHANISMS AND THE WAY FORWARD

Public Sector Funding

As stated earlier, both the NARIs and FCAs derive their primary funding from public budgets which faced severe diminution consequent upon the implementation of an economy-wide Structural Adjustment Programme (SAP) since 1986. The SAP sought to privatise several aspects of the economy, including the knowledge infrastructure in NARIs and FCAs. According to Byerlee and Echeverria (2002), budget austerity as a result of SAP results from competing needs which put the public institutions in financial crisis with high fixed costs that squeezed operating budgets, with the severe consequence being a downwards pressure on the real salaries of scientists. In particular, the research institutes were unable to respond to budget austerity through cost-cutting measures and diversification of fund sources.

The funding situation with the NARIs and FCAs is further compounded by the frequent instances of poor budget commitments whereby the actual amounts released are far less than the amounts allocated in the budget. In such instances, government exhibits a tendency to meet recurrent commitments at the expense of capital allocations, as found by ASTI survey. Moreover, even in more recent times, budgetary allocations to NARIs and FCAs are lopsided in favour of recurrent items at the expense of capital allocations (see Tables 2 and 3).

Table 2: Federal Budget Allocations to Agricultural Research Institutes in 2006

SN	NARI	Total Allocation	Personnel Cost – General	Capital
1	NATIONAL CEREALS RESEARCH INSTITUTE, BADEGI	473,969,988	466,113,043	-
2	NATIONAL VETERINARY RESEARCH INSTITUTE, VOM	1,043,070,122	1,043,070,122	-
3	NATIONAL ROOT CROPS RESEARCH INSTITUTE, UMUDIKE	1,031,127,435	828,518,607	-
4	NIFOR, BENIN CITY	861,383,935	771,523,954	30,000,000
5	INSTITUTE OF AGRICULTURAL RESEARCH, ZARIA	503,109,220	488,370,063	-
6	NATIONAL ANIMAL PRODUCT RESEARCH INSTITUTE, ZARIA	355,555,544	313,792,652	15,112,005
7	NATIONAL HORTICULTURAL RESEARCH INSTITUTE, IBADAN	529,591,275	442,395,125	74,300,000
8	LAKE CHAD RESEARCH INSTITUTE, MAIDUGURI	258,763,547	160,739,656	85,846,151
9	NIOMR, LAGOS	289,514,374	277,837,506	-
10	COCOA RESEARCH INSTITUTE, IBADAN	409,984,600	388,955,412	-
11	INSTITUTE OF AGRICULTURAL RESEARCH AND TRAINING, IBADAN	468,733,845	393,824,456	59,391,000
12	RUBBER RESEARCH INSTITUTE, BENIN	442,162,901	374,773,974	18,857,000
13	NATIONAL INSTITUTE OF FRESH WATER FISHERIES, NEW BUSSA	389,874,192	275,887,642	40,192,000
14	AGRIC. EXTENSION RESEARCH SERVICES, ZARIA	238,043,010	238,043,010	-
15	NIGERIAN STORED PRODUCTS RESEARCH INSTITUTE, ILORIN	392,469,341	275,098,150	93,220,000

Source: National Bureau of Statistics

Table 3: Federal Budget Allocations to Federal Colleges of Agriculture in 2006

SN	FCA	Total Allocation	Personnel Cost – General	Capital
1	FEDERAL COLLEGE OF HORTICULTURE DADINKOWA GOMBE	187,483,733	147,548,200	19,000,000
2	FEDERAL COLLEGE OF ANIMAL HEALTH AND PRODUCTION TECHNOLOGY, IBADAN	198,732,859	148,665,796	30,000,000
3	FEDERAL COLLEGE OF AGRICULTURE AKURE	182,206,693	131,232,268	29,780,000
4	FEDERAL COLLEGE OF AGRICULTURE MOORE PLANTATION, IBADAN	194,931,497	149,799,174	25,000,000
5	COLLEGE OF AGRICULTURE, ISHIAGU, UMUDIKE	264,086,227	247,838,987	-
6	FRESHWATER FISHERY COLLEGE, NEW BUSSA	114,043,942	85,327,942	39,000,000
7	CO-OPERATIVE COLLEGE, IBADAN	87,981,488	63,437,923	-
8	CO-OPERATIVE COLLEGE, KADUNA	68,844,621	47,312,489	-
9	CO-OPERATIVE COLLEGE, OJI RIVER	67,356,269	38,190,027	26,200,000
10	FEDERAL COLLEGE OF LAND RESOURCES TECHNOLOGY, OWERRI	75,206,862	64,097,106	8,400,000
11	FEDERAL COLLEGE OF LAND RESOURCES, KURU JOS	104,523,944	53,747,917	40,768,000

Source: National Bureau of Statistics

Private Sector Funding

Notwithstanding the predominant public good nature of research results, the profit maximization motive compels private enterprises to embark on research and development (R&D) in respect of their products. This is often pursued through sponsorship of researches that promote productivity on the farm and enhance efficiency in the market. Indeed, sometimes the boundaries between public and private good character of research products are unclear, giving rise to the notion of some form of research outputs as having a 'hybrid good' nature that makes these belong in either or both of the two domains of funding.

Thus, private buy-in to research exists in specific cases within two main contexts. Firstly, it pays the private concerns to fund research for continuous improvements of their goods and services in order to operate at the cutting edge of technology and to be able to create special market niches for themselves. In this case, private companies tend to use own funds in conducting research into agricultural technologies that are substantially exclusive to individual companies, in a bid to prevent the possibility of free riders tapping from the results obtained. A copious example of exclusive research of this nature frequently occurs in seed industry, wherein patent rights may be obtained by private companies that sponsored the breeding research, or the companies involved tend to reserve the improved variety for their exclusive use. The recent certification of new sorghum varieties sponsored by the Nigerian Breweries Plc. is a case in point.

Secondly, private companies are under an obligation to demonstrate social responsibility by contributing part of their incomes to research in their spheres of operation, in which case the notion of public-private partnership applies. A familiar instance of this is farmer financing of agricultural research through commodity taxes or levies. In the colonial and immediate post-independence era, commodity taxes on commercial crops were combined with income stabilization to build and operate the knowledge infrastructure required for agricultural research. Though the erstwhile commodity board farmers agreed to pay involuntary taxes in return for mechanisms by which their income would be stable over time, the idea created an opportunity of financial reserves purportedly used to fund research and other agenda. Commodity levies specifically for research may also be extended to the agro-based and agro-allied industries that use agricultural products as raw materials.

The era of involuntary agricultural tax approach came to an end in 1986 when the commodity boards were abolished at once (Idachaba and Ayoola 1991). This led to an era of no mechanisms for recouping from the farm population some of the expenditures on agricultural research financed with public budgets. Yet, the theoretical argument holds true that it is equitable for the beneficiary of research to pay proportional to production, in order to increase total resources available and for sustained funding. Although a set of three commodity development and marketing companies were subsequently established, their modus operandi are not clear in terms of contribution to research in the respective commodity arenas. Nonetheless the main limitation of the approach pertains to absence of a cost-effective means of collecting the levy, especially in agriculture that the products pass through informal marketing channels. It can be shown in most cases that the cost of the levy is proportionately distributed between producers and consumers of agricultural products.

By and large, case studies have shown that research institutes funded by commodity groups generally have more funding, better-paid scientists, sufficient operating costs, and better management. The main problem in Nigeria is, What the best model to apply? Some of the major elements of a good model of voluntary or mandatory levy include the following:

- Enabling legislation to make the levy mandatory in order to avoid the free-rider problem;
- Specification of a matching grant from government proportional to production;
- Joint administration of levy by government and private sector;
- Use of levy for specific purposes in aid of research, e.g. commodity research station either as a completely private entity (such as sugar cane research stations in Colombia and Zimbabwe) or as a parastatal with considerable autonomy (as in Uruguay);
- Levy funds may be administered by an autonomous R & D public corporation or specialized institution that allocates the funds to research institutions through contracting and competitive bidding.

Such partnerships often take the form of contractual arrangements in which the roles of each party are specified. A basic requirement for effective PPP is an irreducible minimum capacity in both the public and private sectors coupled with a set of clear policies and procedures to ensure transparency and accountability. In this connection, it is observed that a comprehensive national policy on agricultural research for Nigeria is not in place, wherein to specify the parameters for PPP, among other components.

Agency funding of agricultural research

Agency funding of research is defined in terms of the substantial activities of local and international bodies corporate that vote money for conducting research as part of their various programmes in the country. At the local level, such bodies include government agencies seeking to investigate issues in their own spheres of operation that require the services of researchers from institutions dedicated for research, such as NARIs, universities, and private research outfits or individuals; some of these include the ministries, namely the FMAWR, FMCI, as well as the parastatals such as RMRDC, NPC, etc.

At the international level, there are several bilateral and multilateral agencies providing official development assistance (ODA), such as USAID, DFID, CIDA, JICA, World Bank, Food and Agriculture Organization (FAO), International Fund for Agricultural Development (IFAD), Common Fund for Commodities (CFC), etc. Also present in the country are several NGOs that are of local, national, regional, or international origin. Some of these provide research funds in focused areas (e.g. CFC) while others operate fund for general purposes. For the most part, bilateral agencies operate funds for general development with direct or indirect components for agricultural research. The requirements for accessing international research money are the openness on the part of agencies to give information to the research community about the research content of their programmes and the dexterity of the research institutions and individuals to access the funds; whereas both aspects are crucial in the efforts to improve the funding of agricultural research in Nigeria, there is presently no alliance between the two sides to give effect to the complementarities.

For the most part, the arena of agency funding of agricultural research lacks proper coordination and organization for the benefit of NARIs. What we have is an army of portfolio researchers penetrating the agencies in the usual tradition of 'any job for the boys'. This has happened for so long that the agencies have almost forgotten the existence of NARIs as the principal actors in the agricultural research market. Thus, it is up to NARIs to rise up and reverse this trend and assert their roles in the research market. The issue here is that for the most part, the quantum of money available through agency funding of research accessible through competition, whether formally through media advertisements or informally by looking over the shoulders of the local and international bodies. This requires that NARIs should build their own capacity for competitive competence in assessing such funds through writing fundable proposals for consideration by the agencies and through a shrewd instinct for fund lobbying from these bodies as others do. The consequence of this not being so is what is already happening to NARIs as backbenchers in the arena of agency funding of research, against the expectation that they should take the driver's seat. We can easily make the following observations to substantiate this statement: That

- whereas government projects such as the Fadama Development Project and others often undertake critical baseline studies and other planning research activities, the NARIs do not play any significant role like this;
- whereas many local agencies and ministries contact research works frequently on issues bordering on agricultural systems and commodities, it is uncommon to see the role of NARIs in such activities;
- whereas several commodity associations with activities bordering on commodity research exist in the country, the NARIs have not established a strong relationship with them for the purpose of servicing their commodity research needs to the exclusion of other institutions;
- whereas the international agencies such as Common Fund for Commodities operate definite commodity research agenda, NARIs have not actively competed for this to a significant extent.

Thus we observe that the posture of NARIs needs to change from a set of traditional research institutions waiting for research funds to come, to a set of modern-day research institutions that are proactive and competitive in the new world market for research funds. This requirement for an attitudinal change leads us to the way forward for NARIs and FCAs in sourcing and revenue generation in order to implement their mandates better in future.

THE WAY FORWARD

The way forward is aggressive fund sourcing and revenue generation in the NARIs and FCAs. The NARIs and FCAs have tried their hands in several modes for sourcing funds and generating revenues to augment the budget provisions, which have not yielded significant results in terms of incremental funds for improving the staffing and infrastructure, let alone in conducting meaningful research.

Fund sourcing refers to deliberate effort of the research institutions to explore sources of fund from the stakeholders rather than depending only on regular budget allocations from government. The scope for sourcing large funds for research exists in the private sector and international agencies in research and development. The response of stakeholders in these categories will depend on several factors, particularly the deliberate effort to harness the capacity of companies and agencies for research funding.

Revenue generation refers to the effort of the institutions to convert their resource base and output of research to viable enterprises. The scope for revenue generation in NARIs and FCAs depends on the degree of entrepreneurship and innovativeness that exist in these institutions. Some potent sources of additional revenue in NARIs and FCAs include making available the slack and idle capacities of staff and facilities in meeting the demand for specialized services from the public, tapping the special professional capacity of scientists for use by government and private sector, sale of products from research and student farms to the consuming public, publishing and marketing of intellectual outputs of scientists such as books and monographs, holding patents and other rights on the intellectual properties emanating from researchers, to mention a few.

The magnitude of revenue generated by these means will depend on the organizational ability of the people at the helm of affairs in the NARIs and FCAs, coupled with the dexterity with which the leadership of the institutions pursues commercialization of research results. Specifically, the revenue-generation activities often work better when they are well organized into formal consultancy outfits and production or marketing enterprises. In this regard, agricultural production which belongs in the real sector of economy harbours a wide range of products for meeting the demand of society for food and specialized services. Besides, it is imperative that research institutions establish such consultancies and enterprises as outlets for demonstrating the best practices for which they are to be known to the general public, in facilitating adoption of new techniques and technologies to further drive the economy.

CONCLUSIONS AND POLICY IMPLICATIONS

The major challenges facing research funding in NARIs and FCAs pertain to frequent budgetary austerity, persistent inadequacy of budget commitment, unmotivated private sector. and unorganized arrangements for agency harnessing of available funds. Thus for a long period of time, the predominance of public funds has exposed the NARIs and FCAs to prolonged financial crisis, which leads to progressive institutional decay and low performance in carrying out the research mandates. The situation is exacerbated by the inability of the institutions to convert their resource base and outputs into viable consultancy services and enterprises to generate revenues to augment budget allocations.

The present retreat of the people at the helm of affairs in NARIs and FCAs is an important step towards the proper analysis of the problem. The retreat will help in identifying the main issues involved in meeting these challenges under the directions of the Agricultural Research Council of Nigeria.

Towards this end, we propose the convocation of National Conference on Fund Sourcing and Revenue Generation for NARIs and FCAs in Nigeria. This would provide the right forum for bringing all stakeholders in the market, R & D financing together to discuss and resolve the issues. The NARIs and FCAs will use the opportunity to present the various dimensions of the funding problem to the public and private sectors and the international agencies. The public authorities will seize the opportunity to improve their understanding of the problem in the various dimensions and justify the need for stepping up budgetary provisions through appropriate means, including considerations about the forthcoming National Agricultural Development Fund (NADF). The private sector will seize the opportunity of the conference to determine the appropriate windows for effectively channelling private funds into research and development in NARIs and FCAs. Also, the international development community to be invited will have the opportunity to research content of their programmes in Nigeria and abroad for the benefit of NARIs and FCAs to log on to them as applicable. Moreover, the major output of the conference will be the consideration of forming an alliance of the stakeholders for regular interactions on the issue of funding of research in the country in order to monitor the honouring of commitments made by stakeholders.

In this regard, we would like to recognize the current initiative for the convocation of the National Conference on Fund Sourcing and Revenue Generation in NARIs and FCAs. The initiative was proposed to the Federal Ministry of Agriculture and Water Resources earlier in the year, by the FIF, *Farm and Infrastructure Foundation*, an organization for the promotion of policy best practices and brokerage services in agriculture and rural development. And with the instrumentality of the Department of Agricultural Sciences, the National Conference was approved by the Honourable Minister of Agriculture and Water Resources, Mallam Adamu Bello (CFR), for it to be held during the first quarter of the year. However, this was postponed at the last minute on the good reasons for such an activity to take place under the auspices of ARCN after settling down properly. Therefore, we suggest that a new date could be considered and determined by participants at this retreat for the conference to be held without further delay.

The policy implication of the National Conference is that it will not only help in setting the agricultural research agenda for the new government but also start up a new programme for the sustainable funding of the NARIs and FCAs in the country, using the budgetary and non-budgetary windows. This relates to the pending issue of National Agricultural Development Fund (FIF 2007), which the current debate on agricultural research funding should specifically focus upon. Moreover, through the national conference, new opportunities will emerge for the domestic and international policies to accommodate innovative ways of revenue generation consistent with the best practices for similar purposes in other parts of the world in these institutions. All these should constitute the elements of a comprehensive *national policy on agricultural research for Nigeria*, which should be articulated at this stage in defining the road map for the new ARCN.

References

Ayoola, G. B. (2002), 'Agricultural Education in Nigeria: Problems and Prospects', In G. B. Ayoola, *Essays on the Agricultural Economy I: A Book of Readings on Agricultural Development Policy and Administration in Nigeria* (Ibadan: TMA Publishers).

Beintema, N. M. and Ayoola, G. B. (2004), *Agricultural Science and Technology Indicators, ASTI Country Brief* (Nigeria: IFPRI-ISNAR).

Byerlee, D. and Echeverria, R. G. (2002), 'Agricultural Research in an Era of Privatization: Introduction and Overview', in Derek Byerlee and Ruben G. Echeverria (eds.), *Agricultural Research in an Era of Privatization* (CABI Publishing).

FIF (Farm and Infrastructure Foundation) (2007), *New Agric. Digest* 1 (4) (July).

Idachaba, F. S. and G. B. Ayoola (1991), 'Market Intervention Policy in Nigerian Agriculture: An Ex-post Performance Evaluation of the Technical Committee on Produce Prices', *Development Policy Review* 9 (3).

The Role Of Agricultural Commodities In Nigeria's Economic Development

Abridged version of a paper presented at the Sensitization Workshop on Common Fund for Commodities (CFC) Activities organized by the Federal Ministry of Commerce and Industry, Kano, Lagos, Calabar (26–30 March 2007).

In Nigeria, as also in all other parts of the world, agricultural commodities play the important role of building blocks of the economic superstructure, which sustains the lives of the people from one historical era to another. Indeed, the country owes much of its precolonial, colonial, and postcolonial history to commodity trade.

The establishment of British foothold in the south of the country emanated from the early forms of commodity relationships between the European merchants who exchanged their own primary commodities such as salt for other primary commodities such as pepper, elephant teeth, palm oil, among other items found along the coast of the Atlantic Ocean. Subsequently, the colonial government engaged in a systematic introduction of new agricultural commodities for planting through the 'travelling teachers of agriculture' to the farm populace, initially from the Botanical Garden established at Ebute Metta, Lagos in 1890 and later from the Forest Department established at Olokemeji near Ibadan in 1900, followed by the Agricultural Department for the South at Moore Plantation, Ibadan in 1900 and another one for the North in Samaru, Zaria, 1912.

At the end of the Second World War, the British government established a set of marketing boards for agricultural commodities, namely cocoa (1947), palm produce (1949), groundnut (1949), and cotton (1949). These boards helped to stabilize supply and incomes, with a view to increasing the raw material requirements of British industries to be resuscitated after the war.

In 1976, the inherited structure of regional boards was changed to another generation of seven commodity boards with national mandates, namely Nigerian Cocoa Board, Nigerian Groundnut Board, Nigerian Grain Board, Nigerian Cotton Board, Nigerian Palm Produce Board, Nigerian Rubber Board, and Nigerian Roots and Tuber Board. All of them were abolished in 1985 to pave way for the Structural Adjustment Programme (SAP).

Status of Nigeria in Agricultural Commodity Trade

The *total merchandise trade* has Nigeria as a small player in the world, with only about 21 million dollars worth of exports accounting for 0.3% and about 9 million dollars worth of imports accounting for 0.1%. Though the country is a relatively big player in Africa, accounting for about 33% of the exports and 18% of the imports of Sub-Saharan Africa, it is regionally inferior to the status of South Africa with 47% and 60% respectively.

In respect of *trade in primary commodities excluding fuels*, the status of Nigeria in SSA was greater as an importer, accounting for about 21%, than as an exporter, accounting for about 3%. That is, in net terms, Nigeria contributes to the dependency of SSA on the rest of the world in primary commodities to the tune of 17%, while the reverse is the case with South Africa, which contributes to the self-sufficiency of the region in primary commodities to the tune of 34%.

The relative commodity status of both countries is similar for *food items* in net terms. While Nigeria contributes to the dependency of SSA on the rest of the world for food items to the tune of 17%, South Africa contributes to food dependency of SSA to the tune of 3% only. The fact remains that SSA is food dependent on the rest of the world, anyway!

As regards trade in *agricultural raw materials*, the net import of Nigeria is about 20% of SSA, while the net import of South Africa is about 60% of SSA, indicating that South Africa has contributed more to dependence of SSA on the rest of the world for agricultural raw materials than Nigeria. Nonetheless, the general picture indicates that Nigeria is a small player both in the world commodity market in absolute terms and in relation to South Africa.

Role of Agricultural Commodities in Nigeria's Economic Development

Role in Internal Economy

Arable agricultural production involves a substantial part of the land resources available. The production of these commodities helps in meeting the food security need of the nation; it represents a major source of livelihoods of the people as a source of family incomes, while it also employs about 70% of the population.

Role in External Economy

Primary commodities comprising agricultural produce contributed only 2.5% of the merchandise exports from Nigeria, but 23% to merchandise imports into Nigeria. Nigeria's share of primary commodities to its merchandise exports was much less than what obtained for SSA, but its share of primary commodities to imports was greater than what obtained in SSA. The share of primary commodity exports in GDP is about 1% in Nigeria compared with 8% in SSA and 2% in the world.

In this capacity, Nigeria did not feature among the 30 principal exporters of primary commodities excluding fuels, or among the 30 principal exporters of agricultural produce in the world. And considering the class of 16 specific agricultural commodities traded in the world, Nigeria featured among the 10 principal exporters of only two commodities, namely cocoa beans (as the 4th in the world and 3rd in SSA) and natural rubber (as the 5th in the world and 1st in SSA). Thus the latent capacity of the commodity sector in contributing to the economy is large, but it is largely responsible for the low level of government revenue and foreign exchange earnings as well as low level of employment.

The Issues in Agricultural Commodity Development

To properly promote the role of agricultural commodities in Nigeria, we need to understand the issues at stake, with particular reference to the economic situation in the country. Three sources of the issues in commodity development will be highlighted.

Production Issues

The cost of production is often complained about for being too high relative to commodity prices that are not under farmers' control. For this reason, government intervention often takes the form of subsidy with the aim of bringing production cost down and to create a larger profit margin for farmers. In Nigeria however, agricultural subsidies have been grossly abused and the farmers have not been effectively reached. Moreover, production subsidy acts like a dog in the manger, by failing to work and also preventing other measures from working. For instance, fertilizer subsidy is accompanied by substantial leakages, and the price signals become distorted in communicating the messages among commodities.

Pricing issues

The recent Cocoa Summit in Abuja revealed the typical price war between producers of primary commodities and the industrial consumers. Producing countries lamented their inability to play a role in price determination for raw cocoa beans produced by them. The ensuing protest led to a proposal to undertake a cost structure study with a view to establishing the basis for fair pricing in future. Besides, the instability index of most agricultural commodities is typically high in value in the free world market for these items.

The volatile prices lead to poor planning of production and marketing activities and frustrations of farm incomes and commodity supplies. It was in a bid to forestall this that marketing or commodity boards came into existence, with a view to stabilizing supplies and incomes. Moreover, the prices of processed products are generally out of proportion with prices paid for agricultural raw materials in the international market as observed by cocoa producers. The matter is complicated by the imposition of discriminatory duties by industrialised countries, in which import duties on processed products are much higher in order to prevent their importation, while the import duties on agricultural raw materials are low in order to facilitate their importation at low prices. The resultant effect is to make processing of raw commodities a higher cost venture in producing countries, thereby blocking their access to semi-finished and processed agricultural products from commodities. Unfortunately, a country like Nigeria cannot influence the trade negotiations in its favour, owing to the volume of output that is too low to make any perceptible impact in the world and to its dependence on a narrow range of tradable commodities.

Value Chain Issues

The critical issue herein is a producer-consumer relation which there has been no proper forum to put right. Thus, proposals for normalizing this relationship entail establishing such a forum for agricultural commodities that incorporates the chain issues from one stage of value addition to another, thereby properly linking producers and consumers together at national and international levels.

Towards a Commodity Development Policy for Nigeria

The solution to the series of agricultural commodity problems of the country requires a refocusing of the commodity sector. This is best achieved through articulation of a commodity development policy for the country. A good commodity policy has these features: It is comprehensive, internally consistent in all aspects combined, crisp and handy, and with the strategy of implementation formulated properly. It is obvious that the agriculture ministry and the commerce ministry are required to collaborate fully in articulating and implementing a commodity policy for Nigeria. The desirable features of a good commodity policy for Nigeria include full participation of stakeholders, all-inclusiveness, openness, and professionalism. The marketing aspects of such a policy are critical, particularly as it is to be properly anchored on the relationship between the agricultural commodity production and marketing companies and the newly established Abuja Securities and Commodity Exchange.

The Nexus Of Food, Drug, Water, And Agricultural Development In Nigeria: The Role Of An Agricultural University

Paper presented at the National Conference on Food, Drug, Water, and Agricultural Development, organized by the International Network of Directories at Merit House, Maitama, Abuja, Nigeria, 23–25 October 2007.

INTRODUCTION

The rationale behind producing a directory on food, drug, and water is predicated on the need to promote development in these areas. It is common knowledge that development actions of government and other bodies have as the central focus reduction of poverty, which presents in the worst forms as hunger, ignorance, and disease and for which food, drug, and water are vital instruments linked to agriculture in the fight against poverty. Therefore, such a directory would help in providing the useful knowledge and information required to contemporaneously address poverty in its major ramifications. This connectivity provides an appropriate framework for the discussion of food, drug, water, and agricultural development nexus as carried out in this paper.

First we establish the poverty character of the nexus under focus; next we highlight the critical linkages within the nexus of food, drug, water, and agricultural development, then we situate an agricultural university within the context of such a nexus, and last, we draw some conclusions and implications for policy.

Poverty Character of the Nexus

The nexus under focus corresponds with the main dimensions of poverty, namely poverty of food, poverty of knowledge, and poverty of health. Poverty of food presents in terms of hunger and malnutrition and has measures for promoting food security as the main approach to fighting it. Poverty of knowledge presents in terms of criminal ignorance among the people and has education, including university education, as the major solution. Poverty of health presents in terms of debilitating diseases and has drugs as the means of combating it. Thus, the relationship between the nexus of food, drug, water, and agricultural development on the one hand, and poverty on the other hand, determines the course and pace of development to a considerable extent.

That poverty is finished if food is out of it is a wise saying which depicts the dominant role of food in the poverty character of the nexus. The hypothetical food problem tree for Africa (Fig. 1) has its taproot deep in poverty and its fibrous roots in several other causal factors, including bad policies, poor infrastructure, poor finance, natural disasters, and socio-economic discontinuities such as incidences of war and conflicts. The shoot system of the food problem tree consists in the main stem and branches comprising poor health condition and poor knowledge situation among the people; the overall character manifests in the leaves as symptoms of hunger, malnutrition, pestilence and squalor, among other miseries of humanity, wherein food, drug, and water play a prominent role in alleviating such symptoms.

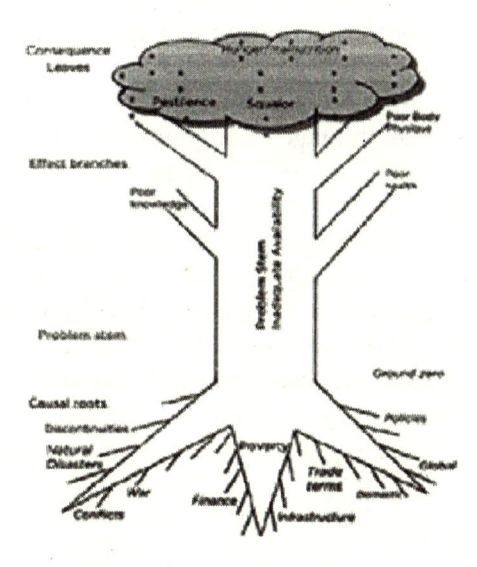

Culled from Ayoola (2003)

Therefore, a constant analytical focus on these diversified aspects of the poverty problem is required. This in turn requires a constant source of renewable information for doing a good job of development. In this regard, the directory would help in the provision of such data for analyzing such information and creating the knowledge required to mobilize the policy instruments and private enterprise towards meeting the demand of the people for good food, safe drugs, and potable water.

Critical Links of Food, Drug and Water to Agricultural Development

Food and water constitute basic needs for human existence; together with drug, which corrects for the negative consequence of poor food and bad water, they are linked to agricultural development as an important approach to uplifting the quality of life of the people. Thus a multilateral approach to providing these needs, integrating the various intervention instruments together, is more appropriate in the fight against poverty.

The central argument in this approach is which is more important for integration purposes, simultaneity or administrative integration in the provision of development services to alleviate poverty. Simultaneous provision of instruments puts the burden on the different ministries—Agriculture, Health and Water Resources—to devise ways and means of delivering their individual services: food, drug, and water respectively. On the contrary, administrative integration puts the joint burden of providing these services on a common administrative establishment. The present effort tends to promote the latter approach by generating information on food, drug and water together in a single directory. The preference for integrated approach is properly justified, given the multifaceted nature of poverty character and the close interconnectedness of the various policy instruments for fighting the menace of poverty. Particularly for rural areas, it is not in doubt that the multilateral or integrated approach to the provision of amenities is superior to deployment of individual instruments in the fight against poverty.

The ADPs (Agricultural Development Projects) illustrate a suitable channel for integrated delivery of poverty reduction strategies. The typical mode of integrating the instruments under the ADP system consists in two aspects as follows:

1. Farm production support through land development services, revamped agricultural extension system, revitalized input delivery, credit supply, etc. and
2. Infrastructure support through rural feeder roads network, rural borehole water supply schemes, rural storage facilities, rural market infrastructure in the form of farm service centres, etc.

However, the ADPs today are only a shadow of the past because the services are no longer available, let alone integrating them effectively. Probably the directory will help in determining the extent of decay of the ADPs, thereby informing the necessary efforts for resuscitating them for the integration of services and better provision of food and water to the rural poor.

Role of an Agricultural University

The concept of agricultural university has its origin in the American land grant colleges designed for addressing problems associated with farm production in an integrated manner using the same institutional arrangement for providing the services. The model forms the backbone of the larger agricultural extension system integrated into the university structure. Following the success of these institutions in achieving the goals of agricultural development, many countries of the world have established agricultural universities (UAs) for the same purpose, notably in Asia and Africa.

In Nigeria, the first generation of UAs was established in at Makurdi and Abeokuta in 1988, followed by another one at Umudike in 1993. The distinguishing features of these universities include their mission orientation towards scientific transformation of agriculture through a tripartite function in teaching, research, and outreach. The central focus of these universities involves the acceleration of the process for the attainment of food security in the country. Thus their staff and programme structures were designed to reflect this vision and mission different from the pre-existing functions of faculties of agriculture in conventional universities.

However, the performance of UAs in the country leaves much to be desired. Thus far, their implementation has featured major departures from the initial road map, which leads the institutions away from the milestones established for them. Some of these include

1. Departure from the provisions of the original statutes establishing the universities: whereas these universities were established by law

which put them under the supervision of the agriculture ministry through an Agricultural Universities Coordinating Agency (AUCA) using parameters that reflect their original mission and mandate, they have been operating the Ministry of Education since 1999 without regard to the law, using the instrumentality of supervision by National Universities Commission that has employed the same parameters as those applied for faculties of agriculture in conventional universities.

2. Departure from original academic structure: whereas the original academic brief of these universities put emphasis on practical work of staff and students in terms of teaching, research, and outreach, which are closely integrated in a multilateral activity structure (e.g. 70:30, 50:30:20, 70:20:10, etc.) and which permit sufficient mental mobility of professional staff between colleges and other organs, these universities have since operated an academic structure not different from the conventional approach of the faculties in conventional universities;

3. Departure from original programme structure: whereas the collegiate system adopted from inception for these universities envisaged considerable administrative autonomy of colleges, these units are run in the same way as faculties with no such independence in programme implementation.

Little wonder then, that after about two decades of their existence, these universities have not performed better than the faculties, when measured against the appropriate parameters that reflect their original mission mandate. With reference to the nexus under focus, the performance status of University of Agriculture Makurdi is briefly described as follows:

1. **Capacity building**—Capacity building relates to teaching of undergraduates and training of postgraduates as follows:

    a. Students' population: presently there are nine colleges in 23 departments and a postgraduate school running six programme areas, namely Agriculture, Engineering and Technology, Science, Veterinary Medicine, Agricultural and Science Education, and Management Sciences; the cumulative student population for about 20 years, from inception to 2006, is 6,985 students made up of 6,493 undergraduate students, 138 postgraduate diploma students, 274 masters students, and 80 PhD students;

b. Practical work: the practical exposure of students is through the Students Industrial Work Experience (SIWES), which is what also obtains in the faculties of other universities under a unified arrangement with the Industrial Training Fund. The expectation of a standard commercial farm operation wherein students will be co-owners and co-managers has not materialized.
c. Linkage activities: linkage activities with local and international institutions are at the lowest ebb, which limits the exposure of staff and students to the outside world.

2. **Generation of technologies**—This relates to research works for constant generation and improvement of agricultural technologies and for increased production and better supply of food, drug, and water. The dimensions of technology generation expected from UAs are as follows:

   a. Technology of land preparation—for ploughing, harrowing, ridging;
   b. Technology of planting—improved seed and seedlings;
   c. Technology of farm management—crop protection products.

Generally the effects of UAs have not been felt in these areas of technology creation to a significant extent.

3. **Outreach**—Outreach work pertains to applying the capacity of the university for the benefit of society at local, national and international levels; this is envisaged to take place in a number of ways, highlighted as follows:

   a. *Extension Services*: The universities are expected to establish and demonstrate model extension systems that are more effective than the traditional systems, to be subsequently adopted by state and federal authorities, but their impact is yet to be felt in this regard.
   b. *Advisory Services*: Policy advisory services to the public and private sectors are crucial to the successful design and implementation of public intervention programmes in the agricultural sector, and this formed a strong basis for establishing the UAs in the first instance; however, in the circumstance in which they operate, their predisposition to performing this role is inadequate. Moreover, policy advisory services of the universities may be strengthened by permitting the growth of centres and foundations of excellence

for the purpose which is presently downplayed and which is why the UAs remain irrelevant to policy actions of the agriculture and other ministries, let alone for them to be part of important policy developments in the sector.

c. *Consultancy Services*: The UAs, like the conventional universities, are expected to establish strong consultancy outfits for revenue generation and institution building. Such revenue flows would come from privatization, commercialization, and cost recovery measures as applicable to products and services from university staff and facilities, namely food items (crop products like yam, cassava, rice, maize, etc.; livestock products, e.g. beef, poultry, pork etc.; fisheries products, e.g. capture fishes, aquaculture, etc.). Furthermore, institutional capacity building results from tapping the slack capacity of staff and facilities available in the universities for entrepreneurial activities, thereby generating employment.

In general, the UAs have not performed to expectation in these areas, so the nexus of food, water, and agricultural development is inconsequential to poverty reduction in the country.

CONCLUSIONS AND POLICY IMPLICATIONS

The nexus of food, drug, water, and agricultural development has a strong poverty character and involves critical linkages that can be explored for uplifting the quality of life of the people of this country. The production of a directory on these elements is a step in the right direction as a source of information and knowledge required for conducting constant analysis of the issues involved. Agriculture has an important role to play in the process, in terms of capacity building, technology generation, and outreach for the benefit of the Nigerian society.

It is recommended that the institutional and policy environment of the agricultural universities should be put right in order to enhance the performance of these important institutions in this country.

References

Ayoola, G. B. (2003), 'The Food Question in Sub-Saharan Africa and the Challenges of Scientific Agriculture' in *Chemistry International*, IUPAC

27 (3) (May–June) also published on the Internet (www.iupac.org/publications/ci/2005/2703/2_crXII.html)

Ayoola, G. B. and **Idachaba F. S.** (1990), 'The Technology and Nigerian Agricultural Development', *Nigerian J. Agricultural Technology*, pp. 23–39.

Relevance Of Horticulture For Economic Empowerment

Invited lead paper presented at the 25th Annual Conference and 30th Anniversary of the Horticultural Society of Nigeria (HORTSON).

INTRODUCTION

The relevance of horticulture to Nigerian development dates back to precolonial times. The initial entry point of the colonial government into the agricultural administration of the country was the establishment of the Botanical Garden at Ebute Metta, Lagos in 1893. Annual reports indicated activities such as systematic introduction and domestication of exotic vegetables (e.g. lettuce) and other horticultural crops in the garden. Other activities involved the planting of grass and flowers as hedges on both sides of the road from the seaport leading to the garden, possibly for the pleasure of Her Majesty the Queen and her proxies.

Subsequently, the Botanical Garden had transformed first into Forest Department at Olokemeji (1900) and later into the Agricultural Department (Southern Provinces, 1910), wherefrom was created the Agricultural Department (Northern Provinces, 1912), both of which were merged into a single Agricultural Department of Nigeria in 1921, and subsequently cleaved into three Agriculture Ministries, Western Region, Eastern Region, and Northern Region (1954), and a fourth Agriculture Ministry (Mid-West Region, 1963), then a Federal Ministry of Agriculture in 1967. Over time, the regional ministries had multiplied at different times following the progressive state creation exercise, from twelve in 1967 to 19 in 1987 and to 21 up to the present 36 including a quasi ministry for the Federal Capital Territory (for details, see Ayoola, 2007).

Thus by virtue of the Ebute Metta garden, the practice of horticulture formally began in its original meaning, *cultivation of gardens*. In practical terms horticulture is defined as the science, skill, or occupation of cultivating plants, especially flowers, fruit, and vegetables in gardens or greenhouses. Otherwise put, horticulture is simple agriculture without many tools *or* a simple form of agriculture based on working small plots of land without using draft animals, ploughs, or irrigation (Encarta).

This paper examines the policy relevance of horticulture to the national economy since inception of this concept on the shores of Nigeria, specifically looking at how seriously horticulture has been taken in the process of economic development and empowerment. First, we examine the concept of economic empowerment in its different dimensions; next, we provide an analytical framework in terms of the nexus of horticulture, agriculture, and the economy; then we trace the preference for horticulture in the policy process with particular reference to policy instruments for enhancing the relevance of horticulture to economic empowerment. The paper is concluded and implicated afterwards.

CONCEPT OF ECONOMIC EMPOWERMENT

Economic empowerment connotes giving power or more power to individual persons, or to individual households, or to society at large, wherein power was hitherto denied in various degrees. Therefore, policies for economic empowerment are focused on the people, at the level of individual persons, individual households and society at large. In reaching individual persons the policy instruments may be directed at the youth, women as well as vulnerable groups such as the widow, the elderly, single mother, nursing mother, disable, etc.

Economic power consists in two dimensions—having a *voice* in the policy process (this we call, for want of another name, *control power*), and having a *vote* in the marketplace for acquiring goods and services (this is generally called *purchasing power*). Thus economic empowerment denominates into:

i. Enhancing the individual persons or households or society at large to participate in policy process with a view to exerting an influence or control on the outcomes or decisions of policy authorities.
ii. Enhancing the incomes of persons, households, or society at large for them to gain title to goods and services in the market.

It is recognized that the distribution of power in the economy determines the accrual of benefits from that economy to persons, to households, and to society at large. This is why the several organs of society engage in deliberate struggle to acquire power in one form or another, namely political power (at times by rigging election), financial power (sometimes through corruptions), social power (sometimes through illicit chieftaincy titles). We posit that in a free enterprise economy, the legitimate means of gaining power is by owning capital and employing it for creating utility of one form or another: We speak of form utility arising from production or processing of commodities, time utility arising from storage and preservation enterprises, and special utility arising from transportation enterprises.

Suffice it to say that horticulture fits into this conceptual framework as a sub-sector of agriculture and an instrument of economic empowerment.

NEXUS OF HORTICULTURE, AGRICULTURE, AND THE ECONOMY

The need arises to specify the role of horticulture in the economy as the appropriate basis to determine its relevance in economic empowerment. In doing this, we construct a role matrix for horticulture. It has the rows defined in terms of agricultural functions and the columns defined in terms of focus of empowerment on the people. Three **rows of functions**, namely food security, income generation, and employment generation are considered; and three **columns of focus**, namely persons, households, and society are also considered. Thus, we have a 3 × 3 matrix in nine coordinates as numbered in Table 1.

Table 1: Role Matrix of Horticulture in Economic Empowerment

Parameters of Agricultural Development	Dimensions of Economic Empowerment		
	Empowerment of Persons	Empowerment of Households	Empowerment of Society
Food Security	1	2	3
Income Generation	4	5	6
Employment Generation	7	8	9

Table 1 provides a framework for interrogating issues in the nexus of horticulture, agriculture, and economy as highlighted in the corresponding numbered cells.

Food security of individual persons: A person is food secure when he eats food of the right quantity and quality to meet the needs of his body in the present state. Individuals have different needs for food at different times; apart from a basic requirement for body maintenance, some people have additional requirements for work, for pregnancy, for nursing baby, and for particular health situations. Horticulture contributes to personal food security in terms of products consumed by individuals for such purposes. Special diets place additional demands for such food items as fruit in different health situations. The issue about personal food security is that most people lack adequate knowledge of what to eat, for what and where. To be more relevant to personal food security status of individuals in this country, horticultural business should generate products that are branded for different purposes.

Food security of individual households: Household food security derives from personal food security of individual members. Nevertheless, the concern of the household in the aggregate sense is how the increase in the food security status of a member results without a compensating decrease in the food security status of another member, which is why rural households will like to maintain a garden around the home, cultivated for vegetable and some fruit trees within the reach of the household members. During the season, they even share some harvests of these items with their neighbours in helping other households to meet their food security as well. This reveals the great relevance of horticulture to the household economy and indicates why such an important role should be further enhanced.

Food security of Nigerian society: Food security at the national level is a different issue altogether. It implies all citizens of the country must have food and have their food entitlements met at all times. This includes the vast majority of the people who will be engaged in works other than keeping homestead gardens. Such people must enter the market to gain that entitlement. Therefore, there must be marketable surplus of horticultural food items (availability), to be distributed far and wide from places where they were produced (accessibility) and to be acquired within the means of the people (affordability). Here comes the three A's of food security (availability, accessibility, and affordability), to which I add a U for utility; the people must be well disposed in good health to properly utilize the food in their body before the importance of food security can be driven home. In this connection, the issue of AIDS makes it imperative for a country to integrate health concerns with food security issues. Horticulture business has many roles to play in such a process, as the source of products with great food and health value.

Income generation by individual persons: Individual persons live for the incomes they earn through productive activities. Horticulture business is such a vast area of productive activities for people to earn incomes. Throughout the length and breadth of this country, individuals engage in selling fruits and vegetables on streets and in markets to receive substantial incomes—oranges, mangoes, tangerine, etc.

Income generation by individual households: Horticulture presents ample opportunity for augmenting income of household. Though the household economy may not completely rely on this, the savings made from harvesting vegetables from the home garden and plucking fruits around the homes may constitute a significant proportion of household incomes.

Income generation by Nigerian society: The potential is there for the country to produce horticultural products for exports through enterprise promotion, e.g. apart from the abundance of tropical fruits and indigenous vegetables, the cut flower enterprise is a multi-million dollar business in Europe which countries of East Africa are presently engaged in.

Employment of individual persons: Opportunities for private employment and wage labour exist for individuals in horticultural enterprises such as plantain farms and in fruit orchards.

Employment of households: Horticulture creates opportunities for members of the household, particularly women and children, to be employed in family businesses involving the production of fruit and vegetables.

Employment of Nigerian society: The country may address the unemployment problems through the development of horticulture such as value addition enterprises in fruits industry.

This shows how relevant horticulture is to economic empowerment through food security, income generation, and employment, at personal, household, and national level. But what preference do we have for horticulture in the policy process to actualize its relevance to economic empowerment? It is answering this question to which we now turn.

Preference for Horticulture in the Policy Process

Sometimes the way society expresses its preference for or against certain items may not be so obvious or totally unnoticed by the general public. Therefore we owe the media a debt of gratitude for recently revealing to us the tall order of preference for house renovation in the national assembly.[58]

---

[58] The story of FIF illustrates this point. When the organization was established in Makurdi in the late 1990s, we occupied a rented bungalow as its headquarters for the initial two years. Then the landlord retired from public service and offered the building up for sale. Although we expressed interest to buy the property, our own offer was lower than the reserve price. The highest bidder who bought it was a politician and who intended to use the building as a beer parlour. The new owner gave us a quit notice and quickly settled down to sell beer and probably reserve some of the rooms for 'quick service' after a drink. One day someone who did not know I was involved remarked that it was *now* the building has found a better use rather than a stupid professor who previously occupied it with books and storytelling. I sat there bemused how a society had preferred development service to drinking alcohol and related services.

Table 2: FEDERAL ALLOCATION TO FEDERAL MINISTRY OF AGRICULTURE AND RURAL DEVELOPMENT IN THE 2005 BUDGET—SELECTED ASPECTS OF RELEVANCE TO HORTICULTURE

0250000	FEDERAL MINISTRY OF AGRICULTURE AND RURAL DEVELOPMENT	Amount Naira
	TOTAL ALLOCATION:	9,286,565,619
Classification No.	EXPENDITURE ITEMS	
2500001000	CAPITAL	5,611,500,000
	Department of Administration	
2500001001	Purchase of Residential Building (FAO Rome)	150,000,000
	Planning, Research and Statistics Department	
2500001005	National Agriculture Data Bank	10,000,000
2500001006	Agric. Sector Macroeconomic and Policy Analysis	5,000,000
2500001007	Food Intelligence, Crop Monitoring and Early Warning System	10,000,000
	Projects Coordinating Unit (PCU)	
2500001008	National Special Programme for Food Security (NSPFS)—PCU	500,000,000
2500001009	Chinese South-South Cooperation Programme—PCU	400,000,000
2500001010	SPFS Adaptive Research and Extension	100,000,000
2500001011	Community Based Agriculture and Rural Development Programme (CBARDP) IFAD	30,000,000
2500001012	Community Based Agriculture and Rural Devt. Programme (CBARDP), ADB	30,000,000
2500001013	Community Based Natural Resources Management Programme (CBNRMP) IFAD, Niger-Delta	30,000,000
2500001014	National Fadama Development Project II(NFDP),World Bank	61,810,000

2500001015	Fadama II ADB	10,403,000
2500001016	Nerica Rice Dissemination Project- ADB	27,787,000
	Federal Department of Agriculture	
2500001017	Presidential Initiative on Increased Rice Production and Export	40,000,000
2500001018	Presidential Initiative on Cassava Production Processing and Export	40,000,000
2500001019	Presidential Initiative on Vegetable Oil Development Project (VODEP)	40,000,000
2500001020	Presidential Initiative on Tree Crops	40,000,000
2500001021	Root and Tuber Expansion Programme (RTEP)	50,000,000
2500001022	PQS Project	7,500,000
	Department of Agricultural Land Resources	
2500001034	Environmental Management Project:	10,000,000
2500001035	Equipping and Upgrading of GIS and Remote Sensing Centre Soil Conservations	50,000,000
2500001036	Erosion Control on Agricultural Lands National Soil Testing Management and Equipping of Laboratories	20,000,000
	Purchase of tractors	
	Department of Rural Development	
2500001048	Machinery/Tractor/Equipment Rehab. and Maintenance (Rehab of 123 Tractors and Implements)	150,000,000
2500001049	Procurement of 37 no tractors and implement, 1 each for 36 states of the Federation and FCT @ N7.5m	300,000,000

2500001050	Rehabilitation of Rural Roads of 20,000 km DFFRI Roads for Rural Access and Nationwide: **Adamawa:** Gaanda to Fotta- 10 km, Kalaá to Kwakwah- 4 km, Kwabaktina to Manzaá- 3 km, Bangshika to Dzumah- 8 km, Kil Nyeekin to Jaba road. **Rivers State:** Sangamabie to Borokiri- 10 km, Bakama to old Bakama 15 km, **Kwara State:** Aran-Orin to Ipetu junction- 3 km, Koro to Eruku- 8 km, Igbonla to Ajase Road- 3 km, Edidi to Alla- 11 km, **Imo State:** Okigwe to Umuna with spur from Umuna to Okwe-Uumuduru Egbeaguru-Arondizogu- 25 km, **Oyo State:** Oko-Oba-Isalekola-Apode, Omi-Adio-Alakoso-Aba-Asa-Bakatari, Ire Saapa-Idi Opele-Arolu-Ilejue- 30 km, Ajawa-Mowolowo-Iwoate-Lagbedu- 20 km, **Kano State:** Tashar Kali-Taranke- 12 km, Gaya-Balare-Gulya Road- 20 km, **Benue State:** Zaki-Biam-Tse Tornajime-Bossua 14 km, Abeda-Dusa 6 km, Harga - Ngokem-Gagu Bur Villee- 7 km	200,000,000
	Rehabilitation to DFFRI standard Rural roads linking Made Iya Oje Igbo Aka-Igbo Ile-Onipaanu, Iwofin/Oke-Asa, Iyelu and Abogunde.	50,000,000
2500001051	Rural industrialization and other productive activities	40,000,000
2500001052	75 Integrated Farm Project Sites	40,000,000
2500001054	Enhancement of Rural Household Food Security and Nutrition	20,000,000
2500001052	World Bank/ADB Assisted Rural Access and Mobility Project (RAMP) - Counterpart funding	200,000,000
2500001053	UNIDO Assisted Ebonyi Salt Project under IP4 CSF-IP-4	1,500,000
2500001054	Implementation of the Rural Development Policy	25,000,000
	Igboye-Ije/Badagri Ikododu Lagos Enclave Project (land preparation including land compensation)	
	Access road and borehole to Badagiri Igboyeiyi Ikorodu Lagos Enclave project	

	Federal Fertilizer Department	
2500001055	Fertilizer Revolving Fund	1,000,000,000
	Agricultural Lime	200,000,000
2500001056	Organic Fertilizer Development and Promotion	10,000,000
	Department of Agricultural Sciences	
2500001057	Completion of Central Agricultural Research Laboratory/Office Complex	60,000,000
0250003	**NIGERIA AGRICULTURAL INSURANCE CORPORATION**	
	TOTAL ALLOCATION:	150,000,000
Classification No.	EXPENDITURE ITEMS	
0250000200 0100	Overhead/Goods and Non-Personal Services—General	50,000,000
0250000200 0101	Premium Subsidy	50,000,000
0250000200 0200	CAPITAL	100,000,000
0250000200 0201	Premium Subsidy Programme	100,000,000
0250601	**AGRICULTURAL RESEARCH AND MANAGEMENT TRAINING INSTITUTE (ARMTI), ILORIN**	
	TOTAL ALLOCATION:	365,668,474
Classification No.	EXPENDITURE ITEMS	
0250000200 0200	CAPITAL	147,265,000
0250000200 0201	Renovation/Rehabilitation of Building	26,700,000
0250609	**NATIONAL HORTICULTURE RESEARCH INSTITUTE, IBADAN**	
	TOTAL ALLOCATION:	429,790,294
Classification No.	EXPENDITURE ITEMS	
0250000100 0001	Personnel Costs (Main)—General	250,161,295
0250000110 0010	Salaries and Wages—General	107,684,051
0250000110 0011	Basic Salary	107,684,051
0250000120 0020	Benefits and Allowances—General	142,477,244

0250000120 0021	Regular Allowances	77,332,846
0250000120 0022	Non-Regular Allowances	42,304,068
0250000120 0024	Social Contribution	22,840,330
0250000200 0100	Overhead/Goods and Non-Personal Services—General	17,378,999
0250000200 0200	**CAPITAL**	**162,250,000**
0250000200 0201	Research into the development of vegetables	5,000,000
0250000200 0202	Research into the development of fruits	4,500,000
0250000200 0203	Research into the development of citrus	6,000,000
0250000200 0204	Research into the farming systems of fruits	5,000,000
0250000200 0205	Dissemination and application of new horticulture	13,000,000
0250000200 0206	Development of Central Horticultural Research	10,500,000
0250000200 0207	Research on the control of major pests	13,000,000
	Fencing of the Institute 350 hectares in phase I	10,000,000
	Rehabilitation of research facilities at Mbato and Bagauda Substation	34,000,000
	Purchase of laboratory equipment and chemicals at HQTR	30,000,000
	Rehabilitation and construction of greenhouses	31,250,000
0250620	**FEDERAL COLLEGE HORTICULTURE DADIN-KOWA GOMBE AFFILIATED TO NIGERIAN INSTITUE OF HORTICULTURE, IBADAN**	
	TOTAL ALLOCATION:	**171,929,303**
Classification No.	**EXPENDITURE ITEMS**	
0250000100 0001	**Personnel Costs (Main)—General**	**131,454,303**
0250000110 0010	**Salaries and Wages—General**	**53,498,401**
0250000110 0011	Basic Salary	53,498,401
0250000120 0020	**Benefits and Allowances—General**	**77,955,902**
0250000120 0021	Regular Allowances	35,629,535
0250000120 0022	Non-Regular Allowances	30,956,957
0250000120 0024	Social Contribution	11,369,410

0250000200 0100	Overhead/Goods and Non-Personal Services—General	12,000,000
	CAPITAL	**28,475,000**
	Research studies	19,975,000
	Purchase of laboratory equipment, chemical, and glassware	8,500,000
0250620	**FEDERAL COLLEGE HORTICULTURE AGBADO, EKITI AFFILIATED TO NIGERIAN INSTITUE OF HORTICULTURE, IBADAN**	
	TOTAL ALLOCATION:	**245,891,251**
Classification No.	**EXPENDITURE ITEMS**	
0250000100 0001	**Personnel Costs (Main)—General**	**131,450,643**
0250000110 0010	**Salaries and Wages—General**	**53,498,401**
0250000110 0011	Basic Salary	53,498,401
0250000120 0020	**Benefits and Allowances—General**	**77,952,242**
0250000120 0021	Regular Allowances	35,629,535
0250000120 0022	Non-Regular Allowances	30,956,297
0250000120 0024	Social Contribution	11,366,410
0250000200 0100	Overhead/Goods and Non-Personal Services—General	5,940,608
	CAPITAL	**108,500,000**
	Building and infrastructure	20,000,000
	Research studies	10,000,000
	Building of hostels	15,000,000
	Building of classroom and offices	15,000,000
	Building of laboratories	25,000,000
	Building of access roads	15,000,000
	Purchase of laboratory equipment, chemicals, and glassware	8,500,000
	Personnel Costs:	9,505,023,207
	Overhead Costs:	1,353,736,333
	Total Recurrent :	10,858,759,540
	Capital:	9,615,950,268
	Total Agric. & Rural Dev.:	20,474,709,808

Economic empowerment is at the core of public policy process for developing the agriculture or horticulture sector. Therefore it is the preference for horticulture in that process that makes it more or less relevant to the process. Table 2 helps to achieve a rough and ready measure of such a preference for horticulture against the preference for other. The benchmark for comparison is the preference for house renovation by the national assembly given the recent political visibility that the matter has attracted. The data used for the purpose were extracted from the 2005 budget as considered by the same national assembly, specifically the allocations to various budget heads for the Federal Ministry of Agriculture and Rural Development.

The decision criterion employed is a stylised *Relevance Index* calculated as N628 million divided by the allocation to a particular budget head to the nearest whole number (sure we know where the figure N628 million comes from). This index represents the number of times that that preference for house renovation for two legislators is greater than the preference for agriculture or horticulture in the federal budget. Thus an index value of unity implies public authorities are indifferent in their scale of preference between horticulture and renovation. So, both are equally relevant to economic empowerment.

In this regard, a few observations concerning some selected budget heads are made below, while members of the audience are free to make several observations on their own.

National Special Programme for Food Security (NSPFS): This programme was implemented all over the country to increase the output of food security crops including horticultural crops. The Relevance Index was less than unity, indicating that such a programme for attaining food security of the citizens was less preferred by public authorities.

Fadama II: The National Fadama project involved building the capacity of farmers and other users for exploiting the potential of water bodies for arable farming particularly for multicropping of vegetables and other horticultural crops, thereby giving opportunity for increased income of rural people. The Relevance Index is much less than unity, indicating that such a project that would have promoted horticultural production was less preferred by public authorities.

Rural Roads Rehabilitation: Many rural roads were earmarked for rehabilitation in six states of the federation to improve access of rural dwellers, including producers of fruit and vegetables to market opportunities. thereby reducing the annual waste of these items. The Relevance Index was less than unity indicating that road rehabilitation is less preferred.

NIHORT is the mandate institute for research into horticultural crops. The Relevance Index was less than unity, which implies that the institute is many times not preferred by public authorities.

Training Colleges: The two training institutions at Dadin-Kowa and Agbado Ekiti were responsible for producing the middle-level manpower for horticulture development, and their products undertake extension services and enterprises in horticulture. The combined relevance index is less than unity, meaning that the colleges were less preferred by public authorities.

Based on the foregoing observations from 2005 budget, it is needless to say that our scale of preference in this country is not right; something is wrong with a country in which the public authorities prefer making huge financial allocations in the budget for such things as renovation of the residence of a few people, to making such financial allocations to a productive sector of the economy such as horticulture. Something is definitely wrong with our leaders who take issues of life and death of the ordinary citizens of this country for granted in preference for things of pleasure and comfort for themselves. Curiously we also observed in the 2005 budget that only N150 million was allocated for the *purchase* of a residential building for the representative of the country at FAO in Rome, and also that a paltry N26 million was allocated in the budget for renovation of buildings at ARMTI (poor ARMTI). Curious also, but for different reasons, the only item on the whole budget of the ministry in that year which had a value of Relevance Index greater than unity is *fertilizer* (N1 billion!).

Given that the scale of preference in the budget will reflect the proper order, policy instruments may be deployed to enhance the relevance of horticulture to economic empowerment of the following kinds.

Research: This is crucial to achieving higher productivity of horticultural produce as well as greater return to farmers' efforts and greater profitability of other horticultural enterprises. The role of NIHORT is called to question in this regard.

Extension: Communicating research results and delivering proven technologies to farmers is crucial to making horticulture more relevant in economic empowerment. This is the domain of ADPs as organs of agricultural extension services in the states, for the delivery of extension services in horticultural practices to the farm population.

Training: The continuous generation of middle-level manpower for agricultural development depends on the training and retraining of the country's agricultural population. The capacity for doing this in horticulture resides in the colleges primarily established for the purpose as well as other institutions for agricultural training in the country.

Business Development: The need to build capacity of people engaged in horticultural enterprises cannot be overemphasized. This is required in driving horticultural business towards higher productivity and greater profitability in the country.

CONCLUSIONS AND POLICY IMPLICATIONS

At issue in economic empowerment are the fundamental rights of the people of this country to the good things of life. These consist in their right to food, right to incomes and right to employment—in their different capacities as individual persons, as households and as a society at large. Suffice it to say that among the classes of fundamental human rights possible, the right to food reigns supreme and is sacred; other rights evolve from and revolve around it. The discussion so far borders on the relevance of horticulture in granting these rights, and the recognition of horticulture in the policy process for doing this. Whereas in theory, the relevance of horticulture in economic empowerment is definitely not in doubt, in evidence, policy authorities have accorded a low priority to horticulture on their scale of preference. The solution lies in matching both theory and evidence together in the policy process through deployment of available instruments, namely research, extension, training, and business development of the horticulture sector.

In conclusion, the right to food is obviously inalienable but not costless. It is high time that we made the political cost of violating people's right to food too high for public authorities to toy with any longer. For this purpose, we probably need to pass a Bill of Rights incorporating right to food in this country. The implication is that, in consideration of the relevance of horticulture specifically or agriculture generally to economic empowerment and food security graduates from its traditional consideration as an ordinary *issue of policy* to its consideration as a statutory *issue of law*. To have the desired effect, it is recommended that such a law granting the right to food to the people should be actionable, justiceable, and remediable. Furthermore, it is recommended that policy authorities should show greater preference for horticulture in the policy process, with a view to making the sector more relevant to economic empowerment of the country.

References

Ayoola, G. B. (2007), *National Council on Agriculture: An Analytical Profile of the Highest Policy Advisory Authority on Nigerian Agriculture and Rural Development* (Makurdi: Farm and Infrastructure Foundation).

—— (2007), 'Social Safety Nets as a Means of Delivering the Right to Food in Nigeria,' Paper Presented at the World Food Day Celebration Symposium, Abuja: Sheraton Hotel and Towers, 5 October 2007. Microsoft Corporation (2006), Microsoft Encarta 1993–2005.

Social Safety Nets As A Means Of Delivering The Right To Food In Nigeria

Paper presented at the 2007 World Food Day Symposium held at Sheraton Hotel and Towers, Abuja, 15 October 2007

INTRODUCTION

The theme of the 2007 World Food Day and Tele Food Campaign is *Right to Food*, which is quite provocative, in the sense that it carries food security concerns beyond the shores of policy to the doorsteps of legality. Hitherto, it is not generally the case to perceive food security concerns of a nation in terms of fundamental human rights, which *ab initio* is an issue of law. Thus the topic represents a paradigm shift from the traditional consideration of food security as an *issue of ordinary policy* to consideration of food security as an *issue of law*, which is a good development to take place at this stage of Nigeria's development. The country has witnessed over one century of agricultural planning and administration through the precolonial, colonial, and postcolonial periods, during which the right to food is almost non-existent.

This paper examines the implications of such a paradigm shift for government of the day in deploying available safety nets to grant the people their fundamental right to food. First we establish the premise that Nigeria has food but not food security. Next, we argue that unless the notion of food security is preconceived beyond mere platitudes of policy to become an issue of human rights law, our effort to implement a Special Programme on Food Security may not produce a quick result. Then, we contextualize the paradigm shift in terms of the nexus between policy and law regarding food security, citing some agricultural case studies. Last, we situate the social safety nets principle within the new rights context of food security.

NIGERIA HAS FOOD BUT NOT FOOD SECURITY!

The first clarification to make is that food production is only a necessary but not sufficient condition for food security. The appropriate definition of food security is that which encompasses the three traditional elements plus one—*availability, accessibility, affordability* and *utility. Availability* pertains to *production* of food items somewhere in the country at particular points in time. *Accessibility* pertains to *distribution* of the food items to the doorsteps of the people who demand them. *Affordability* pertains to the mechanisms of *exchange* that enable the people to acquire their individual food entitlements. *Utility* pertains to the disposition of people to the *consumption* of food in good physical and mental health status so they can assimilate the food nutrients for efficient body use. Suffice it to say that a nation is food secure at that point in time when the sequence of the four elements can be empirically substantiated.

This is the analytical template for us to evaluate the debate on the validity of official report of impressive growth of agricultural production to the tune of 6–7% in recent years. Given that such a growth of food production has taken place but was not accessible to the people, affordable for them, nor was it utilizable by the people, it merely indicates that Nigeria has food but not necessarily food security.

This point can be demonstrated with data from Federal Bureau of Statistics in respect of the last administration which, in collaboration with FAO, launched a vigorous Special Programme on Food Security (SPFS). Consider a hypothetical basket of nine major food items that represent the main sources of energy, namely maize, cassava, rice, millet, guinea corn, yam, groundnut, beans, and cocoyam: Total production of this basket in 1999 was about 84 million metric tonnes or about 40 million MT in grain equivalents amounting to 140 billion calories of food energy available. In the life of the last administration (1999–2007), the annual average of output of items in this basket is 128 million or 78 million MT in grain equivalents amounting to about 273 billion calories. Based on recommended energy food intake per output of 250 kg per year (or 2,500 calories per day), 120 million people living in Nigeria at that time would require only 30 million MT grain equivalents. That is, first, both the before and after energy food supply situations exceed normative requirements. Next, the energy food production has increased during the last administration, from 40 million to about 78 million grain equivalents per year, thereby closing the supply gap to the tune of 38 MT. Nevertheless, this result answers the availability question only and says nothing about food accessibility, food affordability, and food utility.

The other clarification to make is that the concept of food security has a dynamic dimension which presupposes that the food demand gap should be filled on a sustained basis into the foreseeable future. The necessary safeguard to offset any discontinuity is a functional food reserve programme. Such a strategic food reserve programme should be properly perceived as a food security instrument, rather than a price regulatory mechanism. That is, the programme saves food for the rainy day, for the country to fall back upon when threatened by food supply shortfalls, not necessarily because the prices of particular food items have risen or fallen. What the policy authorities often fail to remember is that when the food reserve is used for price adjustment downwards, it pleases the consumers of farm products as much as it hurts the farmers who are the producers of food items.

WANTED—A BILL OF RIGHTS INCORPORATING RIGHT TO FOOD!

Insomuch that rights issues are issues of law that bind vertically (on individuals) and laterally (on the organs of state), food security in its practical meaning and conduct largely remains an issue of policy, but not an issue of law. For it to become an issue of law, the rights of people to their food entitlements should be justiceable, remediable, and enforceable. That is, both the government as policy authorities and other stakeholders in the food system have obligations to observe the dictum of the rule of law.

The rule of law will help to gradually dismantle the protection enjoyed by public authorities and entities in violating people's right to food at no political cost to them. We need to make the political cost of rights violation of food entitlements too high for government to bear. This can only be achieved by passing a bill of rights incorporating right to some irreducible minimum level of food entitlements of the citizens of this country. The bill should give right to any citizen of this country whose fundamental right to a critical minimum food intake is compromised by the act of omission or commission of policy authorities, to seek remedy in the court of law.

In general, we notice a central tendency to take the right to food for granted in the international community, as indicated in the highlights of bills of rights passed in different parts of the world.

1. In the United States, the 14th amendment to the constitution introduced miscellaneous rights of the citizens to all conceivable forms of freedom, rights to freedom of speech, religion, peaceable assembly, keeping and bearing of arms, protection from unreasonable search and seizure, due process, double jeopardy, etc.
2. In Canada, the charter of rights and freedoms as entrenched in the constitution Act 1982 guarantees the rights and freedoms of the people subject to 'such reasonable limits prescribed by law as can be demonstrably justified in a free and democratic society'
3. In Europe, the necessity for greater cooperation among the nations after the Second World War and the compelling need for the reconstruction of Europe towards reawakening of democratic heritage led to the International Rights Convention of 1950, which guaranteed certain rights of the citizen, namely right to life, from torture, from arbitrary arrest, privacy, etc.
4. In the United Kingdom, the Human Rights Act (ushered in with white paper on Rights Brought Home) underscored the unsatisfactory status of existing constitutional guarantees to human rights, whereby rights and liberties were restricted under the case laws such as the Public Order Act of 1986 and Criminal Justice Act 1994, to the effect that if any person's civil liberties have been infringed upon by such domestic laws, the convention offers him a remedy if none was available in a British court.

Though we make an important observation from the various human rights laws of different countries, such was the degree of seriousness about the fundamental right to freedom among the nations of the world. First, necessary safeguards have emerged over time that make such rights inalienable and difficult to be denied by government. Second, the Bill of Rights statements of different countries reveal diversity to be accounted for by reference to the political and cultural contexts in the law in addressing pressing problems of society. The point to make here is that our own society is presently as much overwhelmed by the magnitude of food insecurity that exists, as it was the case with the magnitude of freedom abuse in Europe and America that created the compelling need for those countries to issue bills of rights and various amendments of their laws to guarantee such rights. Hence, we herein make a clarion call for a Bill of Rights incorporating right to food in Nigeria! Invariably this denominates into the following kinds of rights:

- Right of the farmers and other rural poor to have a voice and a vote in the policy process;
- Right of farmers to knowledge about production through extension service;
- Right of farmers to information about commodity markets and prices through organized delivery system.
- Right of farmers to access farm input market without being short-changed in the policy process by the avalanche of rent-seekers and other unintended beneficiaries of government policies;
- Right of people to critical minimum level of their food entitlements.

NEXUS OF POLICY AND LAW REGARDING FOOD SECURITY

We perceive policy and law as two sides of the same coin, the latter giving an expression of force to the former. Indeed, policy is essentially law made by government unto itself. Thus when government is not implementing the agricultural policy properly or at all, it is simply disobeying its own laws, which is what only an irresponsible government would delight in doing. Thus, as far as government is concerned, both the statements of policy and statements of law are one and the same thing, meant to be respected and obeyed. Given this nexus of policy and law, an environment is created for agricultural development wherein both of them are instruments of government for achieving food security and other objectives.

Therefore policies should be properly articulated with full participation of all stakeholders, in a professional manner, so as to make the expected roles of all stakeholders clear. That is to say, in a democratic economy such as ours, explicit policies are preferred to implicit policies. This is why we pushed for the proper articulation of policies on paper and for publication of policies well in advance. When it is done in this way, policy statements become explicit rather than implicit, such that the position and direction of government is clear to all stakeholders in advance. This makes the system predictable and policy environment stable, thereby reducing the frequent occasions of midstream reversals of critical policy commitments of government and to reduce policy inconsistencies and policy somersaults, which are common in the administration of Nigerian agriculture.

Recently we celebrated policy stability in agricultural administration of this country, based on the long term of service of the immediate past minister, Mallam Adamu Bello, who spent over six years in office against the average length of about one-year stay of his predecessors since independence. That inspired me to write a book in honour of him, which he generously launched at the last NCA meeting here in Abuja. Thus, the policy environment of Nigerian agriculture has begun to show a degree of stability in recent years in that respect. We can only expect that this trend will be sustained under the new administration.

Another source of system instability is the frequent reorganization of the ministry which usually takes place when new political leadership assumes office. As indicated in my 1990 paper in an Indian journal, there are two manifestations of this instability. One is the merging and demerging of organs of the ministry, and the other is the changes in leadership positions at three levels—political, administrative, and professional. Such organizational instability reflects in the splits/mergers or personnel turnover in key positions, which produce adverse effects on the management of food security programmes. Such changes lead to alterations in the content and duration of the programmes as also in the pace of their implementation so as to reflect the new political, philosophical, and occupational biases of the people involved. Consequently, the food security pathway becomes uneven. Therefore, making food security an issue of law with a view to stabilizing the system merits consideration.

Stability can be introduced into the policy environment for achieving food security by legislation, whereby certain conditions are prescribed that will make political leaders more cautious or less zealous in causing changes to the organization of the ministry. This is not to say that ministers do not sometimes have their good reasons for taking such radical steps on assumption of office, but as Professor Idachaba puts it in his latest book on agricultural development policy administration of Nigeria, 'good intentions are not enough.' Nonetheless, as law helps in giving an expression of force to government policies, the right to food will be promoted when food security at individual, household, and national level is made an issue of law.

Of course we assume that the rule of law will prevail at all times; the concept of the rule of law is crucial to the nexus of policy and law, which has two critical ingredients. The first ingredient is that law rules, not men. This implies that law automatically supersedes policy in the event of any conflict between them. In applying the rule of law principle to food security discussions in this manner, we need to grasp the real import of law which goes beyond the traditional perception as a means of settling controversies and disputes or resolving conflicts to a method of safeguarding expectations of the people, including their expectation of food consumption in adequate quantity and quality, and as an expression of ideals and values such as the right to food entitlement of citizenry.

Let us examine a few cases to test the compliance of policy authorities in agriculture to the rule of law in agriculture.

1. In the immediate post-independence period when there was no ministry of agriculture at the federal level and the regions were enjoying an immense commodity boom—remember the cocoa West, oil palm East, and groundnut North. A conflict of policy interest ensued, whereby the federal government wanted to create a Federal Ministry of Agriculture, which was not a lawful thing to do by virtue of the constitution that provided for agriculture on the residual legislative list. The regional governments resisted this attempt in jealously guiding their constitutional right to agricultural development. But what did the federal government do to circumvent the law at that time? First, the federal government invited the FAO to conduct a study to recommend the urgent need for such a ministry at the federal level. Yet the regional governments remained opposed to implementing the FAO recommendation, for lack of constitutional provision for doing so. Then the federal government went ahead to create a pseudo ministry of agriculture, namely Federal Ministry of Natural Resources and Research (*Official Gazette* 1965) which carefully avoided the word 'agriculture' in the name so as, according to Alison Ayida, 'not to offend the political sensibilities' of the regions. Of course, following the military takeover of Nigeria and the subsequent suspension of the constitution, the coast became clear for the creation of a full-fledged Federal Ministry of Agriculture and Natural Resources by administrative fiat under General Yakubu Gowon (*Official Gazette* 1967).

2. Again during the second republic when the Shagari government was finishing its first term and election was approaching, another issue of conflict of policy interest arose. Six UPN-controlled state governments from the south (Oyo, Ogun, Ondo, Lagos, Bendel, and Kwara) took the federal government to court over the activities of the River Basin Development Authorities in their domain, claiming that the law did not permit the RBDAs to embark on construction of rural roads and sinking of boreholes and mounting rural electricity projects in their own areas of jurisdiction and that the court should stop them. The fact was that as the election approached, the NPN-controlled federal government had quietly stepped up the activities of RBDAs in these areas by adding 'rural development' to their names, with a view to capturing votes of rural people living in these areas.

The governors of these states could not look on as the general election fast approached.
3. From the distant past till the present time, the provision of the constitution has not been strictly adhered to, pertaining to the federal-state relationships in agricultural development. Agreed, the present position of law is that agricultural development is on the concurrent responsibility of both the federal and state governments, yet the constitution provides for a clear line of division of labour between the two tiers of government in making the policies for food security and other agricultural objectives. According to the Second Schedule of the present constitution (see Part II, at Sections 17(c) and (d)), the extent of federal powers in agricultural development administration is limited to 'establishment of *research* centres for agricultural studies' and 'establishment of institutions and bodies for the *promotion and financing*' of agricultural development. Therefore we make bold to ask, *ipso facto*: Where does the federal government derive its power and authority to engage in fertilizer procurement on annual basis through tenders and distribution of fertilizer to states through contract awards, under our constitution?

The second ingredient of the rule of law principle is that policy authorities will follow the law at all times. *Au contraire* the government does not have a good reputation in respecting the administrative laws governing food security programmes in this country; we also have a few instances to cite in recent past to illustrate the flagrant non-implementation of existing laws and convenient bypass of agricultural laws.

One example is the case of NACB. No sooner had the Third Republic commenced than President Obasanjo ran into troubled waters with some of the reorganizations of some agricultural policy institutions. Specifically it will be recalled that one of the grounds (17 of them) of impeachment that would have swept that government away was that the reorganization of NACB to become NACRDB was not carried out according to law. It is interesting that one of the issues had to do with an institution such as NACRDB, which is for achieving food security of the nation, notwithstanding the good intentions of the former president in carrying out the reorganization of the ministry at that time.

Another example is the case of non-implementation of agricultural laws. It was in 2003 when we were working on the establishment of a fertilizer regulatory system for Nigeria that we discovered that a body had since existed by virtue of a law on fertilizer since 1967 and had never been implemented. At the same time, we discovered that the National Agricultural Seed Council provided in the subsisting national seed law was not implemented as well. Moreover, the draft amendment of both fertilizer and seed laws which have been drawn since 2003 at the instance of government are yet to see the light of day.

The last example to cite is the seemingly innocuous case of agricultural universities law, which provides for them as agencies under the ministry of agriculture but operate under the ministry of education at the moment. The last administration inherited a debate over the appropriate location of agricultural universities at Makurdi, Abeokuta, and Umudike, and simply took administrative decision to transfer them from the ministry of agriculture to ministry of education without reference to the law.

Suffice it to say that the consequences of disregard to law or circumvention of the law are as grave for a society governed by the rule of law as a society operating without laws. My sense is that if the right to food is given the expression of law that is well respected by the government of the day, the pathway to food security will become more certain, more steady, and more stable. The bill of rights incorporating the right to food should be crafted in such a way that an individual can challenge the government or its functionaries when his or her right is or is to be violated as a matter of fundamental human right in such cases, for example, as unwarranted reorganizations, circumvention, manipulation, and avoidance of the law governing administration of food security in the country.

SOCIAL SAFETY NETS FOR ACHIEVING FOOD SECURITY IN NIGERIA

The principles of social safety net as a means of achieving food security borders on granting help to members of Nigerian society who are in difficulty in gaining their title to food in sufficient quantity and quality. Under this principle, food security is considered as an objective of government within the social service context and, as such, enters as a critical explanatory variable in the social welfare function. This function makes it an obligation of government to maximize the welfare of individual members of society using all instruments at its disposal. A few safety net instruments in Nigerian agriculture are highlighted below:

> **Food and nutrition programmes**: An example is the so-called *home-grown school feeding programme*. Such a programme is not new to the country, as it belongs to the generation of primary schoolchildren who enjoyed a free midday meal of rice and cocoa

drink. An extension of this is the targeting of vulnerable groups such as indigent students at lower and higher education levels by issuing meal coupons for them to buy raw food items from stores to be redeemed by government; this type of safety net has the advantage of generating higher demand for food so as to stimulate foods production, and probably the financial budget required for such a scheme in the states is much less than the money required to renovate houses for one or two people in the national assembly!

Strategic food reserve programme: A normative food reserve level to meet the calorie needs of Nigerians for at least one year in the event of scarcity should be maintained. The intuitive reasoning behind this is that within one year of any attack by food enemies such as weather problems, external shocks, or even war, Nigeria should be able to survive in terms of energy food items required to fight back and ward off such an attack; that is, some 30 million MT grain equivalents during the last administration. In this regard, the theoretical maximum capacity of Nigeria's food reserve programme (as officially reported to be installed at the 2004 meeting of the National Council on Agriculture) is about 360,000 MT capacity, which is 1.2% security level; the actual installed capacity at is about 300,000 MT, which is 1% security level. Furthermore, the highest stock level so far is less than 100,000 MT, which is about 0.3% security level. Even if we discount the food items for the energy available from other crops and livestock items or drinks, we are still significantly far behind in meeting the necessary conditions for *dynamic* food security of the country.

Rural infrastructure programme: To be food secure implies to derive maximum utility from the food intake, which borders on availability of clean potable water to drink along with food, suitable physical and mental health to utilize the food in the body, and self-creation of temporal and spatial utility with the available food items. This requires the assistance of government in providing rural infrastructure, including rural physical infrastructure, rural social infrastructure, and rural institutional infrastructure. For details, see Idachaba et al. (2000).

Subsidies: The cases for and against subsidy have been well canvassed (Ayoola 2003). The balance of the argument is that subsidy policy is good for our farmers who are generally resource-poor. Of course, some people argue that even the richest nations subsidize their agriculture, albeit without regard to the fact of the matter: that such nations can easily afford it at much lower opportunity cost in terms of alternative public expenditures foregone, and that the mode of implementing such policies does not present an avalanche of unintended beneficiaries and consequences. It seems to me that we are not fair to the teeming farm population of this country with the knowledge that fertilizer subsidies are not reaching them properly and we have failed to change the strategy of delivery for a long time.

CONCLUSIONS AND POLICY IMPLICATIONS

Right to food as theme of this year's World Food Day campaign has provoked our thought towards the consideration of food security of Nigeria as an issue of law. The fact that Nigeria has food but not food security is generally not in doubt; also that food entitlement of the citizens is their fundamental human right, is intellectually incontrovertible, and that they deserve to enjoy their right to food in large measure is morally inalienable. In delivering the right to food in Nigeria, the government of the day is obligated to take two practical steps: First is to make food security an issue of right as a policy; second step is to make the right to food security at individual, household, and national level an issue of law by legislation.

The implication of this is that the social obligation of government in rendering agricultural services comes under the law. It is required that such a law granting the right to food to the people should be *justiceable*, demandable, and remediable. It is also required that policy authorities should accord greater regard to agricultural laws of the land, and to properly deploy the available social safety nets as a means of delivering the right of all Nigerians to food and nutritional security in the shortest possible time.

PART 3

THIRD PHILOSOPHICAL DECADE
(2008 - 2018)

4o

Transparency, Corruption, and Sharp Practices: The Policy Analyst's Viewpoint[59]

A keynote address / lead paper delivered at the 24th Annual National Congress of Nigerian Rural Sociology Association (NRSA), at Ladoke Akintola University of Technology, Ogbomoso, 12 October 2015

Introduction

In this address or paper, I set out to illuminate the action words in the title as given to me—Transparency, Corruption, and Sharp Practices—then to circumscribe the mindset of the so-called policy analyst, with particular reference to the policy economist like me. Nonetheless, the point must be made from outset that policy analysis is not an exclusive preserve of the economist as widely believed. It is the convergence point of all professionals interested in the analysis of what government is doing or not doing and how it is being done. It may as well be that the policy economist is more concerned in this task than other professionals, probably owing to the universality of the economy in itself and the procession of the appropriate tools of analysis and the diction to communicate the results.

First is a stylized definition of corruption as the misappropriation of public properties, resources and facilities to oneself or others for private gain. Though the policy concerns about corruption are not traceable, some old Indian manuscript of about 2,500 years cited corrupt practices as a policy menace (Daniel Kaufmann), which was perceived in terms of popular clones

---

[59] A keynote address / lead paper delivered at the 24th Annual National Congress of Nigerian Rural Sociology Association (NRSA), at Ladoke Akintola University of Technology, Ogbomoso, 12 October 2015

of the word: dishonesty, double dealing, fraud, misconduct, wrongdoing, adulteration, debasement, graft, etc. In present-day public life in Nigeria as also elsewhere, corruption manifests in several modes, which include paying or receiving bribes, peddling influence, exaggerating outcomes; the adverse effects on economy or society being increased transaction costs of governance, higher price of products and services than normal; society turns on its head by featuring discrimination, inequality, market distortions; misallocation of resources, to name just a few.

Second is a stylized definition of sharp practices, which connotes sneaky or cunning behaviour apparently within the rules or law but deceitful or exploitative somewhat. The variants of this present in terms of unethical behaviour, fraud, dishonesty, misconduct, taking undue advantage of someone and situation, cutting corners, etc. That is, sharp practices and corrupt practices are opposite sides of the same coin.

Last is a stylized definition of transparency, as a process or behaviour that is easy to see for others what actions are being performed; that is, shedding light on decisions and transactions involved open, thereby promoting responsibility, accountability and due process.

Next, we highlight the concerns of a policy analysis about corruption and its variants as illuminated above, which analysis represents a convergence point of all professionals; that is, the outcome of policy analysis depends on the professional spectacles the analyst wears at a particular point in time. Nonetheless, the domain of policy analysis is dominated by economists or agricultural economists as the case may be, whose concerns reflect his application of the economic principles to the analysis of policy decisions (i.e. what government does / does not do, has done / has not done or intends to do / does not intend to do). Hence the emergence of a special knowledge area spuriously known as economics of corruption, which implies that a market exists for corruption to take place (otherwise known as black market) which has a demand side and a supply side, and whereby a price prevails in that market in terms of what society loses to corruption. The key issues in a black market economy pertain to rent-seeking, back-of-the-hand dealings, fake products/substandard products, etc.

Analytical mindset of policy economist

The policy analyst's viewpoint stems from an analytical focus on the market for corruption or sharp practices. The demand side of this market comprises actors such as political office holders, civil servants, and other public service providers, etc., while the supply side comprises the general public, contractors, job seekers, students, etc. The policy analyst then puts his sharp analytical focus on two frameworks: (a) policy cycle, which defines a sequence of technical steps from problem identification to design/formulation, to appraisal or verification, and to implementation, including supervision, monitoring, evaluation, and impact assessment; and (b) policy process, which defines a sequence of administrative steps, from articulation of the problem, to verification, to adoption, authorization, publication, and possibly legislation. The issue of corruption can emanate from either approach, whereby which the instrumentality of policy economist may be deployed to analyze and address the problem using the market analysis framework.

In this framework, corruption in high places may be viewed with three lenses, namely public sector, private sector, and academia. The critical forms of corruption in the public sector involve the entrenched rent-seeking behaviour of public servants and influence peddling by job and favour seekers from the public officers. This is the reason for the NGOs to exist to serve as partners to government and watchdog of government at the same time. Farm & Infrastructure Foundation (FIF) is an example of such NGOs in the agriculture sector, promoting policy best practices in agriculture and rural development devoid of corrupt and sharp practices, which its lessons of practical experience in performing the policy advocacy role poses two questions:

- Who funds policy advocacy as a public good and service? Which suggests a role perception for government, non-government, and international community.
- When the policy advocate runs into trouble with policy authorities or violators in the course of fighting corruption, curtailing sharp practices or promoting transparency, who bails him out? Which suggests a role perception for civil society, media, and international community.

As regards the private sector, the critical sources of corruption involve the different situations of imperfect market, in terms of monopoly, monopsony or monopolistic competition, against which the need arises for regulatory bodies such as NADAC (National Agency for Drug Administration and Control), SON (Standards Organization of Nigeria), or professional associations. Lastly, the manifestation of corruption in academia pertains to academic dishonesty, academic fraud, academic impunity, etc.[60] These malpractices create the need for oversight bodies to be established in the academic sector, such a body

---

[60] For more on academic corruption, see Ayoola (2015).

being NNMA (Nigerian National Merit Award), which was established to maintain an ethical focus on the sector, with a view to promoting academic best practices and rewarding excellence in the knowledge society.

Corruption and Sharp Practices in the Agricultural Sector

The policy analyst's viewpoint about corruption derives from a dichotomous typology, namely: Random/Systematic, as to whether predictable or not; Deterministic/Stochastic, as to whether the parameters of the probability distribution are observable or measurable; Localized/Systemic, as to whether the effect of corruption is restricted to an organ of the system or the whole system; Transient/Chronic, whether temporary or permanent; Mild/Severe, as to whether the effects whether debilitating or life threatening; etc.

Against these definitional backgrounds, we examine the agricultural sector that we are familiar as a fertile ground for wanton corruption or sharp market and non-market practices to thrive; this we do with particular reference to the fertilizer sector; this is based on my many years of experience as a member of the National Fertilizer Technical Committee (NFTC) at the instance of the agriculture minister under successive administrations.

Three fertilizer policy eras may be identified. The first is the era of FPDD (Fertilizer Procurement and Distribution Division), which was established in the Federal Department of Agriculture in 1976 and existed till about 2002. The division enjoyed the monopoly of fertilizer importation in the country, which it distributed to all parts of the country through a network of fertilizer depots and distribution points, for delivery to the states for onward sale to farmers. A high but uniform subsidy rate was applied at all locations in the country, and all the quantities of fertilizer used by farmers were so subsidized, which peaked at about 1.2 MT in 1992 or so.

However, as the rate and volume of fertilizer subsidy was high, so also was the extent of corruption and sharp practices it fostered. In the multiple roles assigned FPDD during this era, the attendant corrupt practices among the actors include (a) rent seeking among public officers who collude with suppliers in round-tripping the same consignment of fertilizers and raise fictitious invoices for payment by the government, and (b) fraudulent practices among transporters and contractors and other handlers of fertilizer consignments from seaport to the hinterland, whereby trailer loads of subsidized fertilizers disappeared en route to be diverted through unintended channels for sale at higher prices in the open market or at the borders with neighbouring countries, etc. The several efforts of government to block the leakages in the subsidy policy failed in quick succession, such that even when policemen were attached to trailers as escorts; the policeman, driver, and the truckloads of fertilizers simply vanished together. Thus, it was during this era that fertilizer became so much politically visible as a source of quick money for corrupt people in the country. Eventually, government liberalized the market by withdrawing from importation.

The second was the era of FFD (Federal Fertilizer Department) created in the ministry in 2002. The department implemented the so-called Market Stabilization Scheme (MSS), which featured lower subsidy rate (25%) and smaller intervention quantity of about 250,000 MT. Yet corruption and market sharp practices persisted in the form of adulteration, short bag weights, and lack of truth in labelling, among others. Moreover, fertilizer round-tripping persisted—collusive practice to take out fertilizer from government warehouses to be resupplied in several rounds to make up the expected quantity awarded to particular suppliers, instead of direct sale to farmers at subsidized prices. During this era, the initial attempts to stop these corrupt or sharp practices in the fertilizer market featured a policy experiment with a voucher scheme designed to bypass the illicit flow of the subsidy payment to wrong channels, followed by a proposal to establish a fertilizer regulatory body NAFRAC (National Agency for Fertilizer Regulation and Control), which did not reach a logical conclusion before a new administration was ushered in, in May 2011.

Third was the era of instinctively FISS (Farm Inputs Support Services) Department, created midstream in the ministry, as the hub for the implementation of Growth Enhancement Scheme (GES), a technologically elegant programme to better eliminate the middlemen in the subsidy delivery channel. To this end, 50% price subsidy was delivered to registered farmers through the use of mobile telephone, to be redeemed by agro dealers. Yet the abuse of subsidy policy persisted in different dimensions, such as buy-back operations (collusive practice whereby an outside supplier buys fertilizer back from registered agro dealer for resale in the open market, for which the agro dealer would later claim the subsidy as if they sold to farmers); impersonation and proxy redemption (whereby fertilizer subsidy flows to people other than the farmers as originally intended); multiple registration that created room for same farmer receiving subsidies many times; elite not rural in nature (network problems, technology failures); supply of fake products; increased rent-seeking behaviour among public officials; emergence of fly-by-night fertilizer and seed producers—all culminating in a heavy debt burden for the succeeding administration to the tune of about N60.5 billion.

Curbing Corruption and Sharp Practices in Agricultural Sector

As pointed out earlier, the three postulates about policy best practices to curb corruption and sharp practices in agriculture sector are transparency, accountability, and policy due process. Three propositions correspondingly emerge therefrom:

- That citizen participation in the policy process is a necessary albeit not sufficient condition for transparency of the policy process for agricultural development

    - The greater the participation of the generality of the people in agricultural policy decisions affecting their lives, the more transparent the decisions become and the fewer the incentives for policy officers to engage in corruption and sharp practices in implementing such decisions;

- That the rights-based policies better engender programme accountability than need-based policies

    - The stronger the foundation of public policies in fundamental human rights, the greater the scope for policy authorities to be accountable to the people who voted them into power, and the less corrupt the public officers will be;

- That policy due process is coded in constitutionality and the rule of law

    - The stricter the public officers follow and obey the extant rules and laws governing the policy process, the more transparent the policy process becomes and the fewer the instances of corrupt and sharp practices in the agricultural sector.

Hence my dominant viewpoint as a policy analyst is anchored on the concept of right to food as the panacea in dealing with the menace of corruption and sharp practices that have slowed down the progress of agricultural sector for many years. In this context, the flagship project of FIF is the National Campaign on Right to Food, which it launched since 2007/2008. Under my close watch, the campaign seeks to change the perception of food by policy authorities and people alike, from the traditional notion of food as a mere human need to the more progressive notion of food as a human right.

Traditionally, food is perceived as a basic human *need*, which implies that the failure of policies in meeting the food entitlements of the people is practically inconsequential, but not as a basic human *right*, which implies that the failure of policies to meet the food entitlement of the people is actionable, justiceable, and ultimately remediable under the law. The difference between the two notions is not farfetched: while the former notion, food as a human need, views the role of government in formulating and implementing food policies as obligatory only (i.e. mere charity or an act of doing the people a favour), the latter notion, food as a human right, views the role of government in this regard as mandatory (i.e. owed as a duty), so as for people to be able to hold the government to account when its policies fail to meet their food entitlement.

Here lies the panacea to fertilizer corruption and other sharp practices in the agricultural policy process. That is, under the regime of right to food, the process is all transparent for the people to see through it, so corrupt officers in government can easily be held to account even in the law court, when corruption and market sharp practices prevent the realization of people's right to food.[61]

Thus, contrary to the popular but wrong notion of right to food, it is not an apology of state socialism that implies that government should provide food for the people free of cost. Rather, the kernel of right to food as an obligation of government, to deal a blow to the menace of corruption and sharp practices in the agricultural sector, is threefold as follows:

1. Obligation to respect the right to food—i.e. to recognize the right of people to nutritious food, which stipulates the state's exercise of power to refrain from acts capable of destroying people's access to food through unfavourable public policies;
2. Obligation to protect the right to food—i.e. to protect the right of people to nutritious food, which entails regulation of the activities of non-state actors or private sector that are inimical to people's food entitlements;
3. Obligation to fulfil the right to food—i.e. to help people in protracted suffering with provision of food at critical times, which entails the provision of food assistance to vulnerable groups and other such people as those that are temporarily displaced from their natural habitats.

---

[61] For more details about FIF's campaign on Right to Food, see Ayoola (2015).

Conclusions and Recommendations

The endemic nature of corrupt and sharp practices in the agricultural sector has its roots in the limited participation of the people in the policy process and the lack of transparency and accountability in the policy process for agricultural development.

In this regard, right to food offers a holistic and philosophical solution approach to addressing the issues of corrupt and sharp practices by subordinating the policy authorities as duty bearers to the wish of the people as right holders, so that the government can become more transparent and be held accountable for its actions and inactions.

References

Ayo, Bamgbose and Gbolagade Ayoola (2015), 'Academic Corruption and the Role of the NNMA in Curbing It', in NNMA, *Proceedings of the 8th Forum of the Laureates of NNOM*, Abuja, 2015 (forthcoming).

Ayoola, G. B. (2015), *Agricultural Policy and Tertiary Education in the Dispensation of Change*, Second Convocation Address of the Landmark University Omuaran (July 2015).

Agricultural Rebirth in Nigeria: Issues of Policy and Institutional Realignment

I. INTRODUCTION

Nigeria is a force to reckon with in the global economy. According to the World Bank, 'the Gross Domestic Product (GDP) in Nigeria was worth 568.51 billion US dollars in 2014. The GDP value of Nigeria represents 0.92 percent of the world economy. GDP in Nigeria averaged 79.89 USD billion from 1960 until 2014, reaching an all-time high of 568.51 USD billion in 2014 and a record low of 4.20 USD billion in 1960'. Presently agriculture contributes about 20% to the economy in terms of GDP, second to industries and third to services. Thus the need for agricultural rebirth of the sector becomes obvious to be able to do better in terms of revenue inflows to the people and the nation at large.

Agricultural rebirth connotes for the sector to be born again, or to be born anew, or better still, to be born afresh. In Nigeria, it is customary to observe a near-perfect correlation between political change and policy change in the agricultural sector whenever one government succeeds another. That itself is not the problem. The real problem is the perennial absence of built-in stabilizers, in terms of appropriate policy and institutional mechanisms, in order to preserve the integrity and stability of public agricultural policy process as political change takes place from time to time. In the ecumenical sense, when somebody is born again, it means 'he is a new creation; old things have passed away; behold, all things have become new'. Even though a total rebirth of the ecumenical type is not possible in the case of agricultural policy, it is nonetheless expected that a major policy change in terms of institutional or other alignments is inevitable, consequent upon a major political change, albeit subject to minimal perturbation of the policy system, in which sense I find the theme of this workshop quite apt—*Agricultural Rebirth: Issues of Policy and Institutional Realignments*.

Fortunately or unfortunately, the present effort towards rebirth is taking place at a bad time for a commodity-dependent economy like Nigeria's. The world market is presently witnessing a major dislocation of its commodity price relativities that is hurting commodity-dependent economies rather badly. Particularly the present downwards gyration of commodity prices is double jeopardy for Nigeria as a commodity-dependent country: on the one hand, the collapse of oil price structure in the world market (from a high of $150/barrel to a low of $28/barrel), a commodity the government of Nigeria nearly totally depends upon, is hurting the Nigerian economy badly; while on the other hand, the sympathetic downwards movement of other commodity prices, specifically agricultural commodities that the rural economy of Nigeria also depends upon, is hurting and threatening the livelihoods and food security of Nigerians badly as well. Thus, the need for rebirth of the agricultural economy at this time is emanating not only from the correlated changes in the polity and policy but also from the world economic downturn presently being experienced.

To start with, let us demonstrate this correlation with episodes in political and policy changes in the past, beginning with political changes since independence: from colonial to civilian (Tafawa Balewa, 1960); from civilian to military (Aguyi Ironsi, 1966); from military to military (Yakubu Gowon, 1966); from military to military the second time (Muritala Mohammed / Olusegun Obasanjo, 1975); from military to civilian the first time (Shehu Shagari, 1979); from civilian to military the second time (Muhammadu Buhari, 1984); from military to military the third time (Ibrahim Babangida, 1985); from military to civilian the second time, but without election (Ernest Shonekan, 1994); from civilian to military the third time (Sani Abacha, 1995); from military to military the fourth time (Abdulsalami Abubakar, 1998); from military to civilian the third time through an election (Olusegun Obasanjo recycled, 1999); from civilian to civilian, same political parties involved (Umaru Yar'Adua / Goodluck Jonathan, 2007); and the latest, from civilian to civilian, different political parties (Muhammadu Buhari recycled, 2015).

This series of political changes correlates nearly perfectly with policy changes in agriculture, in terms of dominant programmes implemented by successive heads of state or presidents, as follows: Nationally Coordinated Food Production Programme (NAFPP, 1972, Gowon); Operation Feed the Nation (OFN, 1976, Obasanjo); Green Revolution Programme (GRP, 1980, Shagari); Directorate of Food, Roads, and Rural Infrastructure (DIFRRI, 1986, Babangida); National Agricultural Land Authority (NALDA, 1990, Babangida); National Programme on Food Security (NPFS, 2000, Obasanjo); National Food Security Programme (NFSP, 2003, Yar'Adua); National Food Reserve Agency (NFRA, 2004, Yar'Adua); and the just-concluded Agricultural Transformation Agenda (ATA, 2011, Jonathan).

I imagine that the rationale behind this policy workshop is to determine which slogan fits the mantra of change under the present President Muhammadu Buhari, which I venture to say 'Agricultural Rebirth Agenda (ARA)'. Nonetheless, the point to make out of this correlation analysis is that the agricultural rebirth that the present administration embarks upon has a strong political economy character, which political character I like to illuminate with some historical antecedent with a view to eliciting the implications of this character for agricultural rebirth, then to present a philosophical foundation for agricultural rebirth, and finally to interrogate the issues of policy and institutional alignments that the rebirth entails.

II. THE POLITICAL ECONOMY OF AGRICULTURAL REBIRTH

Political economy of agricultural development means the interplay between economics, law, and politics, and how institutions emerge for the continual formulation and implementation of public agricultural policy. The political economy of Nigeria's agriculture consists in a series of episodes in three time horizons—precolonial, colonial, and postcolonial eras of agricultural administration. In the precolonial era, the exploratory trading activities of the British and Portuguese merchants along the Atlantic Ocean during the 16th century, which involved the exchange of salt and other foreign products for pepper and palm oil from natives, and the subsequent dominant control of mercantile trade in forest and agricultural products under the Royal Niger Company, underscores the importance of such products at that time, and a trigger factor in the eventual emergence of colonial governance of the area (Ayoola 2007). In the colonial era, the deliberate effort of the British government to establish institutions for agricultural research and administration underscores the joint roles of technology and policy as the pillars for agricultural development. This was done through the notorious but applicable philosophy of surplus extraction pursued by metropolitan Britain in its colonies during that period. The philosophy of surplus extraction of colonial masters naturally changed to one of self-reliance and self-sufficiency, which put the initial emphasis on agriculture as the mainstay of the regional or state economies, thereby demonstrating role agriculture was expected to play in the smooth take-off of the country, and with the benefit of hindsight, the present lessons of experience indicating that the country could only afford to neglect at its own peril.

The recent statement credited to National Bureau of Statistics clearly illustrates the political economy of Nigeria's agriculture that:

> GDP From Agriculture in Nigeria increased to 4816519.15 NGN Million (N4.8 Trillion) in the third quarter of 2015 from 3477845.24 NGN Million (N3.5 Trillion) in the second quarter of 2015. GDP From Agriculture in Nigeria averaged 3583037.05 NGN Million (N3.6 Trillion) from 2010 until 2015, reaching an all-time high of 4816519.15 NGN Million (N4.8 Trillion) in the third quarter of 2015 and a record low of 2594759.86 NGN Million (N2.6 Trillion) in the first quarter of 2010.

We infer from this statement that in 2015, being an election year in Nigeria, agricultural GDP decreased from a higher level in the first quarter, before the presidential election to a lower level in the second quarter, during the presidential election, and later increased to a higher level in the third quarter, after the presidential election. Thus in all likelihood, the presidential election, a political event, has a pronounced effect on the output of the agricultural sector, albeit not of a direct cause-and-effect type.

In light of this, hereunder I like to highlight certain instances in the agricultural history of Nigeria wherein the implications of political economy of agricultural development were manifestly demonstrated, which border on resource control behaviour among the states and boundary maintenance practices between the Federal Government and the States. In the first case of resource control, the political elite of a region or state exhibits a strong tendency to appropriate the benefit of agricultural development to itself alone and to subject the benefit stream therefrom to its own jurisdictional control. In the second case of boundary maintenance practices, a state government draws an artificial boundary in its agricultural development activities for the federal government not to trespass. To such instances we now turn as observed during each Republic.

First Republic (1963–1966)

The constitution that ushered the country into independence in 1960 classified agricultural development as a residual item on the legislative list, which, going by a special clause in the constitution meant that agricultural development was solely a regional responsibility, and as such, each region maintained its own inherited Ministry of Agriculture and Natural Resources (MANR). Soon after independence, the new Federal Government moved to establish a federal ministry for agricultural development at the centre, which all the regions were united to oppose, in jealously guiding their constitutional rights in that regard. This led to a pronounced constitutional crisis over agriculture, culminating in the invitation extended to Food and Agriculture Organization (FAO) to intervene through a study (FAO 1966). But, despite the recommendation in the FAO study that a ministry of agriculture should be established at the centre, the regional governments did not budge or shift ground an inch on the matter. In frustration over the constitutional crisis, the Federal Government forcibly established a pseudo agriculture ministry in 1965, and carefully named it Federal Ministry of Natural Resources and Research (FMNR), which, as Alison Ayida put it, the word 'agriculture' was carefully avoided so as not to 'offend the political sensibilities of the regions' which were constitutionally responsible for agricultural development at that time (Ayida 1973). Nonetheless, this pseudo-ministry operated until the military sacked the Balewa government in 1966, thereby bringing the first republic to an end. And as soon as General Gowon settled down, the constitution having been suspended, and by way of an administrative fiat, he created a truly Federal Ministry of Agriculture and Natural Resources (FMANR), in 1966.

The point to make here is that the action of regional governments reflected the commodity boom of that period that yielded huge revue inflows to the regions from exports of farm produce at that time, which the regions jealously protected from being snatched by the federal government. Therefore, the present administration under President Buhari should note that agricultural rebirth in Nigeria has a political economy character and that the process of rebirth is not necessarily immune to, and it is actually susceptible to a constitutional crisis about who has a superior control over agricultural resources between the federal government and state governments in due course. And per adventure someone thinks that the present 1999 constitution would prevent such a crisis from happening again, by providing for agricultural development on the concurrent legislative list, then he needs to note what happened during the second republic as presented next.

Second Republic (1979–1984)

The legal action about the activities of River Basin Development Authorities (RBDAs), filed by the six UPN-controlled state governments (Oyo, Ogun, Ondo, Lagos, Kwara, and Bendel) against the federal government during this republic testifies the manifestation possibility of resource control or boundary maintenance behaviour even when agriculture is on the concurrent list. Displaying the federal might previously, President Shehu Shagari had increased the number of the RBDAs from the original eleven (11) to nineteen (19) (reflecting the nineteen-state structure of Nigeria at that time). He also expanded their operations from the original focus on the management of water bodies to the boundary infrastructure provision in rural areas, so their name to River Basins and Rural Development Authorities (RBRDAs). Basking in the euphoria of this development, the RBRDAs quickly swung into action and gathered a greater momentum with increased funding from the federal government: sinking boreholes in rural areas, providing rural electricity, constructing rural feeder roads, etc. However, it soon dawned on the opposition state governments if the RBRDAs continued in that tempo, the forthcoming 1983 governorship elections would be lost to NPN candidates in the rural areas. So as the 1983 general elections approached, the opposition governors became jittery, in the fear that ruling party NPN would win the next gubernatorial elections in their states. Thus the fear of RBDAs became the beginning of political wisdom! The six state governments dragged the federal government to court to challenge the right of the federally owned RBDAs to undertake rural development in their own states, thereby claiming trespass. (Can you imagine?) The res of their case even bordered on 1979 constitution, which, despite the provision for agricultural development as an item on the concurrent legislative list (so a joint federal and state responsibility), they claimed it naturally belonged in the jurisdiction of the states.

In all likelihood, this court case informed the amendment introduced in the 1999 constitution, whereby the status of agriculture was retained as an item on the concurrent list, but in addition, a division of labour was established between the two tiers of government. Under this division of labour, the federal government was assigned the responsibility for the establishment of centres for agricultural studies (i.e. research), and for promotion or financing of agricultural projects (Second Schedule Part II, Section 17). Thus, for a successful rebirth of the agricultural sector to take place, it is now left for the present administration to determine whether the roles that federal government is presently performing in the states' agriculture actually belongs in its jurisdiction according to the 1999 constitution. Specifically as a precondition, the PMB administration must engage the states in an exercise in role definition, to jointly determine those implementation activities of the federal government that overreach the state governments in order to discontinue them. Then the rebirth programme can concentrate on research and promotion activities only, as stipulated in the constitution.

Last, under this Republic, another manifestation of resource control or boundary maintenance practices between the state and federal government was the action of old Undo State Government, which challenged the activities of the defunct Technical Committee on Produce Prices (TCPP) during the days of marketing boards (Idachaba and Ayoola 1991). The TCPP was established under the aegis of Central Bank of Nigeria as part of an intervention institution to fix prices at which the commodities would be bought by the different marketing boards in operation at that time—seven of them, namely Cocoa Marketing Board, Groundnut Marketing Board, Cotton Marketing Board, Palm Produce Marketing Board, Grains Marketing Board, Roots, and Tubers Marketing Board (which was later de-scheduled for lack of produce to buy). It was on record that the Ondo State governor, Chief Adekunle Ajasin or so, lodged a complaint to the federal government that the prices of cocoa, an agricultural produce in which Ondo was the leading producer, were deliberately fixed at a low level, insinuating this as a political instrument to reduce the purchasing power of farmers in his jurisdiction, thereby predisposing his party to losing the next election. (Maybe!)

Third Republic (1993–1995)

This republic was short-lived, or better still, it died on arrival, when the head of the military government at that time, General Ibrahim Babangida, having declared the presidential election inconclusive, merely stepped aside to usher in a pseudo-civilian government under Chief Ernest Shonekan. In that circumstance, therefore, no sample was taken in political economy of agriculture before that republic came to an abrupt end.

Federal Executive Council and subsequently to be tabled in a hurry before the National Assembly for its passage into law in a hurry. However, under the intense heat of electioneering campaign of early 2014, this proved to be a very tall ambition, never to be realized. Therefore the output of the policy working group represented an ex-post policy document, a mere medicine after death.

Moving forward, the Agricultural Rebirth Agenda (ARA) of the present administration has taken the right foot forward by organizing a policy workshop, as the starting point of the policy process to subject the document to a critical review and revision, with a view to rearticulating it in line with the new mantra of *change* and against the need for agricultural rebirth. Then we can expect an ex-ante policy document to be used as guidance note for implementing ARA. To this end, seven issues of policy and institutional concern arise for interrogation as identified and highlighted hereunder.

1. *About philosophical direction of agricultural rebirth*: The starting point of agricultural rebirth for this administration is quick determination of the philosophical pathway to follow, i.e. whether it wants to go the way of the last administration that operated on the philosophy of 'agriculture is a business, not development sector', which led to pro-business policies that put more money in the hands of the elite in the organized private sector rather than the farmers; *or* it wants to chart a new philosophical course anchored on freedom from hunger or right to food, which will lead to pro-poor policies that put the emphasis on farmers and other rural dwellers. While the proposed school feeding programme and conditional cash transfer are steps in the latter direction, such pro-poor policy instruments urgently require to be integrated fully in the ARA as part of a philosophy of freedom from hunger and right to food.
2. *About resource control and boundary maintenance behaviour*: These political economy issues are germane to the successful rebirth of the agriculture sector. Suffice it to say that resource control behaviour of the states will not stop as long as the federal government carries on with the excess luggage on its head beyond what the constitution stipulates. The issue is what to do so that the states take up their constitutional responsibility for agricultural development in full force, without looking up to federal government first, and also what to do for the federal government to uphold the division of labour established in the constitution.
3. *About marketing and other institutions*: The last administration spent so much of its energy on providing subsidy on the farm inputs but

failed to develop the commodity market where the products would be sold and income earned therefrom. This is like asking someone to yawn and sneeze at the same time. In the report of a market development study group commissioned by former minister of agriculture Dr Sayyadi Ruma (with me as team leader), the market was likened to the hub or midfield in a football match, where the game is won and lost. If the midfield is weak, pressure quickly builds in the goal area, leading a basket of goals scored against the team with a weak midfield. Similarly the role of market in agricultural rebirth cannot be overemphasized, as the hub of transactions for exchanging farm produce for money. Thus, greater production can be stimulated from the demand side of the market, not only on the supply side asced the last administration did by creating numerous value chains but keeping a weak market structure for commodities. This indicates that we have learnt nothing from the British colonial agricultural administration that used the market development as the principal policy instrument to pursue its surplus extraction policies. It is also questionable if the numerous value chain groups should have been at the pre-existing commodity focused research institutions or stood on their own as they did during the last administration.

Further, the issue about market development goes further to ask what to do with the set of marketing institutions presently existing but not functioning, namely the commodity development and marketing companies—Arable Crops Development and Marketing Company, Tree Crops Development and Marketing Company, and Livestock and Fisheries Development and Marketing Company. What do we do with these companies, which, like their predecessor, the marketing or commodity boards, have failed to meet their goals? Also, what do we do with the Abuja Security and Commodity Exchange that fails to generate desired impact in the market for agricultural produce? What lessons of experience can we learn from the old marketing or commodity boards, and what lessons from the contemporary institutions to re-establish the agricultural market? Surely the present situation does not augur well for a successful agricultural rebirth, whereby there is no institution to serve as an organized market outlet at the national level.

4. *About infrastructure as backbone*: Infrastructure is the backbone for agriculture, which the success story of the ADPs attests to. However, the debate about the poor state of infrastructure in the country is focused on the national grid in electricity supply, without a serious consideration of rural infrastructures that forms the backbone for

as a good thing, if not the best thing, for the sustenance of life on earth that is worth living. Therefore, by my own intuitive reasoning, I have this to say on the subject matter: that unless and until a people become free from hunger, freedom from any other thing is mere platitude.

Indeed, it is based upon this foundation of freedom that the whole world became born again after WWII, through the establishment of the United Nations organization, which holds that all humans are born free of certain encumbrances, and that freedom from these should be guaranteed in terms of rights; and consequent upon the universal declaration of rights, the world body quickly put in place the UN Convention on Social and Economic Rights (to which Nigeria is a signatory). This covers several freedoms as to be entrenched in the constitution of member countries, namely freedom of expression or the right to speak, freedom of information or the right to know, freedom from diseases or the right to health, etc. Unfortunately we observe that in Chapter 4 of the 1999 constitution of Nigeria wherein several rights are been guaranteed, the right to food is conspicuously missing. Thus, it is not and cannot be a mere coincidence that those countries of the world that have observed the tenets of freedom of their people, including their freedom from hunger, otherwise known as right to food, are also those that have made meaningful economic and social progress to make life worth living for their people.

Suffice it to say that agricultural rebirth in Nigeria should be properly anchored on the political philosophy of freedom from hunger and the right to food. When we speak of right to food, most people misconstrue this as an apology of state socialism, whereby government is called upon to provide food for everybody. This understanding of the notion of right to food is totally wrong, and agricultural rebirth cannot be premised on it; not operating anywhere in the free world where the able-bodied people must work to earn their rights to anything including food. That is, the right to food campaign (being championed by FIF) is largely targeted at policymakers, who should be held to account when public policies fail to help the people earn their right in spite of working. Therefore, right to food is properly conceived against the need for a paradigm shift in the mindset of policy authorities and the people who traditionally perceive food as a mere human need instead of perceiving it as a fundamental human right of the people. On the one hand, food as a human need, as presently included in Chapter 2 of the constitution, presupposes that if the policies fail to meet that need, the policy authorities cannot be held accountable for doing a bad job, thereby inconsequential. And this is the fundamental flaw in our policy process for agricultural development, whereby policy authorities cannot be held accountable for the failure of successive government policies to achieve their goals. On the other hand, food as a human right, as intended to be inserted in Chapter 4 of the constitution, presupposes that if the policies fail to guarantee that right, the policy authorities are liable and can be held accountable by the people. Thus, under a regime of freedom from hunger, the right to food of the people is sacrosanct; that is, inalienable, actionable, justiceable, and ultimately remediable.

Thus the obligation of the government as duty bearers under the right to food is threefold as follows:

- **Obligation to *respect* the right of people to food:** This stipulates the limits of state's exercise of power, which includes refraining from destroying people's access to food through unfavourable public policies; that is, government owes the people an obligation to implement policies that do not offend or violates right to food.
- **Obligation to *protect* the right of people to food:** This entails regulation of activities of non-state actors that are inimical to people's food entitlements (e.g. certain activities of the private sector).
- **Obligation to *fulfil* the right of people to food:** This entails the provision of food assistance to vulnerable groups and other people in protracted suffering when necessary.

IV. ISSUES OF POLICY AND INSTITUTIONAL REALIGNMENTS

These types of issues are usually relegated to the background and often addressed last by government, a habit akin to cooking Maggi soup. In cooking Maggi soup, it doesn't matter much to combine the wrong quantities and quantities of different ingredients together. The bad cook puts the pot on fire and the soup is done, only to find that the taste was not right; then the cook hurriedly reaches for a Maggi cube, a condiment to be sprinkled on the soup for it to taste better, nonetheless a bad soup still. At the eleventh hour, in September 2014, the last administration suddenly recognized the need for a policy document which would have been produced at the outset of ATA three years ago. Similar to Maggi soup approach, at the eleventh hour of the administration, the former agriculture minister, Dr. Akinwumi Adesina hurriedly assembled a number of experts and inaugurated them into a policy working group for ATA. The group worked in a hurry and produced a policy document in the same hurry, meant to be hurriedly approved by

Fourth Republic (1999–date)

Instances of resource control or boundary maintenance practices also abound during this period. I recall an instance when the Kaduna State government issued a notice in the press during a period of food scarcity, to ban the movement of food products by what it called 'out-of-the-state buyers', a statement so preposterous in the context of a single national market for food items! Another but less noticed was the action of the Benue State government under Rev. Moses Adasu, to instigate his people against the federal government over the establishment of a University of Agriculture in the state, when a general university was preferred! The case of agricultural universities also presents an instance of lawlessness prevailing in the sector, whereby these specialized institutions were established by law (Act 48, 1992) and nurtured by the preceding Babangida administration as parastatals under the ministry of agriculture, specially for practical exposure of students and direct policy relevance to the ministry, were abruptly transferred by administrative fiat by the succeeding Obasanjo administration, without recourse to the enabling law. And these universities operate like that (somewhat illegally!) which is the reason why they presently face a serious crisis of identity and run programmes in clear departure from their original mission and mandate. Thus, in the context of policy and institutional alignments envisaged under the Agricultural Rebirth Agenda, the movement of these institutions to the ministry of education, well intended as it might have been in the eyes of the past Obasanjo administration, should be quickly reversed by the present Buhari administration, before it is too late.

Perhaps the dominant instance of resource control behaviour in agriculture since the Second Republic till now revolves around fertilizer subsidy. Originally, the introduction of subsidy on fertilizer since 1976 was aimed at stimulating its adoption and use among the farmers, which aim had been achieved since the days of World Bank–assisted Agricultural Development Projects (ADPs). However, the subsidy policy was and still is fraught with corruption, whereby unscrupulous intermediaries and rent-seeking middlemen in the supply chain perennially diverted fertilizers meant for farmers to unintended channels— including fertilizer merchants, transporters, civil servants, and traditional rulers (Figure 1). All efforts at stopping this failed in quick succession over time, even as the budget burden on the government mounted beyond measure and it became obvious that the subsidy programme could not be sustained anymore. Yet, fertilizer has gained so much political visibility that it could not be stopped even under the best of intentions.

Under the growth enhancement scheme of the last administration, the strategy involved the use of mobile phone technology to deliver subsidy to farmers through an electronic wallet, following a registration exercise. The last administration, through its minister of agriculture Dr Akinwumi Adesina, had claimed that it had succeeded in stopping fertilizer corruption in Nigeria to the tune of N870 billion in public saving therefrom, a claim fiercely refuted by Malam Adamu Bello, a former minister of agriculture during the Obasanjo regime. Even as the quantity supplied under the previous Market Stabilization Scheme (586,145MT) was greater than the quantity supplied under the latter GES scheme (536,095MT)

What do all these add up to? An agricultural economy in sluggish growth over time, a nation lacking in food security of its people, let alone its people enjoying their fundamental right to food to a meaningful extent. Therefore, at this juncture, the APC-led federal government, in its present effort towards the rebirth of Nigeria's agriculture, cannot afford to look away, as the last administration substantially did, from these political economy implications for the agricultural economy. In this regard, I am to urge the present administration to shake the pre-existing policy and institutional mechanisms to the roots and to the philosophical foundation of agricultural rebirth.

III. PHILOSOPHICAL FOUNDATION FOR AGRICULTURAL REBIRTH

I am most enthused to extol the philosophy of *development as freedom* as sponsored by Amartya Sen (Sen 1999), on which formed a basis for me to uphold a derivative philosophy of *agricultural development as freedom from hunger* .Upon this is then to be established a philosophical foundation for agricultural rebirth as a necessary but not sufficient condition for growing the agricultural economy of Nigeria at this time. It is the people of Liberia, I think, who have this to say as their national motto: 'The love of freedom brought us here', while the Americans have totally imbibed an enduring culture of liberty or freedom as a political philosophy behind their very existence. Of course, freedom begets rights, particularly the rights of their people to many good things of life. And going by its critical importance to life, food appears to me agriculture. Indeed, it appears that FMARD has lost sight of the word 'rural' in its name, yet it harbours a Rural Development Department to be responsible for initiating and monitoring the stock and flow of rural infrastructures in the country. In this regard, a critical issue confronting the agricultural rebirth agenda is the need for systematic data and information for successful planning and implementation of rural infrastructure project in the country. The Rural Infrastructure Survey project that was established for that purpose remains in a lull for many years now, so there is presently no systematic empirical basis for FMARD to undertake the planning and implementation of rural infrastructure projects, as done in the past with the ADPs, the DFRRI, and the FEAP projects. FIF has submitted a memorandum at the last NCA meeting on the need to resuscitate the survey project, which was adopted but not implemented till date. Meanwhile, FIF has digitalized the old database through the INFRASTAT, an open-source and one-stop-shop domain to obtain such information required to properly plan and implement rural infrastructure project in the country. Figure 2 demonstrates the utility of this domain in accessing information such as the distribution of specific infrastructural facilities by states and local governments, e.g. federal roads in a state or LGA. How will this effort be supported and moved forward under the ARA?

5. *About participation and inclusion*: The hallmark of a democratic agricultural economy is full participation of the agricultural public in the policy process for agricultural development, as well as the social inclusion of all stakeholders in the process. This tenet ensures that what comes round goes around, so the democratic dividend of agricultural rebirth can spread more evenly and widely. All over the world, the mode of achieving this is through associations, cooperatives, and groups that people belong to—NGOs, CBOs, CSOs, etc. In particular, the business case for cooperatives is strong as private enterprises for galvanizing the financial and intellectual resources of farmers and other rural dwellers in the interest of their members; as also the policy case for the associations is equally strong as advocates of policy decisions of government that favour their members as well. The NGOs stand out as partners to and watchdogs of the government in discharging its mandate functions. So these bodies have critical roles to play in the agricultural rebirth process. To my utter dismay, this instrumentality of participation and inclusion was totally ignored by the last administration, thereby making the benefit streams from agricultural policy ineffectual. It was at the University of Agriculture

Makurdi that the rebirth of an apex farmers' association was born in the early 1990s, during the time of Dr Shetima Mustapha as the minister of agriculture. We traversed the length and breadth of the country to consult with different associations in existence about the need for an apex body. At the instance of Dr Mustafa, we organized a workshop and held the first election of officers for what we then named FOFAN (Federation of Farmers Associations in Nigeria), which transmutated into AFAN (All Farmers Association of Nigeria) at the instance of former president Obasanjo. Suffice it to say that AFAN and other non-state bodies have a critical role to play under the rebirth agenda. Then the question is how to integrate these bodies into the rebirth agenda, as institutions working in partnership with government?

6. *Multisectorality of agricultural sector*: Another institutional concern for the rebirth agenda is the multisector nature of agriculture, which issue emanates from the complex nature of commodity value chains. Typically, a value chain starts from production at one end through storage (on-farm/off-farm storage), overlapping with processing (on-farm/off-farm processing), and dovetailing into marketing and consumption at the other end. From end to end, the typical agricultural value chain in the real sector of the economy traverses several institutions in both the real and service sectors, which collectively determines the performance of agriculture. The interdependence among the various policy institutions with the agriculture ministry may be illustrated as follows:

o Foreign exchange policies (Central Bank of Nigeria, CBN)—these set the average levels and variances of foreign exchange rates and their consequences for the foreign and domestic prices of agricultural commodities including inputs and outputs;
o Monetary policies (CBN)—these often have inflationary implications for the domestic terms of trade between the agricultural and non-agricultural sectors, the term-structure of interest rates and the supply and demand for loanable funds in agriculture;
o Fiscal policies (Federal Ministry of Finance, FMF)—namely taxes, tariffs, inflationary deficit financing, etc., these have pronounced consequences for domestic terms of trade which govern the exchange of agricultural products for products from non-agricultural sectors;
o Incomes policies (Federal Ministry of Labour, FML)—namely national minimum wages, equity, etc., these are usually targeted at

that boosting aggregate demand holds the key to increased production and commercialization of agribusiness which is much in line with Keynes's theory. This takes place through the enhanced capacity of people to reveal their preference for appropriate policy measures and to hold policy authorities accountable for unsatisfactory outcome of policy implementation. However the problem pertains to how to change the way policymakers in Africa traditionally think about food as a need rather than as a right of the people, and how to persuade or compel the policy authorities to adopt and implement a right-driven policy option. The twin problems put the policy innovation at risk, given the long time that the basic needs approach is fully entrenched and the preconceived idea that right to food would exerts pressure on public officers for programme accountability, public probity and policy due process, all of which they generally abhor. Thus the disposition of public officers to right-based approach as opposed to need-based approach to food policy in Africa is inherently negative and the concept cannot be successfully introduced with conventional methods of theoretical or ex-ante research.

It is in light of this problem that the study seeks to demonstrate a suitable methodology for introducing a right-based policy innovation and for institutionalizing the same for faster development of agro-industry in Africa. Therefore a 'policy action research' strategy (PAR) was designed and implemented in Nigeria through the Farm & Infrastructure Foundation (FIF), an organization for promoting policy best practices in agriculture and rural development. The paper presents the implementation of this methodology and the empirical progress observed, which helps in identifying the institutional constraints and in encountering other challenges in the system facing the adoption of the policy innovation on the continent.

Food as a right

Food is a natural right of all citizens of the world in the sense of the need to empower the people to express demand for policies that respect, protect and fulfil it at all cost; and the realization of which should, ipso facto, be inalienable, actionable, justiceable, and ultimately remediable under the law. Unfortunately, it is not case in Africa where the perennial failure of government policies to transform the food sector is perpetually inconsequential, as the demand side of policy marketplace remains dull and not proactive enough to hold policy authorities accountable.

Therefore we subscribe to the notion of right to food as anchored on the philosophy of freedom (Hegel 1821), which is consistent with a contemporary worldview of development as pursuit of freedom (Sen 1999), and also compatible with new thoughts about agricultural development as freedom from hunger (Ayoola...). This view confers certain rights on human beings, including the rights to food (freedom from hunger), to life (freedom from wilful death), to know (freedom of information), to speech (freedom of expression), to list a few. This provides the intuitive reasoning to postulate a vertical extension of Sen's postulate to the regime of rights, hence a stylized definition of right to food, as the irreducible minimum degree of freedom from hunger to support a dignified life, which definition is consistent with poverty reduction as a theme of human development. Indeed universally the Right to Food Campaign (otherwise called Freedom-from-Hunger Campaign) is a mandate of United Nations given to Food and Agriculture Organization (FAO) towards the food security of world's people, using the instrumentality of its many conventions and protocols endorsed by African countries.

The list of international laws backing up right to food include the Universal Declaration on Human Rights (1948), the International Covenant on Economic, Social, and Cultural Rights (1966), Rome Declaration on World Food Security (1996). Specifically, Resolution UN 23 was adopted on 11 April 1998. The United Nations Commission on Human Rights had stated that hunger constitutes an outrage and a violation of human dignity, and therefore requires the adoption of urgent measures at national, regional, and international levels for its elimination. The commission also reaffirmed the right of everyone to have access to safe and nutritious food, consistent with the right to adequate food and the fundamental right of everyone to be free from hunger so as to be able to develop fully and maintain their physical and mental capacities. By and large, right to food is explicitly guaranteed in the constitution of many countries already, namely Brazil (Art. 227), Congo (Art. 34), Ecuador (Art. 19), Haiti (Art. 22), Nicaragua (Art. 63), South Africa (Art. 27), Uganda (Art. 14), Ukraine (Art. 48), Guatemala (Art. 51), Paraguay (Art. 53), and Peru (Art. 6).

In practical terms, the obligation of government under the right to food policy is dimensioned as follows:

- Obligation to respect: This stipulates the limits of state's exercise of power, which includes refraining from destroying people's access to food *through public policies*.
- Obligation to protect: This entails protection from non-state actors in the fields of food safety and nutrition, protection of the environment and land tenure systems by regulating their conduct through *legislations and sanctions*.

Above all, there is no gainsaying the fact that a strong political courage is required for the successful implementation of Agricultural Rebirth Agenda in Nigeria.

References

Ayida, A. A. (1973), 'Business Issues in Financing the Nigerian Agriculture in the Seventies' in *Proceedings of the National Agricultural Development Seminar*, Federal Ministry of Agriculture and Natural Resources (Caxton Press).

Ayoola, G. B. (1997), *Toward a Theory of Food and Agriculture Policy Intervention for a Developing Economy with Particular Reference to Nigeria*. Working Paper 97-WP 187, Centre for Agricultural and Rural Development, Iowa State University, Ames, IA, USA. 21 pages.

FAO (Food and Agricultural Organization) (1966), *Agricultural Development in Nigeria 1965–1980*, Rome, 1966.

FMARD (Ayoola, G. B. and ors) (2009), 'Towards a Sustainable Development and Marketing of Agricultural Commodities in Nigeria—A Benchmark Report (Study Group on Agricultural Commodities Development and marketing Companies 2009

Government Press (1992), The Federal Universities of Agriculture Act (No. 48 of 1992).

Galor, Odded and Daniel Tsiddon (1996), 'Income Distribution and Growth: The Kuznets Hypothesis Revisited', *Economica* 63: 250. Supplement: Economic Policy and Income Distribution (1996) pp. S103–S117.

Idachaba, F. S. and Ayoola, G. B. (1991), 'Market Intervention Policy in Nigerian Agriculture: An Ex-post Performance Evaluation of the Technical Committee on Produce Prices', *J. Development Policy Review* 9 (3): 285–299.

NAD (Nigeria Agriculture Digest) (2015), *Curbing Agricultural Corruption in Nigeria. What is the Opportunity Cost of Fertilizer Subsidy?* Farm & Infrastructure Foundation, Abuja, July–September 2015 edition.

Sen, Amartya (1999), *Development As Freedom* (New York: Anchor Books).

Rethinking Africa's Food Policy in Terms of Rights

Introduction

In Africa, many countries inherited a colonial economy from metropolitan Europe wherein agricultural development was anchored on the 'basic needs' approach. The present poor performance of agriculture on the continent is a product of this approach dating back to over one century ago for some of the countries, but failing to sustain a steady growth path towards food security of the peoples let alone create a viable commercial agribusiness and agro-industrial sector. Thus it is high time policy economists began to rethink this approach and to consider a paradigm shift that better reflects the peculiar features of traditional African society.

The economics of basic needs is premised on the supply side of farm input and output markets, whereby farmers and other rural dwellers look forward to government for price and other support in terms of chemical fertilizer, improved seeds, and machinery as well as provision of rural infrastructure, while the demand side of these markets are generally taken for granted and farmers are assumed to be automatically responsive to policy incentives without constraints. Following the many years of implementing the basic needs approach without meeting the goal of a commercial food sector, it presently appears as a fallacy of the facts about the true situation in Africa, in which case the basic rights approach lends itself to serious consideration as an alternative policy option.

Food as a right, contrary to popular opinion of this notion as an apology of state socialism, is based on the intuitive economic reasoning

- the urban labour force wages but have consequences for rural labour supplies, issues of equity, regional income distribution, etc.
- National industrial policies (Federal Ministry of Industry, Trade, and Investment, FMITI)—these make the provision of agricultural raw materials cheap for the purpose of agricultural industrialization in competition with exports;
- International trade and balance of payment policies (FMITI)—these generally reflect the protocols and agreements reached in the WTO, which often hurts developing economies and set the terms of external trade in favour of developed countries.

Based on the foregoing, the FMARD cannot afford to work in isolation, nor can the other institutions also afford to work in silos. In particular, the institutional relationship between the FMARD and FMITI should be properly examined as an important item on the rebirth agenda. For too long, the two ministries have carried on like two wives of the same husband (or one wife of two husbands), featuring as explosive rivals or bomb waiting to detonate at the slightest prompt. The usual claim by FMITI is that in principle the boundary of FMARD stops at the farm gate as the logical endpoint of production of agricultural commodities, which the latter does not accept in practice. The rivalry came to a head during the last administration when we had a minister at FMARD who crossed this boundary far into the other territory to the point that the minister at FMITI sued for trespass several times at the meetings of the Federal Executive Council. This is an institutional problem for the rebirth programme to address once and for all. Truly there are many overlaps there, but boundary maintenance posture is probably not the way to go in resolving this as an institutional or policy *embrace* problem.[62] Probably the way to go under the rebirth programme is to work out a policy framework for strong institutional collaboration between FMARD on the one hand and the FMITI and other instrumental entities on the other hand.

7. *Agricultural corruption*: The change mantra of the present administration is anchored on eradicating corruption. Therefore an

---

[62] For a theoretical treatment of a policy embrace problem, see Ayoola (1997), wherein conceived a dual sector (agriculture-industry) interrelationship, such that one sector depends on the other for reciprocal thrusts and feedbacks (inputs and outputs). 'Deathly embrace' describes a situation whereby the two systems wait for inputs or outputs from each other that are not immediately forthcoming, and then the entire system gradually slows down until it stagnates altogether (Figure 3).

agricultural rebirth programme tolerant to the rot in the sector is shallow or empty, to say the least, and cannot achieve much. Over the years, the breeding spot of corruption in agriculture sector is fertilizer subsidy, which the GES of the last administration wrongly claimed to have been addressed. It would appear that what the last administration succeeded in doing was not stopping fertilizer corruption in any perceptible way, but in substituting one form of corruption for another through the GES.

As reported by *Nigeria Agriculture Digest*, a brand-new set of corrupt practices have emerged in the aftermath of GES, namely impersonation, which led to substituted persons benefiting from subsidy through alternate or the same phone numbers; new modes of round-tripping of products whereby the agro dealers just paid a token amount (say N1,000) to entitled farmers instead of issuing any bag of fertilizer, but record was established for the agro dealer to collect full value of subsidy free from government later; an upsurge of fly-by-night agro dealers comprising new and old companies formed secretly by or in collusion with government officials, who were registered as emergency fertilizer and seed suppliers without experience, thereby perpetrating fraudulent practices to their gain. Apart from financial corruption, the inherited issue of other fraudulent practices in the fertilizer market is equally important—short bag weights, violation of truth-in-labelling norms, outright adulteration, etc. Yet the last administration did not take practical actions to establish a quality assurance system in the market, nor even hold one meeting of the statutory body, the National Fertilizer Technical Committee, in the entire four years. Which way for rebirth?

V. CONCLUSIONS AND POLICY RECOMMENDATIONS

The message of my paper is, first, that in a democratic agricultural economy, policy and politics are Siamese twins or opposite sides of the same coin, so the political economy of agricultural rebirth is very important. Second, that a meaningful rebirth of agriculture sector calls for an overarching philosophical direction to be agreed by policy stakeholders, which the notions of freedom from hunger and right to food fit well the Nigerian situation. And last, that a number of policy and institutional issues are involved in agricultural rebirth to be addressed, namely the situation with marketing and other institutions, the role of rural infrastructures, framework for policy participation and inclusion, multisectorality of agriculture, and the menace of agricultural corruption, among others.

- Obligation to fulfil: This involves facilitating access of vulnerable groups in society to food-producing resources, and even directly providing food if need be *through social services*.

These obligations do not connote in the least the notion of right to food as an apology of state socialism, in which food will be provided free of charge. Rather it connotes a system of empowerment of the people to take public policy authorities on food to account and to expressly demand tenets of policy best practices in terms of programme accountability, responsibility, transparency, and policy due diligence. Unless these virtues and tenets are enshrined in Africa's public policy process for agricultural development, the effort towards commercialization of the food sector will remain but a mere platitude.

Policy experimentation to domesticate the right to food

The research design consists in the sequence of activities of FIF to implement a nationwide campaign in Nigeria for five years at least (2008–2013). The National Campaign on Right to Food has as its unique selling point a methodology for active engagement of policy stakeholders in a practical learning and action process to introduce the concept of right to food as a policy innovation and to follow through its adoption and implementation closely. Deploying the professional and social capital of an independent organization like FIF in this way depicts the postulation of 'development projects as policy experiments', according to Rondinelli (1993). Thus, the various stakeholders in the policy process to be engaged by FIF under the campaign include the following:

(1) The Government (federal, state, and local)—their ministries of agriculture and other related agencies responsible for policy;
(2) Development communities comprising national and international agencies, including bilateral and multilateral institutions engaged in agricultural development in the country;
(3) Academic institutions, such as universities, polytechnics, and colleges responsible for training the manpower required for agricultural and rural development as well as regenerating technologies for the transformation of agricultural enterprises;
(4) Clientele groups comprised of farmers' organizations, commodity associations, market associations, youth associations, and women's associations, which represent potential sources of policies that affect their lives.

The instruments of data generation and collection of qualitative and quantitative types are constitutional amendment, subsidiary legislation, institutional cum policy reforms, and creation of public knowledge or awareness. Respectively, rights matters are indicted deep in the Nigerian constitution, which lacks any provision for right to food; the formulation of strategies and specification of policy instruments for implementing right to food policy is best carried out as a farm bill to be passed into law (as done in the United States in a cycle of six years); of course, the buy-in of policy authorities is inevitable, and the generality of the people must be carried along from outset. It is obvious that this type of (policy) experiment cannot be carried out through a public institution, as the work involves a substantial degree of policy advocacy beyond the casual recommendations inside journal articles and commissioned study reports or through the usual expert advice not binding on the principals in the policy system, which provides justification for FIF as an independent organization as driver of the policy experimentation. Indeed, public policy authorities are negatively disposed to support, let alone undertake, the implementation of right-driven strategies of development like this one.

Thus the specific objectives of the campaign were:

- Propagate the notion of right to food in the general public
- Entrench right to food in the constitution
- Mainstream rights instrument in government policies on food security
- Formulate a Bill of Rights to Food for passage into subsidiary law

The FIF's model involves a number of activities to engage the policy stakeholders, including the following: engagement of policy authorities (executive, legislature, and judiciary); engagement of general public; consultations/collaboration/networking/partnerships; publication and media exposure; legislation; policy debate and dialogue; etc. Table 1 presents some details about these activities.

Table 1—Policy experimentation—Activity Matrix of FIF's National Campaign on Right to Food in Nigeria

Component/Activity	Specific actions/other details	Measurable output/outcome/impact
Engagement with pro/principal stakeholders in the policy process		
General public	Presentations/speaking engagements—Invited paper on right to food presented at World Food Day Symposium on annual basis; own inaugural lecture at University of Agriculture Makurdi	General awareness an sensitization
Executive	Communications with presidency (Exchange of letters to Mr President and agriculture minister)	Government was put on notice from outset
	Sensitization of policymakers—visits to different policymakers at federal, state and local level Participation at policy-related events (meetings, workshops, etc.)	The objective is to arouse the consciousness of the general people about the need for a right to food bill and to invite their participation in the right to food movement generally.
	Sponsorship of memoranda at statutory meetings such as National Agriculture Development Committee (NADC) and the National Council on Agriculture (NCA)	Record created in the files of the national assembly about right to food National assembly was put on notice about the right to food movement

Table 1—Policy experimentation—Activity Matrix of FIF's National Campaign on Right to Food in Nigeria

Component/Activity	Specific actions/other details	Measurable output/outcome/impact
Legislature	Communications with National Assembly (exchange of letters with president of Senate and Speaker of the House)	National assembly was put on notice about the right to food movement Development of rapport for the benefit of the bill
	Presentation of memorandum on Right to Food to the National Assembly Committees for the review of the constitution—Senate and House of Representatives (National sittings, Abuja) Active participation in the constitutional debate and review process—Presentation of memorandum on Right to Food at the zonal sittings of the constitution review committees—six zones: NE (Kano); NW (Bauchi); NC (Makurdi); SE (Enugu); SW (Lagos); SS (Port Harcourt)	Sensitization of legislators
	Sponsorship of legislative bills on right to food—Right to food Bill in the House; Right to food Bill in the senate	Legislative Bills gazette and passed through the first reading
Judiciary	Communications with National Human Rights Commission (exchange of letters with executive director)	Moral support

Development community	Communications with Food and Agriculture Organization FAO (exchange of letters with director general)	Development community was put on notice
Consultations/Collaboration/Networking/Partnerships		
	Collaboration with relevant agencies and organizations—National Human Rights Commission (NHRC), National Orientation Agency (NOA), All Farmers Association of Nigeria (AFAN), Federation of Agricultural Commodity Associations in Nigeria (FACAN)	Increased positive opinion about right to food
	Partnership with Oxfam on the Voices for Food Security platform	Financial support from Oxfam
Publication/Media exposure		
Print media	*Nigeria Agriculture Digest*—a specialized publication of FIF	Awareness on the concept of Food as a Right increased.
	Publications of IEC materials *(production and distribution of IEC materials on Food as a Right for the purpose of public engagement)*	Systems and structures for the implementation of Food as a Right strategy strengthened

Electronic media	*Radio jingles of food security in collaboration with NOA (Conduct of sensitization and jingles in the media to inform rights holders and duty bearers about the concept and meaning of the right to food* Appeared on radio and television programmes at different times. I also featured in a documentary on Independence Anniversary programme of the FMARD discussing issues about right to Food. Sponsored discussions on Right to Food on live and pre-recorded programmes (e.g. Radio Link). Sponsored appearances on other radio programmes at prime times (e.g. Eagle Square)	Public awareness and discussions generated
Social media	Facebook postings, Twitter exchanges	People entered into discussion about right to food
Education/Training		
Workshop	*Training and education of legislative aids to MHR and senators on the agric. committee (strengthen identified partners to constitute a Policy Innovation Platform (PIP) for active engagement through advocacy trainings)*	Knowledge about right to food created for the support of legislative actions

Policy debate/dialogue	Organized school debate; public debate around right to food using telephone and Internet services;	Interest created in right to food across the strata of society
	Public debates on food as a human right conducted across the six geopolitical zones of Nigeria): 20 copies of the CDs produced for distribution to radio and TV stations. public debate was also optimized on YouTube for website use on FIF website.	
	Policy Innovation Platform established for active engagement of policy drivers	

Results and discussion

The active engagement of stakeholders has led to significant sensitization of actors in the food policy system. As the campaign gathered momentum, the general public was sensitized to their role as right holders, and it is expected that in due course they be able to make demand for their right to food to be observed as desired. The public authorities were also sensitized as duty bearers, with the expectation that in no distant future, the government would begin to consciously recognize, protect, and fulfil the right of people to food in its business.

The effect of engagement of legislature was clearly the most considerable, mostly in terms of record creation as yet but accompanied by a bright prospect of concrete action in future. The draft bill on right to food was formally tabled before the House of Representatives in 2010 at the instance of FIF, which as gazette and published in the Journal of National Assembly when it scaled through the first reading stage (No. 20 Vol. 7 of 12 April 2010). The bill sought to amend chapter two of the constitution (Directive Principles of State Policy) to give explicit recognition to right to food alongside food security. However, the bill did not go to the second reading stage before the expiration of the (sixth) assembly, so it needed to be reintroduced according to rules of the House. Meanwhile in 2013, another bill was introduced at the instance of FIF which has also scaled through the first reading. This sought the amendment of chapter four of the constitution (Fundamental rights), with a view to making the right to food actionable and justiceable under the law.

An important lesson learnt is the need to incorporate uncertainty factor about the political environment into the work plans of a policy-induced campaign such as FIF's. The organization faced a major challenge pertaining to the unpredictable political environment that sometimes stalled the process and put things in abeyance, such as the delay in commencing the constitutional amendment process by the national assembly and the distortion introduced by the general elections of 2007 and 2011. In the first case, the delay was due to fight for supremacy between the House and the Senate over which of them should preside over their joint meetings for the purpose of constitutional review. In the second case, the president of the federal republic was indisposed for a number of months during the year 2010 and he was abroad for treatment on a prolonged basis, before he eventually passed on. This introduced delay in normal legislative business as the attention of legislators was focused on resolving a crisis arising from lack of clear provision in the constitution for succession in such a situation.

Both events generated intense public and legislative debates around the issues of chairmanship of joint meetings and succession arrangement, which led to stepping aside of ordinary bills, including the one on right to food. Furthermore, as the general elections of 2011 approached, a lull ensued in the activities of legislators occasioned by a long period of political campaigns, which lasted until the end of that assembly. Thus, the project was implemented in an external environment of considerable distractions of both the executive and legislative arms of the government from normal work, in order to address certain vexed issues.

The implementation of the campaign has led to some structural and lasting changes in society, which includes the change in the mindset of people as influenced by records of events created. Such official records created and the mindset influenced, attributable to the campaign, was towards notion the new of food that is gradually emerging, from the previous notion of food as a basic need to the present notion of food as a basic right in the country. Some evidence of these includes the records created by FIF at the level of political authority through systematic communications with the president on the subject matter of food as a right, through letters that were passed down for treatment by the agriculture minister, thereby changing the mindset of politicians gradually. Specifically in a recent speech, the language of the minister has started to feature the notion of food as a right along with the previous language of agriculture as a business.

Similar records were created by FIF at the level of policy authority through official memorandum to the National Council on Agriculture (NCA), which is the highest policy organ of the federal government on agricultural development and which is comprised of high-level official delegates from the state and local level of agricultural administration of the country, including the members of the international development commitment. Thus the notion of food as a right has been spread from top to the grass roots, capable of changing the mindset of the general public on a lasting basis.

Conclusions and Recommendation

The transformation of African agriculture in terms of productivity, commercialization, and market development requires an appropriate philosophical framework in formulating the food policies. In this regard, it is argued that policymakers in Africa need to think out of the box and give a serious consideration to the right to food approach in their respective policy domains. Towards this end, the outcomes of policy experiment performed in Nigeria suggest that proactive advocacy is superior to mere policy recommendation and advice to make policy changes happen in the food sector. The strategic and operational lessons learnt are useful in resolving the constraints faced and in meeting the challenges posed in the course of implementation.

The way forward is to reinforce the campaign instruments and to maintain the advocacy pressure on the policy authorities until the policy change happens as desired, particularly the policy effort to mainstream rights instruments in the relevant programmes and projects of government. Finally, the capacity of FIF and similar organizations should be improved so that we can stay the course and sustain the effort towards achieving the goals of the campaign.

References

Hegel, W. F. (1821), *Elements of Philosophy of Right*, Oxford World Classics.

Rondinelli, Denis A. (1993), *Development projects as social experiments. An adaptive approach to development administration* (London and New York: Routledge).

Sen, Amartya K. (1999), *Development as Freedom* (Oxford University Press).

Using Smart Subsidies to Support Small Scale Farmers in Africa—Lessons from Nigeria's Growth Enhancement Scheme[63]

Symposium paper in response to the call for papers on: Using Smart Subsidies to Support Small Scale Farmers in Africa organized by African Association of Agricultural Economics (AAAE) during the 29th International Agricultural Economics Association (IAAE) Conference, Milan, Italy, 10–14 August 2015.

INTRODUCTION

1. In Nigeria, the high political visibility of fertilizer owes to the issues surrounding the administration of a price subsidy as a public policy instrument for promoting the production and use of farm inputs. This is with a view to attaining food security of the nation in general and increasing farm incomes and improving livelihoods of the people in particular. However, after many years of policy intervention by the federal government, the situation with fertilizer sector leaves much to be desired in terms of local production that permanently lags behind importation, coupled with the massive leakages, sharp practices, and other frauds associated with subsidy policy, hence the attendant demand and supply gaps in the fertilizer market, culminating in perennial low yield of crops and non-remunerative incomes accruing to small-scale farmers.

---

[63] This paper draws from a recent report at the instance of World Bank (see Keyser et al. 2014)

2. The question that this presentation is centred around is, What incremental lessons do we have to learn from the recent application of smart subsidies in Nigeria's agriculture, which was implemented as a Growth Enhancement Scheme (GES) between 2011 and 2015?
3. The methodology is twofold, namely qualitative situation analysis of the sector, and industry position analysis through participant observation technique of Public-Private Dialogue (PPD).[64] The paper is structured into four sections, with the view to: first, deepening the background knowledge about fertilizer sector in different aspects (legal/regulatory, sector players, and Quality control); next, analyzing the challenges, constraints, and opportunities that characterize the sector that GES was designed to address or explored; then, determining the position reached on different issues about the fertilizer sector as well as the recommendations made by stakeholders at the dialogue; and finally, drawing the policy implications and identifying the way forward in promoting policy advocacy role in the fertilizer sector.

BACKGROUUND OF THE FERTILIZER SECTOR

4. The evolution of Nigeria's fertilizer sector has its background in a process of deepening public intervention over time. At independence (1960), when there was no ministry for agriculture at the federal level, the regional governments engaged in disparate importation of mineral fertilizer for distribution to farmers in their respective jurisdictions. Later, following the creation of a ministry of agriculture at the federal level (1967) and the ensuing civil war (1967–1970), the federal government assumed full powers to embark on agricultural development. Thus, the Federal Department of Agriculture (FDA) was established in 1970 with a special board created by law to deal with fertilizer promotion (Fertilizer Board Act, 1972), and later an organ of the ministry for the same purpose (Fertilizer Procurement and Distribution Division, FPDD) established in 1976. At this stage, the FPDD was assigned the previous responsibility of the regional governments as sole organ of government for importing and

---

[64] The Public Private Dialogue was recently organized by the DFID ENABLE (*Enhancing Nigerian Advocacy for a Better Business Environment, ENABLE*) Project in collaboration with the Fertilizer Producers and Suppliers Association of Nigeria (FEPSAN), and anchored by the author.

supplying mineral fertilizers to the Nigerian market, thereby starting a long period of public monopoly in the market (1976–1997). During the period, the assistance of the World Bank was obtained in terms of two loans for the purpose of massive importation and distribution of fertilizers at subsidized prices, which facilitated a protracted process of market development for agricultural inputs through a policy-driven or, more specifically, subsidy-induced mode of public intervention to promote fertilizer use in the country.

5. Thus, at issue about Nigeria's fertilizer sector is the sustained implementation of a subsidy policy that, after its initial success to stimulate demand for the products from farmers, then constituted a problem in itself in terms of fiscal burden, huge financial leakages, and diversion of policy benefits to unintended channels or unintended beneficiaries. During the period, the strategies employed for delivery of fertilizer subsidy which failed in quick succession include the following: distribution solely by Federal Ministry of Agriculture, though the designated depots; distribution solely by Federal Ministry of Agriculture, though the local governments; distribution by a task force or special committee appointed by the Ministry of Agriculture; distribution through the Agricultural Input Supply Company of the State; distribution through the Agricultural Development Project (ADP) of the states; and distribution through hybrid arrangements of two or more of the above systems.

6. Further to reforming the sector, the department initiated a Fertilizer Market Stabilization Scheme (FMSS) involving the seasonal procurement and distribution of *limited* quantities of fertilizers (in the order of 250,000 MT), with a 25% subsidy by the federal government. For this purpose, a voucher system was introduced which involved distribution of vouchers as promissory notes to farmers to buy fertilizer from agro dealers at a discount of 25% (subsidy), to be redeemed by the sellers in full afterwards. This was implemented on a pilot basis in selected states and positive signals were showing already, when the immediate past administration introduced a modified scheme—Growth Enhancement Scheme (GES)—that replaced the voucher with electronic wallets to deliver subsidy to farmers through mobile telephones.

7. The participating farmer, having been preregistered, tenders the wallet to an agro dealer to collect one bag of fertilizer and get one free. The price subsidy is redeemed from government through a bank account.

8. A description of the fertilizer sector in Nigeria consists in a number of aspects, as highlighted below.

Legal and Regulatory Context

9. The broad framework of the sector for market regulation and quality control by government is anchored on the subsisting policy statement on the sector—*National Fertilizer Policy for Nigeria*. The legal and regulatory context of the fertilizer sector was established in the broad objective of the policy, i.e. to facilitate farmers' timely access to adequate quantity and quality of fertilizers at competitive but affordable prices in Nigeria. Towards this end, notwithstanding the intention of the federal government to pass a law for establishing a fertilizer regulation and control body, as indicated in the policy statement and which has been in the pipeline since 2002 or so, the legal framework for implementing this remains uncertain. Acting under the pressure of farmers who were at the receiving end of poor-quality fertilizer in the market season after season, government embarked on a process to establish an agency for fertilizer regulation and control by commissioning an inventorization study of the sector, followed by production of a framework of regulation and control incorporating a draft bill for the purpose (Ayoola et al. 2002). The draft bill, which was produced for an Act of the National Assembly to establish the National Agency for Fertilizer Regulation and Control (NAFRAC), had passed through the various stages, from technical drafting stage by NFTC, to participative validation stage by a stakeholders' workshop, and then to the administrative appraisal stage by the ministry, and lastly to the official adoption stage by National Council on Agriculture (NCA), before entering into a lull till date.
10. However, in the absence of a strong political will to create the agency from one administration to the next, the draft bill was subsequently revised as the National Fertilizer Quality Control Act, 2012, which is presently before the Federal Executive Council for approval and eventual transmission to the National Assembly for passage into law. The latest draft bill seeks 'to regulate the manufacture, importation, distribution and quality control of fertilizer in Nigeria', albeit without a provision for the establishment of NAFRAC or another implementing agency anymore. Thus the present draft Bill supersedes the previous one and automatically repeals the extant

(1972 and 1991) laws. In place of this, the present draft bill entrusts the minister with its implementation and enforcement, of course with the instrumentality of FFD. Furthermore, unlike the NAFRAC Bill, the draft bill fails to recognize the ECOWAS rules on fertilizer, which supersedes country laws where there is any incurable conflict of provisions or where national laws are inherently defective.

Sector players

11. Among the institutional actors, FFD is the agency of government responsible for the general policy administration and oversight of the fertilizer sector, including regulation of the market and control of quality. In performing this role FFD (formerly FPDD) is secretariat to National Fertilizer Technical Committee (NFTC), which is a statutory advisory body of experts in the subject matters relating to fertilizer as a farm input and the fertilizer economy as a whole.
12. The operational base of NFTC is the National Fertilizer Development Centre (NFDC) located at Kaduna with a laboratory complex for testing of fertilizers and soil samples as well as related scientific investigations about the fertilizer industry generally. In the days of FPDD, the NFTC was responsible for producing annual fertilizer slates for importation, evaluating the technical recommendations based on fertilizer trials conducted by research institutes under the nationally coordinated fertilizer trials, and collaborating with Standards Organization of Nigeria SON (a parastatal of Federal Ministry of Industry, Trade, and Investment) to produce fertilizer standards in consonance with different fertilizer formulations and in subjecting new formulations to laboratory or field test. However, the NFTC is totally inactive in the present time, so issues about fertilizer standards and recommendations have been put in abeyance for a long time now. In the absence of a separate agency to enforce fertilizer rules under the aegis of FMARD, the proxy agency for this task is NAFDAC (a parastatal of Federal Ministry of Health), which acts as the proxy agency for quality maintenance and enforcement of standards for chemicals and other products in the whole economy. But the performance of this role by NADAC leaves much to be desired.
13. The actors on supply side of fertilizer market are comprised of manufacturers, blenders, importers, and other suppliers of products for the replenishment or amendment of the soil to improve the

nourishment and integrity of soils, thereby increasing productivity of crops. The demand side of the market is comprised of the farmers who require these products for application to the soil to increase yield of crops planted by them. In between the two sides are other participants, which include private agro dealers or government bodies acting as middlemen to distribute or sell fertilizers to farmers along with seed and other inputs. The two major fertilizer manufacturers of fertilizers in Nigeria are NOTORE Chemicals (formerly NAFCON) as sole producer of urea and TAC Nigeria Limited (formally Federal Superphosphate Fertilizer Company FSFC) as the sole producer of phosphates, both formally federal government-owned but now privatized. There are quite a number of blending plants scattered all over Nigeria and also a number of importers of finished fertilizer products or raw materials. The umbrella body of participants on the supply side is FEPSAN (Fertilizer Producers and Suppliers Association of Nigeria) while the umbrella body of actors (farmers) on the demand side is AFAN (All Farmers Association of Nigeria), both striving to influence policy on fertilizer in favour of their members from time to time. It is hoped that sooner than later two big plants will come on stream, which are Indorama and Dangote.

14. Recently another participant was added to the market based on political expediency rather than transaction motives, i.e. State Security Service (SSS), whose participation was a sequel to the current security challenges facing the country, whereby subsidized fertilizer became useful for making bombs and explosives. Thus, in order to mitigate this practice of militants or insurgents, the SSS was brought in to issue clearance or end-user certificate to fertilizer importers. Next is the set of state agencies for agricultural development, namely Agricultural Development Programmes (ADPs). Each state of the federation has an ADP for undertaking agricultural development activities, particularly extension and input distribution, including rural infrastructure provision. Some states operate agro-input supply outfits that produce and/or distribute fertilizer (and other inputs) to their farmers.

15. And, last is the presence of NGOs in the sector which their activities cover the agriculture sector generally, and which act in various ways as partners to government to give voice to the farm population while also promoting the accrual of benefits of programme implementation to smallholder farmers. An example of such organizations is Farm & Infrastructure Foundation (FIF), an organization for promoting policy best practices in agriculture and rural development, and a

member of a larger body of NGOs in the agriculture sector by the name Voices for Food Security (VFS), a consortium of organizations numbering about 20 working together to influence the policy process for agriculture in specific ways. Presently, FIF implements a national campaign on right to food, which has sponsored two draft bills in the national assembly for constitutional amendments to this effect. In terms of brokering services in the fertilizer sector, FIF was instrumental to the articulation of several policy documents and studies for addressing issues towards repositioning the industry; such documents include the fertilizer liberalization study (1994), fertilizer inventorization study (2002), fertilizer socioeconomic study (2006, 2012), fertilizer policy of Nigeria (2006), to mention a few.

Fertilizer supply and demand

16. The fertilizer sector of Nigeria is a multibillion-naira market even with a large gap between the actual supply and potential demand. Prior to policy reform of the 1990s, fertilizer supply had reached a peak of 1.2 million tonnes in 1992, but this dropped to a paltry 56,706 metric tonnes in 1997. Two years later, in 1999, subsidy was reintroduced at a level of 25%, in order to shore up supply but subject to a limited procurement operation by Federal Government (in the order of 250,000 tons). However, the low level of supply continued, a situation that obtained till 2012 when the Agricultural Transformation Agenda (ATA) was launched that has GES scheme as a key component focusing fertilizer subsidy delivery. Apart from subsidy withdrawal as a causative factor, supply dropped to a low ebb also on account of the two granulating plants that went out of function for a protracted period of time, thereby stifling the local blending plants of their source of raw materials for producing the different NPK formulations, i.e. urea and phosphate products. Meanwhile, federal government had withdrawn from importation at the outset of reform, and the capacity of private sector to import fertilizer was limited by scarcity of foreign exchange.
17. Supply of fertilizer in the market comes from both domestic production and importation sources. Domestic production of fertilizer is anchored on the two granulation plants, which were initially public companies but later privatized. The outputs of these plants, urea and phosphates respectively, combined with imported potash, constitute

the critical raw materials required by local blending plants to produce the various formulations of fertilizer found in the Nigerian market.
18. According to official sources, the volume of fertilizer supply in the Nigerian market during the decade preceding GES together with estimated cost of subsidy is as follows:

   a. 2001 (615,000 MT; N1.7 Billion)
   b. 2002 (340,746 MT; N1.5 Billion)
   c. 2003 (511,841 MT; N1.9 Billion)
   d. 2004 (560,150 MT; N2.5 Billion)
   e. 2005 (600,000 MT; N1.8 Billion)
   f. 2006 (709,000 MT; N3.5 Billion)
   g. 2007 (990,000 MT; N4.9 Billion)
   h. 2008 (691,153 MT; N14.2 Billion)
   i. 2009 (371,062 MT; N11.0 Billion)
   j. 2010 (586,145 MT; N22.3 Billion)

19. In the present dispensation under GES, in 2013, 4.1 million small-scale farmers (out of the annual 5 million targeted) were registered. In 2013 also the GES featured 41 fertilizer companies as suppliers under the scheme, whereby subsidy support was provided directly to registered farmers at 50% of the price per bag (N5, 500/bag) of fertilizer subject to purchase of two bags at designated redemption points within the local government premises. The total supply under the scheme could not be immediately disaggregated by sources into local production and importation yet, but there were no noticeable instances of fertilizer export, formal or informal, in the market in recent times. It appears that the leakage of fertilizer supplies to neighbouring countries occasioned by price subsidy has stopped with the introduction of electronic wallets to deliver subsidies to registered smallholder farmers.
20. The channel of supply of fertilizer in Nigeria is two-pronged – public (official) and private (open) market. There is a third segment of the market, which is the black market due to the actions of arbitrageurs in between the two recognized market. The official (subsidized) channel is meant for small-scale farmers cultivating one hectare or less to be subsidized, while the private channel is for farmers cultivating land greater than one hectare which is not subsidized. The supply is non-discriminatory by type of fertilizer provided in different agroecological zones. A uniform formulation of NPK and

urea was provided in all areas. notwithstanding the crop or soil conditions prevailing in these areas.

21. Much of the farmers' demand for fertilizer in Nigeria is latent or pent-up in nature, that is, apparent only but not realized. According to a recent socio-economic survey (FFD 2012), the proximate determinants of fertilizer demand in Nigeria include land area cultivated as modal factor. Moreover, demand for mineral fertilizer in Nigeria is generally policy-induced (supply-driven), which reflects the promotional role of federal government, through provision of subsidy, as assigned by the constitution. Thus the potential agronomic demand of fertilizer in Nigeria was initially estimated at about 7 million tonnes (APMEU 1990), which has since increased to about 9 million with increased release of higher yielding and fertilizer consuming crop varieties (FFD 2006). However, the actual demand was estimated at about 1. 7 million tonnes at the historically low consumption rate of below 13 kg per ha., which an early evaluation study of GES indicates a projected increased actual demand estimate of about 2 million tonnes in 2012 at a plausible enhanced rate of 20 kg/ha. (FIF 2013). Nonetheless, both the historical and recent estimates of actual demand have never been satisfied by the supply side of the fertilizer market in Nigeria.

Fertilizer quality control

22. The policy direction is towards ensuring good quality fertilizer products as the primary responsibility of government, and towards encouraging the private sector to monitor the quality of its products to ensure that they conform to the provisions of the existing legal and regulatory framework, particularly truth-in-labelling practices.
23. The draft bill provides the specifications of a good-quality fertilizer in terms of investigational allowances and actual values of primary nutrients obtained from the results of a tested sample. For example, in the case of macronutrients, compound NPK fertilizer with minimum guaranteed actual values of 20-10-10, the investigational allowances should not exceed 0.73% for nitrogen (N), 0.72% phosphate (P), and 1.08% potash (K). In the case of macronutrient straight (single-nutrient) fertilizer such as urea, or the superphosphates (DAP, MOP, MAP), the actual value should not be less than guaranteed value by more than 2%; and in the case of secondary and micronutrients fertilizers, the investigational allowances should not exceed the

following percentages of guaranteed amounts for some nutrients as follows: Calcium 5%, Magnesium 5%, Sulphur 5%, Boron 15%, Cobalt 30%, Molybdenum 30%, etc.

24. Even though the control measures in the market are practically non-existent at the moment, the envisaged mechanism involves the aspects of inspection and enforcement, for which penalties for their violations have been explicitly stipulated in the draft bill, as follows:

   i. Power of entry—that an authorized officer may (a) enter premises, buildings, vehicles, plants, etc. used in the manufacture, importation, sale, storage or transportation of fertilizers; and (b) take samples from the fertilizer found during an inspection for laboratory analysis, for the purpose of ascertaining the quality of the fertilizers
   ii. Display of permits or certificate of registration—that a manufacturer, importer or distributor should display the original permit or certificate of registration at his company or point of sale
   iii. Enforcement procedure—that the minister has power to stop sale of deficient fertilizer;
   iv. Fertilizer weight—that deviation of the weight of fertilizer bags on the label beyond a stipulated limit may attract sanction of minister
   v. Labels—that a set of information should be provided as specified

25. Quality control also involves the application of fertilizer suitable for different soil conditions, which borders on the work of NFTC, for which purpose the NPFS conducts soil surveys so as to generate information on soils from different locations and various crops grown. Also, new fertilizer types are to be tested under the Nationally Coordinated Research Project (NCRP) for one or two seasons before they are introduced to the market, not only to confirm the composition but also the efficacy on Nigerian soils so farmers are not deceived about quality of fertilizers. However, the NFTC and NCRPs are presently inactive, pending their resuscitation under WAAPP soon. The hope for quality control in the fertilizer market rests on the fate of the draft Fertilizer Bill in the presidency, seeking an Act of the National Assembly for the purpose, but the rate of progress is so slow.

26. The capacity for conducting laboratory tests of soils and fertilizer samples is large as presently existing in the National Agricultural

Research System (NARS). The NARS of Nigeria comprises 15 commodity-based National Agricultural Research Institutes (NARIs), three specialized Universities of Agriculture (UAs) as well as about 100 federal or state conventional universities with Faculties of Agriculture (FAs). In addition, there are private labs where quality tests and other analyses may be performed. The original intention of government was to establish a number of reference laboratories but this has not materialized. Nonetheless the National Fertilizer Development Centre (NFDC) located at Kaduna was established, incorporating a laboratory for the conduct of soil and fertilizer tests, which NFTC used to rely very much upon. Recently, mobile test kits for soil samples at the farm level was introduced under the FAO-supported National Project for Food Security (NPFS), but the adoption is still at a low level.

CHALLENGES, CONSTRAINTS, AND OPPORTUNITIES

27. The presence of subsidy was responsible for a long period of chaos in the fertilizer market, which featured heightened corruption, product diversion, and several leakages in the economy. For these reasons, prior to GES it was estimated that only 10–30% of the subsidized fertilizer actually reached the smallholder farmers for whom it was intended, while average consumption remained at a low ebb of less than 10 kg/ha. compared to world average of 100 kg/ha.

28. Thus the analysis of fertilizer sector borders largely on the subsidy issue, which accounts for its political visibility in the Nigerian economy for a long time till now. Originally, subsidy offered an opportunity to increase fertilizer demand by smallholder farmers who have limited purchasing power. This demand, once induced, was expected to be further stimulated and sustained on their own into the future together with the other green revolution technologies, thereby generating both consumer surplus and producer gains far into the future economy along with the growth of private sector in capacity to meet this demand. However, the lessons of experience from the operation of direct subsidy programmes suggest that the policy became self-defeating in many ways, which include the following features of the fertilizer subsidy schemes in Nigeria:

(a) A dependency mentality has developed among farmers on government subsidy, thereby making it a permanent obligation of government and subsequently a huge fiscal burden;
(b) There is absence of proper allocative role of price as a result of distortions created by subsidies, as production decisions of the farmers were not directed along the lines of efficient resource utilization;
(c) The level of morale is low among the private sector participants to build up their stock and make substantial enterprise initiatives in the distribution of fertilizer, owing to price rigidities associated with subsidy, which ultimately produce a limiting effect on agricultural GDP;
(d) There is presence of substantial externalities whereby farm subsidy benefits illicitly flow to unintended channels within and outside the economy;
(e) The subsidy programme faces frequent exploitation to serve the selfish motives of civil servants, politicians, and other elite groups, to the detriment of the ordinary farmer.

29. Moreover, the message of derived nature of fertilizer demand is lost in the subsidy scheme whereby the input is wrongly focused by the subsidy policy rather than the farm commodities that it is used to produce. The implication of this is that farmers with genuine demand for fertilizers find ways around the scheme to collect the subsidized products and deploy them to other channels instead of their farm. This is more so in the face of abject poverty among farmers that will push them to act in favour of one Naira today being better than one naira tomorrow, thereby selling the products in the open market at higher prices in real time.

30. Generally, with respect to these challenges, it appears that government lacks in capacity to oversee the fertilizer sector in terms of the political will to terminate its policy intervention in the fertilizer market through subsidy provision that had overstayed its usefulness and is now akin to a dog in the manger, which fails repeatedly to address the accompanying problems of fiscal burden, huge financial leakages, and diversion of policy benefits to unintended channels or unintended beneficiaries, while also preventing these problems to be addressed by other policy instruments. It also appears that government lacks in the capacity to control the market and rid it of bad products and sharp practices, as no effective system is presently in place for the purpose, and despite the historically rampant episodes of adulteration of products, short bag weights, and lack of

truth-in-labelling practices in the market, there were practically no observations of anyone arrested, let alone prosecuted to enforce the extant laws, hence the consensus among experts at this stage that the liberalization of fertilizer sector should be completed by scrapping the subsidy scheme altogether and considering alternative but more effective modes of farmer support.

31. Apart from the challenge of ineffectual subsidy administration, there is no gainsaying that absence of effective quality control also poses a major constraint to the growth of Nigeria's fertilizer market in terms of structure, conduct, and performance of the market. The net effect of widespread adulteration is that population of farmers demanding fertilizers will dwindle with time as they gradually exit the market upon the very first experience of crop failure attributable to bad quality of fertilizer applied. Eventually the volume of supply will also decrease as companies exit from the market when their turnover consequently becomes low. Also the conduct of the market suffers from lack of truth in labelling, which systematically erodes the confidence of genuine sellers and buyers and predisposes them to exiting the market sooner or later. Thus, growth of the fertilizer market is retarded in terms of price distortions and low volume of trade, which exacerbates the poor performance of the market, thereby discouraging new entrants from home and abroad. As observed, such distortions manifest in terms of inefficient pricing such as the observed higher selling price of urea through the official market (N5,500) than what obtains in the open market at the same time (N4,000), which raises the issue about subsidy illusion among farmers, meaning that effect of GES is probably apparent only but not real.

32. Meanwhile, in principle, the ATA is generally perceived as a pro-business approach to intervention for further inducing the participation of private companies in the market and eliminating the annual wasteful process of tendering for fertilizer purchase by government, with a view to improving the incidence of subsidy on small-scale farmers. In this regard, government has also put in place a programme of support to private fertilizer companies and agro dealers to raise capital from banks at low interest rates and with guarantees against default, in order to enhance their capacity to build stock and expand their operations to grass-roots level. This complements the ATA under another programme of government at the instance of Central Bank of Nigeria—Nigeria Incentive-based Risk-bearing Scheme for Agricultural Lending (NIRSAL), which

contemporaneously stimulates farmers to seek loans and incentivizes the banks to lend at lower risks. In spite of these apparent outcomes of ATA or GES on the positive side, the influence of federal government still weighs too heavily on the market than what is desired, not just as provider of price subsidy directly to farmers instead of through the states as constitutionally assigned, but also as a visible actor in the market through the communication technology platform established for distributing electronic wallets to farmers.

33. In its own right, the dominant role of federal government poses a constraint to maximizing the role of states in agricultural development generally and ATA or GES particularly, with special reference to fertilizer. This happens as producers focus the federal government to meet the narrow range and low volume of products demanded from that single source instead of focusing the 36 states to meet the wide range and high volume of demand from diversified agroecologies of the country. In this regard, the natural response of states to overbearing posture of federal government in agricultural development beyond the constitutional stipulation is their observed complacency that robs the system of substantial growth of the market in terms of product types and quantities. This situation is a major concern to civil society organizations in the agricultural sector, such as FIF, which presently advocates a policy change to properly realign the roles of federal and state governments with the provision of the constitution, as a necessary measure towards resolving the role confusion between the tiers of government genuinely engaged in agricultural development of the country.

34. Last is the quality issue, which at the moment has no effective arrangement in place. In general, this pertains to inadequacy of nutrient composition whereby the active ingredients are less than what the standards stipulate. In this regard, discussions with farmers indicate rampant cases of poor quality, which are not visible to the naked eye except when their yield expectations have not been realized. A variant of the quality challenge arises sometimes when the bags are filled with foreign materials such as sand, ash, etc. and sold to farmers as fertilizers, which represents a form of market sharp practices, or malpractices such as underweight bags whereby a supposedly 50-kg bag is found to weigh less, or lack of truth-in-labelling in order to short-change other suppliers in the market. Given the long-standing nature of this problem when the combination of SON and NAFDAC were unable to address them, a stakeholders' workshop had pushed for the establishment of a stand-alone quality assurance

and regulatory agency specifically for fertilizer under the auspices of FMARD, which was accepted by the ministry and adopted by NCA since 2004 but not implemented till date.

35. The subsidy and quality challenges notwithstanding, the large number of participants in the fertilizer sector in diverse areas, including about 120 million farmers, augurs well for a good structure, conduct of the market, as a strength factor for ensuring that there are appropriate role players of the various functions in critical areas, and also as a strength factor for ensuring that the policy environment is conducive for fertilizer development in favour of market conduct. The diverse functions covered by these participants include Research and Development, Fertilizer Production and Marketing, Quality Control, Import and Export, Fertilizer Extension and Promotion, Fertilizer Pricing, Private Sector Development, and Human Capacity Development. The diversity and coverage of participants implies that opportunity for trade abounds in the sector in favour of new companies that have their confidence built to operate in the market in good conduct.

36. However, the capacity and efficiency of some of these participants leaves much to be desired, thereby posing a constraint to new and existing companies in the fertilizer market to maximize output and make profit. As the number of companies undertaking actual manufacturing is few, only one for urea manufacture and the other one for phosphate manufacture, while the remaining companies deal with blending, importation, or transportation, the tendency towards monopoly in the manufacturing industry is quite strong in Nigeria. Although the number of supply companies has increased under GES, the capacity of NARIs and the production companies to experiment with new formulations for different crops in different agroecological zones of the county is quite weak.

LESSONS LEARNT

37. The dialogue presents opportunity for identifying a number of lessons of experience from a first attempt to implement a smart subsidy regime in Nigerian agriculture through the GES.

The Issues and Industry Position

38. By its very nature, the demand for fertilizer, like other inputs of agricultural or other production inputs, is a derived demand, in the sense that it is not purchased for its own sake but for the sake of producing other (farm) products. This implies that the success of policy efforts to increase fertilizer demand and use depends on the success attained in increasing farm outputs generally and in operating the different commodity value chains specifically. The subsidy policy for increasing fertilizer demand and use is presently targeted at the small-scale farmers who are generally resource-constrained. However, the application of subsidy at the production level could also address the challenge of dual price regime in the same local market with the resultant removal of price distortions. Furthermore, it was recognized that fertilizer and seeds are complementary farm inputs, whereby the former is used to boost the productivity of the latter and many issues under focus affect both of them jointly, and many a time, joint resolution of such issues is required. In this regard, the implementation of GES was commendable, but it lacked a meaningful policy instrument for effective fertilizer and seed quality regulation and control, which created room for product faking, adulteration, false labelling practices, and other sharp practices in the fertilizer market. Specifically the fertilizer quality control bill pending at the National Assembly was inconclusive and the National Fertilizer Technical Committee (NFTC) was in a lull throughout the period.

39. The cost-effectiveness of GES is obviously in question, as the possibility exists that government may have spent more money on the technology aspects than the money saved in curbing fertilizer corruption during the period, while a new generation of unintended beneficiaries of subsidy policy has also emerged even among agro dealers and public servants in charge of fertilizer distribution under the scheme, who abandoned their duty posts to act as agro dealers and sources of huge overpricing and over-invoicing in order to highjack the benefits of subsidy policy in the fertilizer supply chain; the various aspects costs incurred to implement GES including technology, platform management, supply management, and other ancillary costs.

40. There is a growing concern about the absence of integrated soil fertility management practices under GES, which reflects a sustained recognition of the complementarity between the role of organic and inorganic fertilizers in addressing the challenges of environmental

pollution and the need for organically produced foods in Nigeria. In this regard, FEPSAN and other stakeholders have an important role to play to improve technical knowledge of farmers on appropriate fertilizer use, in terms of development and deployment of appropriate technologies as well as enhanced capacity building of farmers. In addition, the deployment of Rapid Soil Testing Kit will enhance farmers' access to soil test results and soil testing laboratories.

41. The visibility of federal government in the implementation of GES was so pronounced that the constitutional role of the state government was practically subsumed. The Nigerian constitution makes provision for agricultural development as an item on the concurrent legislative list, implying that it is a joint responsibility of both tiers of government. Nonetheless, the constitution also provides for a definitive division of labour between them, whereby the federal government was assigned the roles in establishing agricultural research centres as well as in promotion and financing, while the practical implementation of projects and other residual functions constitute the role of the states under the constitution. In this regard, it was obvious that the implementation of GES ascribed an overbearing activity to the federal government that their performance overreached the states in several ways. This non-observance of constitutional roles by federating unit presents a unique political economy issue in Nigerian agriculture, leading to poor role performance and manifest role confusion between the federal and state governments in the agricultural sector of the country.

42. Even though the operation of NIRSAL has sufficiently addressed the problem of access to finance by fertilizer dealers, the benefit of this to small-scale farmers was not maximized, owing to their inability to meet the requirement for N75,000 (i.e. 25%) equity contribution to be made by farmer. Moreover NIRSAL's mandate requires the agency to intervene in the case of the accumulated interest rate accruing to banks on loans borrowed by fertilizer suppliers and agro dealers, when the redemption of subsidy vouchers fail to take place on time. There is ample room for NIRSAL to do more in deploying relevant policy instruments to support farmers and agro dealers. In particular, the timely implementation of the guarantee component of NIRSAL would prevent the accumulation of interest rate on loans borrowed by fertilizer suppliers and agro dealers, arising from non-payment of bills by the FMARD.

43. The massive importation of fertilizer into Nigeria on annual basis has a dampening effect on domestic production of the input which,

given the infant status of the fertilizer industry in Nigeria, besets the local producers with competition with subsidized imported products. Therefore, a trade restriction on fertilizer importation would boost local production, and portends a ripple effect in job creation. Therefore, the need for all fertilizer importers to register with FEPSAN was recognized as a necessary step towards effective fertilizer quality control, monitoring, and tracking.

44. The claim of increase in fertilizer consumption from 13 kg/ha. to 60 kg/ha. by the ministry is spurious, as there is no system in place to get a correct estimate of this, owing to paucity of data for the purpose. Specifically the need for a comprehensive evaluation and impact assessment of GES was considered in view of the need for a clear road map for the new administration for determining the way forward on the notorious fertilizer subsidy issue.

Recommendations

45. Based on the foregoing, the following recommendations emerged from the dialogue:

On Fertilizer Production and Supply:

a. That fertilizer importation should be discouraged using appropriate policy instruments, while the local producers should be incentivized and given the right to bridge their capacity gap by importing during shortage only following a time-bound plan for their upgrade production.

b. That the National Fertilizer Technical Committee (NFTC) should be revitalized as an advisory body for monitoring the entry of new products in the fertilizer market and maintenance of the approved fertilizer slate for guided importation. Also the Fertilizer Quality Control Bill pending at the National Assembly should be resuscitated and pursued to a logical conclusion in quick time.

c. That steps should be taken towards improving the agri-input dealer network across the states of the federation and agro-dealer selection process under GES, through redemption centres at state level, good technology companies as service providers, tracking of GES products (fertilizer and seeds) with security tag and protective seal or GESS mark on bags, prosecution of erring marketers without a bail option,

constant review of the farmers' database in the country, engagement of both commercial banks and development banks such as BOI and BOA as loan providers.
d. That a fertilizer regulatory and quality control body should be established.

On Access to Finance and Subsidy Provision through GES:

e. That the equity contribution per farmer under the NIRSAL should be reduced to 10%, while a timely implementation of guarantee component of NIRSAL should be ensured to address the issue of accumulated interest rate on loans granted agro dealers.
f. That the fertilizer subsidy should be applied at farm output end, while the current subsidy cost sharing formula by federal and state governments should also be extended to local governments.
g. That government should take the responsibility of paying the extra interest accruing to banks as a result of delay in payments to agro dealers by government to redeem the subsidy vouchers.
h. That lending to private investors should be increased.

On Fertilizer Use and Quality Control:

i. That the need arises to create sufficient market demand for agricultural produce as a means of stimulating demand for fertilizers on a larger scale in the country.
j. That efforts should be made to improve the available data on fertilizer use in the country, in order to properly estimate the consumption of fertilizer at the moment and with a view to conducting an impact assessment of GES.
k. That the government should undertake necessary price stabilization policies through guaranteed minimum prices for all crops, with a view to increasing farm output and fertilizer consumption.
l. That soil test kits should be included in the package of farmer support, while also supporting farmers with qualified front-line extension staff by government and other extension services by private and sectors including agro dealers and entrepreneur farmers.

References

APMEU (Agricultural Project Monitoring and Evaluation Unit) (1990), Fertilizer Demand Study, FMARD, Abuja.

Ayoola G. B., V. U. Chude, and A. H. Abdusalam (2002), Towards a Fertilizer Regulatory and Quality Assurance System for Nigeria: An Inventorization of the Fertilizer Sector. Federal Fertilizer Department, Federal Ministry of Agriculture and Rural Development, Abuja. June 2002.

FFD (Federal Fertilizer Department) (2006), Fertilizer Socioeconomic Study, Federal Ministry of Agriculture and Rural Development (FMARD), Abuja.

FIF (Farm and Infrastructure Foundation) (2013), Fertilizer Socioeconomic Study, Federal Fertilizer Department (FFD), Abuja.

John C. Keyser, Marjatta Eilittä, Georges Dimithe, Gbolagade Ayoola, and Louis Sène (2014), Towards an Integrated Market for Seeds and Fertilizers in West Africa, World Bank.

44

Academic Corruption and the Role of NNMA in Curbing It

Paper with Emeritus Professor Ayo Bamgbose, presented at the 8th Annual Forum of NNOM Laureates, Organized by Nigerian National Merit Award, Merit House, Abuja. 1st December 2015.

INTRODUCTION

It is an indisputable fact that corruption pervades both public and private sectors in Nigeria. Against the background of an incoming government that makes the fight against corruption one of its cardinal principles, the choice of eradication of corruption as the theme for the 2015 NNMA Laureates' Forum is a most opportune way of sensitizing the public about the urgency of this task.

Academic corruption is any action that erodes academic standards in an institution. Specifically in the university context, it strikes at the very root of the ideal of excellence and integrity of the university and devalues the quality of the products of the university system. In this paper, an attempt will be made to examine the manifestations of academic corruption in universities in Nigeria, indicating the pre-existing internal remedial measures, and to consider how the NNMA can meaningfully join in the fight against academic corruption through other measures that are external to the individual institutions.

Virtually all stakeholders in the university system may be involved in academic corruption. From the student who cheats in an examination, to the lecturer who fiddles with marks and even the secretary who tampers with admission lists—all are involved in actions that are bound to erode academic standards and consequently mar the image of the university. Thus, in the multi-stakeholder context of academic corruption, the creation of NNMA in 1979 as a vote for meritocracy cannot be gainsaid.

The paper is structured into two main parts. The first part deals with typology of academic corruption that the universities grapple with, namely examination malpractice, admission, handouts and irregular demands, postgraduate research, plagiarism, and promotion. The second part highlights the philosophy behind the NNMA establishment, followed by the promotional or other roles the agency can play in constellation with other bodies and institutions in order to curb the menace of academic corruption.

TYPOLOGY OF ACADEMIC CORRUPTION

Examination Malpractice

The most visible form of academic corruption is examination malpractice, the purpose of which is to give an undue advantage to a candidate who otherwise would not have performed so well in an examination. Examination malpractice can occur before, during, and after an examination has been taken. Because the prevalence of such malpractice has been largely in secondary certificate and matriculation examinations, it will be useful to start with them as an illustration.

When examination malpractice occurs before an examination takes place, it usually requires collusion between a candidate and some other stakeholder in the examination chain. The most common form of such malpractice is leakage of examination questions, which in the Nigerian parlance is called expo (presumably a shortened form of the word 'exposition', meaning putting something on public view). Live papers may be leaked by any of the stakeholders having access to it, including even those who have access to where question papers are stored. The effect of such leakage is to make some undeserving candidates score high marks and, if discovered, cancellation of the question papers involved, with resulting stress to innocent candidates.

When an examination malpractice occurs during an examination, it may be entirely the fault of the candidate, who may have contrived to find some way of cheating in the examination. In the past, the most common method of cheating is spying from the script of a fellow candidate, which is known as the giraffe method. Since then, many more ways of cheating have been devised, including smuggling material into the examination hall through a variety of clever means. With the advent of electronic devices, methods of smuggling in material have become even more sophisticated. Candidates may also be assisted to cheat by other persons. Impersonation used to be one method. Another favourite way of giving such assistance is through the establishment of what has come to be known as miracle centres. These are examination centres where those supposed to be administering an examination freely give assistance to candidates who have paid a handsome fee for the purpose.

Post-examination malpractice can be carried out either through collusion between the candidate and a stakeholder or by the candidate acting on his or her own. The former is the familiar case of altering of marks or grades as a result of gratification given by a candidate or his or her parents. There are apocryphal stories of parents or guardians who have expressed surprise that their wards have been given grades lower than what they paid for. Bad eggs in the ICT department of examining bodies have been known to have fiddled with marks fed into a computer. In the case of a candidate acting by himself or herself, the usual thing is altering grades on statements of result and passing them off as genuine. For instance, I understand that a grade of E is easily altered to B by a clever artist.

Examining bodies such as the West African Examinations Council, the National Examinations Council (Nigeria), and the Joint Admissions and Matriculation Board (JAMB) have waged a vigorous war on examination malpractice in order to maintain the sanctity of their examinations and certificates. Consequently, the gross types of examination malpractice have been eliminated. For example, security on movement of live question papers is tighter, identification of candidates is more rigorous, alternate papers on the same subject are given to candidates sitting next to each other, and above all, the introduction of computer-based tests has eliminated a long chain of stakeholders, since marking and results are electronically done.

Why is examination malpractice so rampant at the level of secondary school certification and matriculation? First, inadequacy in teaching and learning in schools has resulted in high failure rates in certification examinations. Against this background, weaker students are tempted to find a way out through examination malpractice. Second, owning to limited facilities for places in tertiary institutions, there is keen competition to gain admission, and a number of candidates are willing to do whatever it takes to obtain good grades, particularly if they are not sure of their ability to do so by their own effort. Third, there is a craze for university education, and those who could have opted for careers and courses that are less demanding academically will make going to a university a do-or-die affair. Fourth, the attitudes of parents and even candidates, in some cases, do not even encourage choice of less prestigious courses. Everyone wants their son or daughter to be a doctor or lawyer or engineer. Consequently, the high grades required for such courses must be acquired by hook or by crook. Fifth, the government is partly to blame for not according greater recognition to products of polytechnics, teacher-education, and technical colleges.

At university level, there is potentially the possibility of examination malpractice before during and after the actual examination. However, such occurrence is usually minimized by strict controls as well as regular sanctions. For example, leaking of questions is now rare, and any lecturer caught in doing this is instantly charged with misconduct and summoned to appear before the University Staff Disciplinary Committee. If found guilty, such a lecturer is subject to dismissal. Similarly, any student who gains admission into the university through the submission of fake results or who is caught in examination misconduct during an examination is brought before the Students' Disciplinary Committee. Depending on the circumstances of the offence, he or she may be dismissed or rusticated, if found guilty. Every year, the University Bulletins of several universities are replete with lists of students who have been asked to withdraw from the university as a result of these offences.

There is a post-examination malpractice that often goes on in higher institutions. This concerns the manipulation and changing of marks. It may occur during marking of scripts or after the profile of marks shows that candidates have performed poorly, which is often rectified in the name of 'upgrading'. Marking of scripts by a principal examiner and a second reader and the use of external examiners may minimize the injustice involved in these practices. There is, however, an aspect of upgrading which is officially sanctioned and which may therefore pass as justifiable. I have heard of course units being waived for students who are short of mandatory course units in order to allow them to graduate. This is nothing short of officially sanctioned examination malpractice. Some proprietors and officers of private universities are said to believe that because of the high fees payable in such institutions, the university has a duty to help the students to pass in good time and not wait to repeat, thereby incurring more expenses. Some universities also believe that they will look good if they can show that their graduating class records several students with 1st class or 2nd class honours (upper division). I have heard of a proprietor of a faith-based university who is said to have frowned at the number of non-graduating students by claiming that children of God don't fail. Such official manipulation of results is nothing short of academic corruption.

Admissions

Prior to the explosion in secondary school population and the consequent pressure on university places, admission to universities was handled by individual universities through an entrance examination. Particularly in the first-generation universities, merit was the yardstick for admission at that time. Unfortunately, the downside of individual entrance examination was multiple admission offered to the same student, thereby denying other students of otherwise available places. It was partly in an attempt to correct this problem that JAMB was established in 1979. With a central examining system and facility for choice of more than one university, it was possible to limit wastage of places. The University Matriculation Examination (UME) now renamed Unified Tertiary Matriculation Examination (UTME) was generally successful, in spite of initial teething problems. The admission aspect of the scheme had always been problematic. Candidates who did very well and are admitted on merit did not present any problems at all, but in addition to such candidates, many others, with lower scores and lower cut-off points, get admitted on other criteria such as catchment area, educationally disadvantaged states, and discretion. It is in such areas that academic corruption often sets in, with the result that less-deserving candidates get admitted at the expense of better-qualified ones. For example, candidates have been known to claim to belong to educationally disadvantaged states, for which being disadvantaged has become a distinct advantage.

Another problem which later arose was the lack of correlation between performance in UME/UTME and result of first session examination of the same candidated in the university Unlike in the early stages of JAMB when performance in UME was predictive of performance in the university, results of first session examinations in several universities showed that many students who came in with high grades in the admission examinations performed badly. Universities then embarked on setting a supplementary admission examination, known as Post-UTME, including interviews, so as to ascertain that high scores in the UTME correlate with actual performance in the Post-UTME.

While administering supplementary admission examinations solves one problem, it does not eliminate the corruption associated with admissions. Because of limited admission quotas, there is tremendous pressure on university teachers and other officials responsible for admissions. I have always reflected on the fact that one of the rare occasions that university teachers are considered useful and important is during admissions season. In my time in the university system, whenever someone I had not seen for many years suddenly turned up in my office and claimed that he was just passing by and would like to say hello, my immediate reply to him would be 'Just tell me. Which of your children is seeking admission to UI this year?'

Although several high-scoring candidates still fail to gain admission as a result of wrong choice of course, I feel bad when I see such students not having a chance of being offered an alternative course because such courses have already been filled with candidates with lower marks within the cut-off points in the courses they have applied for. Besides, lobbying for places has continued to make it possible for less-endowed students to come in through discretionary places reserved for some university officials and such considerations as biological affiliation with members of staff. The problem with admissions to universities has not escaped the attention of influential Nigerians. Recently, the Nigerian Senate passed a resolution asking a committee to look into the admission policy of JAMB, particularly irregularity in admissions based on what it called favouritism.

Some years ago, I listened to a vice chancellor of a state university, who was defending a policy he had introduced about admissions in his university. He claimed that because of the high demand for courses in medicine and law, he had introduced a fee for those who might fall below the cut-off points by a few marks. Why he did not see the implications of his policy was something that baffled me. A few marks below a cut-off point may mean many candidates on several pages of the computer printout. Besides, only those who were able to pay the deficit surcharge could take advantage of the offer. This might mean the lowest-scoring candidates. In effect, the policy amounts to admission for sale, which is nothing short of academic corruption. Much as I condemn this, it is at least an open sale, for official receipts are issued for any payments made. What is even worse than this is accepting bribes for admission.

Considering the problems associated with admissions in our universities with their attendant avenues for corruption, is it possible to consider a radical departure from our present system in favour of a merit-based system? I recall that when the University of Ibadan conducted its preliminary entrance examination in the early days, marks were collated and a red line was drawn at the bottom of the list of highest scorers. No one below that red line could be admitted unless one of those offered admission failed to turn up. In that case, the next highest scorer below the red line was selected progressively down the list until all the places were filled.

Handouts and Irregular Demands

Persistent reports of lecturers making it compulsory to pay for handouts or making irregular demands have surfaced in many universities. Fees payable in universities are itemized and receiving units are clearly identified and receipts issued. Individual lecturers are not allowed to demand or accept any fees. When they give out handouts, the proper procedure is to give a copy to the class monitor, who will then arrange how the material is to be photocopied. Unfortunately, what tends to happen is that some lecturers insist on selling handouts, and not only that, a list of those who have bought handouts is kept and anyone whose name does not appear on the list is assured of failure in the course. This is nothing short of academic corruption.

A variant of this practice is for a lecturer to bring copies of his book to the class and advise every student to buy a copy with the threat that anyone who does not have a copy may not pass in his course. As usual, a list of those who have bought the book is kept and no sharing of one copy or photocopying of the book is acceptable.

Other methods of extortion include asking students to pay for assisting them with computer work, asking for indirect donation, for example, when the lecturer is involved in such events as the mother's funeral. One funny case I was told about was that of a lecturer who asked his class whether they knew the cost of a certain brand of Nokia phone. On the face of it, it would appear that he was only seeking information. However, his students got the message that what he really wanted was that they should tax themselves to buy him the phone in question.

One type of irregular demand which is generally frowned upon and visited with severe sanctions is sexual harassment. In general, it is associated with asking for sexual gratification in return for favours such as help with examination questions or practicals, award of undeserved marks, coaching, and lending of books and other course materials. Students are at the mercy of their lecturers, particularly in compulsory courses. Hence, weaker students who are not confident that they will pass in compulsory courses tend to fall prey to predatory lecturers. As mentioned earlier, a system in which a student's fate does not rest in the hands of a single examiner will minimize the occurrence of this type of academic corruption. In addition, zero tolerance for sexual harassment through prompt reporting and imposition of heavy punishment will also be a deterrent. Unfortunately, where a student is a willing participant, the offence may go unreported.

Postgraduate Research

Under the British system inherited by most Nigerian universities, postgraduate research is carried out under the supervision of a supervisor. The success of the research depends on a cordial relationship between the research student and his or her supervisor. Complaints frequently heard include laziness on the part of a supervisor who would hold on to drafts of a student's dissertation or thesis for long periods before returning them, exploitation of the research student by asking him or her to carry out assignments related to the supervisor's own work, and lack of proper guidance or supervision. One subtle but disgraceful type of exploitation is the appropriation of findings by a research student by his or her supervisor. It can take the form of incorporating such findings in a paper published by the supervisor or by the supervisor insisting that his name should be added as co-author in a work entirely done by the research student. In either case, it is nothing short of academic corruption.

Research students also have their own shortcomings. Some of them feel that any work already done by their supervisor can be appropriated by them without due acknowledgement. An even worse practice is plagiarizing previous works such as theses on similar topics or of similar titles. Cases have occurred in which tables and statistical calculations found in one thesis have been copied and incorporated into another one or in which data from fieldwork done by an earlier researcher have been modified and passed off as original research by another student. A few years ago, it was discovered that some unscrupulous persons had set up a syndicate for writing theses for lazy postgraduate students. All the student had to do was forward to the syndicate the topic or title of his or her proposed thesis and a made-to-measure draft would be ready for payment of a fee. Obviously, such a practice could not have succeeded if supervision had been diligent and thorough. Fortunately, there are internal mechanisms for detecting and penalizing the malpractices associated with postgraduate work.

Plagiarism

It is the nature of knowledge that it grows by accretion. Old knowledge forms an input into new knowledge and every new researcher learns from the findings of earlier researchers and may then modify or advance such findings. Consequently, there is nothing wrong in someone writing and referring to earlier works. What is unacceptable is for someone to use material by someone else without acknowledging that such material comes from the work of another writer. In academics, this is plagiarism, and it is a cardinal sin. Plagiarism may take a number of forms. For example, material belonging to one writer may be incorporated into the works of the writer plagiarizing without its being conventionally quoted with the use of inverted commas and the accompanying name of the author, publication, date, and page reference. In its gross form, the plagiarist may simply take work published in a distant place or in an obscure journal and simply append his or her name to it as the original writer. No doubt, plagiarism is a serious form of academic corruption which deserves the severest sanction.

Because plagiarism tends to find its roots in postgraduate research, it is important that supervisors should guide their students about its danger and consequences. If research students grow up engaging in it, the chances of its carry-over to publications when they become lecturers are quite high indeed. As we shall see later, the pressures on lecturers to publish for career advancement will give more incentives to a practised plagiarist.

While plagiarism is a condemnable act of academic corruption, a major problem with it is that it is not always easy to detect except when it occurs in its grossest form. I once received a project from a one-year MA student which I believed was of a much higher quality than could be expected from a student at that level. In spite of my suspicion, I could not find any publication from which the project could have been plagiarized. I concluded that it was probably not a publication, but an unpublished material to which the student had access.

In the US, a software called turnitin has been developed for matching any publication with a large database of publications in order to find out the degree of similarity with a previous publication. Postgraduate students in some universities have been obliged to submit their PhD dissertations for verification with this software in order to certify that what they intend to seek approval for was original work. Several students have protested against this practice on the grounds of violation of their intellectual property rights as well as the presumption of guilt involved in the procedure. In fact, a few students have gone to court, and the university senate has had to retreat from the requirement. In short, the software has become controversial mainly because of its inability to distinguish between plagiarized material and previous work properly acknowledged by way of quotation and references.In the Nigerian context, supervisors, external examiners, and assessors should be advised to be on their guard to fish out plagiarized material submitted to them. This they can do by their familiarity with publications in their fields as well as searches on the Internet for suspected similarities. Needless to say, any proven cases of plagiarism should be severely dealt with.

Promotions

Promotions in the university system are a likely source of academic corruption. This is largely because of the 'publish or perish' syndrome. The two relevant aspects of promotions are the publications submitted by the candidate and the assessment of such publications.

Although publications are defined as scholarly works that have appeared or been accepted for publication in refereed journals or books, candidates for promotion engage in all sorts of malpractice such as setting up of publishing houses and acquisition of ISSN labels for non-reputable journals, submission of self-published papers or books not refereed by any scholars, duplication of publications, submission of fake letters of acceptance, and joint publications in which several authors submit the same publication for their promotion. At the University of Ibadan, this has been given the name *the esusu system* (named after the Yoruba thrift system in which several people contribute and take out the monies collected weekly or monthly in turn). One of the worst cases I encountered in my time at the Appointment and Promotions Committee was the case of a candidate who had neatly submitted an offprint, claiming that it had come from a certain volume and number of a certain learned journal. Unfortunately, for the candidate, the editor of the journal cited happened to have been a member of the committee. He dashed out briefly and brought back the number of the journal in question and the paper submitted by the candidate was nowhere to be found in the said journal. Of course, the candidate in question was referred to the Senior Staff Disciplinary Committee for appropriate action. Where the assessment mechanism works, most of these malpractices are likely to be discovered.

The assessment procedure begins with peer assessment at the departmental and faculty levels up to the central committee level and reports by external assessors. Failure to carry out the assessment objectively either on account of favouritism or prejudice can lead to a distortion of the outcome. Well-known malpractices in relation to assessment include choice of assessors to favour or impede a candidate's chances, disclosure of the identity of external assessors to a candidate, contact with an assessor either by the candidate or by someone acting on his or her behalf to ensure speedy or favourable consideration, and manipulation of assessors' reports either by withholding an unfavourable report in order to shop for a more favourable one or by unduly delaying sending favourable reports on a candidate to the Appointments and Promotions Committee. These malpractices are no doubt due to the human element in the system. They need to be exposed wherever they occur and visited with appropriate sanctions.

NNMA AND ACADEMIC CORRUPTION

Now that we have seen manifestations of academic corruption in our universities, the question about what the NNMA can do to curb it will first of all require us to look at internal mechanisms in universities for dealing with corruption. Each university has its statutes and laid-down procedures for discipline of students and staff. In relation to students, universities are *in loco parentis* 'in place of a parent'. This confers some rights to discipline students according to regulations. Hence, most of the malpractices already noted will be handled by university authorities through their Students' Disciplinary Committee and Senate. Reference to outside bodies will not be considered necessary unless there is a criminal offence. Such punishments as rustication and expulsion are routinely handed down for most of these offences. In the case of academic staff, the Senior Staff Disciplinary Committee also investigates any misdeeds. Penalties to be approved by council may even include termination of appointment or outright dismissal. Universities usually insist that internal processes must be exhausted before recourse to any external bodies, including even the courts.

Given the internal processes and autonomy of universities, punitive and deterrent measures for curbing academic corruption are outside the scope of all external bodies, including not only the NNMA but also even the National Universities Commission (NUC), which is the regulatory body for academic standards. However, when we think of attitudinal change and orientation, there is a wide scope for relevant external bodies to participate in the task of curbing corruption. In the case of the NNMA, although its major philosophy is to identify, recognize and reward intellectual and academic excellence by awarding the National Order of Merit (NNOM) for outstanding academic output and innovation, it is also enjoined to promote excellence in the academic system so as to ensure the emergence of suitable scholars deserving recognition. In this connection, a case can be made for the involvement of the NNMA in curbing academic corruption, which is antithetical to academic excellence. Before making the case, however, it is necessary to examine the concept of meritocracy, the philosophical context for the establishment of NNMA.

Literally defined, meritocracy means the power of merit, as decoded from a combined Latin/Greek vocabulary meaning 'merit' (to earn) plus '-cracy' (power). In its original evolution, meritocracy has evolved as a political philosophy which holds that power should be vested in individuals almost exclusively according to merit; that is, in someone who earns it through merit. This implies that advancement in such a system is based on intellectual talents measured through examination and/or demonstrated achievement in a particular discipline, thereby ascribing a central role to the academic system as the producer, enabler, and enhancer of intellectual talents in the society. Unfortunately, the academic system itself is ridden with corrupt practices, which, ipso facto, threatens the legitimacy of the system in properly playing the role as ascribed.

Thus, even though the explicit motive behind the establishment of NNMA was the continuous making of the Nigerian National Order of Merit (NNOM) Award, its implicit motive was premised on the need to save the academic system from the threat of legitimacy that loomed and still looms so large on it, as it also looms large on the society in terms of corruption. Thus, the NNMA Act not only assigns the agency the responsibility for identifying and rewarding Nigerians for their intellectual achievements, but also mandates the agency with the responsibility to promote merit and excellence, with a view to preserving the integrity of the academic system in the long run; the main elements of this integrity including the following: academic discipline, academic honesty, academic accountability, transparency, and due process. Incidentally, these elements of academic integrity also constitute the desirable values of the Nigerian society as demonstrated by the present administration's resolve to fight corruption to a standstill. Therefore, the mandate of NNMA goes beyond sanitizing the academic system specifically, to freeing the Nigerian society from the scourge of corruption generally, under the new dispensation.

Somewhat euphemistically, corruption is like love; everyone knows what it is and experiences it, but no one can define it precisely. To buttress this, we recall the recent outpouring of invectives on the former president, Goodluck Jonathan, when, in a particular definitional context on his mind only, he said that stealing was not the same thing as corruption! On hearing this, not a few people at home and abroad thought he had goofed against the nation by exhibiting a seemingly aberrant notion of corruption as he understood the word, thereby depicting a wrong notion of corruption in such an outlandish statement, indeed perceived as one of the most embarrassing goofs of his tenure as a president on the eve of general election, which eventually carried a political cost for him to pay when he eventually lost that election to the opposition candidate. Such is the high political value of corruption that NNMA cannot take its statutory role in curbing academic corruption with levity. The truth, here, is that the boundary at which a practice to be described as corrupt is drawn depends on the ethical spectacles one wears at a particular place and a particular point in time—that is, whether professional ethics, political ethics, religious ethics, etc. Truly, insofar as Jonathan's statement was outlandish, it practically illustrates the terminological inexactitude inherent in the notion of corruption. That is, whichsoever spectacles one wears; anyway, the claim by Jonathan suggests that people are not on the same page about the proper definition of corruption—and that indeed there might be as many definitions of the viral word as there are analytical spectacles to wear.

For NNMA to do a good job as a watchdog of the academic system against corrupt practices, however, the proper analytical spectacles to wear are ones with lenses focusing on the actors of the system, the goal being to rid the system of the myriad of such practices that presently exist. By some intuitive reasoning, my sense is that the NNMA Act which mandates the agency of federal government to promote intellectual excellence also assigns the agency the onerous duty to maintain an ethical focus on the academic system wherein the intellect dwells as well. This implies that NNMA represents an organ of the federal government for safeguarding the academic system from being corrupt, which, by extension of that reasoning, the role of the NNMA as an anti-corruption agency is implied and in no way conflicts in any perceptible way with the role of the National Universities Commission situated next door to the Merit House. Probably, the government planners who situated both agencies of the federal government side by side in Maitama District of Abuja had this complementarity of their functions in mind *ab initio*. And so, rather than viewing my reasoning as some kind of institutional overreaching of sorts, it is better viewed as a form of division of labour between the two agencies of federal government focussing the same academic system for the overall good of the system, one regulatory and the other promotional. Even at that, such a division of labour is warranted by the need not to concentrate power on a single body for the fear of implicit bias in the performance of different functions; that is, we should not expect the NUC doing a good job of carrying out an oversight of the universities (planning, monitoring, and evaluation) or other bodies doing the same for polytechnics and colleges, to also be able to do a good job as watchdog or ethical cleansing of the academic system, which is why and where the promotional role NNMA comes into play, i.e. to raise the bar of ethical standards among the actors—teachers, researchers, administrators, etc. Moreover, there is so much work to do by everybody that the sky is large enough for many birds to fly without jamming each other in space.

In performing this quasi-academic role, however, the intellectual horizon of NNMA itself must be broadened beyond the present situation with the agency. This informs the decision of the governing board of NNMA in recent past to gradually move the agency from a purely civil service structure and function closer to the academic system so that a better job of ethical oversight can be done. Although this gradual movement is still in its early stages, featuring the appointment of its own staff as different from the inherited set of civil servants, and appointment of a professor-secretary, as well as a few other changes presently under way, the early signs indicate that the promotional mandate given to NNMA by its Act is practically doable.

Next, as a promotional agency itself, NNMA is estopped from encouraging academic corruption, let alone its governing board condoning any form of corruption from its members or staff of the agency. That is, promoting merit and excellence requires that the people involved conduct themselves above the board and are free of blemish. This serves two purposes, namely to preserve the quality and integrity of the NNOM as an order of dignity, and to impart credibility to its promotional efforts in the eye of the academic population and general public, hence justifying the good intention of the governing board of NNMA to reposition the agency outside the mainstream civil service structure, towards both administrative and financial autonomy and perfect standard status of a Grade A parastatal of the federal government it is.

Next is the necessity for NNMA to mount a National Campaign on Meritocracy, as a flagship programme in quality assurance in the academic system. This involves engaging the institutional stakeholders and the entire knowledge society to identify the various forms of academic corruption possible and launching a strong media advocacy, and even legislative advocacy if required, in order to curb the scourge of corruption in the system. Particularly, this campaign is not limited to giving recognition and reward to innovators and inventors only but also should engage the entire educational system from bottom up, towards a mind and policy change required to reverse the trend of corrupt practices at primary, secondary, and tertiary levels in the long term.

Last, NNMA can play an important role in addressing specific corrupt practices such as the rampant abuse of intellectual property right as well as the lack of transparency in teaching and research system among other types, with a view to promoting intellectual excellence and assuring high quality of service in the academic system. As stipulated in the NNMA Act, we need to show a workable institutional relationship with relevant institutions—NUC, NOTAP, ETF, etc.

In light of the foregoing, there are four main areas in which the NNMA may usefully join in curbing academic corruption: sensitization, mobilization, collaboration, and leadership by example.

In terms of sensitization, the NNMA has done well by choosing as one of its sub-themes for this year's Laureates' Forum the issue of academic corruption. By exposing the problem for public discussion, attention will be focused on it. In spite of internal mechanisms that exist for dealing with the malaise, not all universities handle offenders with the gravity it deserves. One often hears of some universities which believe that by admitting that academic corruption goes on in their institutions, they are giving themselves a bad image. By exposing it through lectures, seminars, and even campaigns, the NNMA will be helping to show how widespread the malpractice is and why there is an urgent need to combat it to avoid an erosion of standards and sustenance of excellence in the university system.

Because the primary constituency of the NNMA is the academic community and the academies, it is in a vantage position to mobilize for concerted action to curb academic corruption. For example, the academies may be charged with the organization of activities designed to draw attention to aspects of academic corruption in their disciplines. Academics are the likely perpetrators or victims of such corruption and targeting them for avoidance of the malpractice is bound to yield positive results.

Curbing corruption is a joint action that involves several stakeholders and agencies. In the case of academic corruption, the NNMA can usefully collaborate with agencies which have been involved in collaboration with universities. For example, the Independent Corrupt Practices Commission (ICPC) has been active in fighting corruption in the university system. The NNMA can start by finding out what they have achieved and where they can reasonably join them in pursuing the objective of curbing corruption. Of particular importance is the NUC, which is the regulatory body for universities. The NNMA will have to liaise and collaborate with this agency in its drive to curb academic corruption.

One area in which the NNMA can forcefully show its commitment in curbing corruption is by ensuring that its own operations are devoid of academic corruption. This means that the processes of assessment by the various panels of assessors must be carried out in an unassailable manner. In this connection, I believe that the use of external assessors in determining who should receive the NNOM award should be strengthened rather than diminished. Members of assessment panels should also be bold to give relevant information that may affect the integrity of the award. One case I remember concerns a candidate who had almost passed through the assessment process when someone told the panel of the candidate's corrupt activities in relation to some of his publications. The more the NNMA produces award winners who are truly distinguished, the more it will be showing that excellence in academic work is a virtue worth pursuing. Fortunately, it is widely acclaimed that the NNOM is one national award that has not been bastardized since it was first awarded in 1979. As long as the NNMA continues to uphold standards and retain its reputation, it will definitely be showing leadership by example.

45

A Multisectoral Analysis of Nigeria's Agricultural Policy in the Interconnected World[65]

The notion of multisectorality is central to the contemporary world of interconnectivity among sectors and countries in addressing policy issues of agricultural development, the resurgence of which reflects popular concerns about the need for multisector coordination to implement MDGs and other global initiatives at country level. Conceptually, the multisector economy takes different formations, the most common of which is the dual-economy structure that situates the institutional actors in several binomial contexts of policy governance such as agriculture-industry, social-economic, real-financial, public-private, rural-urban, etc. Analytically, the Deathly Embrace Model provides a framework for investigating the effect of constraints and bottlenecks on the multisector economy and to determine appropriate policy instruments for de-embracing the system. Results of empirical application to World Bank portfolio in Nigeria indicate certain challenges facing the multisector economy in terms of project constellation and overlaps, project integration and coordination, among other issues in the policy process for agricultural development.

Introduction

The role of agriculture as the mainstay of Nigeria's economy and as the cornerstone to reducing poverty, creating wealth, and improving food security of the country, is definitely not in doubt. In the performance of this role, the agricultural sector presently contributes about 40% to the country's GDP, and

---

[65] The role of World Bank office in Nigeria as the source of fund for the study is acknowledged, and the role of ministry staff and project managers in providing the information and documentation required for the conduct of the study.

engages about 70% of the people in gainful employment while also securing their livelihoods in terms of incomes and household food security. However, the context and trends of the country's agriculture indicate that growth in food production (3.7%) has not kept pace with demand (6.5%), which has led to structural deficit in major staples such as milled rice (2 million MT deficit) that situates Nigeria as the highest world importer, worth about US$3 billion annually. The underlying causes of this trend are rooted in the low productivity of agriculture and pervasive poverty, which combined effect is incident on the people's livelihoods in terms of massive unemployment notably among the youth and a latent food insecurity that looms large.

Since Nigeria's independence in 1960, the policy question looms much larger on the agricultural economy of Nigeria than the technology question, which borders on the role of government in the sector and how properly that role has been or is being performed. Indeed the view is widely held among experts that, even though technology matters, policy matters more for effective transformation of agriculture in Nigeria in the medium to long term. Thus it is increasingly recognized that unlocking the potential of the agricultural sector holds the key to transforming the economy, this being a multifaceted undertaking and which, by its very nature, requires a multisectoral approach.

The traditional aspects of agriculture include crops, livestock, and fisheries, which are stand-alone sectors in their own right; whereas there are several other sectors of varying conceptual or definitional scope constellating with agriculture such as health, education, water and sanitation, environment and climate change, and others. Therefore agriculture is a multisectoral problem area which requires multisectoral solutions, implying that no single institutional actor (or sector) could stand alone and work effectively for agricultural development, and that the solution approach to hunger and malnutrition, as the underlying problem, should be multisectoral accordingly.

The major constraints to addressing the negative trend (of food production growth lagging behind population growth) include the perennial lack of sustained policy effort to guarantee predictable return on private sector investment such as coherence tax regime, trade instruments as well as competition and investment promotion policies and an enabling investment environment. Moreover, investment in R&D is low and not demand-driven, which, together with comatose agricultural extension system, has negative implications for the generation and dissemination of farm technologies. Business support services are either poorly developed or do not exist which further debilitates the agricultural economy in terms of poor infrastructure and logistics for bulk storage and aggregation, and adds to lack of competitiveness due to high transaction costs and low access to finance to further retard growth of agricultural sector. Moreover, irrigation is underdeveloped, underperforming, and underutilized, whereby less than 1% of irrigation potential of Nigeria is developed and only about 54% of developed irrigation areas are actually cultivated, so crop intensity could be as low as 10% for some of the schemes during wet season and 0% during dry season. Superimposed on these constraints is the presence of weak supporting institutions and infrastructure for market development and growth. Thus, the critical need to seek solutions to agricultural problems from multisectoral sources is quite obvious.

In dealing with these constraints, in 2011 the Federal Government of Nigeria launched an Agricultural Transformation Agenda (ATA) which was anchored on the development of commodity value chains with design and implementation that cut across the traditional boundaries of the sector. The objective of ATA was stated in terms of assuring food security with additional 20 million MT of foods by 2015. This was premised on a policy philosophy of 'agriculture as a business but not a development sector', whereby focused growth, wealth creation, and employment through the value chain development approach and public-private partnership enabled investment drives. Other key features include mainstreaming gender and youth in agriculture, institutional restructuring and capacity enhancement of Federal Ministry of Agriculture and Rural Development (FMARD) and reform of sector policies, agricultural extension, among other elements.

Lastly, the rationale for multisectoral approach also predicates on the multidimensional nature of the macroeconomic policy environment governing project management in the country. This makes policy governance an integral part of managing the agricultural portfolios of the World Bank and other donor agencies, including the aspects of programme accountability, financial transparency as well as policy due process. In this regard, the different policy regimes that jointly determine the performance of agricultural projects but which are outside the sphere of agriculture ministry or sector, include Foreign exchange policies (e.g. depreciation of the local currency against the major currencies as a result of recession of world economy, which have led to high prices of food imports, fertilizer, and other farm input imports); Monetary policies (e.g. inflationary implications for the domestic terms of trade between the agricultural and non-agricultural sectors, the term-structure of interest rates, and the supply and demand for loanable funds in agriculture); Fiscal policies (namely, taxes, tariffs, inflationary deficit financing, etc., with pronounced consequences for domestic terms of trade governing the exchange of agricultural products for products from non-agricultural sectors, thereby contributing to the perennial high transaction costs of agricultural exports and high farm gate prices of farm input imports; consequently making Nigerian agriculture uncompetitive in the world market); Income policies (particularly national minimum wages, equity considerations, transfer payments, etc. that are usually targeted at the urban labour force wages but have consequences for rural labour supplies, issues of equity, regional income distribution, etc.); National industrial policies (which make the provision of agricultural raw materials cheap for the purpose of agricultural industrialization in competition with exports. Although such import substitution or backward integration policies may reduce the volume of imports in the near term, they often fail to reward farmers with remunerative prices for their efforts, thereby discouraging farm production in the long term); and International trade and balance of payment policies (these generally reflect the protocols and agreements reached in the WTO, which often hurts developing economies and set the terms of external trade in favour of developed countries).

Based on the foregoing, it is obvious that a single-sector approach will not address all those policy issues in favour of agriculture in the contemporary world of interconnectivity of sectors and countries with one another; which further strengthens the case for multisectorality as a desirable policy of the World Bank and other donor bodies for increasing the transformational impact of their project portfolio. Therefore, a number of issues in point need to be investigated conceptually and resolved empirically:

i. How has the notion of multisector played out in individual agricultural projects?
ii. How sensitive to multisectoral character of agriculture at every stage of the project cycle are the policy institutions – government, World Bank, and other donor agencies?
iii. How does each project fit in the policy process for agricultural development of government in a multisectoral pattern?
iv. What multisectoral constraints and bottlenecks has each project encountered and how has the project responded in each case?
v. What is the outcome of these constraints in practical terms? (Whether there are instances of sector-sector embrace?); and
vi. What suggestions can be made to resolve multicultural issues in agriculture; e. g. What should the donor bodies do differently on the supply side of development assistance? And what should government (state/federal government) do differently on the demand side?

The resurgence of multisectoral approach in the World Bank not only reflects the recent client demand for change in Nigeria (with an implicit reference to the recent AgDPO and SCPZ project), but also the emerging trends in global practice and in the delivery of assistance to developing countries by international institutions in development, trade, and finance generally. Previously, the client demand was geared towards ensuring that donor assistance helps to achieve multisectoral cooperation between project components, which under the FADAMA III (a community development project) the demand for integrated service came in terms of agriculture integrated with transport, energy, finance, agro-industry, etc. Nonetheless the client-driven nature of this demand ensures that projects fit into the development process of the country in a multisectoral pattern. For instance, upon demand by the federal government, immediately ATA was launched in 2011, The World Bank was favourably disposed to restructuring FADAMA and CADP in the upstream segment of agricultural value chain to fit the pre-existing projects into the new agenda by scaling them up and sharpening their focus on the targeted crops. Also, the preparation of SCPZ project commenced in response to the ATA programme in the downstream of agricultural value chain, thereby equilibrating the activities of production and industry sectors.

Furthermore, some global initiatives have considerable demonstrative value on the policy institutions to adopt the multisectoral approach, which include:

a. <u>The resurgence of multisector coordination as a good practice for the successful implementation of MDGs</u>. Hitherto assistance projects in agriculture cum health, water resources, etc., generally followed the single-sector approach. And even though these may have been designed with their multisector nature in mind, they were often implemented in silos of staff activities and in substantial isolation from one another without a conscious effort at introducing mechanisms for multisector coordination. However, the implementation of MDG has underscored the need for efforts to coordinate the roles of many sectors together in achieving individual development goals, so the country offices of the World Bank or other institutions involved were expected to adjust their operations accordingly. Unlike other strategies whereby the development assistance is usually segmented by sector, the MDG has changed all that with new partnerships established, participation by private sectors and non-government organizations emerging stronger, and allocation of funds increasing or more diversified. Specifically, the Asian clients have been clamoring

for multisectoral approach in their development process from the World Bank, in terms of an integrated programming approach instead of individual project for better optimization of Bank's work and to build synergy across development sectors (agriculture, energy, transport, etc.).
  b. New concerns about issues such as overlap of activities and need for more effective allocation of resources. For instance in recent times the ODA reached a ceiling due to world recession which made stronger institutional collaboration across sectors inevitable. Also, as the danger of food insecurity loomed large on developing countries occasioned by soaring world food prices or by the recent boom and bust cycles of developed countries, the role of rural infrastructure as a cross-cutting area intensified the need for multisectoral approach.
  c. Other instances that induced the multisectoral approach include The Zero Hunger Challenge launched by the United Nations, which is all-encompassing in nature; the stepping up of the FAO's mandate of right to food including the recent launching of a partnership with ECOWAS to implement a project in that direction; the efforts of some agencies to take multisectoral approach to agriculture, such as the Forum for African Agricultural Research (FARA) that launched the Integrated Agricultural Research for Development (IAR4D).

Specifically, the rationale for multisectoral approaches implicates the World Bank which primary mission is poverty reduction and promoting economic growth in the poorest countries of the world, many of which have agriculture as the mainstay of their economies. Thus, through its investments in multiple sectors, the World Bank is well positioned to support a multisectoral approach to reducing poverty through agriculture, and as such, provide support to its client countries in acting across sectors. Examples of poverty reduction development objectives that are easily combined with agricultural objectives to be addressed through multisectoral programmes are strengthening community institutions, improving agricultural techniques, providing access to basic education and health care, nutritional programmes, improved housing, including access to safe water and sanitation, income generating activities, and child care arrangements, among others.

1. Conceptual Framework and Analytical Model

In general usage, the term *sector* is about the structure of the economy to break the system down into its functional parts or 'institutional actors'.

Thus, the categorization of *multisector* economy takes different formations, of which the most common is the dual-economy formation or two-sector economy structure that situates the institutional actors in several binomial groups of activities. Specifically, the binomial formation describes the dual relations of the horizontal and vertical nature between sectors at a given level of governance (e.g. national level) or at different levels of governance. In the horizontal case, the various binomial formations of a multisector economy include: agriculture versus industry, oil versus non-oil, social versus economic, real versus financial, public versus private, production versus market, rural versus urban, etc. In the vertical case, the dual-economy structure describes the relations of a sector at a given level of economic governance (e.g. federal level) to the same or other sector at a different level of economic governance (e.g. state level). So we have national agriculture sector as different from the state agriculture sector, or possibly Nigeria agricultural sector as different from ECOWAS agriculture or African agriculture sector. In both cases, monomial formations are also possible, whereby the economic activities are grouped together into interrelated functional components such as food sector, water and sanitation sector, housing sector, fuel, security, etc.

In these categorizations, the general misconception is that a ministry is synonymous to an economy sector. Rather, a particular sector of the economy consists not only in the activities of a single ministry but also of such other ministries, departments, and agencies and their activities that relate to one another, extensively or remotely in one way or another; and which the sequence of the interrelationships is captured in the value chains for different commodities. That is, the agriculture ministry is no more an institutional actor in the upstream segment of the value chain (production, storage) than the industry and trade ministry operating in the downstream segments of the same value chain (processing, marketing, consumption), both constituting structural and functional parts of the agricultural sector. Moreover, the structural or functional formations may apply to a sector within a sector, otherwise referred to as subsector, such as the different commodity groups representing the crop sector, livestock sector, and fisheries sector, which constitute the composite agricultural sector of the economy; and also as when the same sector is conceptualized in terms of functional categories such as research and operations, production and market sectors, inputs and output sectors, government and nongovernment sectors, social and economic sectors, among other formations. Sometimes more specific functional categories even apply which may overlap with one another variously, e.g. road sector, marine sector, aviation sector (or generally transport sector), among other possible formations.

Indeed the scope for 'sectorization' of the economy is quite large, and the concept of multisectoral approach refers not only to organization of the system but also to the broad range of actors and roles involved in growing the sectors individually or collectively. The main challenge, of course, is how to coordinate or integrate the efforts of all stakeholders in each sector so that their activities complement but not contradict one another; or given the inevitability of sectors overlapping with one another, how to coordinate the activities of project managers on both sides of the policy divide so that their planning and implementation activities complement but not contradict each other.

The goal of multisector analysis is not just to identify redlines of ministries or other institutional actors against their boundary maintenance practices, but also to establish a framework for understanding the workings of the economic system in terms of the functionality of its structural parts, so as to determine the optimal modes of policy intervention. In this regard, the 'deathly embrace' framework is applicable in analyzing the multisector economy in order to identify appropriate project instruments required to deal with structural and functional failures of institutional actors (Ayoola 1999). In this framework, any two sectors of the economy engage each other in a mutual 'embrace', whereby one sector 'waits' for the other in terms of thrusts and feedbacks (inputs and outputs) required from each other but which reciprocally fail to flow as expected. As a result, the system gradually slows down into an 'embrace situation', until the point of 'deathly embrace' is reached—the limit when the economy stagnates completely. Indeed a 'multiple embrace situation' of the agricultural economy is a theoretical possibility, whereby the projects for growing individual sectors are not properly synergized together so they fail reciprocally to make necessary contributions to one another at the planning and implementation stages. Hence, according to Idachaba (1997), 'it is not only ill-health in agriculture that holds down industry and vice versa, but that ill-health in the transportation sector, social services, education sector, and others lead to more severe illness of the agricultural sector'. The major challenge then is how to resolve the logjam of a multisector economy in constant embrace, which entails first identifying the critical points of embrace and, second, applying appropriate policy instruments at these points in order to 'de-embrace' the system somewhat.

Thus, the deathly embrace model (DEM) provides a framework for analyzing the effect or incidence of constraints and bottlenecks on the multisector economy and to determine appropriate project instruments for de-embracing the system. this illustrates the concept of multiple embrace with the focus of public policy intervention of the government and its development partners on the agricultural value chain, as they implement programmes and projects to impact the performance of the chain at different points or sectors. For instance under the ATA, the federal government presently implements GES, SCPZ, FADAMA III, CADP, and other schemes in order to impact agricultural production in the upstream of the chain, while the World Bank as a development partner provides project support to the sector in terms of financial investments, technical advice, and policy operation. Thus, one sector embraces another in the extent that the inputs and feedbacks fail to flow in desired magnitude and/or speed. To address a particular embrace situation, it is required to identify the particular point or sector of the chain it occurs and apply necessary project instruments to de-embrace the system accordingly.

2. Empirical Illustration

The project portfolio of World Bank in Nigeria provides an empirical basis for conducting a multisectoral analysis of agricultural policy in Nigeria, wherein the entry point to supporting the ATA was the approval of an Agricultural Development Policy Operation (AgDPO) project as a composite and policy-driven intervention within the Sustainable Development (SD) sector of the Bank. The project, which is domiciled with the SD sector of the Bank, represents a development policy operation (DPO) for the structural and functional adjustments of pre-existing portfolio of projects, thereby constituting the trigger factor the Bank requires to take a multisectoral view of its agricultural portfolio and to generate a multidisciplinary and holistic response to leverage transformational impact. However, in putting multisectoral approach to work fully, it is not clear how the other sectors of the Bank's operations outside the SD fit into the management of agricultural portfolio, or how the activities of task team leaders (TTLs) of SD projects integrate with the activities of TTLs of the other operational sectors, namely: Human Development sector (HDS), Finance and Private Sector (FPS) as well as Governance and Economic Reform (GER).

Thus, the sensitivity of the Bank to multisectoral considerations in designing and implementing projects, though quite significant, is in question, both in depth and in scope over time. In the past, the Bank was not so sensitive to multisectoral issues in its operations; individual project managers focused on or were more concerned with the narrow aspects of the portfolio as managed by different experts in isolation of one another. More recently, however, even though the design of projects in the portfolio of the Bank sometimes benefited from the pool of expertise available in different sectors and multidisciplinary specializations, which indicates some multisectoral concerns, coupled with the presence of certain measures or instruments in project documents to strengthen project management for enhanced multisectoral effects, the extent to which the multisectoral design of projects is operationalized on the ground was considerably lax, amounting in most part to mere 'thinking multisectorally but acting sectorally'.

The goal of the analysis is to examine the different multisectoral dimensions of World Bank's agricultural portfolio in Nigeria and to interrogate the emerging issues therefrom while also illuminating the constraints and opportunities therein. This is with the view to helping the Bank develop a perspective about projects constellating with one another in the multisector space, and about leveraging multisectoral resources particularly expertise and knowledge for maximum transformational impact. Towards this end, specifically three issues arise as to: how projects within the portfolio overlap with one another, how projects respond to multisectoral constraints or bottlenecks faced in the agriculture sector, and what the Bank should do differently in realizing or deepening the felt need for a multisectoral approach to its operations in Nigeria. Towards this end, the perspective study involves a content analysis of project documents in order to examine the pre-existing projects in the portfolio of the Bank and diagnose them in terms of objectives and components. This was followed by a series of interactive sessions with the Bank's staff from different sectors, namely, SD, HDS, FPS, and GER. In addition, a number of discussions were held outside the Bank with civil servants that had good institutional memories of the ministry, and also with the national coordinators of active projects

3.1 *Project constellation and overlaps*

The portfolio of the Bank may be viewed as a group of interrelated activities in rendering assistance to its clients. This makes multisectoral approach a compelling need if synergy is to be achieved as these activities constellate together at pre-investment, investment, and post-investment phases of projects. The trends and pattern of project constellation are discernible, whereby the Bank is one among the technical partners working with the government to impact the agricultural value chains through assistance projects, which in the process overlaps occur severally in terms of complementarities or sometimes clashes of components and objectives. Thus, a multisectoral approach would help to eliminate or at least minimize the degree of wasteful duplications resulting from project overlaps at different stages, while also maximizing the complementarities therefrom.

At the moment, when projects in the portfolio are many and more diversified both spatially and functionally so the desire for synergy is presently more pressing, the need for multisectoral approach is both desirable and urgent. In this wise, the World Bank is able to assist ATA along the value chains through new and existing projects, each deploying a mix of project instruments for achieving multisectorality in varied proportions as highlighted below.

   d. Ongoing lending operations: US$250 million FADAMA III adjustment loan for farmer registration and SCPZ activities; US$150 million - IDA credit for restructuring the Commercial Agricultural Development Programme (CADP) to support marketing corporation and SCPZ authority and to initiate the preparation of a new project to support implementation model of SCPZs; US$45 million - for WAAPP supporting extension and NARS; US$200 million - additional financing facility for FADAMA III to support on-farm productivity in SCPZ catchment areas; and US$200 million - AgDPO for policy and institutional reforms.
   e. Pipeline operations: These involve two investment lending amounting to US$500 million, namely, the US$400 million - Nigeria Irrigation and Water Resources Management Project; and the Joint IFC-World Bank SCPZ support project with expected IDA funding of US$100 million.
   f. Technical assistance: These include the Warehouse Receipt project and the Agricultural Insurance Monopoly project under discussion.

This illustrates the inherent complementarities or overlaps in the portfolio as the projects presently constellate in the multisector space, whereby the objectives and components of these projects were benchmarked against one another (A-G). First, the FADAMA III project complements with the AgDPO in Policy Area 1 (*Farmer access to improved technologies*), in its national coverage and focus on on-farm productivity supports to smallholder farmer access to both the input and production components of priority value chains under the ATA. Hence, the consideration for additional finance for the FADAMA project to support government efforts to increasingly focus extension services and knowledge transfer on the application of climate-smart agriculture practices, as well as extend those to cover both production and consumption of appropriate foods and diversified diets by vulnerable groups, particularly women and young children. Second, the WAAPP (West Africa Agricultural Productivity Programme) project also complements AgDPO in Policy Area 1 (Farmer access to improved technologies), by (a) supporting the upstream input supply segment of the prioritized value chains under the ATA, and addressing constraints associated with the production of breeder and foundation seed; (b)contributing to technology development, public and private, and to the regional transfer of technology, and (c) supporting the institutional reform process, thereby accelerating the development of private sector-led fertilizer and seed industry. Third, the CADP project complements AgDPO in Policy Area 2 (Market development), by supporting the ATA's agenda on investment in postharvest management and processing, with reference to envisaged opportunities to support agro-processing zones by investing in rural roads, energy, markets, and the establishment of out-growers schemes.

Thus the three projects exemplified—WAAPP, FADAMA III, and CADP—logically complement one another as well, thereby reinforcing the synergy among them during implementation. Similarly other projects overlap with each other whereby CADP project complement those of FADAMA III and GEM (Growth and Employment Market) in the areas of rural infrastructure provision and commercial enterprise promotions, respectively. In turn, the components of GEM overlap with those of FADAMA III in the area of business support services, and all three projects overlap with AgDPO in its different policy areas. Even though the multisectorality of agriculture implies that these overlaps create the possibility of duplication of activities, unless they occur on a large scale they may not necessarily lead to wastages of project resources. Therefore the real issue is not the presence of complementarities or overlaps as projects constellate in the commodity space per se. Rather the issue is about the present ineffectiveness of the project management to deliver the multisectorality of projects at the planning and implementation stages to the extent desired.

Therefore, the presence of complementarities and overlaps among project objectives and components pose a major challenge in terms of the need for coordination of project activities in their multisectoral contents. Whereas effective coordination of projects across the constituent sectors requires TTLs to hold field meetings with project staff and among themselves occasionally, to discuss emerging issues and share knowledge across projects as they border on multisectorality or lack of it in project management process, such coordination meetings are only feasible when projects are location-specific but not feasible when projects are geographically dispersed. This not only serves as a model for linking projects to different points on the agricultural value chain, but also presents a joint intervention framework for TTLs to share knowledge and information across sectors and to coordinate the project activities under the sustainable development network.

3.2 Multisectoral challenges

The Bank pursues two approaches of multisectoral coordination in order to maximize project impact. One of these is the internal arrangement for joint action of the Bank staff from different sectors to share knowledge and information and to coordinate the project activities through the TTLs, as well as contribute to planning and implementation of projects. The other approach is the external arrangement to strengthen capacity of project management (advisory/steering) committees with membership from various ministries, departments, and agencies. Each of these approaches poses its own peculiar challenges which translate to the two main sets of barriers or bottlenecks to multisectorality, which correspondingly emanate from the project integration and coordination process as well as from the policy process.

3.2.1 Challenge of Project Integration and Coordination

This challenge pertains to the internal efforts of the Bank in terms of project integration and coordination through STLs and TTLs in the planning and implementation processes. The challenge here is that project managers would require a special incentive in terms of career progression and/ or financial inducement to participate in projects other than their own and to express their individual professional capacities fully in the affairs of such projects. This is natural as staff turnover is determined by deliverables coming from 'own projects but not others' projects', so people would normally engage in implicit evaluation of his/her performance in terms of what he/she would be judged on before what he/she would not be judged on. Hence, the natural response of Bank staff to simply play down on multisectoral assignments while prioritizing his/her own sector commitments during planning and implementation of projects.

This challenge is exacerbated by the incremental cost in money and time required to deploy many bank staff on the same mission at the same time. The feasibility of such missions is also doubtful as project handlers might be operating from different locations. An example is the Lagos-Abidjan corridor project, a regional project in West Africa, which combined the components of health sector (HIV/AIDS control) with components of transport sector. The TTL of the project was based in Washington DC, USA, while the co-TTL was based in Lome, Togo. There are several other challenges facing the internal coordination effort of the Bank which include difficult procurement processes for different multisectoral requirements and the need to deal with multiple countries and agencies at the same time. In the case of Lagos-Abidjan corridor project cited above, the procurement processes were different for health and transport sectors which introduced complications to the work of procurement experts as many people would be required to give advice to procurement specialist which invariably introduce delays in delivering on indicators of individual sectors combined in one (multisectoral) project that eventually created the need for extension of the project's life.

Furthermore, there are differences in the views of project managers about multisectoral nature of projects, depending on the definitional formation of multisector applicable to what they are doing. In this regard, it was observed that the FPS managers think about multisectorality of their projects in terms of constellation of services available in the business environment rather than the general idea of seeing sectors as institutional actors. Thus, in the GEM project, the emphasis is on the logistics sector, energy sector, etc., as the different sectors for doing business, but not on the light manufacturing, ICT, and other components of the project as different sectors. The variability of conceptual views about multisector approach implies that the project managers are not necessarily on the same page so they would be enthused in different degrees about internal arrangement for multisectoral coordination of projects in relation to project managers in SD, HD, or GER sectors of the Bank.

The second approach of the Bank to foster multisectorality of its portfolio is the external arrangement to strengthen capacity of project management (advisory/steering) committees with membership from various ministries, departments, and agencies. This is also faced with critical barriers associated with public sector management such as wasteful bureaucracy and rent-seeking behavior that characterize the participation of staff outside the line ministry of projects. This arises from inevitable reliance on staff from other ministries whose primary allegiance is owed their own ministries so they prioritize the project assignments much lower when domiciled in another ministry. Even when the *outside* project enjoys the right priority and tasks are performed promptly, multisectorality is still constrained by the absence of an incentive scheme as the annual staff appraisal exercise does not take cognizance of their role in such projects, which leads to a lackadaisical attitude of outside staff to such projects in actual practice.

Power relation among public functionaries is also important, which is usually encountered at the stage of implementation, bordering on the reluctance of agricultural professionals to transfer power over the project to the authorities or staff of other ministries. In this regard the usual questions pertain to 'Who is in control of this and that project?' or put differently, 'Who is buying what or spending what amount of money under the project?' Thus, the power of budget holding is particularly attractive to project and non-project staff of the sector as conferred by domiciliation. Even when some line items are provided in the budget for activities based in other sectors, the lead agency assumes express powers to approve and disburse budget fund which is seemingly objectionable to participants from other agencies. This poses the challenge of managing civil service staff engaged in a multisectoral project, who feel implicitly interested in sharing the financial power under the project.

3.2.2 Challenge of Policy Process

In addition to operational challenges associated with Bank efforts in multisectorality, other more generic challenges exist which emanate from the external environment wherein the multisectoral approach will work in practice. The major ones are the barriers or bottlenecks originating from the policy process, consisting of the sequence of steps for the government in articulating a policy whereupon bankable project ideas germinate, starting from problem definition (policy conception) through technical drafting (policy design) to stakeholder validation and official verification (staff appraisal), formal consideration and adoption (policy authorization), and eventual approval and publication (policy declaration) prior to implementation, supervision, monitoring, evaluation, and impact assessment.

The challenge of policy process represents the proximate determinants of the Bank's capacity to address multisectoral challenges in providing its support to ATA. This borders on three main process issues with considerable multisectoral implications, namely, policy philosophy, policy governance, and policy sustainability. In the first case (policy philosophy), the ATA is premised on a notion of 'agriculture as a business (but not a development sector)', which the former phraseology is intellectually plausible but the latter is philosophically absurd in the Nigerian case, for the reason that the pro-poor character of the agricultural sector is concealed (but not revealed) in the latter phraseology, which is of course also offensive to the World Bank that poverty forms the central focus and primary concern under its international mandate. That is, while there is no problem with the notion of 'agriculture as a business' per se, this represents only a necessary but not sufficient condition as a philosophy of agricultural development for the country wherein poverty is pervasive. The implication of this for the Bank as a global agency for poverty reduction is the situation whereby the management of its multisectoral project portfolio takes place in a philosophical vacuum, thereby lacking in a definite orientation in that direction. This poses a challenge to the Bank in the policy present operation project (AgDPO) wherein the Bank is responsibility-bound to advise its clients about the need for a suitable overarching philosophy governing the agricultural sector that properly reflects the poverty character of agriculture without necessarily downplaying the need to grow the sector as a business. Such an appropriate philosophy is the right to food which not only reflects the poverty character of the sector as a global practice but also represents a strategy to drive food production from the demand side of the market with active participation of private sector.[66]

In the case of policy governance, the danger of poor policy governance looms large on the multisectoral economy much in terms of a strong tendency for MDAs to operate outside the explicit provisions of the extant laws governing the different sectors in the country. With particular reference to the provision for agriculture, the constitution of the Federal Republic of Nigeria stipulates that agricultural development is on the concurrent legislative list, implying that it is a joint federal and state responsibility. Nonetheless, in order to avoid role confusion, the constitution also establishes a division of labour between the two levels of agricultural governance, but which is generally overlooked and constantly violated by way of federal trespass beyond the set boundary and also by way of states complacency or by way of passive acquiescence. In the constitution, the federal government was explicitly assigned specific roles in research, promotion, and finance; whereas the state governments were assigned generic roles bordering on the actual implementation of projects on

---

[66] The right to food is not an apology of state socialism that the government is required to provide food for the people free of cost. Rather it is a philosophy of poverty reduction through a policy and practice change that sees food as a fundamental human right and not as a mere human need. The rights approach is definitely superior to the needs approach, the philosophical difference being that the former is practically consequential in cases of violation of the right while it is consequential otherwise.

ground.[67] Thus, taking agriculture at federal and state levels as different sectors in the vertical sense, the provision of the constitution for the relationship between the agricultural sector at the federal and at the state level is observed only in breach, which has serious implications for project performance by deepening the negative effects of managing the multisector agricultural economy.[68]

Lastly, in the case of sustainability of the policy process, the challenge manifests in terms of frequent changes in the organization of ministries that lead to incessant perturbation and instability of the system in which multisectoral projects are implemented, thereby putting the Bank's capacity at risk in leveraging the maximum impact of its portfolio. For instance the implementation of ATA has featured considerable organizational instability and high rate of personnel turnover in the political, administrative, and professional leadership positions which creates adverse effects. As such, instability is usually accompanied by particular programmes to be given more or less policy emphasis, redesigned, re-introduced or the implementation pace speeded up or slowed down, so as to reflect the new political, ideological and occupational biases of the new people involved. Thus, the agricultural development pathway consequently becomes uneven, to the effect that the overall health of projects in the portfolio is frequently jeopardized.

It appears that the Bank is generally averse to responding to generic multisectoral challenges of the policy process types, bordering on policy philosophy, policy governance, and policy sustainability. This probably reflects a cautious attitude of the international organization not to delve into sensitive political economy issues affecting agriculture at large. In this connection, even though the Bank should not be dragged into politics at the state or federal level, the risk of which looms on the Bank sometimes, the agriculture sector oftentimes features some political economy issues that place the Bank in a vantage position to assist its clients to resolve such multisectoral

---

[67] Section 4 of the constitution (Concurrent Legislative List) specifically assigns the role of federal government as follows: "... (c) the establishment of research centres for agricultural studies, and (d) the establishment of institutions and bodies for the promotion and financing of industrial, commercial and agricultural projects" (emphasis mine).

[68] In any event, the explicit provision for this division of labour in the 1999 constitution has its background in the political economy of agriculture sector which has featured concrete instances of constitutional crisis at some points in time past (Ayoola 2002).

issues about boundary maintenance practices and others in policy governance, policy sustainability, among others.

The general outcome of these challenges when not resolved, is missed opportunities whereby some activities are underperformed or unperformed altogether, so the benefit stream of projects are not maximized. Thus, as initially postulated, the embrace of different sectors is both apparent and real, whereby in the presence of several bottlenecks or constraints and coupled with the challenges of the policy environment and gaps in project design, the different sectors fail to meet their reciprocal needs for one another so the agricultural economy is sluggish on its growth path and tends towards a deathly embrace situation in the limit. This has implication for the Bank to meet its clients' need for appropriate instruments in the projects for de-embracing the system. In this regard, the candidate instruments to consider include building synergy across sectors and achieving multisectoral harmony of projects, through joint implementation and multisectoral collaboration.

3. Conclusion and Recommendations

The operationalization of multisectorality owes as much to the internal processes of the World Bank in managing the projects on the one hand, as also to also to external processes of clients on the other hand. On the Bank's side each sector undertakes activities in planning and implementation of projects which their design concept did not give adequate consideration to multisectoral elements ab. initio. Also, on the client's side each institutional actor carries out its mandate activities in planning and implementing the projects either as co-development partners to the government or in financial partnership with the Bank, or ordinarily as statutory bodies in the multi-sector space for managing the agricultural portfolio. Thus a multisectoral action plan is required for the comprehensive redesign of these projects, which involves identification and integration of appropriate multisectoral instruments with the current designs and implementation of the projects.

The critical milestones to be reached on this roadmap include the following:

a. Integrating agricultural objectives and target groups into each related sector and incorporating agricultural component in the related sector programmes where one is presently lacking;
b. Sensitizing policy-makers and planners in non-agriculture sectors to agricultural issues;

c. Adopting a culture of multisectoral coordination or partnership among public functionaries; and
d. Motivating the Bank staff and project management staff specially to put multisectorality to work as a deliberate policy.

Therefore the MAP represents a holistic effort to leverage transformational impact of World Bank-assisted projects in Nigeria. Towards this end, the immediate actions required involve defining roles and relations of sectors, identifying the types of collaboration feasible under individual projects, and determining how this will be supported in each case. In the present case of agricultural portfolio, the different collaborations possibly cuts across traditional sectors such as agriculture and water resources collaboration, agriculture and trade sector collaboration, agriculture and investment sector collaboration, agriculture and industry collaboration, and agriculture and social security collaboration, among others, wherein their full specifications and dimensions constitute the main elements of the MAP. In consideration of the cross-relations, the need arises to identify projects that their designs are amenable to multisectorality differently from those that are not, thereby making the multisectoral deliverables of project managers easier to achieve. Such a multisectoral design requires that each sector relevant to the project is adequately represented while the senior managers particularly STL and TTLs should sign off the results agreement of such projects incorporating multisectoral assignments of the staff.

The MAP provides the basis for the next steps which involve determining what the Bank should be doing differently in reinforcing the multisectoral approach to manage its portfolio. A step in this direction is the multisectoral design of projects that their contents may be greater in scope than what is possible to be handled by a sole government organization, thereby requiring direct involvement of counterpart agencies. In this regard, it is recognized that the growing awareness about ATA among various stakeholders, including politicians and chief executives, presents an opportunity for redesigning the programme, deploying the multisectoral approach, and for integrating it into agriculture and other sectors or development programmes. This could also lead to inter-sectoral participation, and collaboration at the national, state or local levels.

In any event, the survival of multisectoral approach in the Bank depends on several factors, including:

i) Necessity of multisectoral collaboration and ex-ante evaluation for potential effects and barriers;
ii) The rationale or scope for collaboration to take place across sectors;
iii) The assessment of the extent that implementation is possible by a single sector and identification of the part of the project plan that requires multisectoral cooperation;
iv) The benefit stream that is perceived by each stakeholder which determines if they will work together and participate in each other's projects effectively; and
v) Sector analysis to identify activities that can be carried out through single-sector approach or activities that cannot be achieved without multisectoral collaboration, and activities that can be achieved better through multisectoral collaboration although achievement is possible through single-sector approach.

In conclusion, the way forward for the Bank and other donor agencies is to take the issue of multisectorality very seriously and beyond the present stage, whereby the staff profess the approach only without operationalizing it on the ground as such. Nonetheless, the lessons of project management experience generally indicate that complex agricultural challenges defy single solution approaches developed in isolation of one another. Therefore, the multisectoral approach to project planning and management entails innovative ways of integrating the interests and actions of all stakeholders to address complex issues emerging at different stages along the agricultural value chains of different commodities or services—production, storage, processing, marketing, and consumption. Despite this complexity, however, interventions that increase knowledge of project management through multisectoral approach can generate incremental impact of World Bank assistance on its clients. By engaging multiple sectors, the Bank can leverage knowledge, expertise, reach, and resources in a manner that permits each actor to do what it does best and to work together in unison towards a shared goal of maximizing project benefits across the board. This represents a strategic investment and project management approach in terms of partnership arrangements, coordination, and funding to promote a more effective use of project resources focused on achieving demonstrable results in the shortest possible time.

However, while it is logical to think and plan multisectorally, actions must follow sector by sector, tailored to the specific context, objectives, and operating environment of each sector. This owes to the traditional practice for appropriating the statutory budget through allocations to institutions and projects at federal and state levels by sectors or ministries, and also to the traditional governance and accountability structures that follow similar sector-based provisions. That is, multisectorality is not cast in stone or sacrosanct that the need for it cannot be questioned. In reality, certain types of project ideas might sometimes fit into single sector more than multisector or vice versa, depending on the nature of projects, their instrumentality, and/ or other factors.

The following recommendations emerge on issue by issue basis for MAP to enhance transformational impact of World Bank's agricultural portfolio in Nigeria.

How projects within the portfolio overlap with one another:

- Clarify roles and responsibilities assumed by each sector in project concepts and designs,
- Verify the effects of multisectoral collaboration, and
- Incorporate agricultural considerations into initial design of projects/policies, including as elements of investments but not necessarily as the primary objective.

How the projects respond to multisectoral constraints or bottlenecks:

- Modify the design and consider alternatives to minimize undesirable outcomes of multisectoral approach and to maximize positive impacts.
- Make sure that the right questions are being asked around project issues throughout the project cycle, as to whether the single-sector or multisectoral approach will lead to attaining the development objectives faster or better.

What the Bank should do differently:

- Produce guidance notes for Bank staff and field actors on the multisectoral approach to its operations.
- Establish a compliance framework to align intervention with constitution.
- Establish a framework for identifying and setting common priorities among sectors.
- Emphasize multisectoral coordination at national, state, and local government levels.

- Set up a mechanism for multisectoral collaboration, e. g. joint implementation committee or task force at each level (intra and extra sector).
- Strengthen the roles of administration at a field level at which collaboration is more successful – local/community.
- Incentivize project managers for cross-sector work in terms of official recognition, career progression and financial inducement.

References

Adekunle A.A. and Fatunbi. A.O. (2012). 'Approaches for Setting-up Multi-Stakeholder Platforms for Agricultural Research and Development'. Forum for Agricultural Research in Africa, Accra, Ghana.

African Development Bank/African Development Fund. (2002). 'Agriculture and Rural Development Sector Bank Group Policy'.

Asenso-Okyere A., Asante F.A., Tarekegn J., and Andam K.S. (2009). 'The Linkages between Agriculture and Malaria'.

Ayoola G. B. (1999), 'Intervention policy in agriculture: A theoretical model of the two-sector developing economy', Petters G.H. and J. Von Brown (eds.), 'Food security, diversification and resource management: Refocusing the role of agriculture'. Proceedings, 23rd International Conference of Agricultural Economists, Sacramento, California, International Association of Agricultural Economists, 10–16 August 1997. Ashgate publishing Co. England (ISBN 1 84014 044 5) 647 pages.

Ayoola, G. B. (1997). 'Toward a Theory of Food and Agriculture Policy Intervention for A Developing Economy with Particular Reference to Nigeria'. Working Paper 97-WP 187.Centre for Agriculture and Rural development, Iowa State University, Ames, Iowa 50011.

Christopher B. and Lynette W. (2004). 'Reaching Communities for Child Health: Advancing Health Outcomes through Multi-sectoral Approaches'. Washington DC, USA.

Garrett J. and Natalicchio M. (2011). 'Working Multisectorally in Nutrition: Principles, Practices and Case Studies'. International Food Policy Research Institute, Washington, USA.

He Changchui. (2008). Technical Meeting of the Asia-Pacific Network for Food and Nutrition on Nutrition Interventions for Food Security – 'Can they work effectively in isolation?' Food and Agricultural Organization (FAO).

http://www.hunger-undernutrition.org/blog/2013/10/sustainable-food-systems-a-call-for-multi-sectoral collaboration-for-empowered-hunger-free-communiti.html

Idachaba, F. S. (1997). Personal communication. In Ayoola, G. B. (1997) op. cit.

International Bank for Reconstruction and Development/ International Development Association or The World Bank. (2013). 'Improving Nutrition Through Multisectoral Approaches.

Issues for Policy, Research, and Capacity Strengthening'. IFPRI Discussion Paper

Jaquelino M., Anina M., Cynthia D., and Almeida T. (2012).Contribution of the Agriculture Sector in Multi-Sector Combat of Chronic Malnutrition in Mozambique: Perspectives from National Seminar about Community Nutrition. Research results from Directorate of Economics, Ministry of Agriculture.

Kerry T., Stavros G., Rebecca C., and Roy B. (2004). 'Economic Valuation of Water Resources in Agriculture'. Food and Agriculture Organization of the United Nations.

Ministry of Health. (2010). 'Multisectoral Action Plan for the Reduction of Chronic Under-nutrition in Mozambique (2011-2015)'.

Nabeeha Mujeeb Kazi. (). 'Sustainable Food Systems: A Call for Multi-Sectoral Collaboration for Empowered, Hunger-Free Communities'.

Namugumya, B.S.(2011). 'Advocacy to Reduce Malnutrition in Uganda: Some Lessons for Sub-Saharan Africa'. Washington, DC: International Food Policy Research Institute.

National Planning Commission, Nepal. (2012). 'Multi-Sector Nutrition Plan for Accelerating the Reduction of Maternal and Child Under-Nutrition in Nepal'.

The World Bank/DFID. (2013). 'Improving Nutrition through Multisectoral Approaches/ Agriculture and Rural Development'.

The World Bank/DFID. (2013). 'Improving Nutrition Through Multisectoral Approaches/ Social Protection'.

UNICEF. (2013). Multisectoral Approaches to Nutrition-Nutrition Sensitive and Nutrition Specific Interventions to Accelerate Progress.

USAID-PVO Steering Committee on Multisectoral Approaches to HIV/AIDS. (2003). 'Multisectoral Responses to HIV/AIDS- A Compendium of Promising Practices from Africa'.

Wegener. J. (2011). 'Multi-Sectoral Perspectives on Regional Food Policy, Planning and Access to Food: A Case Study of Waterloo Region'. PhD Thesis for University of Waterloo, Ontario, Canada.

World Health Organization (WHO). (2010). 'Meeting on Multi-Sectoral Interventions for Non-Communicable Diseases (NCD) Prevention'. Japan.

46

The Human Rights Approach to Ensuring Food Security in Nigeria

Introduction

The problem of food looms large in Nigeria as also in several other parts of the world. According to Jean Ziegler (Vice president, UN Human Rights Council), 'In a world overflowing with riches, it is an outrageous scandal that more than 1 billion people suffer from hunger and malnutrition and every year over 6 million of children die of starvation and related causes. (Therefore) we must take urgent action now.' Generally, seven out of every ten poor individuals lived in households where agriculture represented the main occupation of the head, and lower average incomes among these households is a constant pattern across all regions and countries (WB, Global Economic Prospect 2009). Nigeria is not an exception since its food problem borders on situation of insecurity of a large proportion of the population who are uncertain about the availability, accessibility, and affordability of what to eat to sustain their strength or health, now or in the foreseeable future. The implication is that such people live in protracted suffering from hunger and in a constant state of fear that is life threatening, thereby, such individuals are unable to contribute to society as desired.

In this paper, I am to *profess* that the food problem of Nigeria is less rooted in the absence of technology or natural failure of the market, than it is rooted in the absence of suitable policies to guarantee the right of people to adequate food. This thesis is properly maintained on the strength of CAP 2 of the 1999 constitution, which expressly assigns the State to direct its policy towards ensuring the provision of 'suitable and adequate food' to the citizens.

It is in this context that the food situation of Nigeria presents a problem at three levels culminating in chronic or temporary food insecurity of the people, as individuals, as households, and as a nation in that order of magnitude. First, at the personal or individual level, the citizens of this country feel a perpetual sense of indignity and loss of personal pride in relation to other citizens of the world, arising from their inability to feed themselves. This situation emanates from the massive importation of food items that makes them permanently dependent on other countries of the world. For instance, the country imported about US$1 billion of rice in 2010, notwithstanding that (a) agricultural resources abound in Nigeria (a land mass of 98.3 million ha, 30% of which is cultivated; an ecology varying widely from tropical forest in the south to dry savannah in the far north, yielding a diverse mix of plant and animal life; as well as large water bodies comprising 14 million ha inland water—rivers, lagoons, etc.—and a coast line of 850 km or 800 nautical miles territorial waters; among others); and (b) over 100 years of technology research and deployment around agriculture, ranging from indigenous to imported technologies; through the series of World Bank-assisted Agricultural Development Projects for two decades, coupled a series of irrigation projects in all parts of the country also since mid-1970s. Notwithstanding these endowments and technology deployments, the citizens of the country remain food poor in a resource-rich country.

Second, at the household level, most families do not meet their food needs in terms of calorie and protein requirements to live a healthy and productive live; with particular reference to *children and women* members who are more vulnerable in the household. Thus, hunger and malnutrition persist on a large scale among rural and urban households, which, incidentally are responsible for the production of a large number of energy-giving crops such as cassava, yam, plantain, maize, rice, to mention a few; as well as a for the production of a wide range of protein-giving livestock animals such as cattle, sheep and goats, chicken, and fishes and other aquatic animals. Notwithstanding these endowments, a large proportion of the 160 million Nigerians are permanently starved, and go to bed hungry and/or malnourished on a daily basis. For instance, there is a high level of malnutrition among children in rural Nigeria across the geopolitical zones. The level of malnutrition was estimated to be 56% for the South West and 84.3% in three rural communities in the northern part of Nigeria. Nationally, the overall prevalence of stunting, wasting, and underweight are 42%, 9%, and 25%, respectively. Furthermore, the caput protein intake of a person in the country was about 45.4 g/person as against the standard 63.8 g/day and while the per caput energy consumption was 1,750 kcal /day as against the standard 2500 kcal/day (FAO 2005–2007)

Third, at the national level, Nigeria given its food dependency status is exposed and vulnerable to the negative effects of socioeconomic discontinuities of many countries of Europe and America, such as natural disasters, price volatility, civil disturbances, and war. Such discontinuities often lead to cutting the supply of food items to other countries suddenly, thereby causing temporary or permanent shocks in the food market, which often compromise the stability of governance and livelihood of the people. Here lies the wisdom in the saying that 'Food security begets national security.' That is, it poses a security problem to Nigeria as a nation depending so much on imported food items and to such a dangerous extent; spending 1.3 trillion naira a year (more than US$8.2 billion) importing basic food items like fish, rice, and sugar (greater than the average annual budget expenditure of federal government in recent years; N1.4 trillion for 2012); and now ranking as the world's second-largest rice importer from countries like China and Thailand.

On the flipside is the plight of Nigeria to the vagaries of climatic conditions of the neighbouring countries and the attendant poaching of the limited food resources in Nigeria, which exacerbated the food problem significantly. Such is the threat of famine looming on the Sahelian region in terms of drought, posing a huge problem and already leading to a large influx of hungry people from Chad and Niger in the north of the country, thereby complicating the security challenges facing the country at the moment. In this regard, it is a social responsibility of the government to maintain a safe level of food items in reserve, which should be sufficient to meet the need of the population while any such socioeconomic perturbation lasts; or in the absence of it, for a period of say one year on normative grounds at least. But the food reserve programme of Nigeria is not anywhere near this norm at any point in time, typically at a low level of about 1 million MT installed silos capacity that was never filled with grains beyond a 100,000 MT level.

Based on the foregoing, the food problem assumes such a strategic importance as to make us not only to begin to do things differently, but also to begin to think differently about the task of eliminating hunger and malnutrition from our society. Specifically, it is imperative that our traditional perception of food as a human need should give way to the contemporary perception of food as a human right. This is the philosophical change or paradigm shift we need to make in order to actualize the good intention of the writers of the 1999 Constitution. Furthermore, the right-based approach is superior to the need-based approach in addressing the problem of food insecurity, owing to the large scope it gives to the populace to participate in the policy process for ensuring food security, thereby promoting policy and governance accountability to the people of Nigeria.

From Food Security to Right to Food

That Nigeria has food but no food security

Food production is only a necessary but not sufficient condition for food security. We have made the point that the nutritional intake of an average Nigerian falls below standard. Now consider a hypothetical food basket for the country, containing ten energy-yielding crops—cassava, maize, rice, millet, guinea corn, yam, groundnut beans, and cocoyam. At the outset of democracy in 1999, the total output of this basket was 84 million MT weight or 40 million grain equivalents (GE). In 2007, the size of the basket increased to 128 million MT weight or 78 million MT grain equivalents; which indicates that (a) the energy food situation apparently improved on an annual basis during the Obasanjo regime; and (b) the energy food produced far exceeded the 30 million grain equivalents required by the population of 120 million Nigerians at that time on an annual basis, yet we cannot infer that food security exists owing to the access question and other issues involved. Thus, the food insecurity situation of the country at that time was concealed and not revealed by the data on food production.

Food as a right – how will it work?

The notion of food as a right is derived from the philosophical theory of *humanitarianism* and *egalitarianism*. Thus the following principles apply:

a. That human suffering is abominable and equality of man to man is inherently sacrosanct.
b. That food entitlement of a human being is the very essence of life worth living.
c. That the freedom of everyone from hunger is central to all fundamental freedoms to be enjoyed.

d. That is, right is for all intents and purposes inalienable, undeniable, actionable, remediable, and ultimately justiciable in a civilized society.
e. Therefore, we envision such a society wherein the sanctity of right to food is respected, protected, and fulfilled.

A common misunderstanding is that the right to food requires the State to feed the people. This is not the case, except in the case of the less privileged members of society, as practiced in countries of the world. Thus, the obligations to respect and to protect the rights to food are normal while the obligation to fulfill applies only to vulnerable groups such as children, women, displaced people, or other people in protracted suffering. Rather, the notion of food as a right implies that, on the one hand, the people holds the right with an obligation to work to meet their need for food; while on the other hand, the State bears a duty to ensure that its policies and programmes make food available, accessible, and affordable for individuals to be able to feed themselves. Thence both the people and the State can be held accountable in the performance of their roles to achieve the right to food; the people for their failure to work and the State for the failure of its policies to work, as differently obligated by reason and by law.

In particular, the obligations of government as duty bearer translate as follows:

- **Obligation to respect:** This stipulates the limits of the State's exercise of power, which includes refraining from destroying people's access to food *through public policies*.
- **Obligation to protect:** This entails protection from non-state actors in the fields of food safety and nutrition, protection of the environment and land tenure systems by regulating their conduct through *legislations and sanctions*.
- **Obligation to fulfill:** This involves identifying vulnerable groups in society and facilitating their access to food-producing resources, and even directly providing food if need be to achieve a critical minimum degree of freedom from hunger, as prescribed in CAP 2 of the constitution.

The rights approach to development is different from the needs approach in terms of the former connoting an **obligation** of the government and the latter connoting a **favour** of the government.

- The rights approach presents as both an opportunity and a challenge to development—an opportunity for growth of the food economy at a faster rate and a challenge to create an enabling policy environment for people to earn their food entitlements.
- Right to food generates incremental growth of the food economy by acting as an instrument of market mechanisms, i.e. as a demand management tool for driving production of food items from the demand side of the market.
- Right to food creates a policy environment conducive to development of food economy by acting as an instrument of accountability for holding government (duty bearers) accountable for the outcomes of their policy actions or inactions to ensure food security of the nation, while at the same time holding the citizens (right holders) responsible for the outcomes of their actions or inactions to provide food for themselves.

The global mandate for food security and right to food

Implementation Frameworks

The United Nations Commission on Human Rights had stated that hunger constitutes an outrage and a violation of human dignity, and therefore requires the adoption of urgent measures at national, regional, and international levels for its elimination. The Commission also reaffirmed the right of everyone to have access to safe and nutritious food, consistent with the right to adequate food and the fundamental right of everyone to be free from hunger so as to be able to develop fully and maintain their physical and mental capacities.

FAO as mandate holder produced a *Voluntary Guidelines* adopted by its council in 2004. This obligates member countries including the Federal Government of Nigeria to observe the tenets of right to food and to adopt the instruments for its domestication as a matter of both policy and law. The *Voluntary Guidelines* fulfill a number of conditions as follows:

- Provides practical guidance on how to implement the right to food in countries towards achieving food security of nations;

- Represents the first attempts by governments to interpret an economic, social, and cultural right and to recommend actions to be taken for its realization;
- Covers a wide range of actions to be considered by governments in order to build an enabling environment for people to feed themselves and to establish appropriate safety nets for people who are unable to do so;
- Are to be used to strengthen and improve current development networks with regard to social and human dimensions;
- Also cover issues ranging from democracy, good governance, human rights and the rule of law, economic development policies, legal framework, food safety and consumer protection, support for vulnerable groups, nutrition, international food aid, monitoring, indicators and benchmarks; and
- Deals with international measures, actions, and commitments.

Legal Regimes of Right to food

International laws:

- THE UNIVERSAL DECLARATION ON HUMAN RIGHTS (1948) - 'Everyone has the right to a standard of living adequate for the health and well-being of himself and his family, including food...'
- THE INTERNATIONAL COVENANT ON ECONOMIC, SOCIAL AND CULTURAL RIGHTS ICESCR (1966) - 'The States parties to the present Covenant recognizes the right of everyone to an adequate standard of living... including adequate food. And agree to take appropriate steps to realize these rights' (ICESCR, Article 11.1).
- ROME DECLARATION ON WORLD FOOD SECURITY (1996) - 'We the heads of State and government... Reaffirm the right of everyone to have access to safe and nutritious food, consistent with the right to adequate food and the fundamental right of everyone to be free from hunger.'

National Constitutions - Right to food is guaranteed in the constitution of many countries: Brazil (Art. 227), Congo (Art. 34), Ecuador (Art. 19), Haiti (Art. 22), Nicaragua (Art.63), South Africa (Art 27), Uganda (Art. 14), Ukraine (Art. 48), Guatemala (Art 51), Paraguay (Art 53), Peru (Art 6).

Case laws and success stories

Brazil

- Signed and ratified the International Human Rights framework providing clear obligations for the State to respect, protect, and fulfill the right to adequate food.
- Has put in place a Zero Hunger Program instituting necessary mechanisms for citizens to hold the State accountable to its obligations in relation to the right.

Venezuela

- In July 2008, adopted a decree with the rank of an Organic Law on Food Security and Food Sovereignty. The law guarantees the right of all citizens to have access to adequate and sufficient food.
- The law focuses on the promotion of food sovereignty which it defines as 'the right of a nation to define and develop its agriculture and food policies according to its specific circumstances'.

Nepal

- On 25 September, the Supreme Court of Nepal issued an interim order according to which the Government of Nepal has to supply immediately food to thirty-two food-short districts. This order was sequel to a public interest litigation to seek a public support for the over three million people suffering from food scarcity based on their right to food.

Kenya

- On Friday, 27 August 2010, the President of Kenya signed the country's new Constitution into law, preceded by a popular approval referendum. Article 43 (1) (c) expressly provides for the right to food within the context of economic, social and cultural rights. It states that: 'Every person has the right... to be free from hunger, and to have adequate food of acceptable quality.'

South Africa

- Some judicial pronouncements have been made by Judge Yaccob as follows:

    a) *In Government of the Republic of South Africa and Others v Grootboom and Others:'*

    - That, *'The State must also foster conditions to enable citizens to gain access to land on an equitable basis. Those in need have a corresponding right to demand that this be done. I am conscious that it is an extremely difficult task for the State to meet these obligations in the conditions that prevail in our country. This is recognised by the Constitution which expressly provides that the State is not obliged to go beyond available resources or to realise these rights immediately. I stress however, that despite all these qualifications, these are rights, and the Constitution obliges the State to give effect to them. This is an obligation that Courts can, and in appropriate circumstances, must enforce.'*

    b) *In Minister of Health and Others vs Treatment Action Campaign and Others*

    - That *'The state is obliged to take reasonable measures progressively to eliminate or reduce the large areas of severe deprivation that afflicts our society. The courts will guarantee that the democratic processes are protected so as to ensure accountability, responsiveness and openness, as the Constitution requires in section 1. As the Bill of Rights indicates, their function in respect of socioeconomic rights is directed towards ensuring that legislative and other measures taken by the state are reasonable.'*

The National Campaign on Right to Food

Background of FIF's campaign

Pertinent observations

- **Philosophy:** That the policy efforts towards food security is not guided by an overarching philosophy based on natural rights of the people, so every government is doing what is right in its own eyes and not necessarily right by the nature of the people.
- **Policy instrumentation:** There is no conscious policy effort to deploy rights instruments in government programmes towards achieving food security (e.g. Seven-Point Agenda, NV20:202, NFSP, etc.).
  Constitutionality: The Nigerian constitution fails to strengthen the food security process as a matter of right, so the people cannot participate in the policy process in the process as required.

15 Objectives of FIF:

- FIF's Overall Goal :
  - Create enabling environment for the promotion of human Right to Food in Nigeria, with a view to making policy and practice changes in favour food as a fundamental human right.

- Specific Objectives of FIF:
  - Analyse to what extent the right to food is already incorporated in domestic legislation.
  - Raise awareness of the human Right to Food and build capacity of strategic stakeholder.
  - Provide guidance, methods and instruments to assist the implementation of the Right to Food at country level.
  - Encourage and strengthen the capacities of civil society for participation throughout the process.
  - Formulate a strategy on right to food and facilitate emergence of legal framework for its implementation.

FIF's engagement process

The process of engagement involves putting the three arms of government on notice (executive, legislative, and judicial authorities) through:

- Communications – President, Senate president, Speaker of the House;
- Collaborations – FAO, AFAN, CSO, NSA, commodity associations, women associations, youth associations, etc.;
- Documentation and records –Memoranda (NADC/NCA), briefs (NPAFS), policies (Vision 202020), study report (Commodity Development and Marketing Companies), FAO/NMTPF, NFSP;
- Sensitization – Media exposure, advocacy, lobbying, street works, and
- Legislation – Draft bill on right to food during the sixth assembly (constitutional amendment of CAP 2), ongoing effort to tackle CAP 4 in that regard.

Outcomes:

- Policy and practice changes – Policy recognition of food as a right increased; public debate on right to food sponsored; public awareness about right to food generated; records of right to food officially created; attitude of people about food as a natural right improved; public opinion in favour of right to food mobilized; right consciousness about food aroused; interest in the issues involved in the right to food process rekindled.
- Critical Milestones – Government put on notice; general public sensitized; the legislature activated.

Policy Implications and Conclusions

Policy recognition of food as a fundamental human right is not just a lofty ideal. The perennial failure of policy and market to deliver on the global mandate for food security in Nigeria has its tap root in the non-recognition of food as a fundamental human right of citizens. After all, the right-based approach is consistent with the economics and philosophy of economic development generally, and the goal of Agricultural Transformation

Agenda in its special focus on agribusiness, value chain development, and private investment drive.

Therefore, the following recommendations emerge for the successful implementation of ATA in Nigeria:

- Policy audit to mainstream the rights instruments in the current policies, programmes, and projects;
- Legal framework through the constitutional amendment and formulation of a Bill of Rights to Food;
- Design and formulation of strategies for implementing the right to food; and
- Capacity building and strengthening of the state and non-state actors for effective delivery of the right to food security in Nigeria.

References

1. Ayoola, G. B. year? Food as Right
2. World Bank. (2009). Global Economic Prospect.
3. Wikipedia – an internet based encyclopaedia

> # Overview Of Fertilizer Demand And Investment Opportunities In Nigeria

Invited technical paper presented at the D-8 Private Investment Workshop on Gene Bank Development and Management and Investment Opportunities in Fertilizer Production, Transcorp Hilton Hotel, Abuja, and 15–17 April 2012.

Introduction

Fertilizer is the mainstay of agricultural economy with particular reference to crop agriculture. Consequently, many of the development interventions in the agricultural sector of Nigeria lend credence to the immeasurable value of fertilizer for the advancement of the country's agricultural sector.

First, empirical evidence from the World Bank-supported Agricultural Development Programme (ADP) has shown that crop response to fertilizer was highest among the class of green revolution inputs for increasing yield. Secondly, the evidence in terms of shares of public budget for fertilizer has shown that the policy response in fertilizer market was highest among the class of intervention instruments to induce demand and supply. In 2010, the fertilizer subsidy amounted to N22, 327,500,000, an equivalent of 70% of the capital expenditure of the rest of the Federal Ministry of Agriculture and Rural Development in the same year (N31, 861,528,219).Thus, among the plethora of political campaign issues, the political visibility of fertilizer was highest among the class of many factors of social and economic patronage. Yet the structure, conduct, and performance of fertilizer market leave much to be desired owing to the inadequate supply coupled with sharp practices and rent-seeking behaviour of actors.

This paper discusses fertilizer demand and investment opportunities in Nigeria. Section 2 describes the nature of fertilizer demand in relation to private investment. Section 3 highlights the role of government in stimulating and meeting fertilizer demand in Nigeria. Section 4 provides information about the trends and patterns in fertilizer demand and investment in Nigeria. Section 5 explores the scope and opportunities for foreign direct investment to meet fertilizer demand in Nigeria. The paper concludes with key recommendations regarding the need to enhance farmers' knowledge of fertilizer technology and the emerging opportunities in policy environment.

2. Nature of Fertilizer Demand

This section deals with the nature of fertilizer and the key factors that drive its demand for agricultural production. This includes chemical nature, derived nature, policy-induced nature.

Chemical nature of fertilizer

The chemical nature of fertilizer exists in its constituent elements and compounds. Table 1 below shows the chemical properties of fertilizer formulations. The fact that fertilizer is a chemical product implies that its use has certain consequences for the environment and health of people. Table 1 shows the various formulations in use in Nigeria, whereas Table 2 shows the quantities of nutrients as produced or imported over the years (Liverpool-Tasie et.al. 2010).

Beyond the environmental consequence is also the economic implication of fertilizer as a chemical product. In this regard, fertilizer as a product from chemical industry will rely on high production technology and would require very huge and lumpy investment capital base to set up the factories and agribusiness around it.

Nonetheless, the chemical nature of fertilizer and high-tech characteristics have brought into focus the untoward product knowledge by the end-users—farmers. It is also important to note that the chemical characteristics of fertilizers are also linked to their demand for alternative uses. The non-agricultural uses of fertilizer in Nigeria include soap manufacture and explosives manufacture. While the alternative line of utilization for non-agricultural purpose is largely of economic consequence, it is also an issue of fiscal importance where these activities constitute leakages of government's fertilizer subsidy policy to boost agricultural production.

Table 1: Properties of fertilizer formulation.

Fertilizer	Parameter	Specification
Urea	Nitrogen content	45-46%
	Moisture content	Not exceeding 0.5%
	Biuret content	Not exceeding 1.5%
	Physical condition	Good granular integrity, free flowing, dust free, non-caking, white crystalline.
NPK 15-15-15	Nitrogen content	15%N minimum, 40% nitrate, 60% ammonia.
	Phosphorous content	15%P_2O_5,100% citrate soluble of which one third is water soluble
	Potassium content	15% K_2O as muriate of potash,100% water soluble
	Moisture content	Not more than 1%
	Screen analyses	Minus 2mm, maximum 5%, between 2–4mm, 90%, plus 4mm maximum 5%
	Physical condition	Good granular integrity, free flowing, dust free, non-caking
NPK 25-10-10	Nitrogen content	25% N, 40% as nitrate, 60% as ammonia
	Phosphorous	10% P_2O_5,100% citrate soluble of which a minimum of 65% is water soluble
	Potassium	10% K_2O, 100% water soluble, moisture content not more than 1%
	Screen analysis	Minus 2mm, maximum 5%, between 2–4mm, 90%, plus 4mm maximum 5%
	Physical condition	Good granular integrity, free flowing, dust free, non-caking
	Moisture content	Not more than 1%
NPK 27-13-13	Nitrogen	27% N, 40% as nitrate, 60% as ammonia
	Phosphorous	13% P_2O_5, 100% citrate soluble of which a minimum of 85% is water soluble
	Potassium	13% P_2O_5,100% water soluble, moisture content not more than 1%
	Moisture content	Not more than 1%
	Screen analyses	Minus 2mm, maximum 5%, between 2–4mm, 90%, plus 4mm maximum 5%
	Physical condition	Good granular integrity, free flowing, dust free, non-caking

NPK 29-10-10+10%Ca	Nitrogen	20% N, 40% as nitrate, 60% as ammonia
	Phosphorous	10% P_2O_5, 85% Citrate soluble of which a minimum of 85% is water soluble
	Potassium	10% K_2O, 100% water soluble
	Calcium	10% total Ca of which 50% is soluble in water
	Screen analyses	Minus 2mm, maximum 5%, between 2–4mm, 90%, plus 4mm maximum 5%
	Physical condition	Good granular integrity, free flowing, dust, non-caking
NPK 12-12-17+2MgO	Nitrogen	20% N, 40% as nitrate, 60% as ammonia
	Phosphorous	12% P_2O_5, 100% citrate soluble of which a minimum of 85% is water soluble
	Potassium	17% K_2O, 100% water soluble
	Magnesium	2% MgO, 50% soluble in weak acid
	Moisture content	Not more than 1%
	Screen analysis	Minus 2mm, maximum 5%, between 2–4mm, 90%, plus 4mm maximum 5%
	Physical condition	Good granular integrity, free flowing, dust free, on-caking
SSP 0-18-0	Phosphorous	18% P_2O_5, 100% citrate soluble of which a minimum of 75% is water soluble
	Potassium	17% K_2O, 100% water soluble
	Moisture content	Not exceeding 3.0%
	Free acid content	0.5% as maximum as H_2SO_4
	Screen analysis	Minus 2mm, maximum 5%, between 2–4mm, 90% plus 4mm maximum 5%
	Physical condition	Good granular integrity, free flowing, dust free, non-caking
Ground rock phosphate	phosphorous	20–36% total P_2O_5, phosphorous expressed as P_2O_5 soluble in mineral acids, at least 55% of the declared content P_2O_5 of being soluble in 2% formic acid
	Particle size	At least 90% able to pass through sleeve with mesh of 0.123mm
Muriate of potash, MOP	Potassium content	60% K_2O, 100% water soluble
	Moisture content	Not exceeding 3%
	Physical condition	Crystals

Derived nature of fertilizer demand

Fertilizer demand has a derived nature, i.e. it is not demanded for its own sake, but derived from the demand for final products that it is required to produce. This implies that demand for food and non-food agricultural products is intricately tied to demand for fertilizer. Thus, any investment in fertilizer sector is integrated with investment in food and non-food commodities.

Policy-induced nature fertilizer demand

The policy-induced nature of fertilizer demand is obvious in terms of government intervention through subsidy provision, procurement operations, distribution networks, and extension services. Generally speaking, fertilizer demand in Nigeria is policy induced rather than market induced. Thus, the market for fertilizer is not perfect in structure as it is also poor in its conduct and performance for the most part of its existence.

Therefore, price signals are weak for analyzing demand in the fertilizer market, which implies that the prevailing policy environment should be factored into decisions to invest in the fertilizer market.

Table 2: Fertilizer production, import and consumption in Nigeria, 2002–2008 (Liverpool-Tasie et al. 2010).

Fertilizer Type	Element	2002	2003	2004	2005	2006	2007	2008
Nitrogen fertilizers (N total nutrients)	Production quantity in nutrients (tonnes of nutrients)	0	0	3,800	4,868	20,821	12,505	12,500
Phosphate fertilizers (P_2O_5 total nutrients)		0	0	2,200	2,779	12,540	6,553	6,500
Potash fertilizers (K_2O total nutrients)		0	0	2,450	3,066	14,314	6,803	6,800
Ammonium nitrate	Import Quantity (MT)	2,849	2,437	0	0	0	0	63,538
Ammonium sulphate		4,709	5,17???	74,420	76,490	78,619	24,260	1,576
Diammonium phosphate (DAP)		5,009	466	0	0	0	19,532	24,438
Monoammonium phosphate (MAP)		219	20	30,000	25,000	22,000	0	19,571
NPK complex 10kg		97,605	25,770	0	0	0	0	439,312
Potassium chloride (muriate of potash)		2,946	10,810	11,000	10,000	11,000	12,219	67,380
Superphosphate other		0	0	8,000	10,000	10,000	14,718	0
Urea		288,252	77,207	39,000	417,900	306,900	75,864	601,870

Nitrogen fertilizers (N total nutrients)	Import quantity in nutrients (MT of nutrients)	94,400	137,603	36,868	211,047	160,104	43,508	370,676
Phosphate fertilizers (P$_2$O$_5$ total nutrients)		41,400	49,432	17,040	14,800	13,240	11,634	60,793
Potash fertilizers (K$_2$O total nutrients)		30,400	42,712	6,600	6,000	6,600	7,331	40,428
All	Fertilizer consumption in kg-hectare of arable land	51,937.5	71,795.94	20,896.36	69,302.86	63,227.5	24,201.1	0

Role of Government in Stimulating and Meeting Fertilizer Demand

The role of government with regards to generating demand for fertilizer use in agriculture in post-independence Nigeria and the concomitant efforts at making fertilizer available may be highlighted in a decade-wise mode.

First independence decade (1960–1970)

The main challenge during the first post-independence decade was how to generate demand for inorganic fertilizer from the farm population.

The various regional governments recognized the potential role of fertilizer as a yield-enhancing instrument, which led to promotion of the input by increasing numbers of farmers that adopted the fertilizer technology in the respective regions. The regional governments intervened by importing fertilizer from abroad and introducing it to farmers under various schemes.

Second independence decade (1970–1980)

Further to the demand for inorganic fertilizer generated during 1960–1970 years, the second-generation issue was how to meet a growing demand for inorganic fertilizer. Obviously, Nigeria lacked the capacity to produce inorganic fertilizer at that time, and it was a period when foreign exchange was in short supply.

However, certain measures and policy innovations emerged with the creation of Federal Ministry of Agriculture and Natural Resources (FMANR) in 1966. These include:

- Establishment of Fertilizer Procurement and Distribution Division (FPDD) in the ministry, which took over importation of fertilizer from the regions;
- Establishment of a network of primary distribution points in the country, together with provision of subsidy;
- Promulgation of a Fertilizer decree with provision for the establishment of a National Fertilizer Board (National Fertilizer Board Act, 1977); and

- Establishment of a fertilizer plant in 1976 at Kaduna for domestic production of phosphate fertilizer towards long-term sustainability of supply in the country.

Third independence decade (1980–1990)

The situation during the third post-independence decade presented the challenge of how to sustain demand for fertilizer through importation and how to improve the efficiency of fertilizer use by farmers across the different agro-ecological zones of the country. These were the main effects of the downturn in the economy associated with dwindling foreign exchange incomes and the ultimate need for a structural adjustment programme of the national economy.

In response to these challenges, the National Fertilizer Company (NAFCON) commenced operation in 1982 for the production of nitrogenous fertilizers. This was followed by the establishment of the National Fertilizer Technical Committee to coordinate the indents from states and recommend product slates for different states based on field trials.

Fourth independence decade (1990–2000)

The fourth post-independence decade of Nigeria featured the challenge of how to stem the mounting subsidy burden on public budget and escalating demand for fertilizer that loomed very large. During this period, the federal government commenced a gradual withdrawal from fertilizer supply operations, beginning with a one-stop withdrawal from direct distribution and subsequently from importation.

Fifth independence decade (2000–2010)

How to further meet the subsisting level of fertilizer demand at the current subsidy level while maintaining quality of the products in the open market at the same time were the main challenges with regards to fertilizer as a key input to agricultural production in Nigeria during the fifth post-independence decade. The liberalization of the national economy during the period was undoubtedly a key factor in addressing fertilizer issues. A study group by the Fertilizer Technical Committee was set up in 2001 to produce a blue print for the establishment of a fertilizer regulatory and quality control system for the country, which the process is inconclusive till date.

Current independence decade (2010–2012)

Lastly and currently, the country is burdened with how to incentivize the private sector to drive production and distribution of fertilizer to meet the demand of farmers in Nigeria. Towards this end, the federal government has commenced implementation of Agricultural Transformation Agenda (ATA) which components include a Growth Enhancement Scheme (GES) for fertilizer and seed supply through participating agro-dealer networks.

Trends and Patterns in Fertilizer Demand and Public Investments

The trends and patterns in fertilizer demand and public investments can be decomposed as follows:

a) Fertilizer demand by type or use

- » Demand for macro versus micro nutrients
- » Demand for liquid fertilizer versus demand for granular fertilizers
- » Demand for arable crops versus demand for tree crops
- » Demand for organic versus inorganic fertilizer
- » Demand for soil placements versus foliar applications, etc.
- » Demand for liming as soil amendments to reduce soil acidity consequent upon fertilizer use

b) Fertilizer demand by agro-ecological specialization

- » Different agricultural zones with different soil characteristics
- » Different ecological zones with different land capabilities

c) Fertilizer demand by other proximate determinants

> » **Knowledge:** The knowledge of farmers is weak with respect to different fertilizer formulations, chemical composition, crop requirement, soil requirement, place requirement, recommended application rates, proper timing of application, level of government support, etc.
> » **Attitude:** The attitude of farmers to fertilizer with regards to price (high/low), subsidy (good/not), yield increase (significant/not)
> » **Practice:** The practices of farmers about use of fertilizers with regrads to crop mixture, availability, accessibility, and affordability factors; types of fertilizer applied; time of application; rate of application; subsidy enjoyed; etc.

This implies a wide scope for investment, that is, wide range of product choice; investment capital to meet diversified needs in terms of agroecological specializations and proximate determinants.

Trends and patterns in public Investments

- FSFC: The first granulation plant established at Kaduna for the manufacture of phosphate fertilizers; now privatized.
- NAFCON: Second granulation plant established at Onne, P/Harcourt; now privatized.
- Blending plants operations: Established by states and private individuals all over the country to diversify the supply base of fertilizer formulations.
- Procurement and Distribution: Port facilities/ Distribution points established.
- Extension services: Heavy investment in agricultural extension service in 1970s/80s through the World Bank-assisted ADPs; farm service centres or agro-service centres (FSC/ASC), etc.
- Fertilizer subsidy: Dedicated account for fertilizer; in 2010 fertilizer subsidy amounted to N22, 327,500,000, an equivalent of 70% of the capital expenditure of the rest of the ministry in the same year (N31, 861,528,219).

Other discernible trends and patterns of fertilizer market

Such other tends and patterns emerge from two studies:

a) IFPRI Study (2008?):

- That fertilizer demand in Nigeria outstrips supply from outset (100% consensus);
- That the main deterrents to fertilizer use are the *unavailability of the product at the right time* and *affordability of the product*;
- That very few farmers have access to the subsidized product;
- That public bureaucracy accounts for the late arrival of fertilizers;
- That low retail density accounts for the general difficulty in accessing fertilizer;
- That the price of subsidized fertilizer sold to states by federal government can be higher than that purchased directly by the states from private sources;
- That the Agro dealers complained of patronage challenges whenever subsidized fertilizer is around;
- That recirculation of subsidized fertilizer in the private market is a common occurrence;
- That private dealers take advantage of arbitrage opportunities due to multiple prices, which prevents development of private market; and
- That the subsidy program (as previously) administered was be exacerbating the problem of fertilizer unavailability by reducing incentives for the establishment of private fertilizer retail outlets.

b) PropCom Study (2010):

- That most farmers applied fertilizer to their crops; (>70%);
- That most farmers applied fertilizer without soil testing;
- That extension visits to farmers was a rare occurrence; and
- That the subsidized fertilizer reached a small proportion of farmers (11%).

Table 3: Fertilizer production in Nigeria (FMARD 2006).

S/No	Name and address	Year established	Installed capacity (MT)	Capacity utilization (%)	Type produced
1	National Fertilizer Company of Nigeria (NAFCON), Onne, Port-Harcourt	1976	750,000 p.a.		NPK, Urea, Ammonia
2	Federal Superphosphate Co. Ltd. (FSFC) Kaduna	1976	100,000 p.a.	20	SSP
3	Kano Agricultural Supply Company Ltd	1993	240 p.d.	62	NPK
4	Golden Fertilizer Co. Ltd. Lagos	1997	350,000 p.a.	43	NPK
5	Sasisa Fertilizers Ltd. Kano	1999	60 p.d.	100	NPK, Urea
6	Gaskiya Fertilizer Coy Ltd. Kano	1999	180 p.d.	50	NPK
7	Morris Nigeria Ltd. Minna	1989	200000 p.a.	30	NPK
8	Bauchi Fertilizer Blending Company Bauchi	1999	60 p.d.	70	NPK
9	Agro Nutrient & Chemical Company, Kano	1993	15,0000 p.a.	50	NPK, Urea
10	Shamrock Fertilizer Co. Ltd. Sokoto	2000	75,000 p.a.	45	NPK
11	Fertilizers and Chemicals Company Ltd., Kaduna	1989	200,000 p.a.	50	NPK
12	Aweba Fertilizer & Chemical Co. Lafia	2003	30,000 p.a.	40	NPK
13	Zamfara Fertilizer Company Ltd. Gusau	1998	35 p.h.	30	NPK
14	Gombe Fertilizer Company Ltd. Gombe	1999	15 p.h.	10	NPK
15	Borno Fertilizer Company Ltd Maiduguri	1999	N/A	N/A	NPK
16	Diamond Fertilizer Nig. Ltd, Kano	2000	N/A	N/A	NPK
17	Yobe State Fertilizer Blending Co. Ltd, Damaturu	2002	15–18 p.h.	30	NPK
18	West African Fertilizer, Okpella, Edo State	1993	100,000	N/A	Limestone granules
19	Nasarawa Fertilizer Blending Plant, Lafia	2003	30,000	30	NPK

20	Cybernetics Nig. Ltd, Kaduna South, Kaduna	1986	5,000MT NPK+AMF/ 200,000 CTNS AMF	25	Micronutirent and NPK
21	Edo Blending Plant, Auchi	2001	30,000	N/A	NPK
22	Ebonyi Fertilizer Co., Abakaliki	2002	30,000	N/A	NPK
23	Fertilizer Additive Mfg. Co., Lagos	1998	30,000	60	Limestone granules, NPK
24	Plateau Fertilizer Co., Bokkos, Plateau State	2003	30,000	35	NPK
25	Quest 11 Enterprises Ltd., Lokoja	1986	100,000	50	Agricultural lime
26	Funtuwa Fertilizer and Chemical Company	199	30,000	35	NPK 20-10-10

Table 4: Supply of fertilizer by the federal government, 2000–2010.

Production season	Target (MT)	Actual quantity (MT)	Cost of Quantity (MT)	Achievement
2000	120,000.00	101,148.75	2,394,736,083.60	84.29
2001	120,000.00	119,089.25	3,530,406,650.00	99.24
2002	50,000.00	45,306.70	1,346,148,348.00	90.61
2002	163,700.00	114,793.65	3,605,662,509.20	70.12
2003	120,000.00	95,994.90	4,620,418,025.00	80.00
2004	248,400.00	202,038.55	11,024,019,200.00	81.34
2005	157,000.00	146,249.25	8,341,772,360.00	93.15
2006	261,000.00	259,838.86	16,258,649,932.00	99.56
2007	528,000.00	297,260.04	19,422,363,970.00	56.30
2008	650,000.00	566,590.03	57,055,503,960.00	87.17
2009	400,000.00	371,062.45	38,050,847,750.00	92.77
2010	900,000.00	586,145.05	58,429,230,250.00	65.13
GRAND TOTAL	3,718,100.00	2,905,517.48	224,079,759,037.80	78.15

Opportunities for Foreign Direct Investments (FDI) in Fertilizer Sector in Nigeria

Such opportunities reside in current programmes of government currently under implementation. Some of these are summarized as follows:

5.1 Agricultural Transformation Agenda (ATA)

ATA is a programme of FMARD with a mission to 'Achieve a hunger-free Nigeria through an agricultural sector that drives income growth, accelerates achievement of food and nutritional security, generates employment and transforms Nigeria into a leading player in global food markets to grow wealth for millions of farmers' The main features of ATA include the following:

- Focus on agriculture as a business
- Create jobs, create wealth, and ensure food security
- Focus on value chains where Nigeria has comparative advantage
- Develop strategic partnerships to stimulate investments to drive a market-led agricultural transformation
- Participative - State and local governments, inter-ministerial collaboration, private sector driven, farmer groups and civil society, sharp focus on youth and women.

The transformational policies include: Fertilizer reforms; Marketing institutions; Financing agricultural value chains; Agricultural investment drive. Specifically for fertilizer and seed, the following reform measures **were include:**

- Revitalize the private sector fertilizer and seed industry
- Government withdraws from fertilizer and seed procurement and distribution
- Private sector to commercialize seeds and fertilizers to reach farmers directly
- Growth Enhancement Support (fertilizer subsidy through telephone

5.2 Nigeria Agribusiness and Agro-industry Development Initiative (NAADI)

NAADI is a programme of FMTI that promotes development of an industrialized, high-growth, and diversified economy that will create jobs, wealth, and provide food security for Nigerian people. The opportunities for investment emerge through: backward integration, import substitution, local patronage, and industrial partnerships.

Specifically relating to fertilizer sector as an import substitution investment area:

6. Fiscal and non-fiscal policy measures for dealing with particular production and trade issues
7. Policy measures for strengthening sectoral innovation systems
8. Policy measures for spreading the impact of FDI on domestic capacity and capability building
9. Specific policy measures for strengthening local and regional chains
10. Policy measures for facilitating the role of non-state actors and civil society in the policy process for agricultural development

5.3 Targeted Policy Incentives (TPI)

The package of fiscal and other incentives is structured as follows:

- With respect to the manufacture of biological fertilizers, organic fertilizers, or soil conditioner, etc., the following incentive would apply:

  1. Variable tax holiday ranging from 10 and 50 years depending on magnitude of investment.
  2. Tax deduction of up to 35% of the cost of providing infrastructure facilities (capitalized during the period of tax holiday) distributed over a five-year period.
  3. The benefits would apply to existing as well as prospective investors.

- With respect to the manufacture of petrochemicals and subject to certain conditions, the following incentive will apply:

    1. Variable tax holiday ranging from 7 to 17 years depending on the magnitude of the investment;
    2. Deduction of up to 150% of the R&D investment made by existing and new industries to develop products and services;
    3. A 2% tax concession on in-plant training expenditure for up to 15 years for Nigerian employees, based on certain criteria;
    4. The benefits will apply to existing as well as prospective investors;
    5. A 50% reduction in corporate income tax on net profit for five years after expiry of tax holiday.

Conclusions and Recommendations

Fertilizer demand is derived by nature, and policy-induced to a considerable extent in Nigeria. These features offer considerable opportunity for investment in fertilizer production and adoption as premised on increasing trend and pattern of use in Nigeria.

In the past five decades, the government has made substantial public investments in the fertilizer sector towards stimulating and meeting demand for the critical farm input, and these have yielded fruits in terms of infrastructure and policy environment conducive for the private sector investment. Furthermore, the present administration has improved upon the policy environment with a view to generating domestic and foreign direct investment. Specifically new opportunities have emerged through ongoing policy measure—ATA and NAADI. The opportunities in the new policy reforms exist in terms of growth enhancement support, marketing reforms, value chains financing and agricultural investment drive as well as several other fiscal and non-fiscal incentives being put in place.

References:

1. Ayoola, G.B. (2001). *Essays on the Agricultural Economy: A book of Readings on Agricultural Policy and Administration in Nigeria*; TMA Publishers; ISBN 978-33454-3-5
2. Liverpool-Tasie, Saweda L. O., Abba Auchan and Afua B. Banful (2010). An Assessment of Fertilizer Quality Regulation in Nigeria. Nigeria Strategy Support Programme Report 09 October 2010.
3. Anonymous (2012). Nigeria Agribusiness and Agro-industry Development Initiative (NAADI) Federal Ministry of Trade and Investment, Abuja Nigeria.
4. *Nigeria Agriculture Digest* – First Quarter 2012 Edition Farm and Infrastructure Foundation.

Emerging Issues for the Formulation of Policy on Agri-Input Delivery System in Nigeria

With **J.B. Ayoola**, In M. Behnassi, S. Draggan and S. Yaya (eds. 2011). *Global Food Insecurity: Rethinking Agricultural and Rural Development Paradigm and Policy.* Netherlands Springer, Chapter ISBN 978-94-007-0889-1, 1st Edition, 2011. XXIV, 40.

In Nigeria, the perennial inefficient distribution system of farm inputs, namely fertilizer, seed and crop protection products, represents a major constraint to achieving food security. The goal of this paper is to determine the key issues involved in the formulation of agri-input policy for the country, and to examine these issues with a view to facilitating policy implementation. The analysis of agri-input delivery system was conducted within the context of a development communication framework, which had as its main elements the specific package of inputs as a policy *message*, originating from a *source*, and passing through a *channel* to be delivered to a *target*. In this framework, the various policy bottlenecks associated with each segment of input flow from factory gate to farm gate were identified as a *noise*, which obstructed the free flow of the agri-input message in its passage from the source through the channel to the target. The negative consequences of these noise elements on the performance of agriculture has been manifested in the status of the country as a net importer of food items combined with input import dependence on other countries of the world. The main issues under focus belonged in the political economy and governance as well as structural and systemic categories. The conclusion reached was that a rule-based, evidence-led, and internally consistent policy articulation and strategy formulation was required for enhancing the performance of the agri-input delivery system of the country.

Key words: Agri-input, policy articulation, strategy formulation, development communication, food security.

1. INTRODUCTION

The production system is expressed in one economic jargon called production function, mathematically written as: $P = f(x)$. In ordinary language, this means quantity of output is determined by quantity of input. This is how economists appreciate the role of inputs in the production process, implying that without inputs there is no output. Therefore, the efficiency of the agricultural production system depends on the efficiency of the input delivery system. And unless the problem with the input delivery system is effectively addressed, there is no basis to expect that the volume of output will grow as desired.

The problem with the agri-input delivery system of Nigeria pertains to inadequate quantity of supply, delays in supply relative to the needs of farmers, and widespread nature of market sharp practices, among others. These problems form the original basis to justify policy intervention in the system. However the situation is compounded by second-generation problems of implementing the policies in terms of perennial abuse of these policies coupled with the lack of sufficient implementation commitments.

In this paper it is argued that poor policy formulation begets poor policy implementation. That is, good formulation is a necessary, albeit not a sufficient, condition for good implementation of agri-input policies. Thus the goal of this paper is to determine the key issues involved in the formulation of agri-input policy for Nigeria, and to interrogate these issues somewhat, with a view to proposing their optimal resolutions. First, after the introductory overtures (Section I), we survey the background of such policy issues (Section II) followed initially by a highlight of the policy formulation process as regards agri inputs (III), and finally by the discussion of the issues as they emerge from previous expositions (IV).

Primer of Agri-input policies

The need for policy intervention in the agri-input market predicates on three important arguments, namely: the market failure theory, which holds that market for agri inputs fails in certain respects that the market on its own is unable to correct for them on its own; equilibrium adjustment, which pertains to the long time of adjustment of the market from one equilibrium position to the next equilibrium position; externalities, whereby there is visible divergence of private course from social course; and the theory of second best, states that once the conditions of *pareto optimality* in attaining the first best equilibrium of the market is violated as is generally the case, there is no basis to pursue the remaining conditions in attaining the second best option (Lipsey and Lancaster 1956).

The agri-input delivery system is conceived in the development communication context, which has as its main elements the specific package of inputs as a *message*, originating from a *source*, and passing through a *channel* to be delivered to a *target*. Within this context we identify the various policy bottlenecks associated with each element as a *noise*, which obstructs the free flow of the agri-input message as it passes from the source through the channel to the target. Elsewhere the forms and functions of the noise elements in the policy pathway have been described in detail (Ayoola 2001). The manifest consequences of these noise elements on the performance of agriculture is reflected in the status of the country in terms of food import dependence notwithstanding its status in terms of input import dependence (Tables 1 and 2).

Now the African green revolution is underway, having gathered substantial momentum in the recent past. We recall the Africa Fertilizer Summit held in Abuja in June 2006, which has generated a 12-point agenda for action among the countries (IFDC 2006). We also observe the follow-up action under the Alliance for Africa Green Revolution in Africa (AGRA). Both developments will definitely focus on the judicious use of agri inputs in turning the situation around on the continent, with a view to attaining food security of the people in the shortest possible time. Thus now is the time to get the policy environment right for agri-inputs delivery in Nigeria with a view to maximizing the advantage of the new programmes.

Table 1. Imports of food into Nigeria, 1991–2003 (tonnes)

Commodity/year	1991	1992	1993	1994	1995	1996	1997	1998	1999	2000	2001	2002	2003	2004	2005	2006	2007
1 Sugar refined	449780	477133	529708	406000	719355	468000	635768	896000	771782	615930	893500	1099200	716500	456577	461060	288972	288972
2 Rice milled	296000	350000	350000	350000	300000	345500	699054	594057	812452	740000	1765500	1232411	1600000	1350000	1040322	963140	963140
3 Wheat	230000	292506	1125196	675282	608609	799520	1068802	1424009	1473940	2219708	2190200	2397839	2217000	2608947	3714683	3243998	3243998

Source: FAOSTAT (FAO Website); NA: Not available

Table 2 – Imports of Farm inputs and machinery into Nigeria (1991-2002)

Farm input	Quantity/Unit	1991	1992	1993	1994	1995	1996	1997	1998	1999	2000	2001	2002	2003	2004	2005	2006	2007	2008
Pesticide	Value (1000 $)	24373	8495	11000	13000	16065	19000	32187	37149	16000	14361	12721	108313	28029	49688	36095	60906	169082	82338
Fungicide	Value (1000 $)	NA	626	667	NA	5523	NA	8467	15660	1955	NA	1085	10325	6537	-----	-----	7586	6170	6170
Herbicides	Value (1000 $)	NA	534	725	NA	1960	NA	3195	2543	3793	NA	7757	50602	7575	-----	-----	48622	90525	47063
Mineral fertilizer	Value (1000 $)	5000	4500	4000	4000	3603	15797	5547	24207	1913	29133	12418	49169	29056	100000	150000	213050	281918	159452
Organic fertilizer	Value (1000 $)	NA	NA	NA	NA	NA	200	76	3619	54	181	189	49169	29056	100000	150000	213050	281918	159452
Agric. tractors total	No. (Numerical)	NA	NA	NA	NA	405	520	360	400	235	589	2309							

Source: FAOSTAT (FAO Website); NA: Not available

3. BACKGROUND TO POLICY INTERVENTIONS IN AGRI-INPUT MARKET

3.1 Policy intervention modes

The principal focus here is on the so-called green revolution inputs—seed, fertilizer, and crop protection products or CPP, in that order.

Seed:

Public intervention in the seed industry dates back to the colonial era when the 'travelling teachers of agriculture' used to carry planting materials from place to place introducing 'new improved' crop varieties similar to the 'New Improved Blue OMO' in the present time. This translated into organized extension work for the delivery of seed and other planting materials to rural dwellers.

At the moment, the National Seed Service (NSS), now the National Agricultural Seed Council (NASC), is the superintending agency for policy interventions in the seed industry. The National Crop Varieties and Livestock Breeds Registration and Release Committee was established by law (Decree 33 of 1987), for the purpose of regulating the activities of stakeholders in the seed industry. Specifically, the committee was charged with receiving and processing applications for the registration, naming, and release of old and new crop varieties, and officially releasing the list of varieties recommended by the technical sub-committee established for that purpose. Subsequently, the National Agricultural Seed Committee was established (National Agricultural Seed Act No. 72 of 1992), leading to the publication of a comprehensive list of all crop variety released and registered in Nigeria (PASS, 2007), and giving mandate to NSS (NASC) as sole source of foundation seed production in Nigeria, in collaboration with the National Agricultural Research Institutes (NARI).

In practical terms, the current key modes of policy intervention in the seed industry are as follows:

- The NARIs and universities as public institutions act as the original source of *breeder seed* of new varieties produced by them; the breeder seed of such public bred varieties should be released to NSS (NASC).
- The NASC is responsible for the production of *foundation seed* from breeder seed in collaboration with NARIs, universities, and private

seed companies; the NASC would release the first stage foundation seed to private seed companies while the second stage foundation seed would be released to ADPs.
- The private seed companies and ADPs are responsible for the production of *certified seed* from the foundation seed provided by NASC, through companies' farms or contracted out to out growers/contract farmers; both the private seed companies and ADPs are also responsible for selling certified seed to farmers to produce commercial grain.

Fertilizer:

Policy intervention for replenishing the soil dates back to the era of regional control through to the first half of the second post-independence decade, when the Regions/States engaged in separate importation of mineral fertilizer for distribution to farmers. At the federal level, the Fertilizer Procurement and Distribution Division (FPDD) was established as the sole agency for supplying mineral fertilizers to the Nigerian market. A loan was obtained from the World Bank in two stages for the purpose of massive importation and distribution at subsidized prices, which lasted till the 1990s. Also domestic supply capacity was enhanced by the establishment of two granulation plants namely National Fertilizer Company (NAFCON) and Federal Superphosphate Fertilizer Company (FSFC) followed by a series of blending plants. The sector has since undergone substantial reforms leading to the re-designation of the FPDD as FFD (Federal Fertilizer Department) and the systematic withdrawal of the government from importation and distribution activities and the reduction of subsidy on fertilizers.

Currently, the main elements of policy intervention in the fertilizer sector in the market stabilization programme (FGN, 2006) are as follows:

- The federal government undertakes limited purchase of mineral fertilizer, typically 250,000 MT through tendering in the local market, on a seasonal basis, and at a uniform price; the consignments would be distributed to the states as indented.
- The state governments should sell the allocations of fertilizer to farmers at 25% subsidy and uniform price throughout the country; the proceeds should be remitted to the federal government.
- The states and local governments are also at liberty to support the farmers with additional subsidies on federal fertilizers or to procure additional fertilizer for distribution to farmers at subsidized prices.

CPP:

Public intervention in the market for crop protection products also has its roots in the defunct regional control era. The old Western Region launched the popular Cocoa Pesticides Scheme in the 1960s. Subsequently, the supply of CPPs has been mainstreamed through several projects of agricultural development to facilitate access and judicious use. At the present time, the market for CPPs is considerably liberalized.

Lessons from the Implementation

The collective lessons gained from the experience of policy implementation are better perceived within the development communication context. This helps in examining the agri-input delivery system by treating (a) the package of agri inputs as a policy *message* to the farm population, (b) the policy authorities as the *source* of the message for its packaging and mobilization, (c) the market cum public extension system as the joint *medium* or *channel* for the smooth passage of the message, and (d) the farmers as the *target* of the message for decoding and eventual use. This framework helps in determining the nature of issues associated with the delivery system as a means of communicating the policy message about agri inputs, thereby facilitating the determination of issues emanating therefrom.

Thus the following lessons of implementation experience can be discerned.

- *Lessons about the Message* – The well-packaged policy message has the property that the **availability** of agri inputs in desired quantity and quality holds the key to the efficiency of the agri-input delivery system. The fertilizer policy intervention accounts for large-scale awareness and adoption of fertilizers among the farmers. Nevertheless, the degree of availability falls short of the demand generated, which leads to frequent crisis in the past. Even at the present time, substantial pent-up demand for fertilizers exist that is not met. The situation is similar with seed whereby limited availability of improved seeds provides the basis for the vast majority of farmers to continue with the traditional practice of using own seed saved from previous harvests, season after season. Moreover, the nature of agricultural input demand is also important; as inputs they are not demanded for their own sake unlike final products. Thus, the demand for agri inputs is a derived type, in the sense that they are demanded for the production of other items than themselves. This implies that for their use to expand, the demands for the products that they are used to produce must first expand. That is, much of the sluggish uptake of such inputs would be explained by the low production of food items in the domestic market possibly emanating from policy-induced competition with food imports (Ayoola 2001).

- *Lessons about the Source* - The role of government as the source of the policy message initially involves the proper packaging of the agri inputs in terms of proper articulation of the policy statement and proper formulation of strategies for implementing them (FMAWRRD 1987, FMARD 2003), followed by sustained commitment of resources during implementation. This is not the usual experience with past implementation of agri-input policies. Budget provisions could no longer keep up with the growing demand for fertilizers (Nagy and Edun 2002) through public distribution while farmers have also become dependent on the subsidy. Similarly, the public system for distributing improved seed failed to meet the need of farmers based on insufficient funding.

- *Lessons about Channel* – The channel or medium of message flow in delivering agri inputs to farmers represents the most crucial aspect, from where most of the policy issues of the communication model emerge. For agri inputs, the channel comprises the market and the public extension system. Indeed the economist's viewpoint is that the extension system is just a parallel market for agri input. In any event, the most important issue about the channel is the magnitude of noise present, which is any factor impeding the flow of the policy message, distorts it, or at least impairs its reception on reaching its destination. Thus in the market for agri inputs, the several noise elements affecting the policy message include (a) the unorganized nature of the market that makes it difficult to reach the agri-input dealers for policy participation and from enjoying the economy of scale in their operations; (b) sharp practices in the market such as short bag weights, adulteration, and general lack of truth in labelling practices, etc. On the other hand, the noise elements in the public agricultural extension system include (a) poorly-trained and immobile extension workers; (b) poor basic infrastructure in rural areas such as rural road networks, rural water supply, rural electricity, etc.

- *Lessons about the Target* – The disposition of the farmers as targets in receiving the policy message depends on their socioeconomic circumstances and possession of voice in the policy process, among other factors. All too easily, the farmers are too vulnerable to cheats and interest groups who corner the part of the policy benefits meant for the farmers. The fertilizer policy is a case in point whereby the inputs arrive too late and too little for the needs of the farmers.

4. POLICY FORMULATION FOR AGRI-INPUT DELIVERY

Getting the policy environment right for agri-input delivery system starts with proper formulation of the policy in the first instance, which is a serious rather than casual analytical exercise to be followed by disciplined adherence to its implementation. The latter is often blamed for policy failures, which is not necessarily true in all cases. In this section, we first make the case for the adoption of a process approach to policy analysis before proceeding to highlight the requirements for policy formulation with reference to the agri-input delivery system.

4.1 Process approach

In carrying out any type of policy analysis such as policy formulation, the choice of analytical framework is crucial, which is between treating policy as a set of discrete events or as a sequence of events in process (Idachaba 2006). The process approach to policy analysis is superior because it brings out the real explanatory factors of policy behaviour much more clearly, including the roles of different stakeholders, and helps in subjecting such roles to efficiency tests while proffering more feasible solutions. The key stages in the policy process analysis are as follows.

- Articulation (of the policy problem)
- Formulation (of the implementation strategy)
- Appraisal (of the strategy document)
- Implementation (of the strategy as appraised)
- Evaluation and feedback

We are presently concerned with the first two stages that comprise policy formulation, i.e. proper articulation of the policy problem and proper formulation of the implementation strategy.

4.2 Articulation of policy on agri inputs delivery

Formal *articulation* of government policies represents the first stage of any aspect of the agricultural *policy process*. This involves the formal recognition and proper definition of the policy problem in all its ramifications and the deliberate specification of the policy directions to follow in addressing the problem. In a practical sense, a policy is formally articulated when it is written down and explicitly and publicly declared in advance, following the due process. Otherwise, we have disparate policy statements about different elements of the policy tucked away in active and closed files or inside some grey literature within the ministry or the agencies at federal and state levels, constituting the 'implicit policies'. It is obligatory for government to articulate its policies in a formal way, and such formally articulated policies become laws, more or less, that are binding on its agencies operating in the sector while also guiding the behaviour of stakeholders in the sector. Suffice it to say that at the moment, a formally articulated policy on agri input integrating all the inputs together is not in existence in this country. And now is the time for one, bearing in mind the looming commencement of the African green revolution.

Policy articulation in this sense is an attempt to answer the question of *what* the government position is on agri-input delivery. The critical elements to be considered should be: the background of the agri-input policy environment, the challenges and objectives of agri-input delivery policy, guiding principles, policy directions, and instruments to be employed. These elements should be fully specified and succinctly presented and then published as small, handy quick reference materials for use by government officials and stakeholders in public and private sector, as recently done for the recently articulated National Fertilizer Policy for Nigeria.

Thus, consistent with the development communication paradigm discussed earlier in articulating the policy on agri-input delivery system for Nigeria, the *what* question borders on: what the position of government is on the issues relating to the different aspects of the agri-input delivery as a communication system—the adequacy and completeness of the package of agri inputs as a message; the role of government as the source of packaging the agri inputs message; the noise level in the agri-input message delivery channel; and the disposition of farmers as the target of agri-input message.

4.3 Formulation of implementation strategy for agri inputs delivery

While policy articulation attempts to answer the *what* question, strategy formulation attempts to answer the *how*; that is, going by the policy statement articulated, *how* will the government deploy the *policy instruments* at its disposal and follow the *policy directions* predetermined in addressing the *policy challenges* identified and in meeting the *policy objectives* stated. The main elements of strategy design and formulation include: elaboration of each instrument and how they will be utilized in various combinations, suitable institutional arrangements for implementation, logical framework, action plans and time phasing, phasing of activities and work programmes, financial plans and fund raising, among others. Thus, strategy formulation results in an elaborate document that contains the full specification of parameters for implementing policy as previously articulated. Again, such a formulation is presently non-existent for agri-input delivery system in the country at the moment, yet one is urgently required preparatory to implementing the 12-point agenda of the Africa Fertilizer Summit.

5. EMERGING POLICY ISSUES AND OPTIMAL RESOLUTION OF ISSUES

The foregoing expositions serve as veritable sources of the several issues to consider in formulating the policy on agri-input delivery system in Nigeria. We shall highlight such issues or pose relevant questions for discussion at this stage only, without attempting to resolve them ahead of the consensus around them to be built by the policy stakeholders themselves and not the analyst.

Some selected issues have emerged to be grouped in two categories as follows.

5.1 Political economy and governance issues

- *Role of Government in the Agri-Input Delivery System* – There are several options in supporting agricultural production, comprising direct and indirect roles of the government. Should government continue to provide direct subsidy in the agri-input delivery system in the presence of widespread abuse by its officials and the perennial leakages of the subsidy policy benefits to non-targeted individuals in the society?

- *Federal-State Relationship in Agri-Input Delivery System* – The division of labour in agricultural development is clearly established in the constitution of the Federal Republic. Is federal government permitted under the relevant section of the constitution (FGN 1999), to embark in market stabilization programme in the agri-input delivery system, like the one in operation for fertilizer?
- *Policy Due Process and Policy Best Practices in Agri-Input Delivery System* - The buzz words of good policymaking include: Inclusiveness, Consistency, Stability, Transparency, Openness, Programme accountability, Participation, Professionalism, and Documentation, to mention a few. What are their elements in formulating the policy on agri-input policy delivery system for Nigeria?

5.2 Structural and systemic issues

- *Regulatory and Legal Frameworks* – The nature of agri inputs is very scientific and technical, implying that unsuspecting farmers can be easily deceived and exploited unless they are effectively protected from the sharp practices of agri-input dealers in the market. Are we satisfied that the existing regulatory agencies such as NAFDAC and SON are doing a good job in this area, or should the Agricultural Ministry establish its own regulatory agency, particularly for fertilizer?
- *Public-Private Partnerships Frameworks* – The private sector operates in the agri-input market purely for profit motives, whereas the public sector operates the agricultural extension system as a complementary policy to fill the gaps in service provision to farmers. What frameworks exist that will maximize this complementarity within the framework of PPP for the smooth working of the agri-input delivery system?
- *Small- versus Large-Scale Farmers* – Both scales are desirable for agricultural development of the country. Is there or is there not a need for discriminatory instruments in the policy on agri-input delivery in respect of each category of farmers?
- *Organized Private Agri-Input Sector* – The effort towards an organized agri-input sector is quite recent, through the IFDC/DAIMINA project. But the observed trend is to the effect that the Agro-Input Dealers Associations have toed the line of traditional commodity and farmers' associations—political interferences, top-down structure rather than bottom-up, etc. What are the necessary safeguards required in the policy on agri-input delivery to eliminate these negative developments?

- *Technical Back-up support services in formulating and implementing Policy on Agri-Input Delivery System* – The need for certain supportive services have been recognized as the responsibility of private sector and NGOs. These include professional services such as policy advocacy and brokering services that are critical to policy best practices and conducting policy (varietal) trials prior to large-scale adoption. How can we promote the private and non-government sectors to render such services to the benefit of the agri-input delivery system?
- *Role of Agricultural Universities in Agri-Input Delivery System* – The best way for agricultural universities to contribute to the functionality of agri-input delivery system as in the case of the American 'Land Grant Colleges', which is our role model for the three agricultural universities in Nigeria, is through the 'Cooperative Extension System' involving resource collaboration between the federal, state, and local governments under a single administrative umbrella of the universities. Are these universities involved in the input delivery system to that extent; if not, why not?
- *Regional dimensions* – There is a serious effort to harmonize the agri-input delivery policies of countries through the regional bodies such as ECOWAS and NEPAD, particularly within the contexts of Africa Fertilizer Summit and African green revolution. The main issue is the extent of participation of Nigeria and the degree of commitment of the Nigerian government to international treaties in regard to policy on agri-input delivery system for the country.

6. CONCLUDING REMARKS

The formulation of policy on the agri-input delivery system in Nigeria depends on the optimal resolution of the several issues raised. Specifically, the proper articulation of the policy statement on agri-input delivery system and proper formulation of an implementation strategy for the system require that the lessons learned from the experience of implementing past policies are considered in resolving the issues, subject to strict adherence to the tenets of policy due process and consistent with policy best practices in other parts of the world.

In conclusion, the most important lesson gained from public intervention in the agri-input delivery system of Nigeria is that there have been certain instances in the past and at present, such as the case of fertilizer, wherein the subsidy policy is good but the implementation strategy failed to work. Suffice it to say that from this point on, we desire a rule-based, evidence-led, and internally consistent policy articulation and strategy formulation for the agri-input delivery system of the country.

REFERENCES

Ayoola, G. B. (2001) 'Effective Communication with Women for Agricultural Development' in G. B. Ayoola *Essays on Agricultural Economy 1: A Book of Readings on Agricultural Development Policy and Administration in Nigeria*. T. M. A. Publishers. Ibadan.

FGN (Federal Government of Nigeria) (1999). *Constitution of the Federal Republic of Nigeria*, Government Press.

FMAWRRD (Federal Ministry of Agriculture, Water Resources and Rural Development (1987). *Agricultural Policy of Nigeria*.

FMARD (Federal Ministry of Agriculture and Rural Development) (2003). *Policy Thrust for Agricultural Development*.

FMARD (Federal Ministry of Agriculture and Rural Development) (2006) *National Fertilizer Policy for Nigeria* Abuja. June 2006.

Idachaba, F. S. 'Agricultural Policy Process in Africa: Role of Policy Analyst' in F. S. Idachaba *Good Intentions are not Enough: collected Essays on Government and Nigerian Agriculture*. University Press Ibadan, 2006.

IFDC (an International Center for Soil Fertility and Agricultural Development), 'Africa Fertilizer Summit: Abuja Declaration on Fertilizer for African Green Revolution' In *IFDC Corporate Report, 2005/06*. Circular IFDC S29, USA.

Lipsey, R. G. and K. Lancaster (1956). 'The General Theory of Second Best', *Review of Economic Studies*, Vol. 24 1956.

Nagy, J.G. and O. Edun (2002). *Assessment of Nigerian Government Fertilizer Policy and Suggested Alternative Market-Friendly Policies* IFDC September 2002.

AGRA (African Green Revolution Programme) Program for Africa's Seed Systems (PASS): Country Report – Nigeria. January 2007.

NASC (National Agricultural seed Council) (2010). National Seed Policy for Nigeria (in progress).

Stocktaking Of The Soaring Food Prices In 2007/2008: Evidence From Nigeria

With Oyin Oyeleke, March 2009, Federal Ministry of Agriculture and Rural Development

INTRODUCTION

1. In mid-2007, the FAO raised an alarm indicating that the whole world was in a food crisis situation, when it was observed that food prices had soared to an unprecedented level of 40% in the past year. And in the first three months of 2008, food prices reached their highest level in real terms in thirty years. It was envisaged that many developing nations, including Nigeria, might not be able to cope. The situation was soon to resonate with a global meltdown in the financial market to trigger the recession of the world economy at large, which poses a major challenge to many countries at the moment.

2. The soaring food prices was attributed to droughts and floods linked to climate change as well as rising oil prices, boosting demand for bio fuels. In particular, the use of land to grow plants which can be used to make alternative fuels—and the use of food crops themselves for fuel—has reduced food supplies and helped push up prices. Changing diet in fast-developing nations such as China was also considered a factor, with more land needed to raise livestock to meet increasing demand for meat. As a result, it was observed that international cereal prices already sparked food riots in several countries, which formed the basis for further food shortage alerts and the call for urgent action to provide small farmers in the affected countries with

improved access to seeds, fertilizers, and other inputs to increase crop production.
3. In December 2007, FAO launched an initiative to boost food production in the short term, namely *Initiative on Soaring Food Prices (ISFP)*. This involved the distribution of seeds, fertilizer, animal feed, and other farm inputs to smallholder farmers against the following season in developing countries. Part of the initiative is advising governments on policy measures in response to the crisis, including a general guideline to provide an overview of different policy responses to higher food prices, their possible effects, advantages, and disadvantages.
4. The ECOWAS Commission convened an extraordinary meeting of the Ministers of Trade, Agriculture, Finance and Economic Affairs, held at Abuja, Nigeria on 19 May 2008, which led to the launching of a 'regional offensive for food production and the fight against hunger'. Prior to this meeting, individual countries had launched some initiatives to deal with the negative effects of soaring food prices and to ward off a looming food crisis in their domains. In this regard, the response of the Federal Government of Nigeria was a mixed type involving an initial panic measure followed by longer term measures.
5. The purpose of this paper is to take stock of events relating to the soaring food prices in Nigeria since commencement in 2007 to date. The terms of reference are as follows:

   o Document the appearance and the evolution of the food crisis;
   o Describe the food supply and demand profile of the country; and
   o Document responses given so far by the government.

NIGERIA'S FOOD SUPPLY AND DEMAND ROFILE

Trend in Commodities and Energy Prices and Food Trade

6. Table 1 and figure 1 depict the movement of average month-on-month percentage movement in prices of all consumer items and food (CPI) between 2003 and 2008 with 2003 as base year. All items and food showed a steady upward movement from the beginning of the second quarter of 2003 up to the end of second quarter of 2004 but with higher rise in all items than food. The period between the middle of 2004 and September 2005 witnessed dramatic surge in prices of all items and food but with greater increases in the prices

of food items. Prices nosedived from the last quarter of 2005 up to the end of third quarter in 2007 with greater increases in all items compared with food. However, the price situation took a dramatic turn of great surge from near 0% in the last quarter of 2007 to a peak of about 22% change by the last two quarters of 2008. The food prices soared far higher than the prices of all items compared.

7. Similarly, for retail consumer prices of major food commodities (rice, beans, maize millet, etc.), table 2 and figure 2, 3, and 4 showed a significant upsurge between June 2007 and September 2008. The same trend is noticed in the retail prices of imported agricultural products(fig 3). Price of imported rice rose from N140 per kg in June 2007 to N165 per kg in Sep 2008, a surge of 50%. Price of agricultural rice (long grain improved variety) increased from N100 per kg to N130 per kg—a 30%-rise within the same period. In the case of local traditional rice varieties, price jumped from N85 per kg to N120 per kg (41% increase). Maize rose to about 30% during the same period. Average prices in figure 4 rose from about N85 per kg to N115 per kg between June 2006 and Sep 2008, that is, a growth of about 35%. Intentional price of corn (major ingredient of ethanol and bio fuels) soared from US$120 per MT in Jan 2007 to US$168 per MT in Jan 2008 or a 40% hike.

8. However, it is noteworthy that retail consumer prices of all staple food items have assumed a slight downward trend from Sep 2008. It is also empirically right to conclude that Nigeria joined the global trend of soaring food prices from the middle of 2007. The general slight downward trend in prices starting from the last quarter of 2008 is a positive respondto various short- and median-term measures taken by the federal, state, and international community.

9. In contrast to the sharp increases in the retail prices of staple food (sorghum, millet, rice, maize, yam, cassava, etc.), producers or farm gate prices remained stable between 2005 and 2007 for grains and tuber crops (table 3 and fig 5 & 6). The margin between the retail consumer and producer prices remained higher and stable between 2006 and 2007, ranging from N30 per kg in Millet to N55.70 per kg bean in 2006. vide table 4 and figure 7. Guinea corn recorded the lowest retail-producer price margin of N28.40 per kg in 2007, whereas the highest margin occurred in yam tuber at N54.60 per kg. High marketing margin between retail or consumer prices and producers or farm gate price is a sign of market failure occasioned by weak marketing institutions, very high transportation, and poor marketing extension services. This situation mostly profits the

middle men rather than farmers who toil hard for many months each year without remunerated prices for their produce. It could also be inferred that soaring global food price had less impact on producer or farm gate prices compared with significant upsurge in retail prices of food items. This is a reflection of the positive impact in the subsidy on major production inputs of seeds and fertilizers by the federal and state government. The only area where the farmer bears the brunt remained in transportation and energy costs, which are major causes of soaring prices in global markets. With stable producer prices, farmers in Nigeria experienced erosion in their net income in the wake of high energy and rising transportation prices.

10. Fertilizer prices in the open markets (Fig. 8) remained steady at an average of N55 and N57 per kg between October 2005 and June 2007. The prices rose astronomically between June 2007 and Dec 2008. In the case of NPK 15-15-15, average price per kilogram rose from N54 in June 2007 to N85.5—a 40% jump. Urea also rose by 65% during the same period. This scenario is a great constraint to the usage of fertilizer which is put at an average of about 8 kg per hectare for an average African farmer (the lowest in the world). However, the intervention of both Federal and State Governments in the form of subsidy on fertilizer prices cushioned the effects on farmers. Fertilizers from Federal and State sources sold between N1000 and N1700 per 50kg bag in most states of the federation. Farmers also adopted various coping methods in direct response to soaring fertilizer prices. Farmers in the grain production zones of the country cultivated more of millet and guinea corn which require less fertilizer instead of maize. According to IFDC sources, global fertilizer prices rose more than 200% in 2007 due to the increase in use of maize products for ethanol and biofuel and rising energy cost. The Arab gulf prices of Urea rose from US$272 per MT in Jan 2007 to US$415 per MT in January 2008—about 52% increase.

11. Transportation cost fluctuated sharply within the months of Jan and Dec 2005 to 2008 (Fig. 9). Average cost of 3km journey surged from N28 in January 2005 to N60 in 2008, an increase of about 115%. Transportation fare jumped from December price of N30 in 2005 to N70 in Dec 2008, a 133% increase. On the other hand, energy prices (Fig. 10) remained high but stable at an average unit price of N400 in Jan 2007 and N425 in Dec 2008. The stable energy price is a reflection of price regulation in electricity and petroleum products as well as subsidy by the federal government on these products.

APPEARANCE AND EVOLUTION OF GLOBAL FOOD CRISIS

Trend in Production

12. Production of key crops (yam, cassava, sorghum, millet, rice, maize, etc.) table 5 portrayed marginal increases between 2006 and 2007. Total output of crops increased from 98.564 million MT to 104.915 million MT within the same period, that is, 6.4% growth see figure 11. However, the total area cultivated for the crops dropped from 27.4 million hectares in 2006 to 26.87 million Ha in 2006 (about 2% drop). In the period under review, yam production rose from 28.890m MT to 30.450m MT, cassava from 40.57 million MT to 43.57 million MT, rice from 2.7 million MT to 2.978 million MT, maize 6.76 million MT to 7.24million MT. The yields [kg/ha] increased from 12.5 MT/Ha to 13.7 MT/Ha for cassava, from 1.18 MT to 1.28 MT for sorghum, from 1.08 MT/Ha to 1.13 MT for millet, 1.91 MT to 2.1 MT/Ha for rice, etc. (Table 4). Growth of over 7% was recorded in the output of cassava, rice, maize, and sweet potatoes. This is closely followed by growth of about 5 and 6% in sorghum and groundnut. A least growth in output of 1% was recorded in cocoyam production. Yield increased significantly in dried cowpeas [41%], rice [12%], cassava [10%], sorghum [9%], bean [40%], millet [4.5%]. Yield dropped in yam [8%], cocoyam [5%] and sweet potatoes [4%]. The drop in the total area cropped with contrasting increase in total output typified an increase in yield or productivity. This could be attributed to interventional measures of improved technology, better varieties seed/seedling, better application of fertilizers, and crop production products (cpp). Also recent emphasis on high value grain crops by the federal government with various interventions and support in the forms of subsidies, credit facilities and extension services accounted for the recorded growth in output and productivity.

Agricultural Trade

13. The summaries of agricultural trade including food import and export, real import and export, percentage growth in import and export, percentage share of food to the total import, structure of import and export, etc., are contained in tables 6, 7 and 8 as well as

figures 12 to 17. The total food import increased from 2006 value of N513.7 billion to N802.6 billion in 2007. In real term, the food import soared from N330.6 billion in 2006 to N480.6 billion in 2007. This translates to 45% growth in real food import during the period, figure 12. Total import in 2007 amounted to N4.12 trillion compared with the 2006 figure of N2.922 trillion or about 41% growth. Food import share of total import rose marginally from 18% in 2006 to 19% in 2007. Figure 13a and 13b give a pictorial view of food import share of total import, growth in food import, growth in the real food import and growth in the consumer price index between 2004 and 2007.

14. Nigeria Food export surged from N28 billion to about N122 billion between 2006 and 2007 or about 300% increase. Total export of the country declined slightly from N7.5 trillion in 2006 to N6.8 trillion in 2007. Food export as a share of total export rose by about 1.6% in 2007 compared with less than 1% in 2006 (table 6).

15. A look at the structure of import in 2007, table7 and figure 14 shows five commodities accounting for about 88% of the total food import. These include cereals 42%, fish and crustacean 18%, dairy products and bird eggs 16%, preparations of cereals, flour (6%) and sugar/confectionery (5%). In 2006, wheat and meslin accounted for about 33% while fish (frozen) represented 16% of the food import. The importation of cereals of rising global prices (rice and wheat) has pulling effects on local retail prices of the imported and local food items. High prices of imported rice and wheat, for instance, influenced higher demand and consumption of local and agricultural rice and other grains (millet, cowpea, maize, etc.) causing prices to assume an upward trend in these substitutes. This is the pattern empirically shown in the retail prices of commodities in figures 2, 3 and 4. A sudden jump in the real value of imported food between 2006 and 2007 is a manifestation of quick response to rising price of food commodities with an objective of dousing the price upsurge through food importation. This is also an indicator of deficient food self-sufficiency ratio.

16. Structure of food export (table 8 and figure 15) clearly shows export concentration around cocoa and cocoa preparation accounting for about 53% of the total food export in 2007. This is closely followed by oil seed, seed, fruit of 13%, edible fruit and nuts (12%), gum, resin, veg. saps (7%). The five categories of export accounted for 87% of the total food export in 2007. The 2006 record showed lac,

natural gums, resins, as accounting for 70% of the export, followed by goat, skin/leather (11%) and natural rubber (6%).

POLICY RESPONSES TO SOARING FOOD PRICES

Short-term measures

17. The federal government first responded in an initial panic mode, by ordering the immediate release of grain reserves coupled with direct importation of rice to the tune of N80 billion Naira. The basis for this mode was the fear that the soaring food prices in the world market could introduce significant shocks into the Nigerian food market, given the status of Nigeria as a net importer of stable food items, wherein rice and wheat predominate among others in the food import bill of $2.8 million per annum.

18. A stakeholders meeting was convened at the instance of agriculture minister, which took place at Abuja on 3 May 2008. The agenda of the meeting was to examine the food situation with the country and to obtain the commitment of stakeholders towards implementing the policy decision. The sole objective was to bring the domestic price of rice down quickly having jumped by about 100% in a couple of months. It was established that rice output in 2007 was 3.4 million MT out of which only 1.4 million MT was milled, which leaves 2.0 million MT rice paddy unprocessed because of inadequate processing capacity. It was also established that the requirement of the country for paddy rice for its 140 million people at 30 kg per caput consumption was 6.5 million MT or 4.2 million milled rice equivalent at 65% recovery rate; and that the harvest of paddy in 2008 was estimated at 3.94 million tones.

19. The following actions were taken:

a. The Federal Government released 65,000 metric tonnes of various grain crops and garri to the public, which led to a reduction of prices of some tradable and non-tradable food items such as maize, sorghum, millet, and garri. The states were enjoined to do the same but this took place at an insignificant level given the low stocks available from the buffer stocks.

b. The intervention buying operation of the Federal Government was aimed at increasing supply of rice within the next three months (May-July 2008) and sustained it for the following three months in the first instance (August-October 2008). This was expected to cause a significant reduction in market price for rice, based on evidence that it had risen from about N6000/50Kg bag (N120, 000/mt) to about N12, 500/50Kg bag (N260, 000/50Kg bag), much faster than the current price at the border with Benin Republic of only about N7, 500/mt or N150, 000/mt at that time.

c. The incremental income would be sold to the genial public at a subsidized price, with a view to bringing the price down to an average of N152,096/mt or N7,606/50Kg bag as it was a couple of months before.

20. The federal government had targeted 500,000MT of rice to be imported from Thailand and other countries into the market by the end of July 2008, through government-to-government intervention. This would be achieved via three sources, namely: mobilizing the organized private sector; mopping the stock available in the domestic market; and opening up of the border trade in rice with Benin Republic for increased importation by private sector. Towards this end, the following pledges were extracted from the stakeholders in the private sector. Messrs Veetee offered 300, 000 from Thailand in 50kg bags (150 000mt parboiled and 150 000 white rice), which was confirmed. The Stallion Group made an unconfirmed offer from USA in bulk supply (i.e. bagging at the port required). A number of offers came from India and Pakistan which were uncertain. The features of the offers include the following:

- **Pricing** - The offer price of Thai rice was N9,400/bag (= $1,600) CIF, ex Lagos) against the current world market is $1,200/mt (FOB).
- **Timeliness:** The offer from India would not arrive earlier than 6months which was outside the short-term range; moreover delivery within 3 months was unlikely from any of the countries (at a maximum load of 30,000mt/ship, and 2 ships per month, then minimum of 4 months is required for the 25 000mt import target).
- **Shipment** – The assemblage of ships from different shipping zones also takes time; shipment is not spontaneous from each country thereby requiring more time is required to assemble rice from different locations in each country.
- **Logistics:** Port activities include clearing/bagging(USA)/loading etc.; inland transportation is feasible by road only, at

30mt/truck thereby requiring 80 trucks to move the whole rice load at once to different parts of the country.
- **Cost implications** - CIF ex Lagos: 250 000mt of rice X $1600 = $400m (N47.2Bill); Internal distribution is N8000/mt = N2Bill.
- **Socioeconomic implications** - The subsidized price would gladden the heart of consumers but dampen the enthusiasms of the farmers in Nigeria. This will lead to reduction of land area under rice in the subsequent season. Furthermore proportion of women rice processors would be displaced form job while others could take to the jobs of packaging.

21. However, public opinion about the proposed importation of rice was not favourable, which led to its discontinuation by the government. The following arguments were canvassed by government and other experts:

- That the country is blessed with a range of climate and soils to produce a wide range of agricultural food and cash crop commodities; therefore the scope of substitution in the food commodity market was large enough to ward off price shocks in the short term.
- That the country is potentially food secure and an exporter of food commodities to its neighbors and the rest of the world.
- That even though the country imports a significant quantity of food items it is essentially not out necessity but out of preference for cheaper foreign commodities, a situation that can be easily reversed with adequate support to domestic production, processing and marketing of such items.
- That at the moment there was a glut of cassava in the market while large quantities of unprocessed paddy rice existed in storage on-farm and off-farm.

22. This helped the federal government to quickly come out of the panic mode in the short term, leading to substitution of the direct implementation option with tariff measures. Hitherto, though there was no total ban on rice importation, the level of tariff on it was quite high, so much so than what obtained in the neighboring countries, thereby causing a lot of smuggling from other countries into Nigeria. It was estimated that about 1.5 million metric tonnes of rice come into Nigeria through smuggling yearly from Chad, and Niger alone. A number of alternate options were considered in the short term as follows:

a. **Mopping up operation** – This involves purchasing the current stock of imported rice in the country form local stores including Messrs Intercity and Stallion Group. It was established that about 110,000mt available to be mopped up at the current price level of N12,000/bag - N240 000/mt, to be sold to the consuming public at subsidized prices. Although this can be achieved with immediate effect the quantity available is too small to generate any perceptible impact.

b. **Distribution of small-scale machines** – The government considered the option to process the paddy in storage as established. The RIFAN (Rice Farmers Association of Nigeria) had claimed that about 4 million MT of paddy would be available in the short term, comprising present stock level in July 2008 (2.5 million) as verified by the ministry plus new harvest (1.5 million MT) in the next season in October 2008. However, the problem of processing was underscored, in terms of low capacity of small-scale processors and poor quality of domestically milled rice in Nigeria. Thus small-scale machines could be distributed to processors and small-scale irrigation pumps to farmers in the short run for the purpose of milling the paddy to be made available in October 2008. Therefore government made an attempt to place an order for small-scale rice processing machines from abroad. About 1000 small-scale milling machines could be purchased which would need one month to install in all parts of the country. This would lead to at least 30% reduction in price, about N140 000/mt. The following estimation was made:

   i. Small irrigation pumps (100/state = 3,700units); @N45000 = N1.67 Bill); Procurement of mini-combined harvesters. 18 states (100No. To be distributed to rice producing states) @N1 Mill. = N100 Mill;

   ii. Procurement of motorized threshers (for remaining, 10 No. per state = 370 No.; 1.8 mt/hour of paddy; @ N672000/unit = N248,640,000;

   iii. Procurement of mechanical dryers (multipurpose); to be imported. 5No./state = 185; @N800,000/unit = N1.48 Billion; Procurement of local dryers. 10/state @N75000 = N27.75 mill; Rapid steam parboiler (locally fabricated); 2No/mill; 800mills @ N100,000 = N1.6 Bill;

   iv. Procurement of milling machines for small-scale milling (300days/year @4mt/day; capacity = 1,200mt/year): 800No. = 960000mt/year. N3m/mill X 800000 = N2.4 Billion; Procurement of packaging machines; 1No. per 5mills: = 160 units; @N3million = N480 Million; Rehabilitation of non-functional processing machines(30-60mt/day of milled rice capacity).

The socioeconomic implication involved creation of jobs for processors; engineers; etc.; also farmers will also be encouraged to produce, which made this option particularly attractive. However, this option failed as a short-term measure as about 2-3 months would be required for importation of machines and putting up the factory buildings.

c. **Tariff waivers** - The federal government approved the suspension of all levies and duties on rice imports with effect from 7th May 2008 to 31st October, 2008, which stimulated the private sector to place an order for rice importation to the tune of 9,651,075 MT valued at USD 543,653,416 or N62,238,409,910. As a result, actual rice importation was 172,518, which led to fall in price to an average of N7000 in August 2008 against N12 000/bag in May 2008 or about 45% reduction in price. In addition cross-border trade in rice probably increased. The major cost implication was loss of revenue, which was expected to reach about N47.2Billion at zero tariff level. The socioeconomic implication was in terms of increased business among rice traders and consumers in the short run, which of course was at the expense of low morale of farmers in the long run.

Medium- to long-term measures

23. Subsequently the satiation of soaring food prices was converted to an opportunity for Nigeria to take advantage **of and to** institute longer term measures **toward** agricultural development. The federal government set up an Implementation Committee comprising Ministers of Agriculture and Water Resources; Finance; Commerce and Industry. A number of medium term measures emerged to deal with the looming food crisis, as follows:

o **Food production** – Dedication of accruals to the Natural Resources Development Fund from 2008 to 2011 to boost domestic production of food crops, development of agro-allied industry, and research and development on seed varieties; this amounts to 1.68% of the federation account or about N240 Billion.

- **Agricultural credit** – Approval of N10 Billion from the rice levy account as a credit scheme at concessionary interest rate, in support of local rice processing capacity in the country. Furthermore the Central Bank of Nigeria resolved to raise N200 billion funds from the commercial banks in two weeks. The fund would be used for commercial farming to be disbursed by select banks.
- **Food reserve** – Decision to complete the outstanding storage projects before end of 2008 in order to increase the national strategic food reserve capacity from 300,000 to 600,000 mt. The silos project had commenced a long time ago which had not been completed, so government provided funds for their completion. Further the state governments were encouraged to step up their buffer stock operations, which involving at least 10% of food output in their respective domains. It was envisaged that up to 2 million MT silos capacity would be required for the country.

24. Subsequently the federal government has produced a food security strategy document which prioritizes a number of measures in the long term. The policy thrust behind this includes a number of desirable attributes namely: value chain approach; commodity focus; visibility of private sector; successor farmer generation; participative policy process; policy advocacy and brokerage; safety net. In this regard, the aspects of the policy response in the medium and long terms include the following aspects:

   a. **Production** – These include:

      i. Land clearing for commercial agriculture – provision not less than 2000ha/state; for commercial farming (370000ha); average of N250000/ha = N18.5 Billion;
      ii. Establishment of farm settlement to attract the teeming population of unemployed qualified youth ((1 settlement/state; 20 houses per settlement (1-bedroom with a store by the side) @N2m; store (N500 000); motor cycle (100000/person = N2m); bore-hole water with overhead tank (N3mill); Stand-by generator (N3mill); 5 tractors + set of equipment @N5mill/set = N25 mill; infrastructure development (bulk provision) N5m/settlement; training (N3mill/settlement); approximately N100mill per settlement excluding the cost of land to be provided by the state (200

ha per settlement). Additional provision of N1.3million for further infrastructure (small clinic; markets; community centre);

iii. Establishment of Farm Service Centres to provide agro-inputs extension advise, tractor and machinery service support as well as producer marketing;

iv. Agricultural land cadastral mapping and certification (Programme to map out all agricultural land and issue certificate of ownership to farmers to enable them use their land as collateral to access credit – N57 Billion for 4 years;

v. Research and development Leading to new technologies for agricultural development: 15 research institutes (N100mill each); support 3 universities of agriculture (N100mill each); 37 ADPs (N50mill each); 30 agricultural colleges (N300mill) = N3.9 Bill);

vi. Training facility for successor farmer generation - provision of integrated farm training facility to prepare the youth for agricultural enterprise in place of the ageing farming population, the encouragement of school garden programme in primary and secondary schools, and practical exposure of youth including farm tours/study tours

b. **Storage** - Completion of 17 silos at various stages of completion (1%-85% completion); Start-up of building of 11 others currently at zero level in various states of the country; the sum of N15 Billion was earmarked for this purpose.

c. **Processing** - The rice milling capacity would be increased by an additional 88,000mt per annum and create about 8000 direct and 110,000 MT indirect job opportunities. The mils would be located in the major rice producing states to take advantage of nearness to raw materials; the local capacity for operation and maintenance of rice mills and fabrication of spares would be gradually built thereby creating employment for the youth.

d. **Marketing** - Physical development of market for livestock and birds; Physical development of grain market; Guaranteed Minimum Price; Association building; etc.

ANNEXES

Table 1. TREND IN COMPOSITE PRICE INDEX (2004–2008; 2003 IS THE BASE YEAR)

		Jan	Feb	March	Apr	May	Jun	July	Aug	Sep	Oct	Nov	Dec
2004	All items	22.4	24.8	22.5	17.5	19.8	14.1	10.7	13	9.1	10.7	10	10
	Food items	11.9	14.5	15.6	14.4	18.1	14.5	12.2	16.3	14.6	15.4	15.1	12.1
2005	All items	9.8	10.9	16.3	17.9	16.8	18.6	26.1	28.2	24.3	18.6	15.1	11.6
	Food items	15.1	18.7	25	20.3	15.7	18	35.7	38.5	29.5	24.6	19.7	15.5
2006	All items	10.7	10.8	12	12.6	10.5	8.5	3	3.7	6.3	6.1	7.8	8.5
	Food items	14.7	10.5	9.3	9.6	8.6	6.2	-3.7	-2.3	4.3	4.7	5.4	3.9
2007	All Iiems	8	7.1	5.2	4.2	4.6	6.4	4.8	4.2	4.1	4.6	5.2	6.6
	Food items	-0.1	3.3	1.7	2.1	2.4	3.2	1.1	-1.2	-0.8	-0.1	3.2	8.2
2008	All items	8.6	8	7.8	8.2	9.7	12	14	12.4	13	14.7	14.8	15.1
	Food items	12.6	8.7	12.4	13.1	14.7	18.8	20.9	18.8	17.1	19.2	18.1	18

Source: CBN reports.

Table 2. Trend in Retail Prices (2006–2008).

	Mar	Jun	Sep	2006	Mar	Jun	Sep	2007	Mar	Jun	Sep	2008
Rice - agric/long grain	105.44	110.8	127.3	106.0	112.2	101.6	104.9	110.9	112.9	121.8	131.5	129.3
Rice - local	92.6	93.2	107.57	88.4	89.19	86.45	89.28	91.89	98.5	109.1	122.5	112.1
Rice - imported	126.47	126.14	135.45	130.78	130.48	122.8	127.52	134.91	144.8	167.3	166.2	158.9
Beans - white	77.1	72.47	88.11	72.18	81.69	67.02	69.46	76.2	94.6	117.4	115.8	97.7
Maize - grain (white)	53.65	49.43	61.1	47.26	46.22	48.48	53.39	53.04	64.9	69.8	70.3	66.1
Millet	52.5	49.04	53.55	50.25	44.93	45.2	47.6	45.42	61.8	62.3	74.1	70.2
Guinea corn	51.53	49.88	52.9	49.22	49.49	46.64	47.59	48.54	62.1	80.4	72.4	75.0
Beans - brown	94.68	85.64	89.37	89.41	89.85	80.27	90.64	92.22	108.9	136.1	143.9	128.6
Gari - yellow	80.18	77.19	76.82	71.01	73.98	72.85	66.43	68.08	76.2	75.0	87.1	83.1
Cabbage	89.22	96.17	115.27	131.09	116.03	111.05	129.72	114.52	133.0	138.2	130.8	141.8
Onion - bulb	80.2	73.88	79.35	116.97	112.51	93.88	77.36	107.93	88.4	115.3	106.8	119.3
Palm oil	152.3	159.62	171.61	180.95	165.84	167.99	186.18	179.04	211.0	219.9	229.8	216.5
Sweet potato	46.91	51.38	52.79	46.1	58.3	51.62	62.13	47.64	62.2	45.5	61.3	53.7
Yam - tuber	62.27	71.85	79.64	70.03	68	73.94	72.08	74.77	84.5	74.5	90.3	83.0
Gari - white	68.76	65.99	69.56	64.44	68.44	62.62	58.5	59.48	66.3	76.3	75.2	69.9
Groundnut - shelled	101.36	112.16	138.3	100.08	187.81	140.23	125.52	113.12	132.8	139.1	148.8	143.5
Simple average	83.4	84.1	93.7	88.4	93.4	86.1	88.0	88.6	100.2	109.2	114.2	109.3

Source: Collated from various volumes of NBS reports.

Table 3:

Commodities	Producer Prices					
	2002	2003	2004	2005	2006	2007
Sorghum	17120	19000	27740	18750	19490	19120
Millet	18210	20150	19970	19790	20540	20170
Rice	23160	27500	26450	25400	25840	25620
Maize	20010	22030	20140	19730	20240	19990
Beans	21520	23550	22905	22260	21820	22040
Groundnuts	46500	46500	47500	48664	49827	50995
Yam	18550	21590	20805	20020	21570	20800
Cassava	17590	20910	19970	19810	20010	19910
Cocoyam	12710	15720	16215	16870	16710	16790

Table 4. Margin Between Retail and Producer Prices.

	2006 Retail Price	2006 Producer Price	2006 Retail Price	2006 Producer Price	2006 margin	2007 margin
Beans - white	77.5	21.8	69.7	22.04	55.7	47.6
Maize (white)	53.1	20.2	48.9	19.99	32.9	28.9
Millet	50.8	20.5	45.4	20.17	30.3	25.2
Guinea corn	51.1	19.5	47.5	19.12	31.6	28.4
Yam - tuber	71.7	21.6	75.4	20.8	50.1	54.6

Source: Computed from CBN/NBS Reports.

Table 5. Production of Key Crops.

Production ('000 metric tonnes) of key output	1990	2002	2003	2004	2005	2006	2007
Yam	13624	21707	21743	24977	27126	28890	30450
Cassava	19043	27938	28546	31067	36583	40573	43575
Sorghum	4185	4649	4627	4657	5039	5251	5561
Millet	5136	3944	3964	4088	3970	4076	4333
Rice	2500	2236	2367	2416	2660	2765	2978
Maize	5768	4424	4483	5001	6203	6767	7247
Beans		1454	1452	1483	1504	1650	1765
Dried cowpeas	1345	1218	1233	1239	1529	1576	1686
Groundnuts	992	2040	1997	2232	2701	2737	2892
Cocoyam	731	2633	2622	2869	2719	2765	2803
Sweet potatoes	143	1108	1154	1248	1453	1514	1625
Total production of above crops		73352	74188	81276	91487	98564	104915
Total area used in production of key output above (hectares, ha)		22872210	23317350	24384893	26494182	27433753	26876000
Growth in total production above (% increase in total tonnes produced)			1	10	13	8	6.4

	1990	2002	2003	2004	2005	2006	2007
Yields (kg/ha) of key output							
Yam	10677	11412	11405	11978	12273	12552	11582
Cassava	12937	12091	12213	12061	12317	12505	13708
Sorghum	1000	1156	1144	1141	1149	1182	1287
Millet	1075	1049	1053	1063	1069	1085	1134
Rice	2070	1857	1874	1875	1884	1916	2142
Maize	1130	1521	1500	1547	1556	1573	1731

Beans	463	679	659	688	698	713	782
Dried cowpeas		502	503	506	540	551	782
Groundnuts	1403	1220	1093	1083	1218	1271	1293
Cocoyam	5184	7458	7559	6871	7210	7305	6939
Sweet potatoes	5107	6343	6357	6369	6183	6368	6139
Production weighted average yield		8449	8507	8736	9006	9266	9692
Growth of production weighted average yield (%)			1	3	3	3	4.5
Area weighted average yield (kg/ha) i.e. Total kilograms/total area^2	2430	3207	3182	3333	3453	3593	3903.6
Growth in area weighted average yields above (%)			-1	5	4	4	8.6
Growth in total production above (% increase in total tonnes produced)			1	10	13	8	6.4

Source: Data from NBS, Nigeria Foreign Trade Summary (Various Years)

Table 6. Agricultural Trade.

	2002	2003	2004	2005	2006	2007
Food imports (millions of Naira)	196,598	287,463	280,260	368,860	513,773	802591
Total imports (millions of Naira)	1,054,076	1,923,099	1,547,371	1,779,602	2,922,248	4127620
Real food imports (CPI deflated, millions of Naira))	211,623	271,294	229,909	256,866	330,613	480593
Food imports share in total imports (%)	19	15	18	21	18	19
Growth of food imports (%)(	6	46	-3	32	39	56
Growth of real food imports (%)	-6	28	-15	12	29	45
Food exports (millions of Naira)	20,481	948	1,427	51	28,118	122303
Total exports (millions of Naira)	2,167,412	3,109,288	5,129,026	6,621,304	7,555,141	6881501
Food exports share in total exports (%)	0.9	0.0	0.0	0.0	0.4	1.6
Growth of food exports (%)	5,153	-95	51	-96	54,573	301
Agriculture exports value (millions of Naira)						
Live animals (HS codes 1-5)	6,260	164	153	0	64	8042
Vegetable Products (HS codes 6-14)	781	190	423	51	25,762	41335
Cocoa (HS codes 1801-1802)		1	121		643	48256
Tobacco (HS codes 2401)			-			4469
Rubber (HS codes 4001)	4	192	607		2,179	24666
Hides (HS codes 4101-4106)	34	774	6,789		3,758	31011
Wood (HS codes 4401-4403)			5		66	5603
Cork (HS codes 4501)						34877
Cotton (HS codes 5201-5203)	98	36	12		87	6035
Total agriculture exports	7,178	1,355	8,110	51	32,560	204294
Growth rate of Agricultural exports (percentage)	1,430	(81)	498	(99)	63,209	

Source: Computed from NBS, Nigeria Foreign Trade Summary (Various Years)

Table 7. Structure of Imports.

2006 Imports			
HS Code	Description	Share	Cshare
1001	wheat and meslin	0.33	0.33
0303	fish, frozen (no fish fillets or other fish meat)	0.16	0.49
0402	milk and cream, concentrated or sweetened	0.15	0.64
1006	rice	0.10	0.75
1901	malt ext, food prep of flour etc. un 50% cocoa etc.	0.06	0.80
1701	cane or beet sugar and chem pure sucrose, solid form	0.02	0.83
2002	tomatoes prepared or preserved nesoi	0.02	0.85
1107	malt, whether or not roasted	0.01	0.86
2106	food preparations nesoi	0.01	0.87
0305	fish, dried, salted, smoked etc., ed fish meal	0.01	0.88
1502	fats, bovine, sheep or goat, raw or rendered	0.01	0.90
ICE			0.82
*Imports were made under 139 HS4 tariff lines			
2007			
10	Cereals	0.42	0.42
3	Fish and crustacean	0.18	0.61
4	Dairy prod, bird eggs	0.16	0.77
19	Prep of cereals, flour	0.06	0.83
17	sugar/confectionery	0.05	0.88
15	Animal/veg fat and oil	0.02	0.91
21	Misc edible Preparations	0.02	0.93
20	Prep of veg, fruits, and nuts	0.01	0.95
11	malts, starches	0.01	0.97
5	Products of animal origin	0.01	0.98

Source: Data from NBS, Nigeria Foreign Trade Summary (Various Years)

Table 8. Structure of exports.

2004 Exports

HS Code	Description	Share	Cshare
4106	goat or kidskin leather, no hair nesoi	0.75	0.75
4001	natural rubber, balata, chicle, prim form, etc.	0.07	0.82
1701	cane or beet sugar and chem pure sucrose, solid form	0.05	0.87
1208	flour and meal of oil seed and olea fruit (no mustard)	0.02	0.89
1806	chocolate and other food products containing cocoa	0.02	0.91
	* Exports were made in 50 HS4 tariff lines		

2005 Exports

HS Code	Description	Share	Cshare
2201	waters, natural, not sweetened etc., ice and snow	0.97	0.97
2302	bran, sharps, etc. from working cereals & leg plants	0.01	0.99
1007	grain sorghum	0.01	1.00
2306	oilcake etc. nesoi, from veg fats and oils nesoi	0.00	1.00
1302	veg saps and extracts: pectates : agar-agar etc.	0.00	1.00
	* Exports were made in 5 HS4 tariff lines		

2006 Exports

HS Code	Description	Share	Cshare
1301	lac, natural gums, resins, gum-resins, and balsams	0.70	0.70
4106	goat or kidskin leather, no hair nesoi	0.11	0.81
4001	natural rubber, balata, chicle, prim form etc.	0.06	0.87
0801	coconuts, brazil nuts, and cashew nuts, fresh or dry	0.04	0.90
	* Exports were made in 55 HS4 tariff lines		

2007 Exports

HS Code	Description	Share	Cshare
1801-06	Cocoa and cocoa preparation	0.53	0.53
1201-12	oil seed, seed, fruit	0.13	0.67
0801-13	Edible fruit and nuts	0.12	0.79
1301-02	Gum/Resin/veg. saps	0.07	0.87
0301-03	Fish and crustacean	0.06	0.93
0901-10	coffee, tea	0.01	0.95
1701-04	sugar/confectionery	0.01	0.96

Source: Computed from NBS, Nigeria Foreign Trade Summary (Various Years)

References

1. CBN (Central Bank of Nigeria), Various reports.
2. (NBS) National Bureau of Statistics, Nigeria Foreign Trade Summary (Various issues)

50

The Hungry and the Rest of Us
A Food Security and Right to Food Manifesto for Nigeria

INTRODUCTION

The **National Campaign on Right to Food in Nigeria**[69] is the flagship project of Farm and Infrastructure Foundation (FIF), an organization for promoting policy best practices in agriculture and rural development. The Campaign was conceived within the policy process context, whereby the successive Governments perennially failed to ensure that the country was food secure, despite the numerous programmes implemented for the purpose – Nationally Coordinated Food Production Programme (NAFPP, 1972, Gowon); Operation Feed the Nation (OFN, 1976, Obasanjo); Green Revolution Programme (GRP, 1980, Shagari); Directorate of Food, Roads and Rural Infrastructure (DIFRRI, 1986, Babangida); National Agricultural Land Authority (NALDA, 1990, Babangida); National Programme on Food Security (NPFS, 2000, Obasanjo); National Food Security Programme National (NFSP, 2007 Yar'adua); National Food Reserve Agency (NFRA, 2007, Yar'adua); Agricultural Transformation Agenda (ATA, 2011, Jonathan), and the current Agricultural Promotion Policy (APP, otherwise known as Green Alternative). It is within this context of perennial policy failure that the Campaign was mounted by FIF in 2008, and is currently being implemented nationwide, in collaboration with Voices for Food Security (VFS), a coalition of civil society organizations working in the agriculture sector, and in association with Nigeria Zero Hunger Forum (NZHF), a platform for facilitating the realization of the Sustainable Development Goal No.2, i.e.

---

[69] The support of Oxfam from the onset of the Campaign is hereby acknowledged.

The latest World Hunger Index published by IFPRI (International Food Policy Research Institute) indicates that as at last year (2017), with an index value of 25.5 Nigeria ranked 84th among 119 countries, in the category of *"Serious Hunger Prevalence"*. This implies that about three quarters (¾) of our population (of about 180 million at the last count) presently suffers from acute (adult) malnutrition, severe child stunting, extreme child wasting and (under-five) child mortality, all linked to food insecurity of the country. Little wonder why, given the widespread hunger in the land and the attendant generally poor physical and mental health conditions of the people, the country is faced with high unemployment, excruciating poverty, runaway inflation, among other debilitating diseases of the economy; a rather grim situation of a people living in protracted suffering, albeit in the midst of plenty – plenty natural and human resources for that matter.

As conceptualized ab initio, the problem reflects the absence of a definite philosophical direction required for the series of food security policies and programmes of Government to endure long enough. This creates the room for policy ad-hocery, thereby permitting the political and policy authorities to act irresponsibly with absolute impunity, as their policies fail repeatedly to meet the food entitlements of citizens in adequate quantity and quality. The absence of a policy philosophy is linked to the absence of political philosophy, both of which are real in the Nigerian situation.

Thus, in addressing this situation our Campaign is anchored on the notion of food as a fundamental human right, not as a mere human need; this notion constituting the policy philosophy required for the progressive attainment of food security in Nigeria on a sustained basis. By and large, the two notions contrasts with each other in the sense that: while the traditional notion of food as a human need does not place any burden of public accountability on the Government of the day when its policies fail to ensure food security of the people, the contemporary notion of food as a human right puts such a burden on Government. The burden is derived from the innate gratification that the people feel when they are empowered as of right to be able to take government to account for its policy failures, and to ask questions at such times of policy failure, thereby putting Government on its toes and helping it to prioritize food security on top of the political and policy agenda. Hence, under this Campaign, the promotion of **Right to Food** as the appropriate policy philosophy for sustained progress towards food security in Nigeria!

MAIN THRUST OF THE NATIONAL CAMPAIGN ON RIGHT TO FOOD

The goal of this Campaign is to bring about food justice as a norm in Nigeria, whereby every citizen is assured an irreducible minimum degree of freedom from hunger for him or her to live a healthy, dignified and productive life (i.e. right to food). The campaign involves mind change on the part of the citizens and policy change on the part of Government, about the notion of food. Towards this end, we have carried out several activities in communication, education and sensitization at national and state level since 2008 when the campaign was launched. In particular, the need for a suitable legal environment for right to food to take a firm root in Nigeria is recognized, which informed the FIF's initiative to introduce and promote a Bill at the National Assembly for the purpose in 2010.

The Bill seeks to amend the constitution for the recognition of food security as a right issue in Chapter 2 (Directive Principles of State Policy), and to insert a clause to that effect in Chapter 4 (Fundamental Human Rights), thereby making it actionable, justiciable and ultimately remediable in Nigeria. The Bill passed the First Reading Stage at the House of Representatives during the Sixth Assembly (2007-2011), and also passed the same First Reading Stage at the Senate during the Seventh Assembly (2011-2015). It is presently before the Constitutional Review Committees of Senate and House of Representatives, during the present Eighth Assembly (2015-2019), having passed the First Reading and Second Reading in both chambers as at June 2016.

On the one hand, in a regime of right to food the citizens are the *right holders*, implying that the people themselves, except the vulnerable members of society and others in protracted suffering, have the obligation or indeed the responsibility, to work in order to earn their individual and collective right to food. On the other hand, the Government is the *duty bearer*, whereby it is the obligation of Government to implement policies for ensuring food security as a sacred duty, not as a favour or charity as is presently the case.

In this regime, therefore, Governments have three obligations for which to be held to account, following:

- Obligation **to respect** the right of citizens to food: This stipulates the limits of state's exercise of power, which includes refraining from destroying people's access to food through unfavourable public policies. That is, the government owes the people an obligation to implement policy decisions or actions if and only if such policy decisions or actions do not offend or violate the right to food of citizens.

- Obligation **to protect** the right of citizens to food: This entails the regulation of activities of state and non-state agencies and bodies that are inimical to people's food entitlements. The common examples of these instances include environmental pollution that destroy the capacity of farmers to grow crops; water pollution that destroys the capacity of water bodies as habitats for fish and other marine products; air pollution that endangers the quality life and jeopardizes the health of people, etc.
- Obligation **to fulfill** the right of citizens to food: This involves the provision of food assistance to vulnerable groups and other people in protracted suffering (children at home or in school, nursing mothers, internally displaced persons, etc.). The obligation of Government to fulfill citizens' right to food underscores the role of agriculture as a social investment sector; not just the business case for agriculture as a private sector.

UNIVERSALITY OF THE RIGHT TO FOOD

The global regime of food as a human right is best illustrated by the weight of international consensus that exists in terms of conventions, protocols and frameworks namely:
- THE UNIVERSAL DECLARATION ON HUMAN RIGHTS (1948) - *"Everyone has the right to a standard of living adequate for adequate for the health and wellbeing of himself and his family, including food..."*
- THE INTERNATIONAL COVENANT ON ECONOMIC, SOCIAL AND CULTURAL RIGHTS (1966) - *"The States parties to the present Covenant recognizes the right of everyone to an adequate standard of living... including adequate food. And agree to take appropriate steps to realize these rights.*
- ROME DECLARATION ON WORLD FOOD SECURITY (1996) - *"We the heads of State and government... Reaffirm the right of everyone to have access to safe and nutritious food, consistent with the right to adequate food and the fundamental right of everyone to be free from hunger."*
- UN RESOLUTION 23 (1998) –The United Nations Commission on Human Rights: *a) states that hunger constitutes an outrage and a violation of human dignity, and therefore requires the adoption of urgent measures at national, regional and international levels for its elimination; and also b) reaffirmed that the right of everyone to have*

access to safe and nutritious food and to be free from hunger is paramount so as to be able to develop fully and maintain their physical and mental capacities.
- *AFRICAN CHARTER OF HUMAN RIGHT (Articles 4, 16, 22)- That:* "The right to adequate food is an individual right that is indivisibly linked to the inherent dignity of the human person and is indispensable for the fulfillment of other human rights ...". That the state has an obligation to "Take the necessary action to guarantee the right of everyone to be free from hunger and to mitigate and alleviate hunger even in times of natural or other disasters; ..."

Furthermore, the universality of right to food also exists in the large number of countries of the world that have not only adopted the concept of food as a fundamental human right but also have passed a right to food Bill into constitutional or subsidiary law. They include the following (with references to their constitutional provisions): **Bangladesh** - Article 15; **Brazil** - Article 227; **Colombia** - Article 44; **Congo** - Article 34; **Cuba** - Article 8; **Ecuador** - Article 19; **Ethiopia** - Article 90; **Guatemala** - Article 51 (minors and elderly), Article 99 (feeding and nutrition); **Haiti** - Article 22; **India**: Article 47; Iran Article 3; **Malawi** – Article 13; Mexico - Art 27; Nicaragua: Art 63; Nigeria Article 16 (wherein not justiciable); **Pakistan** - Article 38; **Paraguay** - Art 53; **South Africa** - Chapter 2, Section 27; **Sri Lanka** - Art 27; **Uganda** - Objective 14; **Ukraine** - Art 48.

A CALL FOR CITIZENS' ACTION

In sum, the quick passage of the Right to Food Bill holds the key to alleviating the suffering of Nigerians in several ways. As citizens become more and more conscious of their fundamental right to food, and move in various directions to claim it, Government then becomes more sensitive and more responsive to the food entitlements of the people and eventually become positively disposed to respecting, protecting and fulfilling the right; this by prioritizing food security higher on the policy agenda thereby increasing the allocation to it in the annual budget. Also, passage of the Bill into law now will empower the citizens to build greater confidence in the food market; thus helping the people to place demand for all conceivable kinds of food items – energy foods, protein foods, minerals, etc. more than ever before now, and helping to increase farm output and farmers' incomes in the medium to long term.

Furthermore, the population will be better fed, including the vulnerable people and the less privileged, thereby increasing the workforce initially and diversifying GDP of the agricultural economy ultimately. Lastly, the passage of the Right to Food Bill is the master key to unlocking or addressing the several issues of environmental and other actions of the private and public sectors that abuse people's right to food at will, so the victims of such actions can seek remedy under the proposed right to food law.

Therefore, at this stage we call on the good citizens of Nigeria in their different capacities to lend their full support to the **National Campaign on Right to Food in Nigeria,** by a) endorsing it generally and mandating their lawmakers at national and state level to support the passage of the Right to Food Bill pending at the National Assembly; and b) thereafter putting a political price on food security of the people by voting in the next general election for candidates, who and only who are ready to give an undertaking that they would support the smooth passage of the right to food Bill now (i.e. before the expiration of this Assembly) and implementation of the Right to Food Bill, before and after May 2019. IT IS DOABLE!

Index

A

Abuja Securities and Commodity Exchange 509
academia 549
accountability 13, 34, 122, 381, 499, 548, 552, 554, 650, 653
ACGS (Agricultural Credit Guarantee Scheme) 15, 80, 121, 224, 236, 246
ADAP (Accelerated Development Area Projects) 98, 294-5, 340
Adasu, Moses 562
ADC (Agricultural Development Corporation) 414-16, 459
Adesina, Akinwumi 563, 565
Adewusi, Samuel 158
Adjbeng-Asem 204, 210
ADP (Agricultural Development Project) x-xi, 11-12, 15, 23, 30, 47-51, 54, 57, 70, 80-1, 90-2, 98-9, 113, 120-2, 129, 133, 135-8, 152-3, 155, 176, 184, 195, 199, 203, 216, 230, 237, 248, 250, 260, 262-3, 266, 269-70, 277, 279, 281-3, 285, 287-8, 290, 295, 328, 337-45, 347, 351, 409-10, 425, 438, 445-6, 470, 478, 512-13, 532, 562, 568, 588, 591, 657, 681-2, 702
AFAN (All Farmers Association of Nigeria) 569, 580, 591, 655
Africa 18, 349, 453-4, 474, 492, 506, 511, 513, 573-4, 576, 584, 586, 644, 689
Africa Fertilizer Summit 443, 450, 453, 677, 686, 689
Agbado Ekiti 531
AgDPO (Agricultural Development Policy Operation) 629, 631-2, 636
agri-input delivery system 442-3, 446, 448, 450-3, 675, 677, 682, 684-9
Agricultural Department for Nigeria 4, 60, 468
agricultural extension 10, 14, 29, 48-9, 54, 69, 79, 89-90, 92, 134-5, 176, 181, 197, 199, 231, 243, 259, 330, 363, 379, 392, 467, 623
Agricultural Policy for Nigeria 5, 35, 52, 139, 233, 244, 404, 414, 433
agricultural rebirth 555-7, 559, 563-4, 566-8, 571
agricultural science 160-1, 426, 466, 527
agricultural technology 28, 86, 161, 170-1, 173, 176-7, 181-2, 184, 200, 209, 211, 371, 391, 498, 515
AIDAs (Agro-Input Dealers Associations) 452, 687

717

AISU (Agricultural Inputs and Services Unit)   315, 321
APMEPU (Agricultural Projects Monitoring, Evaluation, and Planning Unit)   39, 216, 281, 288, 295, 337
APMEU (Agricultural Projects Monitoring and Evaluating Unit)   11, 23, 70, 81, 120-1, 140, 152, 156, 191, 323, 327, 344, 346, 418, 470, 594
APPC (Agricultural Produce Price Commission)   304-5
ARA (Agricultural Rebirth Agenda)   557, 562, 566, 568, 572
ARCN (Agricultural Research Council of Nigeria)   198, 491, 503
ARSN (African Regional Soyabean Network)   251
ASTI (Agricultural Science and Technology Indicators)   491, 495
ATA (Agricultural Transformation Agenda)   557, 565, 592, 598-9, 623, 625, 629, 631-2, 635, 639, 655, 666, 671, 673, 711
AU (African Union)   454

B

basic resource theory   42, 45
Beatle, A. G.   60
Bello, Adamu   503, 538
Bendel State   55-6, 213, 215, 228, 337, 364, 368
Better Life   154, 249
Blinder, Alan S.   233, 389
Botany (Oliver Indian)   158
Bourdillon, Bernard   160
budget illusion   463
Buhari, Muhammadu   559, 562

Bureau of Public Enterprises   354

C

CADP (Commercial Agricultural Development Programme)   625, 629, 631-2
CAPL (Chemical and Allied Products Limited)   327, 333, 335
CBN (Central Bank of Nigeria)   83, 121, 193, 236, 246, 426, 472, 561, 569, 598, 701, 710
CFC (Common Fund for Commodities)   439, 500-1, 505
CIB (Cooperative Intervention Buyer)   305
classical-neoclassical theory   42, 45
Cocoa Summit   508
Colonial and Welfare Acts   160
commercialization   196-7, 200, 202-10, 212, 353-4, 408, 421, 502, 516, 574, 576, 584
commitment   30, 137, 262, 266, 315, 338, 340, 342, 345, 413, 418, 452, 469, 479, 482, 503, 620, 651, 688, 696
commodity boards   8, 98, 100, 105, 124, 236, 246, 298-9, 302, 304-6, 311, 315, 355, 384, 387, 408, 411, 415, 420, 425, 471, 499, 506, 508, 567
corruption   34, 125, 385, 413, 520, 547-53, 562, 571, 606, 610-11, 616-20
counterfactual   457, 460, 462
CPP (crop protection product)   439, 444, 446, 477, 515, 675, 680, 682

D

Dadin-Kowa 531
DEM (deathly embrace model) 621, 628
DFRRI (Directorate of Food, Roads and Rural Infrastructure) x, 12, 26, 32, 47, 49-51, 92, 98, 121, 135, 153, 230, 261, 265, 280-1, 296, 372, 383-4, 397-8, 425, 472, 568, 711
diffusion model 43, 48
DPO (development policy operation) 629
dual-economy model 42, 46

E

Ebute Metta Botanical Station 60, 135, 158, 176, 197, 518
Economic Development Plan 19, 62, 433-4
economic empowerment 518-20, 523, 530-3
economy:
 multisector 621, 627-8
 national 72, 420, 519, 665
ECOWAS (Economic Community of West African States) 452, 626, 688, 691
Elementary Botany (Silver) 158
esusu system 615
export-led growth model 42, 46

F

FACU (Federal Agricultural Coordinating Unit) xi, 11, 23, 52, 64, 70, 73, 120, 128, 133, 264, 269, 281-3, 288, 306, 344, 358, 378, 470

Fadama II 530
FADAMA III 625, 629, 631-2
Falusi, A. O. 351
FAMAS 440-1
FAO (Food and Agricultural Organization) 4, 10, 22, 35, 45, 52, 61, 64, 69, 73, 78, 96, 101, 192-3, 197, 211, 251, 279, 299, 306, 351, 357, 375, 378, 438, 454, 468, 500, 531, 535, 540, 559, 572, 575, 580, 626, 643, 647, 650, 655, 690-1
FARA (Forum for African Agricultural Research) 626
FAs (faculties of agriculture) 56, 161, 187, 368, 492, 513-14, 596
FASCOM (Farmers' Supply Company) 27, 99, 101, 123, 225, 233, 390, 416
Faulkner, O. T. 60
FCAs (federal colleges of agricultural) 491, 493
FDA (Federal Department of Agriculture) 22, 250, 365, 438, 470, 525, 550, 587
FDARD (Federal Department of Agriculture and Rural Development) 23, 52
FDI (foreign direct investment) 658, 671-3
Federal Grain Storage Scheme 228
FEPSAN (Fertilizer Producers and Suppliers Association of Nigeria) 587, 591, 602-3
FFD (Federal Fertilizer Department) 445, 527, 551, 590, 594, 681
FIF (Farm and Infrastructure Foundation) xv, 440, 480, 482, 488, 503-4, 523, 549, 552, 564, 568, 574, 576-7, 580, 583-5, 591-2, 594, 599, 654, 711

719

First National Development Plan 5-6, 19, 63, 433
FLD (Federal Livestock Department) 361, 363-5
FMA (Federal Ministry of Agriculture) 4, 20, 35, 52, 61, 73, 96, 98, 101, 103, 131, 135, 139, 198, 233, 244, 279, 306, 346, 414, 470, 480, 518, 540, 588, 664
FMANR (Federal Ministry of Agriculture and Natural Resources) 4, 20, 61, 73, 101, 103, 131, 135, 153, 198-9, 370, 469, 540, 559, 572, 664
FMARD (Federal Ministry of Agriculture and Rural Development) x, 61, 103, 131-2, 530, 568, 570, 572, 581, 590, 600, 602, 623, 657, 669, 671, 683, 690
FMAWR (Federal Ministry of Agriculture and Water Resources) 4, 61, 135, 198, 472, 478, 480, 500, 503
FMAWRRD (Federal Ministry of Agriculture, Water Resources, and Rural Development) 4, 7, 20-1, 35, 52, 61, 103, 123, 128, 132, 135, 138, 198, 229, 233, 279, 303, 362, 389, 399, 404, 408, 414, 683
FME (Federal Ministry of Education) 198
FMF (Federal Ministry of Finance) 569
FMITI (Federal Ministry of Industry, Trade, and Investment) 570
FML (Federal Ministry of Labour) 569
FMNRR (Federal Ministry of Natural Resources and Research) 20, 61, 103, 469, 540, 559
FMSS (Fertilizer Market Stabilization Scheme) 588
FMST (Federal Ministry of Science and Technology) 198, 248
FOFAN (Farmers Organizations in Nigeria) 89-90, 569
food deficiency 82-4, 88-9, 93
food production 8, 14-16, 18, 21-2, 27, 29-30, 32, 34, 86, 91, 108, 118-19, 121-2, 145-6, 156, 211, 230, 370, 382, 389, 391, 397-8, 403-4, 423, 535, 622, 648, 691
food security 67, 82, 100-1, 117, 119, 443, 454, 456, 465, 473, 478-9, 481, 507, 511, 520-3, 530, 533-6, 538-9, 541-2, 544, 556-7, 563, 573, 577, 581, 583, 623, 642-3, 648, 650, 652, 654-6, 671-2, 675-7, 712-13, 716
food self-sufficiency 12, 18-19, 33, 65, 82, 84, 112, 117-19, 126-7, 343
Food Strategies Mission 24, 35, 48, 69, 73, 306, 340, 346
food sufficiency 82, 85-9, 93, 115
FOREX (foreign exchange market) 26, 121, 261, 280, 306, 354, 383, 472
FOS (Federal Office of Statistics) 78, 219, 249, 255
Fourth National Development Plan 5, 21, 433
FPDD (Fertilizer Procurement and Distribution Division) 350-1, 353, 357, 445, 471, 550, 587, 590, 664, 681
FPS (Finance and Private Sector) 629-30

FSFC (Federal Superphosphate Fertilizer Company) 350-1, 412, 415, 445, 667, 669, 681
FSS (farm settlement scheme) 5, 10-11, 14, 21, 25, 47, 54, 69, 79, 135, 260, 278, 375, 388, 469
FTE (full-time equivalent) 492

G

GDP (Gross Domestic Product) 9, 19, 47, 192-3, 208, 226, 385, 473, 507, 555, 558
George, Cornelius 158
GER (Governance and Economic Reform) 629-30
GES (Growth Enhancement Scheme) 551, 563, 571, 587-8, 593-4, 596, 598-601, 603-4, 629, 666
GFS (Graduate Farmers Scheme) 24, 312-14
Glaxo Nigeria Limited 205
GMPs (guaranteed minimum prices) 241, 298-9, 387-8, 604
Gowon, Yakubu 540
green revolution 22, 24, 35, 73, 80, 83, 120, 306, 346
Gross margins 252

H

HDS (Human Development sector) 629-30
high-input pay-off model 43, 47
horticulture 519-23, 528-33

I

IAR4D (Integrated Agricultural Research for Development) 626

ICB (International Competitive Bidding) 263, 284
ICESCR (International Covenant on Economic, Social and Cultural Rights) 651, 714
IFAD (International Fund for Agricultural Development) 500, 524
IFPRI (International Food Policy Research Institute) 491, 643, 712
IITA (International Institute of Tropical Agriculture) 55, 176, 201, 205, 211, 213-14, 217, 248-9, 251, 255, 368
IMF (International Monetary Fund) 21, 83, 224, 381, 396
IMF loan 84, 381
Imo State 55-6, 98, 186-7, 213, 215, 363, 368, 526
Imperial College of Tropical Agriculture 160
industrialization policy 50
Integrated Agricultural Development Programme 11, 70, 434
internal combustion theory 42, 45
intervention boards 9, 355, 402
ISFP (Initiative on Soaring Food Prices) 691

J

JAMB (Joint Admissions and Matriculation Board) 608, 610-11
Jimeta, Mohammadu Gambo 90
Joint Federal Grain Storage Consultative Group 228

K

Kaduna State 27, 55-6, 61, 98-9, 129, 186-7, 213, 215, 295, 363-4, 368, 390, 562
Kaduna State Agricultural and Marketing Company 410
Katsina 159, 347, 363
Kwara State 55-6, 123, 187, 190, 213, 215, 314, 364, 368, 526

L

LCB (Local Competitive Bidding) 263, 284-6
LIB (Limited International Bidding) 263, 284-5
LIBC (Livestock Investigation and Breeding Centre) 363

M

Mackie, J. R. 60
Makurdi 56, 89, 127, 155-6, 161, 165, 187, 215, 247, 255, 298, 368, 396, 471-2, 513, 523, 542, 569, 579
Makurdi Declaration 89
MAMSER (Mass Mobilization for Social and Economic Recovery) 141, 154
MANR (Ministry of Agriculture and Natural Resources) 4, 60, 141, 197, 314, 438, 468, 559
market mechanism 59, 331, 423-5, 650
MLAE (Middle Level Agricultural Education) 491, 493
Mosher, A. T. 259, 269
MSADP (multi-state ADP) 98, 287, 340, 342, 347
MSS (Market Stabilization Scheme) 551, 563
multisectorality 621, 624, 631-5, 638-41
Mustapha, Shetima 102, 127, 132, 569

N

NAADI (Nigeria Agribusiness and Agro-industry Development Initiative) 672-4
NAC (National Council on Agriculture) 89, 350, 478, 480, 543, 578, 584, 589
NACB (Nigerian Agricultural and Cooperative Bank) 121, 379, 416, 421, 541
NADAC (National Agency for Drug Administration and Control) 549, 590
NADC (National Agricultural Development Committee) 420, 578
NADF (National Agricultural Development Fund) 503-4
NAFCON (National Fertilizer Company) 353-4, 422, 445, 461, 591, 665, 667, 669, 681
NAFDAC (National Agency for Food & Drug Administration & Control) 451, 590, 599, 687
NAFPP (national accelerated food production project) xi, 14, 22, 48, 54, 79, 99, 104, 135, 199, 223, 236, 246, 470-1, 557
NAFRAC (National Agency for Fertilizer Regulation and Control) 551, 589
NAIC (Nigerian Agricultural Insurance Company) 121

NALDA (National Agricultural Land Development Authority) xi, 93, 100-1, 106, 121, 123, 370-4, 376-7, 379, 425, 472, 557, 711
NARI (National Agricultural Research Institute) 444-5, 490-3, 496, 500-1, 596, 600, 680
NARP (National Agricultural Research Project) 437, 490
NARS (national agricultural research system) 28, 184, 230, 391, 477, 491, 595-6, 631
NASC (National Agricultural Seed Council) 541, 680-1, 689
National Agricultural Seed Committee 444, 680
National Campaign on Right to Food 552, 576, 711
National Examinations Council 608
National Fertilizer Board 664
National Fertilizer Quality Control Act 589
National Food Security Programme 434, 472, 478, 480, 557, 711
National Rolling Plans 433
NBAE (National Board of Agricultural Education) 166
NCA (National Council on Agriculture) 89, 350, 478, 480, 488, 543, 578, 584, 589, 600
NCRI (National Cereals Research Institute) 55, 186, 213-14, 224, 251, 368, 496
NCRP (Nationally Coordinated Research Project) 248, 595
NDE (National Directorate of Employment) 24, 69, 314, 317, 388-9, 397-8
NEPAD (New Partnership for Africa's Development) 452, 688
Newberry, M. R. J. 159

NFDC (National Fertilizer Development Centre) 590, 596
NFRA (National Food Reserve Agency) 472, 478, 557, 711
NFSP (National Food Security Programme) 472, 557, 654-5, 711
NFTC (National Fertilizer Technical Committee) 350, 550, 571, 589-90, 595-6, 601, 603, 665
NGB (Nigerian Grains Board) 100, 124, 224, 227, 236, 241, 246, 308
NGO (non-governmental organization) 437, 452, 482, 488, 500, 549, 568, 591-2, 688
NGPC (Nigerian Grains Production Company) 224, 227, 236, 246, 411, 415
NIFOR (Nigerian Institute for Oil Palm Research) 55, 174-5, 213
Nigeria Agriculture Digest 571-2, 674
Nigerian livestock policy 360, 367
NIHORT (National Horticulture Research Institute) 55, 213-14, 368, 492, 531
NIRSAL (Nigeria Incentive-based Risk-bearing Scheme for Agricultural Lending) 598, 602, 604
NNMA (Nigerian National Merit Award) 550, 606-7, 616-20
NNOM (Nigerian National Order of Merit) 616-17, 619-20
NPFS (National Project for Food Security) 557, 596, 711
NPN (National Party of Nigeria) 91, 98
NRSA (Nigerian Rural Sociology Association) 547

723

NSA (National Soyabean Association) 248, 250-1, 255, 655
NSPFS (National Special Programme for Food Security) 524, 530
NSS (National Seed Service) 201, 250, 444-5, 680
NUC (National Universities Commission) 164, 198, 514, 616, 618-20
NZHF (Nigeria Zero Hunger Forum) 711

O

Obasanjo, Olusegun 541, 557, 562, 569, 648, 711
ODA (official development assistance) 500, 626
OFN (Operation Feed the Nation) 14, 24, 54, 79, 83, 113, 224, 236, 246, 471, 557, 711
Ogun State 56, 93, 122, 187, 215, 228, 364, 368, 410, 416, 488-9
Ogunwolu, E. O. 202
Okoye, Mokwugo 125, 128
Olayide, S. O. 48, 69, 74
Olufokunbi, B. 112, 125, 128
Oyaide, O. F. J. 343, 346
Oyo State 55-6, 186-7, 213, 215, 251, 313, 315-17, 319-21, 368, 526

P

PCU (Projects Coordinating Unit) 470, 524
perturbation 30-1, 64, 85, 183, 262, 268, 412
pesticide marketing 323-7, 331-4
PFA (Price Fixing Authority) 241, 298-9, 304, 309
Philips, Nathan 158

PMU (project management unit) 340-1
PPD (Public-Private Dialogue) 587
private sector xv, 33-4, 67, 166, 225, 229, 354, 372, 377, 421, 450-2, 472, 478-9, 485, 488, 490, 493, 499, 501-3, 515, 549, 553, 565, 592, 594, 596, 606, 625, 636, 666, 671, 685, 687-8, 697, 700-1, 714
privatization ix, 353-4, 410-12, 421-3, 425, 461, 516
process technology 86, 200, 203, 248
product technology 86, 200
public sector 33, 451, 485, 549, 687, 716

R

RAIDS (Rural Agro-Industrial Development Scheme) 248
RBDAs (River Basin Development Authorities) 12, 14, 24, 54, 70, 80, 91-2, 97-8, 100, 104, 121, 135, 224, 227, 230, 236, 246, 260, 268, 271-2, 291, 372, 375, 379, 397, 434, 470-1, 540, 560
RIFAN (Rice Farmers Association of Nigeria) 699
Right to Food 534, 553, 581, 648, 654, 711-12, 716
RMRDC (Raw Materials Research and Development Council) 491, 493, 500
roads, candidate Palm Produce 263
Ruma, Sayyadi 567

S

SAP (Structural Adjustment Programme) 6, 11, 21, 52, 64,

83, 128, 220, 224, 232-3, 238, 242, 244, 297, 302, 305, 349, 352, 354, 380-4, 386-7, 393, 396-7, 399, 402-4, 420, 424, 426, 471-2, 495, 506, 665, 695, 710
SAR (Staff Appraisal Report)   282, 340-1
SD (sustainable development)   3, 101, 370-1, 377, 629
Second National Development Plan   5, 20, 53, 63, 117, 433
Second World War   433, 505, 537, 564
Sen, Amartya   357, 563
Shagari, Shehu   91, 556, 560
Shonekan, Ernest   556, 561
Social and Economic Rights   564
SON (Standards Organization of Nigeria)   451, 549, 590, 599, 687
SSS (State Security Service)   591
subsidy policy   453, 544, 550-1, 562, 586, 588, 597, 601, 688
sustainability   xi, 91, 113, 137-8, 146, 154-5, 181-4, 225, 253-4, 325, 343, 371, 387, 398, 422, 424, 637

T

TAC Nigeria Limited   591
Taraku Limited   249, 251
TCPP (Technical Committee on Produce Price)   241, 298-301, 304, 311, 561
technology   20, 28, 42, 45-6, 52, 56, 63, 71, 86, 88, 107, 112, 120, 128, 136, 161, 164, 170-1, 173, 175-6, 178-83, 187, 189, 197, 206, 208-10, 215, 225, 233, 244, 269, 281, 292, 391, 426, 465-6, 473, 476-7, 498, 502, 514-15, 532, 547, 558, 601, 632, 645
Ten-Year Plan of Development and Welfare   19, 62, 433
TES (Techno-Economic Surveys)   491
Third National Development Plan   5, 20, 63, 433
Thomas, Josiah   158
Titilola, T.   74, 125, 128
Tropical Agriculture (Nicholl)   158
Tropical Oilseed Journal   251

U

UAM (University of Agriculture Makurdi)   248, 250, 491, 514, 568, 578
UN (United Nations)   279, 564, 575, 626
unified extension system   133, 137-8
United Nations Commission on Human Rights   575, 650, 714
UPN (Unity Party of Nigeria)   92, 98
urban industrial impact theory   42, 47
UTME (Unified Tertiary Matriculation Examination)   610

V

VFS (Voices for Food Security)   580, 592, 711
Vom   55, 134, 160, 213, 360-1, 368, 496

W

WAAPP (West Africa Agricultural Productivity Program)   595, 631-2

725

West African Examinations Council 608

World Bank x, 11, 23, 39, 67, 70, 74, 83, 104, 114, 116-17, 137, 176, 185, 193, 199, 234, 260, 262-3, 269-70, 281-5, 292, 334, 338, 340-2, 344-5, 348, 353, 364, 380-1, 395, 437, 439, 445, 458, 460, 470-1, 500, 524, 555, 586, 588, 621, 623-6, 629-31, 636, 638, 641, 643, 656, 681

Z

Ziegler, Jean 645

Xlibris

ISBN 978-1-5434-0180-6

90000